THE TECHNOLOGY AND APPLICATIONS
OF ENGINEERING MATERIALS

THE TECHNOLOGY AND APPLICATIONS OF ENGINEERING MATERIALS

Martyn S. Ray

Lecturer in Chemical Engineering,
Western Australian Institute of Technology,
Bentley, Western Australia

Formerly lecturer at
The Polytechnic of Wales,
University of the West Indies
and Huddersfield Polytechnic

Prentice/Hall PHI International

Englewood Cliffs, N.J.   London   Mexico   New Delhi
Rio de Janeiro   Singapore   Sydney   Tokyo   Toronto

Library of Congress Cataloging in Publication Data

Ray, Martyn S., 1949
  The technology and applications of
  engineering materials

    Bibliography: p.
    Includes index.
    1. Materials.   2. Welding.   3. Engineering design.
    I. Title
    TA403.R37   1987   620.1′1   86–5088

    ISBN 0–13–902099–3
    ISBN 0–13–902081–0 (pbk.)

British Library Cataloguing in Publication Data

Ray, Martyn S.
  The technology and applications of
  engineering materials
  1. Materials
  I. Title
  620.1′1   TA403

  ISBN 0–13–902099–3
  ISBN 0–13–902081–0 Pbk

Prentice-Hall Inc., *Englewood Cliffs, New Jersey*
Prentice-Hall International (UK) Ltd, *London*
Prentice-Hall of Australia Pty Ltd, *Sydney*
Prentice-Hall Canada Inc., *Toronto*
Prentice-Hall Hispanoamericana S.A., *Mexico*
Prentice-Hall of India Private Ltd, *New Delhi*
Prentice-Hall of Japan Inc., *Tokyo*
Prentice-Hall of Southeast Asia Pte Ltd, *Singapore*
Editora Prentice-Hall do Brasil Ltda, *Rio de Janeiro*

Printed and bound in Great Britain for
Prentice-Hall International (UK) Ltd.,
66 Wood Lane End, Hemel Hempstead, Hertfordshire, HP2 4RG,
at the University Press, Cambridge

1 2 3 4 5   91 90 89 88 87

ISBN 0-13-902099-3
ISBN 0-13-902081-0 PBK

Contents

Preface

This book should satisfy the demands of the syllabus for those parts of an engineering course related to materials and their applications. The book is written for use in engineering degree courses and in pre-degree (TEC) certificate and diploma courses. It is also suitable for use by students in their last years at school, to provide an appreciation of the uses of materials and related engineering activities.

Most engineering degree courses contain units concerned with materials science and related topics. It is in this area that the book will provide a full coverage. Despite the abundance of textbooks published in this field, the majority have one or two disadvantages. Either the material is presented from a scientific or metallurgical basis, e.g. 'Science of Materials' courses, and engineering applications are too sparse, or the emphasis is firmly towards engineering but important topics such as fabrication methods are omitted.

The choice of topics is intended not merely to satisfy the demands of a syllabus, but also to provide a wider view of the importance and scope of materials in many aspects of engineering. With this aim in mind topics such as basic applied mechanics and the concepts of engineering design are included, and also details of relevant BS and ASTM standards. This book should occupy a prominent place on the bookshelf, first as an undergraduate textbook and later as a reference source for the young graduate engineer.

Materials should be considered and taught as an integrated topic in engineering, i.e. its importance in many areas should be recognized and emphasized. The applications of materials are particularly relevant to the 'Engineering Applications' units now required in engineering degree courses. The relationships between topics presented in different chapters of the book are emphasized and there is frequent cross-referencing. Relevant theory and the explanation of effects are related to particular applications, e.g. the theory of dislocations, cold working and the effect on mechanical properties. The relevance of design to engineering and to materials is discussed in Chapter 9.

Each chapter begins with a short list of objectives and ends with exercises, complementary activities, keywords and references. It is hoped that the reader will try to establish the important areas within each section, and will evaluate how successfully the chapter objectives are fulfilled by the material as study proceeds.

Keywords should be referred to frequently when reading a chapter, to establish whether important concepts and terms have been appreciated. Definitions or detailed explanations of keywords are not given as they should be identified by the reader within the chapter, as part of the learning process. For these reasons keywords are presented in the order in which they first appear in a chapter, not in alphabetical order. Exercises at the end of a chapter are intended to establish whether the material and the course of study have resulted in an increased awareness of engineering applications, rather than just an assimilation of facts. References for further reading are included in a bibliography at the end of each chapter. They are chosen from the large amount of literature available within this field.

Complementary activities are included for each chapter, the 'doing' being considered as important as the reading. The students will learn a great deal from performing relevant experimental work, especially if *they* are required to plan the experiments to be performed.

Very many self-assessment exercises are included throughout each chapter, relating directly to the preceding topics. These exercises are meant to be a direct and immediate revision aid, and generally do not include questions outside the particular text material. However, the exercises at the end of each chapter may require some additional reading and understanding. They are intended to test the student's wider appreciation of materials topics and applications.

I hope that on reaching the end of this book (or the end of a materials course), the reader will have gained not only knowledge but also an understanding and appreciation of how different topics are related. The reader should be aware of the factors that influence the selection of materials, that being one of the many tasks of the graduate engineer. The following are actual, and typical, answers to problems on the selection of materials – given even after studying a materials unit!

A cheap and light material – polymer.
A strong and light material – aluminum alloy.
A universal panacea for all materials problems – stainless steel!

I hope that after reading this book the student will have a more realistic (and safer!) approach to materials' selection and design.

Notes
   (a)   The material presented in Chapter 9 was originally published (in greater detail) in M. S. Ray *Elements of Engineering Design: An Integrated Approach*, Prentice-Hall International (UK) Ltd, London (1985).
   (b)   The male gender is used through this book for reasons of consistency and readability (i.e. avoiding he/she). Wherever the male gender appears it should be taken to include both sexes.
   (c)   Where column headings in tables are written as units multiplied by a factor of 10, this means that the values presented in the table *have been* multiplied by that factor of ten. For example:

   Column heading: Pressure $(N/m^2 \times 10^{-3})$;
   Sample value (from column): 5.79

   Therefore, actual value is $5.79 \times 10^{+3}$, i.e. 5790 N/m^2

I would like to acknowledge the support, time and help contributed by my wife, Cherry. Without her encouragement this book would not have been completed.

Martyn Ray

PERIODIC TABLE OF THE ELEMENTS

Group:	IA	IIA	IIIA	IVA	VA	VIA	VII	VIII	VIII	VIII	IB	IIB	IIIB	IVB	VB	VIB	VIIB	O
	1 H 1.008																	2 He 4.003
	3 Li 6.940	4 Be 9.013											5 B 10.82	6 C 12.010	7 N 14.008	8 O 16.000	9 F 19.00	10 Ne 20.183
	11 Na 22.997	12 Mg 24.32											13 Al 26.98	14 Si 28.06	15 P 30.98	16 S 32.066	17 Cl 35.457	18 A 39.944
	19 K 39.10	20 Ca 40.00	21 Sc 45.10	22 Ti 47.90	23 V 50.95	24 Cr 52.01	25 Mn 54.94	26 Fe 55.85	27 Co 58.94	28 Ni 58.69	29 Cu 63.54	30 Zn 65.38	31 Ga 69.72	32 Ge 72.60	33 As 74.91	34 Se 78.96	35 Br 79.916	36 Kr 83.8
	37 Rb 85.48	38 Sr 87.63	39 Y 88.92	40 Zr 91.22	41 Cb 92.91	42 Mo 95.95	43 Tc 99	44 Ru 101.1	45 Rh 102.91	46 Pd 106.4	47 Ag 107.88	48 Cd 112.41	49 In 114.82	50 Sn 118.70	51 Sb 121.76	52 Te 127.61	53 I 126.92	54 Xe 131.3
	55 Cs 132.91	56 Ba 137.36	57 La* 138.92	72 Hf 178.5	73 Ta 180.88	74 W 183.86	75 Re 186.31	76 Os 190.2	77 Ir 193.1	78 Pt 195.09	79 Au 197.2	80 Hg 200.61	81 Tl 209.39	82 Pb 207.21	83 Bi 209.0	84 Po 210.0	85 At 210	86 Rn 222
	87 Fr 223	88 Ra 226.05	89 Act 227															

Atomic number

Chemical symbol

Atomic mass

Transition metals

Refractory metals

Precious metals

Metallic bonding

Covalent bonding tendency

Nonmetals covalent bonding

Inert gases

*Lanthanum series:

58 Ce 140.1	59 Pr 140.9	60 Nd 144.2	61 Pm 145	62 Sm 150.4	63 Eu 151.9 BCC	64 Gd 157.2 CPH	65 Tb 158.9 CPH	66 Dy 162.5 CPH	67 Ho 164.9 CPH	68 Er 167.3 CPH	69 Tm 168.9 CPH	70 Yb 173.0 CPH	71 Lu 175.0 CPH
			—	—									

†Actinium series:

90 Th 232.0	91 Pa 231	92 U 238.0	93 Np 237	94 Pu 244	95 Am 243	96 Cm 247	97 Bk 247	98 Cf 251	99 Es 254	100 Fm 257	101 Md 256	102 No 254	103 Lr —
				—	—								

CONVERSION FACTORS

Quantity	From	Exact conversion To	Multiply by	Engineering approximation
length	inch	cm	2.540	25 mm = 1 in.
	foot	m	3.048×10^{-1}	300 mm = 1 ft
	mile	km	1.609	
	m	foot	3.281	
	cm	in.	3.937×10^{-1}	
	angstrom(Å)	m	1.00×10^{-10}	
	mil	m	2.54×10^{-5}	
	micron (μm)	m	1.00×10^{-6}	
	micron (μm)	mil	3.937×10^{-2}	
mass	ounce	g	2.835×10	
	pound	kg	4.536×10^{-1}	2.8 g = 1 oz
	ton (2240 lb)	kg	1.016×10^{-3}	1 kg = 2.2 lb
	ton (2000 lb)	kg	9.072×10^{2}	
	g	ounce	3.527×10^{-2}	
	kg	pound	2.2046	
density	lb/ft³	kg/m³	1.602×10	16 kg/m³ = 1 lb/ft³
	kg/m³	lb/ft³	6.245×10^{-2}	
temperature	°F	°C	(°F−32)/1.8	0°C = 32°F
	°C	°F	(°C × 1.8)+32	100°C = 212°F
				−40°C = −40°F
area	in.²	mm²	6.452×10^{2}	645 mm² = 1 in.²
	ft²	m²	9.290×10^{-2}	1 m² = 11 ft²
	m²	in.²	1.550×10^{3}	
	m²	ft²	1.0764×10^{-1}	
	hectare (ha)	m²	1.0×10^{4}	
	hectare	acre	2.47	
volume	in.³	mm³	1.639×10^{4}	16400 mm³ = 1 in.³
	ft³	m³	2.832×10^{-2}	1 m³ = 35 ft³
	m³	ft³	3.531×10	
	m³	liter	1.0×10^{3}	
	m³	gal (UK)	2.1998×10^{2}	
	ft³	gal (US)	7.481	
	gal (UK)	gal (US)	1.201	
	gal (US)	gal (UK)	8.326×10^{-1}	
force	lbf	N	4.448	
	kgf	N	9.807	
	lbf	kgf	4.535×10^{-1}	4.4 N = 1 lbf
	N	lbf	2.248×10^{-1}	
	N	dyn	1.0×10^{5}	
pressure	Pa	N/m²	1.0	
or stress	lb/in.² (psi)	kPa	6.895	1 MPa = 145 psi
	atm	kPa	1.01325×10^{2}	
	atm	mm Hg (0°C)	7.60×10^{2}	
	mm Hg (0°C)	Pa	1.333×10^{2}	
	mm Hg (0°C)	mbar	1.00	
	mbar	in. WG	4.02×10^{-1}	
	bar	Pa	1.0×10^{5}	
	Pa	lb/in.²	1.45×10^{-4}	
torque	in.−lbf	N−m	1.130×10^{-1}	1 N−m = 9 in.−lbf
	ft−lbf	N−m	1.356	
	N−m	ft−lbf	7.376×10^{-1}	1.4 N−m = 1 ft−lbf
energy	Btu	J	1.0551×10^{3}	1 Btu = 1.1 kJ
	cal	J	4.187	1 kW−h = 3.6 MJ

| Quantity | Exact conversion | | | Engineering approximation |
	From	To	Multiply by	
	Btu	ft−lbf	7.78×10^2	1 Btu = 29 MW−h
	J	electronvolt	6.24×10^{18}	
	kW−h	J	3.6×10^6	
	kW−h	Btu	3.412×10^3	
	J	erg	1.0×10^7	
power	W	J/s	1.0	
	Btu/s	W	1.055×10^3	
	kW	horsepower	1.34	
	horsepower	W	7.46×10^2	
thermal conductivity	W/(m−K)	Btu/(h−ft−°F)	5.7803×10^{-1}	
	Btu/(h−ft−°F)	W/m−°C	1.731	
heat transfer coefficient	Btu/(h−ft²−°F)	W/(m²−K)	5.678	
viscosity (dynamic)	Pa−s	N−s/m²	1.0	
	Pa−s	poise	1.0×10	
	lb/(s−ft)	N−s/m²	1.488	
viscosity (kinematic)	stoke	m²/s	1.0×10^{-4}	
surface tension	N/m	lbf/ft	6.854×10^{-2}	
quantity of electricity	ampere-hour	coulomb	3.600×10^3	

SI DERIVED UNITS WITH SPECIAL NAMES

| Quantity | SI unit | | | |
	Name	Symbol	Expression in terms of other units	Expression in terms of SI base units
frequency	hertz	Hz	—	s^{-1}
force	newton	N	—	kg.m/s²
pressure	pascal	Pa	N/m²	kg/(m.s²)
energy, work, quantity of heat	joule	J	N.m	kg.m²/s²
power, radiant flux	watt	W	J/s	kg.m²/s³
quantity of electricity, electric charge	coulomb	C	A.s	A.s
electric potential potential difference, electromotive force	volt	V	W/A	kg.m²/(A.s³)
capacitance	farad	F	C/V	A².s⁴/(m².kg)
electric resistance	ohm	Ω	V/A	kg.m²/(A².s³)
conductance	siemens	S	A/V	A².s³/(m².kg)
magnetic flux	weber	Wb	V.s	kg.m²/(A.s²)
magnetic flux density	tesla	T	Wb/m²	kg/(A.s²)
inductance	henry	H	Wb/A	kg.m²/(A².s²)

SI PREFIXES, DECIMAL MULTIPLES AND SUBMULTIPLES OF SI UNITS

Factor	Prefix	Symbol	Factor	Prefix	Symbol
10^{-18}	atto	a	10	deka	da
10^{-15}	femto	f	10^2	hecto	h
10^{-12}	pico	p	10^3	kilo	k
10^{-9}	nano	n	10^6	mega	M
10^{-6}	micro	μ	10^9	giga	G
10^{-3}	milli	m	10^{12}	tera	T
10^{-2}	centi	c	10^{15}	peta	P
10^{-1}	deci	d	10^{18}	exa	E

SI BASE UNITS

Quantity	Name	Symbol
length	meter	m
mass	kilogram	kg
time	second	s
electric current	ampere	A
thermodynamic temperature	kelvin	K
amount of substance	mole	mol
luminous intensity	candela	cd

VALUES OF SELECTED PHYSICAL CONSTANTS

Quantity	Symbol†	SI units
Planck constant	h	6.626×10^{-34} J.s
Boltzmann constant	k	1.381×10^{-23} J/K
Avagadro constant	N_A	6.023×10^{23}/mol
Electron mass	m_e	9.109×10^{-31} kg
Proton mass	m_p	1.672×10^{-27} kg
Charge of electron or proton	e	1.602×10^{-19} C
Faraday constant	F	9.649×10^4 C/mol
Gas constant	R	8.314 J/(mol.K)
Molar volume at STP‡	V_m	22.41×10^{-3} m³/mol
Permittivity of vacuum or dielectric constant	ϵ_0	8.854×10^{-12} F/m
Permeability of vacuum or inductance	μ_0	$4\pi \times 10^{-7}$ H/m
Stefan–Boltzmann constant	σ	5.670×10^{-8} W/(m².K⁴)
Speed of light in vacuum	c	2.998×10^8 m/s
Standard gravity (free fall)	g	9.807 m/s²
Base of natural logarithm	e	2.718

† Symbol often used, not unit symbols.
‡ STP ~ 273.15 K and 101.325 kPa.

1
Materials Science

---- CHAPTER OBJECTIVES ----

To obtain an understanding of:
1 the structural characteristics of materials, the forces holding them together and the defects which may be present;
2 the relevance of equilibrium diagrams in the treatment and selection of engineering materials;
3 the effects of changes in temperature and composition on the structure of alloys.

---- IMPORTANT TERMS ----

chemical bond	close-packed hexagonal	alloy
covalent bond	coordination number	solid solution
ionic bond	atomic packing factor	equilibrium (phase)
metallic bond	atoms per unit cell	diagram
hydrogen bond	stacking sequence	liquidus line
van der Waals forces	dislocation	solidus line
crystallization	point defect	eutectic
space lattice	line defect	peritectic
unit cell	bulk defect	eutectoid
body-centered cubic	phase	intermetallic compound
face-centered cubic		

┌─────────────────────── NOTE ───────────────────────┐

A glossary of terms associated with materials is included in Section 1.1. It is
not intended to be exhaustive, nor is it meant to provide absolute definitions
and detailed descriptions of the terms included. It is intended to provide a
useful and quick *aide de memoire*. The reason it has not been included at the
end of the chapter or as an appendix is that hopefully the reader will notice its
presence early in his reading of the book, and will remember (more easily)
where to locate this reference source.

└───┘

1.1 GLOSSARY OF TERMS AS APPLIED TO MATERIALS

Abrasive – A hard particle with sharp edges which removes softer material by scratching; used for grinding or cutting; commonly made of a ceramic material.

Activators – Components added to start a reaction.

Addition or **chain polymerization** – Polymerization by sequential addition of monomers.

Age-hardening – Hardening with time by incipient precipitation.

Aging – Process of age-hardening

Aggregate – Coarse particles used in concrete, e.g. sand and gravel.

Alloy – A metal containing two or more elements.

Alloying elements – Elements added to form an alloy (sometimes referred to as alloys).

Alpha iron – Iron with a body-centered cubic structure which is stable at room temperature.

Amorphous – Noncrystalline and without long-range order.

Anisotropic – Having different properties in different directions.

Annealing – Heating and cooling to produce softening.

Antioxidant – Inhibitor to prevent oxidation of rubber and other organic materials by molecular oxygen.

Asbestos – A naturally occurring fibrous silicate mineral.

Atactic – Lacking long-range repetition in a polymer (as contrasted to isotactic).

Ausforming – Process of strain-hardening austenite prior to transformation.

Austempering – Process of isothermal transformation to form bainite.

Austenite – Face-centered cubic solid solution of iron and carbon.

Axis (crystal) – One of three principal crystal directions.

Bainite – Microstructure of discrete rounded carbide dispersed in ferrite, obtained by low-temperature isothermal transformation.

Body-centered (crystal) – A unit cell with center positions equivalent to corner positions.

Boundary (microstructure) – Surface between two grains or between two phases.

Brass – An alloy of copper and zinc.

Bravais lattices – The 14 basic crystal lattices.

Brazing – Joining metals at temperatures above 500°C (932°F) but below the melting point of the joined metals.

Brinell – A hardness test utilizing a spherical hardened steel indenter. The hardness is measured by the load/area of the indentation.

Bronze – An alloy of copper and tin (unless otherwise specified – e.g. an aluminum bronze is an alloy of copper and aluminum).

Burgers vector – Displacement vector around a dislocation. It is parallel to a screw dislocation and perpendicular to an edge dislocation. It describes both the magnitude and direction of slip due to external stress.

Carbon steel – Steel in which carbon is the chief variable alloying element (other alloying elements may be present in nominal amounts).

Carborundum – A trade name for silicon carbide.

Carburizing – Introduction of carbon to the surface of a steel to change the surface properties.

Case-hardening – Hardening by forming a case of higher carbon content.

Casting – The process of pouring a liquid or suspension into a mold, or the object produced by this process.

Cast iron – Casting of iron–carbon alloys or iron–carbon–silicon alloys containing 2–4% carbon and 0·5–3% silicon.

C-curve – Isothermal transformation (I–T) diagram or time–temperature–transformation (T–T–T) curve.

Cement – A material (usually ceramic) for bonding solids together.

Cement, Portland – The main constituents are di- and tricalcium aluminate; hydration produces a crystalline structure.

Cementite – The intermetallic compound iron carbide (Fe_3C).

Ceramic (phase) – Compound of metallic and non-metallic elements.

Chain polymerization – *see* **Addition polymerization**

Charpy – One of two standardized pendulum impact tests (the other is **Izod**) using notched specimens.

Cleavage – Plane of easy splitting.

Close-packed – Structure with the highest possible packing factor.

Cold working – Permanent deformation below the recrystallization temperature.

Concrete – Agglomerate of an aggregate, water and a cement.

Condensation polymerization – Polymerization by chemical reaction which also produces a by-product.

Constituent (microstructure) – A distinguishable part of a multiphase mixture.

Continuous cooling transformation – Transformation during cooling (as opposed to isothermal transformation).

Coordination number – Number of closest atomic neighbors.

Copolymerization – Addition polymerization involving more than one type of mer.

Corrosion – Environmental deterioration of a material or its properties.

Covalent bond – Atomic bonding by sharing electrons.

Creep – A continuous slow deformation at constant stresses below the normal yield strength.

Cross-linking – The tying together of adjacent polymer chains.

Crystal – A physically uniform solid, in three dimensions, with long-range repetitive order.

Decarburization – Removal by diffusion of carbon from the surface of a steel.

Deep-drawing – The process of forming cup-shaped articles out of sheet metal by punching.

Degree of polymerization – Number of mers per average molecular weight.

Delta iron – Body-centered cubic iron, which is stable above the temperature range of austenite; also used to describe body-centered cubic solid solutions of iron and carbon at high temperatures.

Dendrite – Skeleton crystal.

Depolymerization – Degradation of polymers into smaller molecules.

Diamond cubic (crystal) – The cubic crystal structure possessed by diamond.

Die – A forming tool.

Dislocation – Lack of coherence. A dislocation moves as a shear front through the material. The type of dislocation is given by the direction of the movement relative to the dislocation line.

Dislocation, edge – Linear defect is an extra half plane of atoms above the slip plane. The Burgers vector is perpendicular to the defect line.

Dislocation, screw – Linear defect with Burgers vector parallel to the defect line.

Ductility – The ability of a material to undergo cold plastic deformation, usually by tension, before fracturing.

Elasticity – Non-permanent deformation.

Elastic limit – Stress limit of elastic deformation.

Elastomer – Polymer with high elasticity due to its coiled structure.

Elongation – The amount of permanent strain prior to fracture.

Enamel (ceramic) – A protective coating of glass on metal.

Enamel (paint) – A paint possessing a gloss like glass.

End-quench – A test for determining hardenability (also called a Jominy test).

Endurance limit – The maximum stress allowable for unlimited cycling.

Equiaxed – Shapes with approximately equal dimensions, e.g. grains.

Equilibrium diagram – *See* **Phase diagram**.

Etching – Controlled chemical corrosion to reveal the microstructure.

Eutectic (binary) – A thermally reversible reaction:

$$\text{liquid} \; \underset{\text{heating}}{\overset{\text{cooling}}{\rightleftarrows}} \; \text{solid}_1 + \text{solid}_2$$

Eutectoid (binary) – A thermally reversible reaction:

$$\text{solid}_1 \; \underset{\text{heating}}{\overset{\text{cooling}}{\rightleftarrows}} \; \text{solid}_2 + \text{solid}_3$$

Extrusion – Shaping operation accomplished by forcing a plastic material through a die.

Face-centered – A unit cell with face positions equivalent to corner positions.

Fatigue – Tendency to fracture by crack propagation due to cyclic stresses.

Fatigue limit – *See* **Endurance limit**.

Ferrite – Body-centered cubic solid solution of iron and carbon.

Fiber structure – Two-phase structure where the strengthening phase is in the form of fibers, macroscopically heterogeneous, and increased fiber alignment increases anisotropy.

Fictive temperature – Transition temperature between supercooled liquids and glassy solids (glass transition temperature).

Fillet – A rounded corner between two surfaces.

Fireclay – A refractory (rather pure) clay.

Firing (ceramic) – High-temperature treatment for agglomeration.

Firing shrinkage – Shrinkage accompanying sintering.

Flakes (graphite) – Two-dimensional graphite sheets in cast iron.

Flakes (steel) – Internal fissures.

Flame-hardening – Hardening by surface heating with appropriate flames.

Forming – Shaping by deformation.

Frenkel defect – Atom or ion displacement to an interstitial site, producing a vacancy and an interstitial atom or ion.

Gamma iron – Iron with a face-centered cubic structure which is stable between $908°C$ and $1388°C$ ($1666°F$ and $2530°F$).

Glass – An amorphous material with three-dimensional primary bonds.

Grain – Individual crystal.

Graphite – Most common phase of carbon (sheet-like).

Graphitization – Carbide dissociation to graphite.

Gray cast iron – Cast iron with flake graphite which was formed during solidification.

Hard-drawn – Cold worked to high hardnesses by drawing.

Hardenability – The property that determines the depth and distribution of hardness induced by quenching, under specified conditions.

Hardening – Heat treatment to increase hardness.

Hardness – Resistance to penetration.

Hole – Vacancy in a crystal or in an electronic structure.

Hot working – Shaping above the recrystallization temperature.

Hyper-eutectoid – A steel with a higher carbon content than the eutectoid composition.

Hypo-eutectoid – A steel with a lower carbon content than the eutectoid composition.

Impact test – A test which measures the energy absorbed during fracture.

Inclusions – Particles of impurities contained by a material.

Induction-hardening – Hardening by using high-frequency induced currents for surface heating.

Ingot – A large casting which is to be subsequently rolled or forged.

Interrupted quench – Two-stage quenching of steel which involves heating to form austenite and an initial quench to a temperature above the start of martensite formation, followed by a second cooling to room temperature (marquenching or martempering).

Interstices – Open-pore spaces between particles.

Isomer – Molecules with the same composition but different structures.

Isotactic (polymer) – Long-range repetition in a polymer chain (in contrast to atactic).

Isothermal transformation – Transformation with time by holding at a specific temperature.

Isotropic – Having the same properties in all directions.

Izod – One of two standardized impact tests (the other is **Charpy**) using notched specimens.

Jominy bar – A heated bar of metal rapidly cooled from one end; – *see* **End-quench**.

Kiln – A furnace for firing ceramic materials.

Lattice (crystal) – The space arrangment of atoms in a crystal structure.

Ledeburite – Eutectic microstructure of austenite and cementite.

Liquidus – The locus of temperatures with compositions above which only liquid is stable.

Long-range order – A repetitive pattern over many atomic distances, a property of a crystal structure.

Low-alloy steel – Steel containing up to approximately 10% (total) alloying elements.

Macroscopic – Visible to the unaided eye (or up to $10\times$ magnification).

Macrostructure – Structure with macroscopic heterogeneities (*cf.* **Microstructure**).

Malleability – The ability to deform plastically, usually by compressive forces, e.g. cold hammering or squeezing.

Malleable iron – Cast iron in which the graphite was formed by solid graphitization.

Marquenching – *See* **Interrupted quench**.

Martempering – *See* **Interrupted quench**, and also **Tempering**.

Martensite – Metastable body-centered tetragonal phase of iron supersaturated with carbon, produced through a diffusionless phase change by quenching austenite.

Matrix – The enveloping phase in which another phase is embedded.

Mechanical properties – Those properties associated with stress and strain.

Mechanical working – Shaping by the use of forces.

Mer – The smallest repetitive unit in a polymer.

Metallic bond – Atomic bonding in metals caused by the valency electrons forming an electron gas or cloud.

Metals – Materials containing elements which readily lose electrons.

Metastable – Temporary equilibrium.

Microstructure – A structure with heterogeneities as shown by a microscope (*cf.* **Macrostructure**).

Mineral – A rock phase or a ceramic phase.

Mixture – Combination of two phases.

Mold – A shaped cavity for casting.

Molecule – Group of atoms with strong mutual attraction.

Monel – An alloy of copper and nickel.

Monoclinic (crystal) – Three unequal axes, two of which are at right angles to each other.

Monomer – A molecule with a single mer.

Nitriding – Introduction of nitrogen into the surface of a steel to increase the hardness.

Nodular cast iron – A cast iron with spherical graphite microstructure.

Normalizing – Heat treatment for homogenization.

Notch sensitivity – Change of behavior from ductile to brittle, due to the presence of stress concentrations caused by a standard notch.

Nucleation – The start of the growth of a new phase.

Ordered (crystal) – Structure with a long-range repetitive pattern.

Orthorhombic – A crystal with three unequal but perpendicular axes.

Overaging – Continued aging until softening occurs.

Packing factor – True volume per unit of bulk volume.

Pearlite – A microstructure of ferrite and lamellar carbide of eutectoid composition.
Peritectic (binary) – A thermally reversible reaction:

$$\text{solid}_1 + \text{liquid} \xrightleftharpoons[\text{heating}]{\text{cooling}} \text{solid}_2$$

Peritectoid (binary) – A thermally reversible reaction:

$$\text{solid}_1 + \text{solid}_2 \xrightleftharpoons[\text{heating}]{\text{cooling}} \text{solid}_3$$

Phase (material) – A physically homogeneous part of a material system.
Phase diagram – Graph of phase relationships with composition and environmental coordinates.
Plain carbon steel – Iron-carbide alloys with only nominal amounts of other elements.
Plastic – Moldable organic resin (*see* **Plasticity**).
Plasticity – Ability to be permanently deformed without fracture.
Plasticizer – An additive to commercial resins to induce plasticity.
Polyethylene – Polymer of $(C_2H_4)_n$.
Polymer – Molecules with many units or mers.
Polymerization – Process of growing large molecules from small ones.
Polymorphism – The existence of a composition in more than one crystal structure.
Portland cement – A hydraulic calcium silicate cement.
Powder metallurgy – The technique of agglomerating metal powders into engineering components.
Precipitation hardening – *See* **Age-hardening**.
Process annealing (steel) – Annealing close to, but below, the eutectoid temperature.
Pro-eutectoid cementite – Cementite which precipitates from hyper-eutectoid austenite above the eutectoid temperature.
Pro-eutectoid ferrite – Ferrite which precipitates from hypo-eutectoid austenite above the eutectoid temperature.
Quench – Rapid cooling to preserve a metastable phase, e.g. martensite or glassy metals.
Recrystallization – The formation of new annealed grains from previously strain-hardened grains.
Recrystallization temperature – Temperature at which recrystallization is spontaneous.
Refractory (ceramic) – A heat-resistant material.
Resilience – The ability of a material to absorb and return energy without permanent deformation.
Resin – Polymeric material, usually a thermosetting resin.
Rhombohedral – Crystals with three equal axes not at right angles.
Rigidity – The property of resisting elastic deformation due to shear.
Rockwell hardness – A test utilizing an indenter; a function of the depth of indentation is a measure of the hardness.
Rubber – A polymeric material with a high elastic yield strain.
Schottky defect – Ion pair vacancies.
Segregation – Heterogeneities in composition.
Simple cell (crystal) – A unit cell with only corner atoms (also primitive cell).
Sintering – Agglomeration of a powder using its thermal surface energy.
Slip (deformation) – A relative displacement along a preferred structural direction on a slip plane.
Slip plane – Crystal plane along which slip occurs.
Soldering – Joining metals below $500°C$ ($932°F$); the joined metals are not melted.
Solidus – The locus of temperatures and compositions below which only solids are stable.
Solute – The minor component of a solution.
Solution treatment – A heat treatment to produce a solid solution.
Solvent – The major component of a solution.
Solvus – Solid solubility curves in a phase diagram.

Spheroidite – Microstructure of coarse spherical carbides in a ferrite matrix.

Spheroidizing – Process of making spheroidite.

Sterling silver – An alloy 92.5% silver and 7.5% copper (approximately the maximum solubility of copper and silver).

Strain hardening – Increased hardness accompanying deformation.

Stress – Force per unit area.

Structure – Geometric relationships of material components.

Supercooling – Cooling below the solubility limit without precipitation.

System (phase diagram) – Compositions of equilibrated components.

Temper (hardness) – Extent of strain hardening.

Temper carbon – In cast iron, carbon that is a product of graphitization.

Temper glass – Glass with residual compressive surface stresses.

Tempered martensite – A microstructure of ferrite and carbide obtained by heating martensite.

Tempering – A toughening process of heating martensite to produce tempered martensite.

Tensile strength – Maximum resistance to deformation (based on original area).

Tetragonal (crystal) – Two of the three axes are equal; all three are at right angles.

Tetrahedron – A four-sided solid.

Thermoplastic resin – A linear-chain structural polymeric material that softens with increased temperature.

Thermosetting resin – A network-structured polymeric material which does not soften, but rather polymerizes further with increasing temperature.

Toughness – The property of absorbing energy before fracture.

Transformation temperature – Temperature of an equilibrium phase change.

Transition temperature (steel) – Temperature (range) of change from ductile to nonductile fracture (i.e. brittle-ductile transition temperature).

Transverse – Section perpendicular to the elongated direction (in contrast to longitudinal).

Triclinic (crystal) – Three unequal axes, none of which are at right angles to each other.

TTT curve – Temperature–Time–Transformation curve (isothermal).

Unit cell – The smallest repetitive volume that comprises the complete pattern of a crystal (sometimes known as a primitive cell).

Vacancy – Unfilled lattice site.

van der Waals forces – Secondary bonds arising from structural polarization.

Vulcanization – Treatment of rubber with sulfur to cross-link the elastomer chains.

Welding (fusion) – A joining operation involving melting part of the metals to be joined.

Welding (pressure) – A joining operation involving heating part of the metals to be joined (below the melting point) and applying pressure.

White cast iron – Cast iron with all the carbon as cementite rather than as graphite.

1.2 THE STRUCTURE OF MATERIALS

1.2.1 Atoms and molecules

There are approximately 100 different kinds of atoms; those with the largest atomic masses have been produced artificially. An atom can be regarded as a positive core or nucleus orbited by a series of negative electrons. The chemical properties of a substance are related to the number of electrons it possesses: when the orbits are not completely filled there is a tendency to become stable by gain or loss of electrons. There are also isotopes, which are atoms with the same chemical properties but slightly different nuclei. We are normally concerned with only 20 to 30 of the most common atoms. Each species of atom is called an element, and all the atoms of a particular element are virtually identical, whereas those of different elements have very different properties.

Atoms of an element can exist either in a pure, homogeneous state as elements e.g. tin or lead, as mixtures or solutions, or as definite chemical compounds, e.g. iron carbide (Fe_3C). Chemical compounds are normally formed from definite proportions of the elements and exhibit distinct, characteristic properties.

Compounds which are formed from the elements carbon, oxygen, hydrogen and nitrogen in living matter or artificial substances, e.g. oil, are broadly classified as organic compounds. Other compounds, such as metals and ceramics, are broadly termed inorganic.

A *molecule* is the smallest unit of a substance which possesses the properties of that substance, e.g. sodium chloride (NaCl), benzene (C_6H_6) or chlorine (Cl_2).

1.2.2 Chemical bonds

Chemical bonds are the electromagnetic forces which hold together atoms and molecules.

(a) Primary bonds joining the atoms within a molecule

COVALENT BONDS
These bonds occur when two atoms share a pair of electrons. They are hard to form but are strong and rigid. They occur within organic molecules and ceramics.

IONIC BONDS
These bonds occur between positive and negatively charged elements or ions, e.g. sodium gives an electron to chlorine so that Na^+ is positively charged and Cl^- is negatively charged. These ions are then attracted to each other:

$$\left. \begin{array}{l} Na - e = Na^+ \\ Cl + e = Cl^- \end{array} \right\} Na^+Cl^-$$

It is common for substances to possess both ionic and covalent forces. Covalent bonds are strongly directional, whereas ionic forces tend to be uniformly distributed around the elements.

METALLIC BONDS
These bonds hold together metals and their alloys when they are not formed as chemical compounds. Some of the outer electrons are not held permanently in orbits but move between atoms (an 'electron sea', gas or cloud), accounting for the good electrical conductivity of metals. Metallic bonds are relatively easy to form, break and reform.

(b) Secondary bonds attaching molecules to each other

Although as a whole a water molecule is electrically neutral, the distribution of charge within a water molecule is not uniform. This force can be attractive towards other similar groups, e.g. hydroxyl (OH^-) ions in organic molecules which attract water molecules. These forces are known as *hydrogen bonds*.

Small local charge variations can result in weak forces between molecules, known as *van der Waals forces*.

1.2.3 Crystallization

When liquids are cooled the molecules tend to adopt a regular structure, this leads to crystal formation associated with the forces existing between the molecules. All solids have a tendency to be crystalline, but not all substances achieve it. All metals and many simple inorganic substances are crystalline. The ideal crystal is used as a standard (*cf.* the ideal gas); no substance actually exists in this state but it represents a useful first approximation to real systems.

When a liquid begins to solidify the unit cells form a regular three-dimensional array known as a *lattice*. A crystal may be as large as 5 cm (2 in.) in diameter, although the normal size is between 0.01 and 0.1 mm (0.0004 and 0.004 in.) diameter. When the solid is composed of many crystals (termed a polycrystalline material), they are all identical except for the directions of the crystal axes, with respect to a reference system. There are usually no void spaces within the solid crystal structure. The individual crystals are called *grains* and they meet at *grain boundaries*.

The grain boundaries have a higher energy than the surrounding grains at lower temperatures. When a specimen is etched, the grain boundaries dissolve quicker and they can be seen as grooves if examined with a microscope. The crystalline nature of a solid is determined by X-ray or electron diffraction methods. The regular planes of crystals diffract a beam of X-rays and crystal sizes can be determined.

Crystals are not ideal because of defects within the geometrical arrangement of the molecules. The properties of a substance can be either structure-insensitive, e.g. density, or structure-sensitive, e.g. electrical conductivity of semiconductors at low temperatures (this becomes structure-insensitive for metals at high temperatures). A small number of defects can change the value of a structure-sensitive property by several orders of magnitude.

Metals are elements that possess a positive temperature coefficient of resistance. They can also be defined by reference to their properties, for example, they

(a) are crystalline;
(b) conduct heat;
(c) conduct electricity;
(d) are malleable and ductile under certain conditions;
(e) possess a metallic lustre;

(f) are opaque;
(g) have high density;
(h) are able to form alloys.

There are seven crystal systems. These can be defined in terms of the geometry of the unit cell on which the crystalline space lattice is built. These seven modes are defined in Table 1.1, with reference to Figs. 1.1 and 1.2.

Crystals are structures composed of atomic arrangements which exhibit a repeating pattern in three dimensions; this structure is known as a *space lattice*.

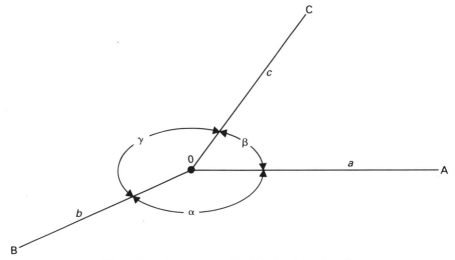

Figure 1.1 Axes and angles defining the unit cell.

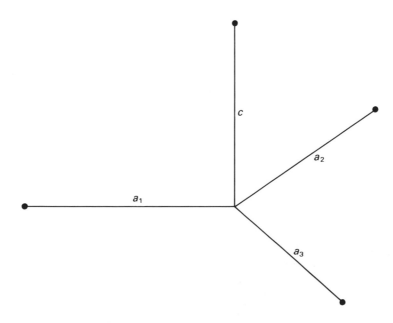

Figure 1.2 Axes defining the hexagonal unit cell.

Crystalline structures may occur in seven main modes, although pure metals usually adopt either a cubic or hexagonal pattern. In this section consideration will be given mainly to metals containing only one kind of identical atom.

Table 1.1   Defining the seven basic crystal systems

System	Axes	Axial angles
Cubic	$a = b = c$	$\alpha = \beta = \gamma = 90°$
Tetragonal	$a = b \neq c$	$\alpha = \beta = \gamma = 90°$
Orthorhombic	$a \neq b \neq c$	$\alpha = \beta = \gamma = 90°$
Monoclinic	$a \neq b \neq c$	$\alpha = \beta = 90° \neq \gamma$
Triclinic	$a \neq b \neq c$	$\alpha \neq \beta \neq \gamma \neq 90°$
†Hexagonal	$a = b \neq c$	$\alpha = \beta = 90°; \gamma = 60°$
Rhombohedral	$a = b = c$	$\alpha = \beta - \gamma \neq 90°$

† The Miller–Bravais nomenclature defines the hexagonal form in terms of four axes, of which three are coplanar as shown in Fig. 1.2, i.e.

$$a_1 = a_2 = a_3 \neq c$$

The three coplanar axes (a_1, a_2, a_3) are inclined at 120° to each other, the other axis (c) is mutually perpendicular to the other three axes.

1.2.4 Crystal modes

Atoms can be packed in a cubic pattern with three different types of repetition: simple, body-centered or face-centered modes. Another common form is an hexagonal unit cell.

The simple (primitive) cubic structure shown in Fig. 1.3 provides a starting point for the consideration of different crystal modes. There are no pure metals of this form because the atoms are not packed in an efficient arrangement.

The body-centered cubic structure (bcc) shown in Fig. 1.4 contains an atom at each corner of the unit cell, and another at the body center of the cube.

The face-centered cubic structure (fcc) shown in Fig. 1.5 contains an atom at each corner of the unit cell and one at the center of each face, but none at the center of the cube. This is a *close-packed cubic structure.*

A simple hexagonal structure is shown in Fig. 1.6. It can be represented as either an hexagonal cell, or as a rhombic structure with a shared corner atom in each plane, thus forming an hexagonal cell. As with the simple cubic structure, pure metals do not crystallize in this form because of the low atom packing efficiency.

A close-packed hexagonal structure (cph) is shown in Fig. 1.7 and provides a denser packing arrangement of the atoms comprising the unit cell. Each atom in a layer is directly above or below the interstices of three atoms in the adjacent layers. Also each atom touches six atoms in its own layer, three atoms above and three atoms below.

Alternative crystal structures. All crystal structures fall into one of the 14 categories shown in Fig. 1.8. These crystal structures can also be defined by reference to their geometry, as shown in Table 1.2.

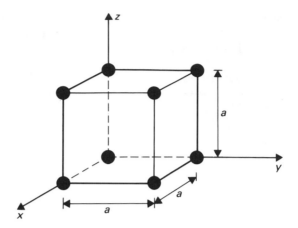

(a) Simple cubic cell. The corners of unit cells are at 'like' positions in the crystal: $a = a = a$. Axes are at right angles.

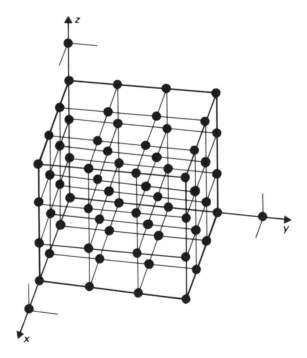

(b) Cubic lattice. Space is divided by three sets of equally spaced parallel planes. The *x, y, z* reference axes are mutually perpendicular. Each point of intersection is equivalent.

Figure 1.3 Simple cubic structure.

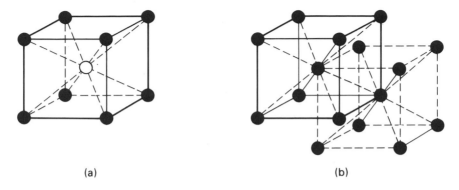

Figure 1.4 Body-centered cubic structure. (a) Unit cell (b) Crystal lattice.

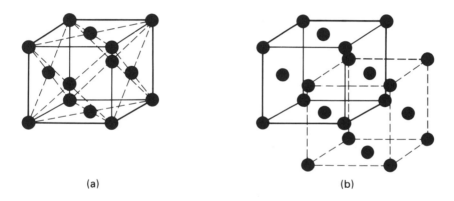

Figure 1.5 Face-centered cubic structure. (a) Unit cell (b) Crystal lattice.

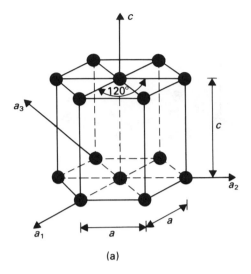

(a)

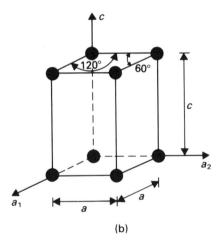

(b)

Figure 1.6 Simple hexagonal unit cells. (a) Hexagonal representation. (b) Rhombic representation. These two are equivalent with $a \neq c$, a basal angle of 120° and a vertical angle of 90°.

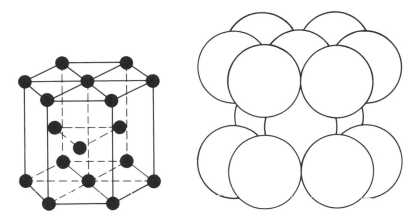

Figure 1.7   Close-packed hexagonal structure.

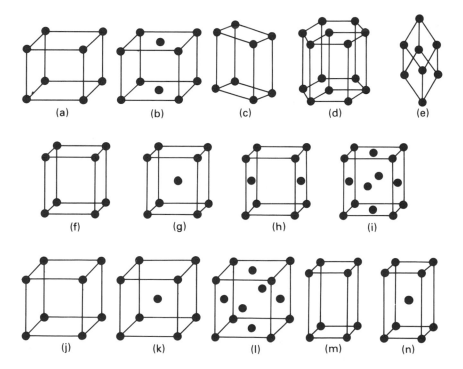

Figure 1.8   Space lattices. These 14 Bravais lattices continue in three dimensions, each indicated point having identical surroundings.

(a) simple monoclinic	(g) body-centered orthorhombic	(k) body-centered cubic
(b) end-centered monoclinic	(h) end-centered orthorhombic	(l) face-centered cubic
(c) triclinic	(i) face-centered orthorhombic	(m) simple tetragonal
(d) hexagonal	(j) simple cubic	(n) body-centered tetragonal
(e) rhombohedral		
(f) simple orthorhombic		

Table 1.2 The characteristics of crystal systems

System	Bravais lattices	Lattice parameters and interaxial angles	Examples
Cubic	Simple, face-centered and body-centered	Three axes at right angles, lattice parameters all equal: $a = b = c$ $\alpha = \beta = \gamma = 90°$	Ar, Au, Ag, Cu, Ca, Al, Pb, Ni, Si, NaCl, LiF
Tetragonal	Simple and body-centered	Three axes at right angles, two spacings equal: $a = b \neq c$ $\alpha = \beta = \gamma = 90°$	In, TiO_2, KIO_4, 'white' Sn
Rhombohedral	Simple	Three axes equally inclined but not at right angles, equal spacings: $a = b = c$ $\alpha = \beta = \gamma \neq 90°$	Hg, Bi, Sb
Hexagonal	Simple	Two axes at 120°, third axis at 90° to these: $a = b \neq c$ $\alpha = \beta = 90°; \gamma = 120°$	Zn, Mg, Cd, NiAs
Orthorhombic	Simple, base-centered, face-centered and body-centered	Three axes at right angles, spacings all unequal: $a \neq b \neq c$ $\alpha = \beta = \gamma = 90°$	Ga, I, Fe_2S, Fe_3C, $BaSO_4$
Monoclinic	Simple and base-centered	One axis at right angles to the other two, which are not at right angles to each other: $a \neq b \neq c$ $\alpha = \gamma = 90° \neq \beta$	$KClO_3$, KNO_2, $K_2S_4O_6$, As_2S_4
Triclinic	Simple	Three axes not at right angles: $a \neq b \neq c$ $\alpha \neq \beta \neq \gamma \neq 90°$	$K_2S_2O_8$, Al_2SiO_5, $B(OH)_3$

* * *

Self-assessment exercises

1 Describe the following terms and provide examples of each:
   (a) ionic bonding;
   (b) covalent bonding;
   (c) metallic bonding.

2 Explain what is meant by dynamic equilibrium as applied to atoms and lattice structures. (This exercise may require additional reading, e.g. Barrett *et al.*, (1973), Chapter 4, and discussion with a tutor.)

3 Specify the crystal systems and Bravais lattices for each of the following crystal descriptions:
   (a) $a \neq b \neq c, \alpha = \beta = \gamma = 90°$;
   (b) $a = b \neq c, \alpha \neq \beta \neq \gamma$;

4 Draw the lattice structures of:
   (a) the simple cube;
   (b) bcc;
   (c) fcc;

(d) simple hexagonal;
(e) cph;
(f) bc tetragonal.

5 Build a model of each of the lattice structures (a) – (f) in Exercise 4.
   Identify the corner atoms and other atoms within the lattice. Build models showing long-range atomic order and the stacking sequences of the atoms.

(*Note*. If a commercial model building kit is not available use table tennis balls, golf balls, glue, plasticine, straws, etc.
The importance of this exercise cannot be overemphasized. A far greater understanding of crystal structure is possible from model building than from pencil-and-paper representations.)

*   *   *

1.2.5 Crystal directions and planes

The unit cell is used as the basis for describing crystal directions and planes. This can be seen most easily by considering the crystal directions on the simple orthorhombic lattice shown in Fig. 1.9.

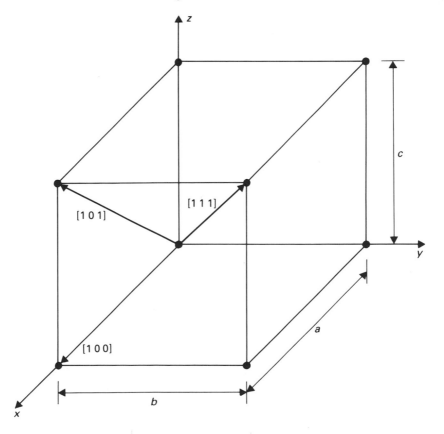

Figure 1.9   Unit cell for a simple orthorhombic lattice.

Unit cell distances, and not intercept distances, are used ($a \neq b \neq c$). The crystal axes are used as base directions.

The direction [1 1 2] is from the origin through the top face.

Miller indices are used to define points and planes within a crystal structure. The atomic spacing is the lattice parameter with respect to an arbitrary origin. A point is defined in terms of its coordinates, and a plane in terms of the reciprocals of its intercepts with the axes.

The (0 1 0) planes are parallel to the x and z axes, and the (1 1 1) planes cut the three crystal axes as shown in Fig. 1.10.

With reference to a unit cell, assume a plane cuts the axes at $3/2$, 2, 2. Take reciprocals of these intercepts, obtaining $2/3$, $1/2$, $1/2$, then rationalize to obtain (4 3 3). These are the Miller indices of the plane.

With reference to Fig. 1.11, where OA, OB, and OC represent the sides of a unit cell, then

$$OH = 2 \times OA$$
$$OJ = 1/2 \times OA$$
$$OK = 2/3 \times OB$$

The plane ADEG intercepts the axes at A, ∞, ∞.

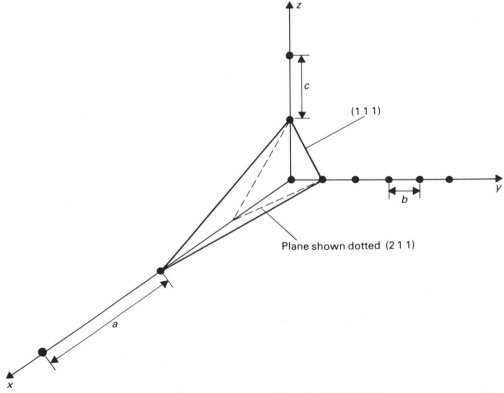

Figure 1.10 Definition of crystal planes using Miller indices.

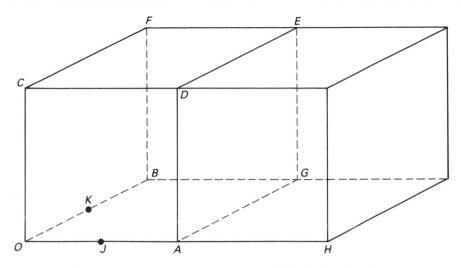

Figure 1.11　Unit cell used to define the Miller indices of a plane.

The Miller indices of the plane are given by the reciprocals of the intercepts, namely

$$\frac{OA}{OA} \quad \frac{OB}{\infty} \quad \frac{OC}{\infty} \qquad \text{or} \qquad (1\ 0\ 0)$$

Similarly, plane BFEG would be (0 1 0) and plane CDEF would be (0 0 1). Planes parallel to the faces of the cube have similar indices; the family of planes parallel to ADEG is represented by {1 0 0}.

The plane CJK intercepts the axes at J, K and C, the indices are given by

$$\frac{OA}{OJ} \quad \frac{OB}{OK} \quad \frac{OC}{OC} \qquad \text{or} \qquad (2\ \tfrac{3}{2}\ 1).$$

After rationalization this becomes (4　3　2).

The plane HKC is represented by

$$\frac{OA}{OH} \quad \frac{OB}{OK} \quad \frac{OC}{OC} \qquad \text{or} \qquad (\tfrac{1}{2}\ \ \tfrac{3}{2}\ \ 1)$$

After rationalization this becomes (1　3　2).

* * *

Self-assessment exercises

1 Show that the Miller indices are (4 6 3) for a plane that intersects the *x*, *y* and *z* axes at:
   (a) 3, 2 and 4 units, respectively;
   (b) 6, 4 and 8 units, respectively.

2 Show that the Miller indices are (3 4 6) for a plane that intersects the x, y and z axes at 4, 3 and 2 units, respectively.
   Show that the direction perpendicular to this plane is [3 4 6].

3 Sketch the following planes in a body-centered cubic lattice: (1 1 1), ($\bar{1}$ 1 1), (1 0 0) and (2 2 0). Show that the number of atoms touching each plane in one lattice is 3, 3, 5 and 0, respectively.

4 Repeat Exercise 3 for a face-centered cubic lattice and calculate the planar atomic packing for each plane.

5 Determine the Miller indices for the specific planes in the {2 1 0} family for a cubic crystal.

6 Determine the direction indices for a line connecting points at ⅓, ½, 0 and ⅔, ½, ½ in a cubic crystal.

7 Determine the direction indices for the specific directions in the <1 1 2> family for a cubic crystal. Show these directions on a sketch or a model.

8 Sketch the plane that cuts the x, y and z axes of a cubic crystal at distances of +¾, −⅔ and +½ respectively from the origin. Draw the direction arrow perpendicular to this plane and state its direction indices, and those of the family of directions.

* * *

1.2.6 Characteristics of crystal structures

For metals there is an interrelationship between the melting point, the elastic modulus (which is a measure of the bonding) and the strength (resistance to deformation). This is shown by the variation in values of a particular property for different metals, and values of the parameters which describe the crystal structures. The properties of some pure metals are given in Table 1.3 (refer also to Tables 3.1, 3.2, 3.27, and 3.28).

COORDINATION NUMBER
For a particular atom in a regular repeating crystal structure, the coordination number is the number of nearest (or first) neighbors of the atom.

ATOMIC PACKING FACTOR
This is the fraction of the unit cell which is occupied by the atoms comprising that structure. The metal atoms are considered to be 'hard balls' of radius r for a cubic structure with unit cell dimensions a.

Examples
(1) *Simple cubic cell*

$$\text{Atomic packing factor} = \frac{\text{volume of atoms}}{\text{volume of unit cell}}$$

$$= (\tfrac{4}{3})\pi r^3 / a^3$$
$$= (\tfrac{4}{3})\pi r^3 / (2r)^3$$
$$= 0.52$$

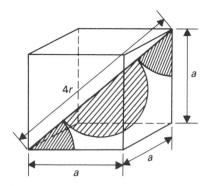

Figure 1.12 Body-centered cubic unit cell. This structure has two atoms per unit cell, one atom at the center and each corner atom is shared by eight unit cells.

(2) *Body-centered cubic*

The lattice constant (a) is related to the atomic radius (r) as shown in Fig. 1.12, such that

$$a = 4r/\sqrt{3}$$

$$
\begin{aligned}
\text{Atomic packing factor} &= 2(\tfrac{4}{3})\pi r^3/a^3 \\
&= 2(\tfrac{4}{3})\pi r^3/(4r/\sqrt{3})^3 \\
&= 0.68
\end{aligned}
$$

(3) *Face-centered cubic*

Using $a = 4r/\sqrt{2}$ from Fig. 1.13, then the atomic packing factor is 0.74.

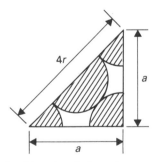

Figure 1.13 Face-centered cubic unit cell. A section across the face of a unit cell. The structure has four atoms per unit cell; each corner atom is shared by eight unit cells and each face-centered atom is shared by two unit cells.

NUMBER OF ATOMS PER UNIT CELL

This is another useful parameter; it conveys information about the density of the atoms in a cell. For a cubic structure each corner atom is shared by eight unit cells, it can therefore be considered that each corner atom contributes one-eighth of an atom to each unit cell, or only one corner atom belongs wholly to each cell.

Table 1.3 Structure and properties of some pure metals

Element and symbol	Atomic number	Atomic mass	Melting point °C (°F)	Density kg m^{-3} × 10^{-3} (lb ft^{-3})	Structure†	Young's modulus E MN/m² × 10^{-3} (psi × 10^{-6})	Tensile strength MN/m² (psi × 10^{-3})	Atomic radius (Å)	Ionic radius (Å)	Most common valence
Aluminum (Al)	13	26.98	660 (1220)	2.699 (168.7)	fcc	70 (10.2)	60 (8.7)	1.43	0.57	+3
Copper (Cu)	29	63.54	1083 (1982)	8.96 (560.0)	fcc	125 (18.1)	160 (23.2)	1.28	0.96	+1
Iron (Fe)	26	55.85	1535 (2795)	7.87 (491.9)	αbcc<908°C γfcc 908–1388°C δbcc>1388°C	206 (30.0)	270 (39.2)	1.28	0.87	+2
Lead (Pb)	82	207.21	327 (621)	11.36 (710.0)	fcc	16 (2.3)	15 (2.2)	1.75	1.32	+2
Magnesium (Mg)	12	24.32	651 (1204)	1.74 (108.8)	cph	45 (6.5)	100 (14.5)	1.60	0.78	+2
Nickel (Ni)	28	58.71	1458 (2656)	8.90 (556.3)	fcc	200 (29.0)	370 (53.7)	1.25	0.78	+2
Tin (Sn)	50	118.70	232 (450)	7.30 (456.3)	bct	40 (5.8)	13 (1.9)	—	0.74	+4
Titanium (Ti)	22	47.90	1660 (3020)	4.51 (281.9)	α cph<880°C β bcc>880°C	114 (16.5)	460 (66.7)	1.47	0.64	+4
Tungsten (wire) (W)	74	183.86	3410 (6170)	19.3 (1206.3)	bcc	400 (58.0)	4500 (652.5)	1.41	0.68	+4
Zinc (Zn)	30	65.38	420 (788)	7.13 (445.6)	cph	90 (13.1)	155 (22.5)	1.37	0.83	+2

† cph – close-packed hexagonal; bct – body-centered tetragonal; bcc – body-centered cubic; fcc – face-centered cubic.

Table 1.4 Comparison of parameters used to describe crystal structures

Structure	Coordination number	Number of atoms per unit cell	Atomic packing factor
Simple cubic	6	1	0.52
Body-centered cubic	8	$^8/_8 + 1 = 2$	0.68
Face-centered cubic	12	$^8/_8 + ^6/_2 = 4$	0.74
Simple hexagonal	8	$^{12}/_6 + ^2/_2 = 3$	—
Close-packed hexagonal	12	$^{12}/_6 + ^2/_2 + 3 = 6$	0.74

SUMMARY

The values of these three parameters for different crystal structures are given in Table 1.4. Calculations show that the atomic packing factor is independent of the atomic size, if only one size of atom is present. If more than one type of atom is present then the relative sizes will affect the packing factor.

STACKING SEQUENCES

Table 1.4 shows that the face-centered cubic and close-packed hexagonal structures have the same coordination number (12), and the same atomic

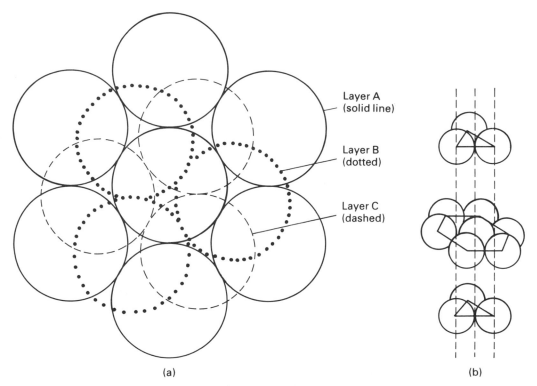

Layer A (solid line)

Layer B (dotted)

Layer C (dashed)

(a)   (b)

Figure 1.14 Atomic arrangements. (a) Stacking sequence. (b) Atomic coordination for a close-packed hexagonal structure (exploded vertically). Each atom is directly above or below atoms in alternate planes.

packing factor (0.74). However, it is necessary to examine their atomic stacking sequences in order to appreciate the differences in their structures.

With reference to Fig. 1.14, for the close-packed hexagonal structure the stacking sequence is ... ABABABAB ... for alternate atom planes. For the face-centered cubic structure the sequence is ... BACBACBAC ...

THE STRUCTURE OF METALS

The crystal structures of some metals are summarized in Table 1.5.

Table 1.5   The crystal structures of common metals

System	Metals	Comments
Body-centered cubic	Nb, Ta, W, Cr, Mo, V	Brittle metals. These are refractory metals with a high melting point, and high solubility for small atoms.
Face-centered cubic	Al, Cu, Pb, Ag, Ni, Au, Pt	Ductile metals.
Close-packed hexagonal	Zn, Mg, Cd, NiAs	

N.B. Copper and zinc have different structures but combine to form a closely related structure, which is important when considering alloys of these metals (*see* Chapter 3).

1.2.7 Polymorphism (allotropism)

Polymorphs are different types of crystal structures which have the same composition (this is analogous to isomers which have the same chemical composition but different molecular structures). The most important example of a polymorph is iron; this property makes possible the heat-treatment of steel to modify its properties. Below 910°C (1670°F), iron has a body-centered cubic structure (coordination number 8, atomic packing factor 0.68). The atomic radius is 1.241 Å at room temperature, but as iron is heated the radius increases due to thermal expansion. At 910°C (1670°F) there is abrupt shrinkage: the atomic radius is 1.258 Å, and the structure changes to face-centered cubic (coordination number 12, atomic packing factor 0.74). The atomic radius when the change is complete is 1.292 Å. At 1388/1400°C (2530/2552°F) the iron again changes to a body-centered cubic structure, and at approximately 1535°C (2795°F) it becomes liquid. An important fact is that these changes are reversible as the iron cools. These changes in structure are also accompanied by a change in properties, e.g. specific gravity.

1.2.8 Amorphous structures

Amorphous materials tend to melt gradually as the temperature is increased, and this distinguishes them from crystalline solids which have a sharp melting point. Amorphous materials may also be considered as containing a large number of defects (see Section 1.3) that make their properties significantly different from those of other substances. Amorphous means 'without form' and this can include gases, liquids and glasses.

Gases have no structure except within their individual molecules; liquids are more comparable to crystal structures and contain short-range order. Common amorphous materials at room temperature are water, oil, mercury, bakelite and glass. At high temperatures glasses are true liquids; below the melting temperature the glass contracts rapidly and continuously as the atoms develop more efficient packing arrangements. Below the glass transition temperature (see Section 4.7) no further rearrangement occurs, the volume contraction is less and is due only to reduced thermal vibrations. This transition temperature marks the change from supercooled liquid to solid.

Most amorphous materials possess some short-range order between the molecules which make up the molecular chain. However, there is no long-range order (or very little) between the chains themselves.

1.2.9 The structure of nonmetallic solids

Ionic crystals are similar to the metallic crystal structures, except that the atoms at the lattice positions are ions. There are two ions (one positive and one negative) associated with each lattice point. Each ion is surrounded by several ions of the opposite sign. Sodium chloride is a common example, based on an fcc structure with one ion situated on the lattice point and the other ion (opposite sign) halfway along the edge of the fcc unit cell. There are no NaCl molecules present: the entire crystal represents a large molecule. However, some ionic crystals have molecules (both ions) at each lattice point.

Covalent crystals are based upon the space lattices already described, but the bonding between adjacent atoms is due to very rigid covalent bonds. The strength of the bond is due to the shared valence electrons. Covalent bonding is typical of ceramic materials which are usually brittle (strong bonding and no atomic slip), hard, inert and electrical insulators (no 'free' electrons). Common ceramic materials are described in Sections 4.21 to 4.24.

Polymers consist of long-chain repeating organic molecules. The atoms within the repeating molecular chain are usually joined by strong covalent bonds. However, the bonding between the chains is due to weaker electrostatic (van der Waals) forces. Some polymers are said to be crystalline, although this is not in the normally understood sense of the term. It is used to describe the condition whereby the long-chain molecules are aligned parallel to each other, so that a regular 'pattern' exists within a particular region. When the chains are entangled they appear amorphous by X-ray diffraction; the degree of crystallinity depends upon the chain alignment. The molecular weight of a polymer is an important property that can be used as an indication of the strength of the material. Since not all chains are the same length, an average value is measured. One example is the melt flow rate based upon the flow of molten polymer through an orifice under standard conditions. Polymer materials are described in greater detail in Sections 4.1–4.20.25.

Composites are combinations of two or more materials, usually formed by coating, internal additives or laminating. The properties of the composite are usually very different from those of the component materials. Examples are wood (a natural composite), ceramic-based materials and polymers, e.g. fiberglass.

* * *

Self-assessment exercises

1 Calculate and check the values of the crystal characteristics given in Table 1.4.

2 Find examples of polymorphous structures other than those given in Section 1.2.7.

3 Calculate the crystal characteristics for iron as changes occur at particular temperatures.

4 Compare the crystal structures of selected metals, polymers, amorphous substances, ceramics, wood and concrete. (Reference should be made to handbooks of material properties or the texts given in the Bibliography, e.g. Brick *et al.* (1977), Lynch (1977), Smithells (1976), or the *Metals Handbook* (1981), published by the American Society of Metals.)

* * *

1.2.10 Strength and structure of materials

Suppose a crystalline solid is subjected to forces that tend to break the atomic bonds. If the bonds are very strong, then cleavage can occur within the crystal and the atomic bonds will rupture along a particular plane. This will happen with brittle metals and alloys, ceramics possessing covalent bonding and also ionically bonded crystals. In metals, the bonds are such that cleavage does not occur at room temperature. Instead yielding occurs during which planes of atoms 'slip', which takes place by means of the movement of dislocations. Types of dislocations will be discussed in more detail in Sections 1.3.1–1.3.4. Plastic deformation in metals (see Section 6.2.1) occurs because of the movement of dislocations. If these movements can be prevented or restricted, e.g. by alloying, cold working or precipitation hardening, then the strength properties will be increased.

The elastic properties of a material are determined under conditions where the force exerted does not cause a permanent deformation (see Section 6.2.1 for a more detailed description). The elastic modulus (Young's modulus) of a crystal is a measure of the force required to displace the atoms of the crystal from their normal equilibrium spacing, and is dependent upon the bond forces between the atoms. It is difficult to modify the elastic properties because of this dependence, although alloying does have some effect. The Young's modulus and shear modulus of an amorphous material are the same (elastically isotropic) whatever the direction of the applied stress. However, cubic crystals are elastically anisotropic and the properties depend upon the direction of loading. In polycrystalline metals, the elastic constants are averages of those of the individual grains if they are randomly oriented, and the metal behaves like an isotropic elastic material.

1.3 DEFECTS IN SOLID CRYSTALS

Defects in crystals are important because they determine the mechanical behavior, such as plastic deformation, ductility, yield stress and fracture stress. They are also the cause of low values of material strength. In a perfect crystal, the atoms would be arranged in a regular pattern and situated in energy wells. In actual crystals, defects can occur which may be difficult to produce, but when produced they are relatively easy to propagate.

Structural defects may be classified as

(a) point defects;
(b) line defects;
(c) interfacial, plane or sheet defects;
(d) bulk defects.

1.3.1 Point defects

In its simplest form a point defect is a vacancy caused by a missing atom, and it may involve more than one site vacancy. Defects may occur due to imperfect packing of the atoms during crystallization, or at high temperatures thermal vibration of the atoms may cause movement from positions of low energy. In compounds this can be a cation or anion vacancy.

A *Schottky imperfection* is a defect composed of both an anion and a cation vacancy, occurring in a compound which must maintain a charge balance.

Within a crystal structure there may be foreign atoms which do not belong to the given lattice; these are *impurities*. They can occur either in lattice sites as *substitutional impurities* or between lattice sites as *interstitial impurities*. Foreign atoms which are small compared to the matrix atoms are often found as interstitial impurities. Larger atoms are most common as substitutes, otherwise they can cause atomic distortion.

A *Frenkel defect* is the displacement of an ion from its lattice position into an interstitial site.

Additional energy is required to transfer atoms to new positions, and close-packed structures have more vacancies and Schottky defects than interstitial and Frenkel defects.

The main types of point defect are shown in Fig. 1.15.

1.3.2 Line defects

These are generally *edge dislocations* or *screw dislocations*. It is possible for mixed dislocations to occur (e.g. the dislocation loop) but these are more difficult to

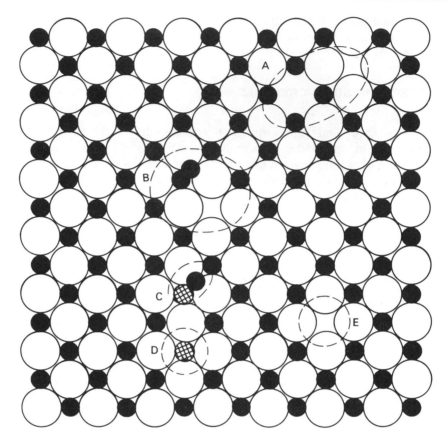

Figure 1.15 Possible point defects in an ionic lattice. (A) Schottky imperfection. (B) Frenkel defect. (C) Interstitial impurity atom. (D) Substitutional impurity atom. (E) Cation vacancy.

describe and are usually represented as superpositions of edge and screw dislocations. A graphical representation has been devised by Burgers and modified by Frank (the Frank–Burgers circuits or vectors).

An edge dislocation is due to an extra plane of atoms within a crystal structure, and this is represented diagrammatically by the symbol $\perp$. Regions of compression exist around the extra half-plane of atoms and tension below the extra half-plane of atoms. These regions are created within the crystal structure, resulting in a net increase in energy along the dislocation. An edge dislocation has its displacement perpendicular to the dislocation line.

A screw dislocation (symbol $\curlywedge$) has its displacement parallel to the defect. The associated shear stresses also cause a net increase in energy.

Edge dislocations occur during crystal formation when adjacent parts of the growing crystal are misaligned, so that an extra row of atoms is introduced or eliminated. With a screw dislocation, extra atoms can be added to the 'step' of the screw.

These types of defect are shown diagrammatically in Figs. 1.16–1.20.

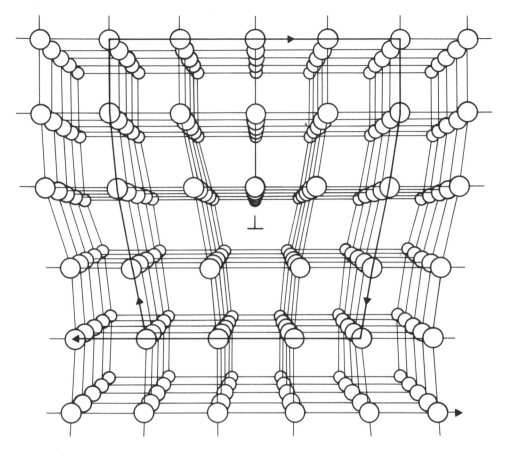

Figure 1.16 Atom arrangements in an edge dislocation. A linear defect occurs at the edge of an extra plane of atoms.

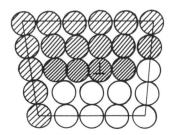

Figure 1.17 Dislocation energy. Atoms are under compression (darker) and tension (lighter) adjacent to the dislocation. The displacement vector (Burgers vector) is perpendicular to the dislocation line.

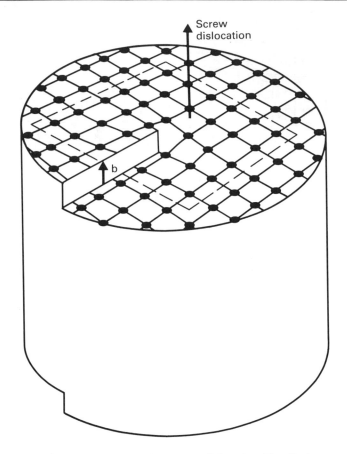

Figure 1.18 Atom arrangements in a screw dislocation. The displacement vector 'b' (Burgers vector) is parallel to the linear defect, and represents the direction and magnitude of atom displacement.

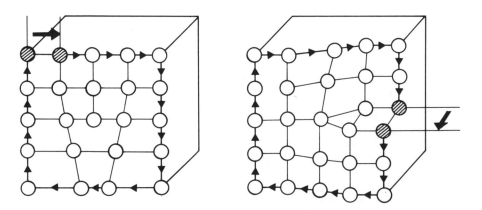

Figure 1.19 Burgers circuits for edge and screw dislocations.

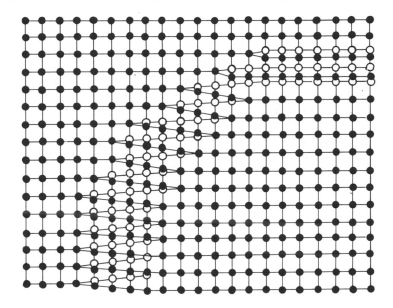

Figure 1.20   Top view of atom arrangements in a mixed dislocation. ○ atoms below plane of page; ● atoms above plane of page.

1.3.3 Interfacial, plane or sheet defects

These crystalline imperfections can occur in a variety of physical situations:

(a) at free surfaces between gases and solids;
(b) at interphase boundaries, where changes in structure and chemical composition occur across the interface;
(c) domain boundaries, at interfaces between regions containing the same atoms;
(d) at grain boundaries.

The atoms which make up the external surface of a crystal structure possess more energy than the internal atoms because these atoms are not entirely surrounded. An example of this state is the spherical shape assumed by liquid droplets possessing the minimum surface area, and surface energy, per unit volume.

Grains are individual crystals which combine to make up the solid metallic phase, and which exist in a variety of orientations. The shape of a grain is normally determined by the surrounding grains. The atoms within a grain have the same orientation and packing, which is determined by the lattice arrangement. At the boundary between two adjacent grains the atoms cannot align perfectly with both crystal lattices. There is a transition layer between the two grains, where the structure is that of neither grain. The grain boundary has a higher energy per unit area than the material in the neighboring grains, and it can be shown that this energy is approximately constant.

1.3.4 Bulk defects

These are normally:

(a) casting defects, e.g. shrinkage cavities or gas holes;
(b) working or forging defects, e.g. cracks;
(c) welding or joining defects, e.g. cracks, gas holes, slag penetration, corrosion regions.
These types of defect will be considered in Chapters 7 and 8.

The sizes of different types of structural defects are important when considering their effect and detection in materials; the relative sizes are shown in Fig. 1.21.

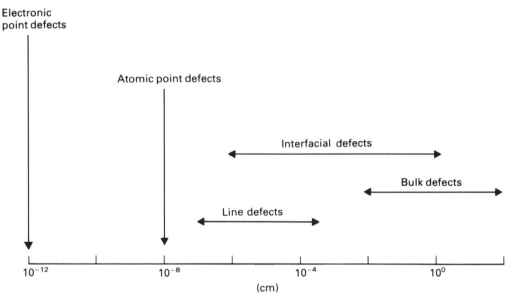

Figure 1.21 Size ranges for different kinds of structural defects.

* * *

Self-assessment exercises

Exercises marked with an asterisk will require additional reading and would be suitable as assignments or group discussion topics.

1 If the lattice points in a bcc unit cell are 0, 0, 0 and ½, ½, ½, determine the lattice points in the following lattices:
(a) simple cubic;
(b) fcc;
(c) simple hexagonal;
(d) cph;
(e) bc tetragonal.

2 Calculate the size of the largest sphere that can fit into a bcc lattice without distortion, if the largest interstitial sites have coordinates of:
(a) 0, ½, ¼;
(b) 0, ½, ½.

3 Repeat Exercise 2 for an fcc lattice with an interstitial site at 0, 0, ½, for example. How many metal atoms surround the interstitial atom?

4 Explain why an interstitial impurity in an fcc metal lattice usually causes an increase in the lattice parameter, whereas a substitutional impurity does not necessarily cause any change.

5*Describe where, why and how defects can occur in crystal structures.

6*Describe the practical effects that are caused by crystal dislocations, e.g. changes in properties. (Their influence on the fabrication of metals is discussed in Chapter 8.)

7*Describe how crystal defects can be detected, and how they can be alleviated.

8 Show on diagrams how atoms in a crystal may move due to the presence of vacancies.

9 Explain how dislocation movement can be affected by grain boundaries.

10 Distinguish between the Burgers vector, the dislocation line and the direction of dislocation movement for:
   (a) an edge dislocation;
   (b) a screw dislocation.

11 Show on a diagram the distortion that occurs around an edge dislocation. Indicate where large and small substitutional atoms and interstitial atoms can be accommodated.

※   ※   ※

1.4 EQUILIBRIUM RELATIONSHIPS

1.4.1 Introduction

A *phase* is a region which is structurally homogeneous, and each phase contains a particular atomic arrangement. A crystalline phase possesses a definite atomic arrangement, with long-range order in the repeated pattern. An amorphous phase has only short-range order, and may be a liquid or a non-crystalline solid. Gases exist only in a single-phase system. A mixture is a dispersion of two or more materials which does not result in the formation of a new substance, e.g. oxygen and nitrogen mixed to form air. The properties of the mixture are determined by the law of mixtures and depend upon the properties of the individual materials and upon the proportions used.

A *compound* is formed when materials are chemically reacted, to form a new substance that has its own distinctive properties, e.g. carbon and oxygen are burnt to form carbon dioxide.

Alloys are metallic substances in which the atoms of two or more elements combine to form a substance with metallic properties. Some engineering applications require the use of pure materials, e.g. copper for electrical wiring, zinc coating on galvanized steel, pure Al_2O_3 in spark plug insulation. However, a pure material often possesses extreme properties, e.g. low tensile strength but good corrosion resistance. Alloys can possess properties which are similar, improved or very different from their constituents. Most alloys are formed in the liquid state, where the constituents completely dissolve in each other forming a homogeneous solution. This is the situation discussed here. However, a small quantity of lead is often added to brass and steel; this does not dissolve in the liquid state and is present as tiny particles when solidified. The advantage is that, although the toughness the material is reduced, it is then easier to machine.

1.4.2 Solid solutions

If an element (the *solute*) is added to a pure material (the *solvent*), and becomes an integral part of the crystal structure when solidified, this is known as a *solid solution*. A *substitutional solid solution* is formed when the solute and solvent atoms have similar electron structures and atomic sizes. The solute atoms are then substituted for solvent atoms in the crystal structure, the crystal pattern being unaltered. An example of a substitutional solid solution is the substitution of zinc atoms for copper atoms to form brass. The atoms are substituted in a random manner. It is possible to obtain an ordered substitutional solid solution where the solute and solvent atoms adopt a specific arrangement. This is more common in solutions formed at lower temperatures where there is less thermal energy available to destroy the order of the substitution, e.g. bcc β-brass below 460°C (860°F) and fcc solid solution $AuCu_3$ below 380°C (716°F). The 'compound' CsCl has essentially perfect order, as has the intermetallic compound $CuAl_2$ (unlike the random distribution of small amounts of copper in aluminum).

An *interstitial solid solution* is formed when the solute atoms are small enough to occupy positions in the interstices between the larger solvent atoms. An example of this type of solid solution is when carbon dissolves in iron. The solubility of carbon is less at temperatures below 908°C (1666°F) when iron has a bcc structure and smaller interstices. At temperatures between 908°C and 1388°C (1666°F and 2530°F) iron possesses an fcc structure and the solubility of carbon is increased due to the unoccupied region at the center of the unit cell. However, the introduction of small amounts of carbon into the iron structure greatly changes the properties of the material, and the alloy properties depend upon the quantity of carbon dissolved. Reheating the solid solution does not cause the iron and carbon to separate. However, the subsequent cooling process causes changes in the alloy structure, and affects the physical and mechanical properties of the alloy. The effects of heating and cooling processes will be considered in more detail in Section 2.19.

An alloy may have a different structure from that of the pure metal, it behaves differently when cooled from the liquid state and solidified. This is shown in Fig. 1.22 for an alloy formed from two metals, A and B.

When a pure metal is cooled, it solidifies or freezes at a definite temperature which remains constant from the onset of solidification until the complete solid is formed. This is shown by the plateau regions of Fig. 1.22(a) and (d). This region of solidification is also accompanied by a change in the

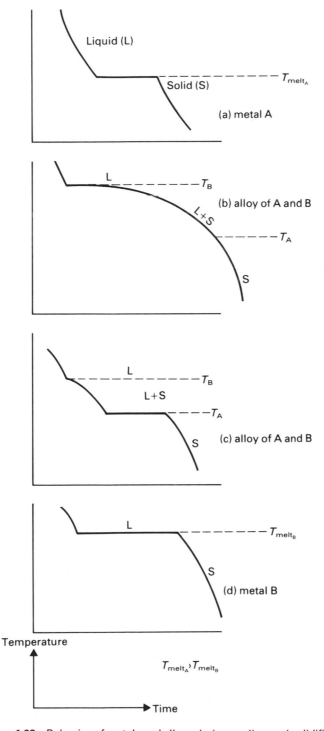

Figure 1.22   Behavior of metals and alloys during cooling and solidification.

specific volume. The cooling curve for an alloy (Fig. 1.22(b)) indicates that the solidification process occurs gradually at a temperature between the freezing points of the pure metals, depending upon the relative amounts of A and B present. The alloy solidification may be completed by a period of constant temperature (Fig. 1.22(c)).

Two metals which form a homogeneous liquid phase, i.e. they are completely soluble in one another, may not necessarily solidify to form a homogeneous solid phase or solid solution. When solidified the two metals may be:

(a) completely soluble in each other;
(b) completely insoluble in each other;
(c) partially soluble in each other;
(d) combined to form an intermetallic compound.

These alternative situations will now be considered, although discussion will be confined to alloys which are composed of only two elements (binary mixtures). Use will be made of diagrams with a base line showing percentage composition of each metal in the alloy, and an ordinate axis of temperature. Such a diagram is known as an *equilibrium* or *phase diagram* (sometimes called a solubility curve). It is a convenient method of indicating the changes in state and structure through which each alloy passes during slow cooling from the liquid state, i.e. under equilibrium conditions. It also shows the temperatures at which changes occur.

1.4.3 Complete solid solubility

When two metals are completely soluble in each other in the solid state, a substitutional solid solution (Section 1.4.2) is formed. Examples are alloys formed from copper and zinc, or copper and nickel (Monel), in which any proportion of the pure metals may be used.

The phase diagram for the copper–nickel system is shown in Fig. 1.23. A general phase diagram is also shown in Fig. 1.23 and will be used to describe the changes which occur when a liquid alloy solution is slowly cooled. Considering an alloy containing $X\%$ of component A, at temperatures above T_1 a homogeneous liquid phase exists. At temperature T_1, solidification commences and the first solid solution crystals which form have a composition given by point P (rich in component B) on the solidus line. The composition of the liquid remaining is enriched in component A. The freezing temperature therefore falls slightly. When the temperature reaches T_2, the crystals of solid solution contain more of component A (by diffusion) as given by point Q, and the composition of the remaining liquid is given by point L (richer in A than the original liquid). At T_3 the composition of the solid solution crystals (point R) is approaching that of the original alloy liquid (along the solidus line), the composition of the remaining liquid (point M) is increasing in component A (along the liquidus line). At T_4 the last drops of liquid (composition at point N) solidify, and the composition of the solid solution (point S) becomes $X\%$ of A. Only one phase is present.

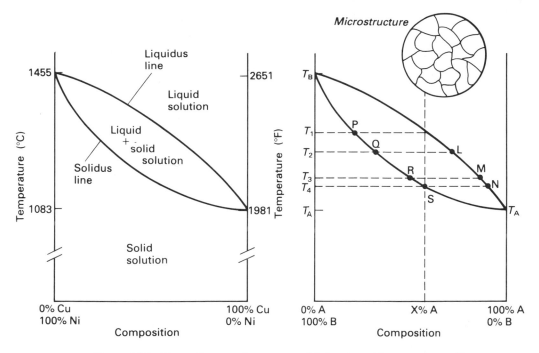

Figure 1.23 Phase diagrams for copper–nickel and any binary system with complete solid solubility.

If the solidification rate were very slow, then the solid solution crystals would be uniform in composition at each stage of the process. If the crystals were examined microscopically it would be impossible to recognize the constituent metals, and the structure would resemble a pure metal. In practice, cooling is not sufficiently slow to allow complete equilibrium to be established at each stage, and the core of each crystal is richer in B than the outer layers. This non-uniformity is known as *coring*. It can sometimes be seen through a microscope as color differences. A uniform structure can be produced by reheating the alloy to a suitable temperature below the solidus line, i.e. below T_4.

1.4.4 Complete solid insolubility

No engineering alloys exist which exhibit complete insolubility in the solid phase, but it represents a useful starting point for explaining the behavior of more complex systems. The phase diagram for this system is shown in Fig. 1.24.

The temperatures T_A and T_B represent the freezing points of the pure metals, A and B; the addition of a second metal causes a depression of the freezing point, as shown in Fig. 1.24. The point E on the diagram is referred to as the *eutectic point*. It defines the lowest temperature (T_E) at which a liquid alloy of A and B can exist, and also the composition of this particular *eutectic mixture* of A and B. This eutectic mixture solidifies at a constant temperature (T_E). If

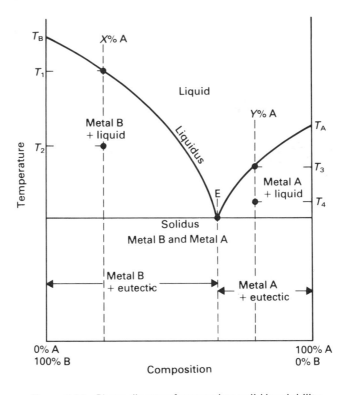

Figure 1.24 Phase diagram for complete solid insolubility.

examined under a microscope it is seen to be composed of laminations of pure metal A and pure metal B, as shown in Fig. 1.25.

Consider a liquid solution containing $X\%$ of A (richer in metal B). When cooled to temperature T_1, solidification commences and solid crystals of metal B begin to form. The liquid is then enriched in metal A and freezes at a lower temperature. As cooling proceeds the liquid becomes richer in metal A, as metal B continues to solidify, and its composition follows the liquidus line until the temperature reaches T_E. Any remaining liquid has the composition given by point E, and solidifies to form the eutectic mixture. When completely solid, the alloy containing $X\%$ A consists of crystals of metal B surrounded by laminar grains of the eutectic mixture.

Similarly, cooling a liquid alloy containing $Y\%$ A (richer in metal A) will result in solidification of metal A, and eventually the final liquid forms the surrounding eutectic mixture. It is important to remember that the composition of the eutectic mixture is always constant, as given by point E in this example.

* * *

Self-assessment exercises

1 Describe how metals freeze to form crystals (see Section 2.4.1, Nucleation and growth of new phases).

2 Draw the cooling curves and describe what happens when a pure metal and an alloy of two metals freeze.

3 Define the terms:
(a) alloy;
(b) grain;
(c) solid solution;
(d) homogeneous structure;
(e) solid solubility.

4 What is the relationship between cooling curves and equilibrium diagrams?

5 Describe the cooling characteristics when two metals are soluble in each other in the liquid state, and when solidified they are:
(a) completely soluble;
(b) partially soluble;
(c) completely insoluble.

*   *   *

1.4.5 Partial solid solubility

A more common situation occurs when two metals are partially soluble, rather than the two cases considered previously. The phase diagram for this situation is shown in Fig. 1.25, which is a combination of Figs. 1.23 and 1.24 showing solid solubility and the formation of a eutectic mixture.

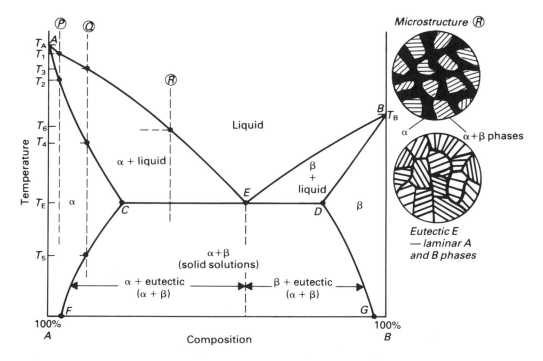

Figure 1.25 Phase diagram for partial solid solubility.

With reference to Fig. 1.25:

(a) line *AEB* is the liquidus and line *ACEDB* is the solidus;
(b) solid solution α is metal *B* (solute) dissolved in metal *A* (solvent);
(c) solid solution β is metal *A* (solute) dissolved in metal *B* (solvent);
(d) lines *CF* and *DG* are called the solvus lines, denoting the maximum solubility of *B* in *A*, and *A* in *B*, respectively.

If a small quantity of either metal is present, then a solid solution (α or β) forms; for intermediate proportions the structure contains laminations of both solid solutions. The eutectic mixture contains both solid solutions, and the microstructure is shown in Fig. 1.25. The solid solubilities of *B* in *A*, and *A* in *B*, both occur at the eutectic temperature (T_E) and are represented by points *C* and *D* respectively on Fig. 1.25. If a liquid alloy has a composition given by point *E*, then when cooled to T_E the eutectic mixture is formed according to the reaction:

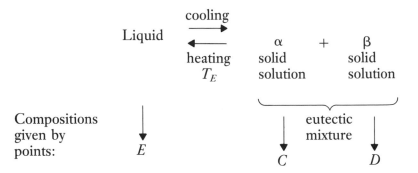

Consider the cooling of three alloys shown in Fig. 1.25 and represented by points *P*, *Q* and *R*. For alloy *P*, solidification commences at T_1 and is complete at T_2, forming a complete solid solution (α).

For alloy Q, solidification commences at T_3 and is complete at T_4, forming a complete solid solution (α). If the temperature is reduced further, then below T_5 the solubility limit of *B* in *A* is exceeded, and excess metal *B* is precipitated from the α solid solution. However, it is not pure *B* but a saturated β solid solution. The precipitated β phase may occur either at the α grain boundaries, or within the α crystals, or at both locations.

For alloy *R*, solidification commences at T_6 and produces α solid solution. The proportion of *B* in the remaining liquid rises, until at T_E solidification is complete and the structure contains α and the eutectic (α + β). Cooling of this mixture below temperature T_E causes the solubility of *A* in *B*, and *B* in *A*, to change and the compositions of the solid solutions (α and β) are given by the points on the curves *CF* and *DG* respectively. The microstructure for this (solid) alloy is shown in Fig. 1.25. Similar results can be deduced for cooling liquid solutions with compositions of metal B greater than the eutectic.

For a particular alloy composition the microstructure can be composed of one or two phases. It is often necessary to know how much of each phase is present. Quantitative information can be obtained from the phase diagram. Referring to the portion of the aluminum–silicon phase diagram shown in Fig.

1.26, consider an alloy containing 3% (by weight) of silicon. Below a temperature of 577°C (1071°F), only the solid phase is present. This solid phase is composed of α solid solution dispersed in the eutectic mixture (11.6% Si). As the temperature increases, the relative proportions of the α phase and eutectic mixture changes (as described in this section with reference to the alloy compositions *P* and *Q* in Fig. 1.25). In Fig. 1.26 at 577°C (1071°F), the first drops of liquid are formed (the eutectic composition).

If this same 3% Si alloy is heated to 600°C (1112°F), it will consist of α solid solution (silicon in aluminum) and a liquid phase. From Fig. 1.26, the compositions of these two phases are

     solid solution (α) contains 1.2% Si and 98.8% Al;

     liquid phase contains 8.3% Si and 91.7% Al.

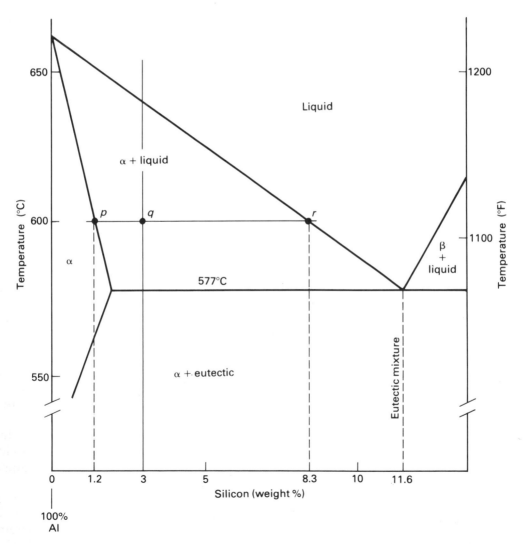

Figure 1.26 Section of the aluminum–silicon phase diagram.

If no material has been added to or removed from the system, the overall composition remains as 3% Si and 97% Al (e.g. 3Si–97Al). If 100 g of alloy were originally present and at 600°C (1112°F) there are a grams of liquid, a material balance can be performed for either component: for example,

$$Si_{alloy} = Si_{liquid} + Si_{solid}$$
$$0.03 \times 100 = 0.083a + 0.012(100-a)$$
$$3 = 0.083a + 1.2 - 0.012a$$
$$a = 25.4 \text{ g of liquid phase.}$$

Therefore, the α phase is 74.6 g.

Alternatively, if there are L grams of liquid and S grams of solid, then

$$0.03(L + S) = 0.083L + 0.012S$$

or
$$\frac{L}{S} = \frac{0.03 - 0.012}{0.083 - 0.03} = \frac{0.018}{0.053} = \frac{18}{53}$$

Substituting general terms for the mass fractions of the phases, i.e. x_a, x_1, x_s representing the mass fractions of the alloy, liquid and solid phases, respectively:

$$\frac{L}{S} = \frac{x_a - x_s}{x_1 - x_a}$$

This equation is known as the *lever rule*. Either component can be used as the basis for calculations, and the lever rule may be expressed in several equivalent algebraic forms. For example, the mass fraction of solid phase in the two phase mixture may be written as

$$\frac{S}{S+L} = \frac{x_1 - x_a}{x_1 - x_s}$$

The results of the previous calculation for the amounts of the liquid and solid phases present at 600°C (1112°F) can be obtained directly from the phase diagram. First, a 'tie line' is drawn horizontally at the temperature in question, i.e. 600°C (1112°F) in the example considered here, across the two-phase region. The points p and r at the ends of the tie line (shown in Fig. 1.26) represent the compositions of the solid and liquid respectively, and point q represents the original alloy.

The lever rule states that

$$\text{percentage of } \alpha \text{ phase} = \frac{qr}{pr} \times 100\% = 74.6\%$$

and
$$\text{percentage of liquid phase} = \frac{pq}{pr} \times 100\% = 25.4\%$$

These relationships are derived directly from the material balances performed previously. The interpretation of Fig. 1.26 is that the 3% Si alloy at 600°C (1112°F) contains an amount of solid α phase represented by qr and of composition given by point p, and an amount of liquid phase represented by pq and of composition given by point r.

* * *

Self-assessment exercises

Use the lever rule or material balances to calculate the amounts of the phases present (given the initial solid alloy composition) at the stated temperatures (the alloy compositions are weight %):

(a) 30Ni–70Cu at 1000°C (1832°F), 1200°C (2192°F) and 1400°C (2552°F);

(b) 80Ni–20Cu at 1400°C (2552°F);

(c) 20Cu–80Ag at 700°C (1292°F), 800°C (1472°F) and 900°C (1652°F);

(d) 0.5C–99.5Fe at 750°C (1382°F);

(e) 3C–97Fe at 1200°C (2192°F).

Reference can be made to Fig. 1.23 for parts (a) and (b), and to Fig. 2.2 for parts (d) and (e). Comprehensive data relating to phase diagrams can be found in the handbooks by Hansen (1958) with supplements by Elliott (1965) and Shunk (1969), Levin *et al*. (1964) and Moffatt (1976–77) listed in the Bibliography.

* * *

1.4.6 Peritectic phase diagram

A *peritectic reaction* often occurs during the cooling of some copper alloys containing zinc, tin or aluminum. It involves the reaction between a solid produced during cooling and the remaining liquid to form another phase. The phase diagram is shown in Fig. 1.27, where

(a) line AEB is the liquidus line;
(b) line $ACDB$ is the solidus line;
(c) FC and GD are the solvus lines;
(d) CDE is the peritectic line;
(e) D is the peritectic point.

The peritectic reaction occurs at temperature T_p when the alloy composition lies between C and E.

When the liquid alloy Q is cooled, solidification begins at temperature T_1; the composition of the α solid solution formed is found from the solidus curve. As cooling proceeds the α solid solution composition approaches C (along the solidus curve) and the liquid composition approaches E (along the liquidus curve). At the limits (temperature T_p) the solid solution (C) and liquid (E) are in equilibrium and react to form a new phase, the β solid solution. The quantity of B is insufficient to produce an entire β phase and the final solid product contains both α and β. With further cooling below T_p, the quantities of α and β are adjusted as the solubilities of A in B, and B in A, vary. At a particular temperature the compositions of solid solutions α and β are given by the points on the curves DF and DG, respectively.

With alloy R a similar situation arises except that the peritectic reaction produces β, and there is excess liquid which remains in equilibrium when all the

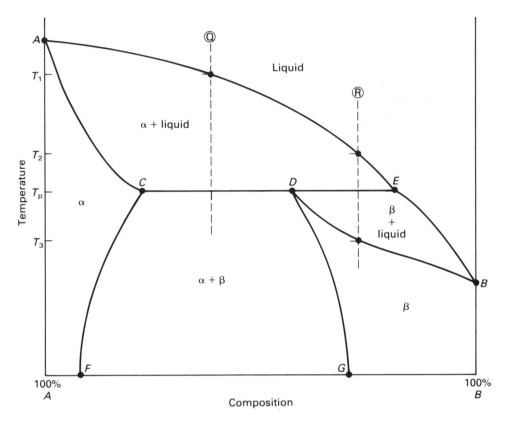

Figure 1.27   Phase diagram for peritectic reaction.

α phase has been consumed. Complete solidification is achieved when the temperature falls to T_3, then only the β phase is present.

For an alloy composition represented by point D and at temperature T_p, the peritectic reaction would produce only phase β according to the reaction:

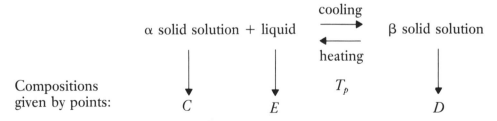

1.4.7 Eutectoid phase diagram

The property of allotropy, or polymorphism, has already been discussed in Section 1.2.7. Diamond and graphite are two allotropic forms of the element carbon. Pure iron has bcc structure at temperatures up to 908°C (1666°F) (α iron), and also at temperatures between 1380°C (2516°F) and the melting point of 1535°C (2795°F) (δ iron). At temperatures between 908°C and 1388°C (1666°F and 2530°F) the iron has an fcc structure (γ iron). These allotropic changes affect the shape of the phase diagram for alloys of the metal. The iron–carbon phase diagram will be considered in detail in Section 2.4, and the general effects of alloying two allotropic metals *A* and *B* will now be considered.

Let metal *A* be bcc at low temperatures and fcc at high temperatures. Metal *B* is close-packed hexagonal at low temperatures and is also fcc at high temperatures. One possible phase diagram is shown in Fig. 1.28.

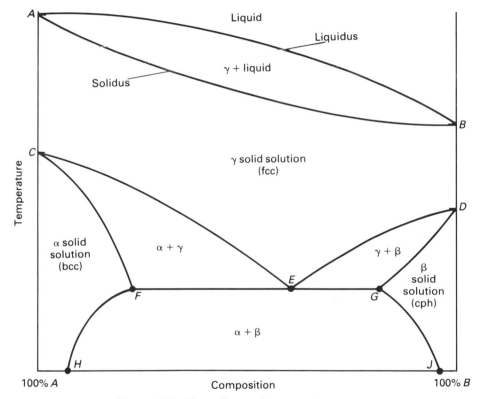

Figure 1.28   Phase diagram for eutectoid reaction.

It is assumed that both metals are completely soluble in each other when liquid, and also when solidified at high temperatures (above that represented by point *C*). Both metals then possess an fcc structure, forming γ solid solution. At lower temperatures only partial solubility is possible because the allotropes have different crystal structures. The phase diagram of Fig. 1.28 below the solidus line resembles the eutectic phase diagram of Fig. 1.25. In Fig. 1.28, upon cooling, the homogeneous solid phase forms two separate solid phases; the lower region of Fig. 1.28 is known as *eutectoid* phase diagram, which literally means 'eutectic-like'. The difference is that when cooled, a homogeneous *solid* phase forms two separate solid phases. In Fig. 1.28,

(a) line *CED* is called the liquidoid;

(b) line *CFEGD* is called the solidoid.

The use of the eutectoid phase diagram to interpret the temperature and composition changes during cooling is essentially the same as for the eutectic phase diagram (Fig. 1.25). It should be remembered that changes which occur entirely in the solid phase take much longer to reach completion than those involving a liquid phase. Rates of cooling, particularly for solid phases, can greatly affect the structure and properties of the alloy. This will be discussed in more detail in Sections 2.15 and 2.19.

1.4.8 Intermetallic compounds

Some metals combine to form intermetallic compounds which may obey normal valency laws, or may be ordered crystal space lattices with specific arrangements of the constituent atoms. Intermetallic compounds are usually hard and brittle; when pure they are of little use for engineering purposes. However, when combined with a solid solution they produce an alloy with increased wear resistance and toughness. Most intermetallic compounds have different crystal structures from the constituent metals and a unique higher melting point. Table 1.6 is an example of the formation of an intermetallic compound.

Table 1.6

	Melting point	*Crystal structure*
Magnesium (Mg)	649°C (1200°F)	Close-packed hexagonal
Tin (Sn)	232°C (450°F)	Body-centered tetragonal
Mg_2Sn – intermetallic compound	783°C (1441°F)	Complex cubic.

Two possible phase diagrams for two metals (A and B) which combine to form an intermetallic compound, A_xB_y, are shown in Fig. 1.29(a) and (b). It can be seen that these diagrams are actually two simple phase diagrams (compare with Figs. 1.24 and 1.25) which have been combined together. For this reason the intermetallic compound is often regarded as a pure metal. It can be seen in Fig. 1.29 that each phase diagram has two separate eutectic mixtures which do not solidify at the same temperature.

*   *   *

Self-assessment exercises

1 Annealing is a heat treatment process used to modify the properties of an alloy. Read Section 2.19.1 for a description of this process.

2 What effects is annealing intended to produce? What are the disadvantages of using this process?

3 Why have you been asked to study this section now? What relevance does annealing have to the material presented in this chapter? This should become clear after reading Section 1.4.9.

*   *   *

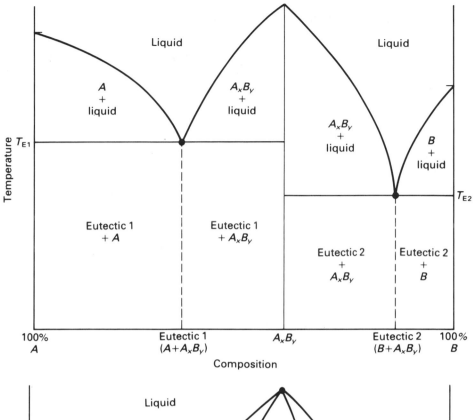

(a)

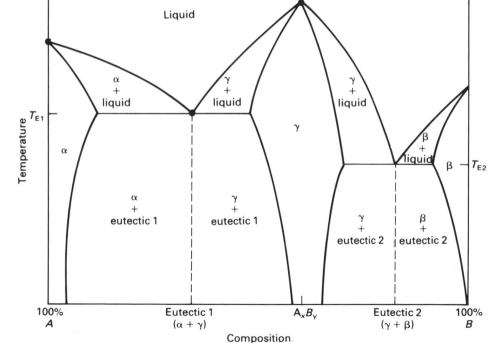

(b)

Figure 1.29 Phase diagrams for intermetallic compound formation. (a) No solid solubility. (b) Partial solid solubility.

1.4.9 Applications of equilibrium phase diagrams

A phase diagram is a graph with axes of temperature and percentage of alloying elements. This graph indicates the boundaries of the regions where particular phases exist. An equilibrium-phase diagram indicates the phases present in a metal alloy when it is *slowly* cooled from the molten state. This slow cooling is a lengthy process and must be sufficient for the required changes to occur throughout the entire phase.

Phase diagrams can be used to determine the percentages of the phases present in a particular alloy at a specified temperature, e.g. using the lever rule – see Section 1.4.5. However, for the designer and the engineer an important use of phase diagrams is to determine the effects produced by particular heat treatment processes. It should also be possible to explain why these effects occur, to predict the microstructure of the phases that are present and to describe the phase changes that take place during heating or cooling cycles.

The main disadvantage of an equilibrium phase diagram is that the data it contains was obtained under equilibrium conditions, i.e. very slow heating or cooling. In most practical situations, process times and hence heating costs must be kept to a minimum. If a metal is heated under non-equilibrium conditions, then the phase diagram can be used with only slight modifications. However, for non-equilibrium cooling the situation is very different from that predicted by the equilibrium phase diagram. In this case a time–temperature transformation (TTT) diagram is used in conjunction with the equilibrium phase diagram, and with a requisite amount of experience, understanding and caution! Descriptions and examples of TTT diagrams for particular steels are presented in Section 2.5.

However, one particular heat treatment process, namely annealing, is performed under conditions comparable to those which apply to the equilibrium phase diagram. That is why the self-assessment exercises preceding this section were directed towards a description of the annealing process (Section 2.19.1). Let us now consider the process of annealing, the reasons why it is performed, the effects that can be produced in an alloy and its relationship to equilibrium phase diagrams.

Annealing involves heating a metal alloy to an appropriate temperature, maintaining that temperature for a certain time and then cooling sufficiently slowly to avoid the formation of an undesirable structure. The cooling stage may be interrupted and the alloy held (soaked) at particular temperatures. The annealing process is described in Section 2.19.1 with particular reference to the phases present in steels; this description will require an understanding of the iron–carbon equilibrium diagram of Fig. 2.2. The discussion in this section will be confined to general situations and effects. There are several reasons why annealing is performed:

(a) to soften an alloy, by producing a change in the crystal structure from a hard brittle phase to a softer, more easily machinable constituent;
(b) to relieve internal strains present within a structure, due to either cold working (see Section 8.2.7) or welding (see Chapter 7), which prevent

further work hardening or give rise to stress corrosion cracking (see Section 5.5.3).

To produce these changes, an alloy must be heated to the temperature where phase transformations will occur. This temperature can be determined from the phase diagram, e.g. Figs. 1.25, 1.28 or 2.2. However, as changes occur the component will 'grow' (the reverse effect of cold-working compression) and this needs to be considered if dimensional tolerances are important.

Consider the stages that occur, i.e. the effects produced, in the microstructure and the mechanical properties as annealing proceeds. Three distinct stages occur, and these will now be described.

Recovery is the first stage, involving rearrangement of dislocations to produce a more stable configuration. Running out or annihilation occurs due to the attractive or repulsive forces that exist between dislocations. The magnitude of the residual stresses in a cold-worked material may exceed the tensile strength of the material, and localized cracking occurs. If a cold-worked metal is heat treated in the recovery range, then this is known as *stress relief annealing*.

The second stage is *recrystallization*, which occurs at higher temperatures and causes a rapid decrease in the hardness of the metal. A *nucleation process* occurs as minute new *equiaxed grains* (the dimensions are independent of the direction of measurement) are formed. These new grains have the same composition and lattice structure as the original undeformed grains, and they appear first in the most deformed region of the microstructure, usually at previous grain boundaries. The higher the temperature, the greater the rate of recrystallization and the less the time required to obtain full (or say 95%) recrystallization.

While recrystallization is taking place, the third stage known as *grain growth* is also occurring. The structure becomes more stable as the grain size increases and the length of the grain boundaries is reduced, thus reducing the total energy of the material.

The following changes in the mechanical properties (see Section 6.2) and behavior of a material are produced during the annealing stages.

Recovery stage:

   (a) hardness and strength virtually unchanged;
   (b) stress corrosion cracking problems reduced.

Recrystallization stage:

   (a) hardness and strength reduced;
   (b) ductility increased.

Grain growth (higher temperatures):

   (a) coarse grains produced;
   (b) hardness and strength (further) reduced;
   (c) ductility increased.

The temperature required for annealing is reduced by an increase in cold working, metal purity, or fineness of grain size.

When alloys are formed that possess complete solid solubility, as shown for example in the phase diagrams for copper–nickel (Fig. 1.23) or copper and less than 37% zinc (Fig. 3.1), then the alloying elements raise the recrystallization temperature. Insoluble constituents or impurities, e.g. the pitch or oxygen content (copper oxide) of pure copper (see Section 3.2), do not affect the recrystallization temperature but give rise to a finer recrystallized grain size.

* * *

Self-assessment exercises

1 Distinguish between the phase diagrams for:
   (a) eutectic reaction;
   (b) eutectoid reaction;
   (c) peritectic reaction.
   Describe actual examples of each type of reaction.

2 Obtain published data for the phase diagrams of particular binary metal combinations, e.g. beryllium–copper, lead–arsenic, cadmium–bismuth, etc.
   Describe what happens when these particular metal combinations are heated or cooled. Describe the phases that are present and the microstructures that are formed.

3 What practical limitations restrict the use of equilibrium diagrams?

4 What is the structure of an intermetallic compound? Describe actual examples.

5 Define the terms:
   (a) peritectoid;
   (b) monotectic;
   (c) hypoeutectoid;
   (d) hypereutectoid.

6 Study the iron–carbon phase diagram in Chapter 2 (Fig. 2.2) and identify particular regions. Describe what happens when particular alloys are cooled.
   Compare these ideas with your answers after studying Sections 2.1 to 2.9.

* * *

Complementary activities

A textbook provides a descriptive body of knowledge relevant to a particular area of study. It should also provide questions, to help the reader ascertain whether a satisfactory understanding has been achieved, and ideas to stimulate further study and actions. Knowledge is useful only if it can be applied in practical situations. Just as the written word is enhanced by a picture, so the descriptive study is enhanced by a practical demonstration or an application. Complementary activities are included at the end of each chapter to provide a few ideas as to what the reader should be 'doing' as well as reading. There are many other practical activities that could be performed. Luckily the field of materials is literally all round us – to see, to investigate, to analyze and to test.

1 Make models of some crystal structures. Introduce impurity atoms, substitutional atoms, etc., and show what effects are produced. How are these models different from polymer structures (Chapter 4)? Do these models help

provide an understanding of the bonding and strength of crystal structures?
If model-building kits are not available, improvise with straws, plasticine, table tennis balls, etc.

2 Arrange a layer of ball bearings on a flat tray and use this model to illustrate the types of dislocations (defects) that can occur in crystal structures. Demonstrate the dislocation movements that can occur in a plane.
Extend the model (or other models) in three dimensions and demonstrate the types of dislocations that can occur, and how they can move.

3 Dissolve some common chemicals such as sugar or salt in water, demonstrate the effects of temperature and concentration upon the solidification and dissolution processes.

4 Measure the melting points of some common pure metals, alloys and other materials. How pure are these materials? What effects do particular impurities produce? How accurate are the temperature measurements?

5 Determine the cooling curves for several common alloys. How are the results related to the phase diagrams for these alloys?

6 Obtain published data related to the phase diagrams of binary mixtures of metals. Prepare a list of binary mixtures that could be investigated when cooled from the molten state. Perform experiments to obtain data for some alloys which illustrate the alternative situations described in Sections 1.4.1–1.4.5. Practical limitations will include:
(a) available equipment, e.g. source of heat and the range of the temperature measuring equipment;
(b) availability of materials;
(c) cost of materials;
(d) melting points of materials;
(e) time available.

7 Examine prepared sections of the solidified alloys from activities 4, 5 and 6, using a metallurgical microscope. Compare these structures with those examined in the complementary activities for Chapters 2 and 3.

*   *   *

KEYWORDS

(See Section 1.1, Glossary of terms)

materials science	space lattice	crystal
atom	unit cell	crystal direction
molecule	cubic	crystal plane
nucleus	tetragonal	Miller indices
electron	orthorhombic	coordination number
element	monoclinic	atomic packing factor
homogeneous	triclinic	lattice constant
mixture	hexagonal	atoms per unit cell
compound	rhombohedral	stacking sequence
chemical bond	axes	polymorphism
covalent bond	axial angles	(allotropism)
ionic bond	simple cubic	amorphous
metallic bond	body-centered cubic	structural defect
hydrogen bond	face-centered cubic	dislocation
van der Waals forces	simple hexagonal	point defect
crystallization	close-packed hexagonal	vacancy

Schottky defect
Frenkel defect
interstitial impurity
substitutional impurity
line defect
   edge dislocation
   screw dislocation
  interfacial, plane
   or sheet defect
  domain boundary
  grain boundary
  bulk defect
   casting defect
   working defect
   joining defect
equilibrium relationship
phase
alloy
solid solution
solute

solvent
substitutional solid
  solution
interstitial solid
  solution
solidification (freezing)
equilibrium (phase)
  diagram
solubility curve
liquidus line
solidus line
coring
eutectic point
eutectic mixture
eutectic temperature
laminations
lever rule
peritectic reaction
solvus line
peritectic line

peritectic point
eutectoid phase diagram
liquidoid line
solidoid line
intermetallic compound
annealing
T T T diagram
soaking
cold working
welding
work hardening
stress corrosion cracking
recovery
residual stress
stress-relief annealing
recrystallization
nucleation
equiaxed grains
grain growth

*   *   *

BIBLIOGRAPHY

Alexander, W. and Street, A., *Metals in the Service of Man*, 8th Ed., Penguin Books Ltd, Harmondsworth, Middlesex (1982).

Anderson, J.C., Leaver, K.D., Alexander, J.M. and Rawlings, R.D., *Materials Science*, 2nd Ed., Thomas Nelson and Sons Ltd, Walton-on-Thames, Surrey (1974).

Barrett, C.R., Nix, W.D., and Tetelman, A.S., *The Principles of Engineering Materials*, Prentice-Hall, Inc., Englewood Cliffs, New Jersey (1973).

Brick, R.M., Pense, A.W. and Gordon, R.B., *Structure and Properties of Engineering Materials*, 4th Ed., McGraw-Hill Book Co., Inc., New York (1977).

Cottrell, A., *An Introduction to Metallurgy*, 2nd Ed., Edward Arnold (Publishers) Ltd, London (1975).

Gourd, L.M., *An Introduction to Engineering Materials*, Edward Arnold (Publishers) Ltd, London (1982).

Guy, A.G., *Essentials of Materials Science*, McGraw-Hill Book Co., Inc., New York (1976).

Hansen, M., *Constitution of Binary Alloys*, 2nd Ed. (1958); 1st Supplement (1965) by R.P. Elliott; 2nd Supplement (1969) by F.A. Shunk; McGraw-Hill Book Co., Inc., New York.

Harris, B. and Bunsell, A.R., *Structure and Properties of Engineering Materials*, Longman Group Ltd, Harlow, Essex (1977).

Higgins, R.A., *Engineering Metallurgy*; Volume 1, Applied Physical Metallurgy, 5th Ed. (1983); Volume 2, Metallurgical Process Technology, 2nd Ed. (1974), Hodder and Stoughton Ltd, Sevenoaks, Kent.

Higgins, R.A, *Properties of Engineering Materials*, Hodder and Stoughton Ltd, Sevenoaks, Kent (1977).

Jastrzebski, Z.D., *The Nature and Properties of Engineering Materials*, 2nd Ed., John Wiley and Sons, Inc., New York (1977).

John, V.B., *Introduction to Engineering Materials*, 2nd Ed., Macmillan Publishers Ltd, London (1983).

Levin, E.M., Robbins, C.R. and McMurdie, H.F., *Phase Diagrams for Ceramists* (1964), two Supplements (1969, 1975) by Levin and McMurdie, American Ceramic Society, Inc., Columbus, Ohio.

Lynch, C.T. (Ed.), *Handbook of Material Science*, 3 Volumes, CRC Press, Cleveland, Ohio (1974).

Moffatt, W.G., *Binary Phase Diagrams Handbook*, 4 Volumes, General Electric Co., New York (1976, 1977).

Pascoe, K.J., *An Introduction to the Properties of Engineering Materials*, 3rd Ed., Van Nostrand Reinhold Co., New York (1978).

Rollason, E.C., *Metallurgy for Engineers*, 4th Ed., Edward Arnold (Publishers) Ltd, London (1973).

Rosenthal, D. and Asimow, R.M., *Introduction to the Properties of Materials*, 2nd Ed., Van Nostrand Reinhold Co., New York (1971).

Ruoff, A.L., *Materials Science*, Prentice-Hall, Inc., Englewood Cliffs, New Jersey (1973).

Smithells, C.J. (Ed.), *Metals Reference Book*, Butterworth and Co. (Publishers) Ltd, London (1976).

Van Vlack, L.H., *Elements of Materials Science*, 4th Ed. (1980); *Materials for Engineering: Concepts and Applications* (1982); *Materials Science for Engineers* (1970); *A Textbook of Materials Technology* (1973), Addison-Wesley Publishing Co., Inc., Reading, Massachusetts.

Metals Handbook, 9th Ed., American Society of Metals (ASM), Metals Park, Ohio (1981).

2
Iron and Steel

nucleation	wear-resisting steel	stress relief
growth	stainless steel:	annealing
TTT diagrams	ferritic	normalizing
critical cooling rate	martensitic	quenching
plain carbon steel	austenitic	tempering
mild steel	BS and ASTM	age-hardening
low-carbon steel	specifications	precipitation
medium-carbon steel	AISI, SAE and ASME	hardening
high-carbon steel	heat treatment	solution treatment
alloy steel	hardening	
high-speed tool steel	recrystallization	

2.1 INTRODUCTION

Iron ore is the most common metal ore and it is often found sufficiently close to the surface to be obtained by open-cast mining. The main types of iron ore found are oxides and carbonates, which are obtained commercially from the USA, Russia, Africa, Sweden, South America and Australia. British ore is low grade but is often mixed with imported material. The majority of the iron produced today is processed into steel, which is one of the most important engineering materials. Steels are alloys of iron and carbon, containing between 0.1% and 1.7% carbon in a combined state. One advantage of steels is that they can be hot worked (see Chapter 8). By controlling the quantities of impurities and by introducing additional alloying elements, a wide range of steels can be produced with very different properties.

2.2 PRODUCTION OF IRON

Controlled amounts of iron ore, coke and limestone are fed to the top of a shaft-type blast furnace. A blast of high-pressure, heated air is blown through the furnace by a series of jets situated near the bottom, and the material inside is subjected to intense heat. The coke acts as a fuel and the carbon monoxide produced reduces the iron oxides to iron. The coke also acts as a source of carbon. The limestone assists the separation of the iron from the waste material (or *gangue*) by making a fusible slag. The liquid iron collects in the hearth (at the bottom of the furnace) and the liquid slag floats on top. Both are tapped off separately at regular intervals.

At this stage, the main impurities in the iron are carbon (approximately 4%) from the coke, silicon (approximately 2%) and manganese (approximately 1%) which are reduced from the ore, and sulfur. Depending on the source of the ore, up to 2% phosphorus may also be present. These impurities are detrimental to the quality of the iron if they are present in larger quantities. They should be removed as waste gases depending upon the furnace temperature, or in the liquid slag depending upon the quantity of limestone added. The iron produced is either fed directly to the steel-making furnaces (as hot metal), or cast into molds and obtained as small ingots or pigs. This *pig iron* can then be used to produce *wrought iron* or a range of grades of *cast iron*. Figure 2.1 shows the melting and process temperatures for different metals.

2.2.1 Wrought iron

Wrought iron is refined pig iron with a low carbon content – less than 1% and usually approximately 0.02%. It is now only produced in small quantities because of the relatively high cost. Wrought iron was traditionally produced in a puddling furnace, using a slag that was rich in iron oxide to act as an oxidizing agent. The removal of carbon and other impurities produced a pasty state, as the melting temperature of the iron was increased. Older furnaces could not maintain the iron in a molten form and, although some slag was removed, the pasty iron contained entrapped slag. The pasty wrought iron was worked, i.e. forged and rolled, causing some slag to be exuded, but the remaining entrapped slag was responsible for the characteristic fibrous texture of wrought iron. Most wrought iron is now produced by adding slag to refined liquid iron.

Wrought iron is very ductile because of its low carbon content (less than 0.1%). It can be easily worked and welded, although machining is more difficult due to the included slag which can damage cutting tools. It is tough and resistant to atmospheric corrosion. It is used when a visual warning of failure is required, such as link chains where overloading causes a considerable increase in length (and a reduction in diameter) of the links. Applications which used wrought iron have now generally substituted mild steel which is easier to machine, cheaper and more reliable.

2.2.2 Cast iron

Cast iron is an iron alloy that contains more than 1.7% carbon, usually between 2.4% and 4.0%. Pig iron is mixed with scrap iron and/or scrap steel; the latter helps control the phosphorus and silicon content and therefore the graphite content. This mixture produces a better quality casting with increased strength. The mixture is then melted in a small furnace. About 90% of all cast iron production uses a cupola, or small blast furnace, which receives alternate charges of metal and coke and is economical to operate. An alternative type of furnace is the air or reverberatory furnace which is a batch melting furnace, but it is less economical. However, the fuel does not come into contact with the charge and therefore better control of composition can be exercised.

The rotary furnace and electric furnace are used only when a special quality cast iron is required. The molten cast iron can then be obtained in the following three basic grades. *Ordinary cast iron* is classified as white or gray. *High duty cast iron* is produced if graphitizers are inoculated into the cast iron in order

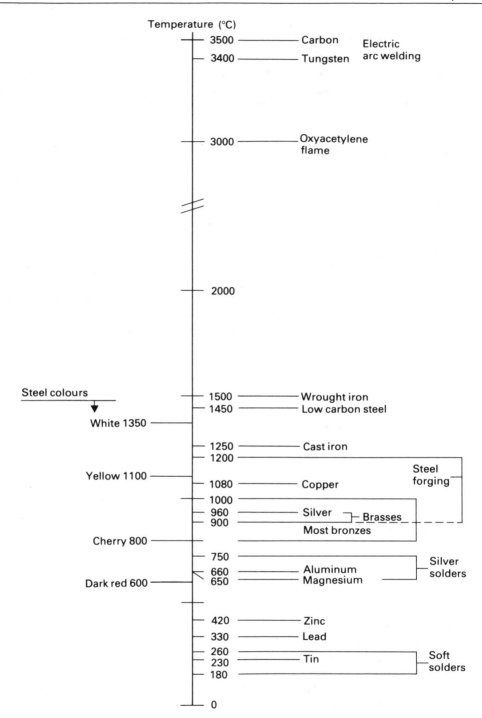

Figure 2.1   Melting points and process temperatures.

to control the form taken by the graphite. This can be done either in the furnace or as the molten metal leaves. *Alloy cast iron* is produced by introducing alloying additions, either in the furnace or as the metal leaves. Alloy iron castings have replaced some traditional steel forged products, e.g. axles and crankshafts, because the equipment can be smaller and it is more economical. Castings can be produced in more complex shapes than with wrought iron, and less allowance is required for machining.

2.2.3 Ordinary cast iron

The carbon present in cast iron is largely responsible for the properties which it possesses. The carbon may be present

(a) as free carbon or graphite, mixed with the iron but not forming a homogeneous phase;
(b) absorbed in an interstitial solid solution with the iron and called ferrite (this accounts for only a small amount of carbon);
(c) combined with the iron and forming Fe_3C (cementite).

Gray cast iron is obtained when cast iron is cooled slowly; the combined carbon then has time to separate out and is present in the free graphite form. Gray cast iron has a low tensile strength because of the presence of the graphite flakes at the crystal boundaries. It is also self-lubricating and absorbs vibrations, has a good resistance to compressive loads and is used for frames that are not subject to shock loads. It can only be used in a cast form and is not suitable for hot or cold working due to the high carbon content. It can be welded if this is necessary.

White cast iron is obtained when cast iron is rapidly cooled and the carbon does not have time to separate out. It is hard and brittle and has few engineering uses because it can only normally be machined by grinding.

Malleable cast iron can also be produced: the three forms available are *whiteheart*, *blackheart* or *pearlitic* (describing the microstructure). White cast iron has a more even distribution of carbon within its structure, and is heated for a prolonged time in the presence of iron oxide to reduce the carbon content. The edges of the graphite flakes act as stress raisers and the heating process changes the size and shape of the flakes, making the material more ductile and malleable. Malleable cast irons are a cheap substitute for steel forgings and are often used for pipe fittings. Pearlitic malleable cast iron has an increased manganese content (about 1%), which acts as a carbide-stabilizing element and improves the shock-resistant properties.

Silicon is always present in cast iron; it acts as a softening agent by promoting free carbon and it is also a reducing agent due to its affinity for oxygen. In gray cast iron, silicon causes brittleness if present in proportions greater than 4% but has little effect up to 2.5%. Phosphorus is also an impurity but in small amounts it is used to keep the iron fluid in the mold, and hence fill any small crevices. It also makes machining difficult. Manganese controls the sulfur and oxygen content and helps keep the iron soft. The sulfur content is usually restricted to about 0.1%.

2.2.4 High-duty cast iron

This type of cast iron has good casting properties and better mechanical properties than gray cast iron. A graphitizing material, such as magnesium mixed with nickel, is added to the molten metal as it leaves the furnace and causes the graphite to assume spheroids rather than flakes. This is known as *spheroidal graphite cast iron*, or *SG iron*. Depending upon the heat treatment process employed, the final product is either a *ductile ferritic SG iron* or a *pearlitic SG iron* (which is stronger). An SG iron can be partially hot or cold worked, machined or welded. It is a good substitute for steel forgings, and its uses include crankshafts, pump casings and gears.

2.2.5 Alloy cast irons

Alloying elements are added to cast irons to produce special properties such as heat resistance (reduced oxide formation and growth), corrosion resistance, improved strength, hardness, etc. Nickel can be added to promote graphitiz-ation, to promote grain uniformity and to improve heat resistance. Typically 5% nickel is added, although 15% to 25% may be added for improved corrosion resistance. Chromium promotes the stability of combined carbon, and is used with nickel to achieve balanced properties, producing hardness without brittleness. A small amount of copper will enter into a solid solution of iron; it improves the corrosion resistance and promotes strength and hardness with ease of machining. About 1% molybdenum, with nickel, produces a needle-like structure with strength and hardness but without loss of machinability.

* * *

Self-assessment exercises

1 List the main iron-bearing ores and where they are found.
2 Describe (briefly) how iron is produced industrially, and list all the materials that are required.
3 What is the difference between wrought iron and cast iron?
4 What are the main differences between the various types of cast irons that are produced?
5 List the main uses of cast irons.

* * *

2.3 PHASES AND STRUCTURES IN STEELS

The general term steel has already been used (Section 2.1) to describe iron–carbon alloys with a carbon content between 0.1% and 1.7%. These alloys may contain other elements, and can be produced in various grades possessing a wide range of properties. In order to understand the structural changes which occur and the phases that are present when iron–carbon alloys are cooled, it is necessary to interpret the various regions of the phase diagram (Fig. 2.2). Certain special names are used to describe the constituents that may be present in different steels. These terms will now be defined and will be used throughout the remainder of the chapter.

2.3.1 Ferrite

Ferrite is the name given to the body-centered cubic (bcc) solid solution of iron and carbon that exists at temperatures below 908°C (1666°F). The maximum carbon that can be present is 0.03%. The name ferrite is sometimes used for the low-temperature bcc allotrope of pure iron, but in this text the term α-iron will be used for the pure metal (the terms α-iron and ferrite are used in an interchangeable manner in some texts for both the pure metal and the solid solution). Pure α-iron is a soft, ductile metal with little mechanical strength, and is a ferromagnetic material at temperatures below 768°C (1414°F, the Curie temperature). The loss of ferromagnetism above 768°C (1414°F) is a reversible process on cooling.

2.3.2 Austenite

Austenite is the name given to face-centered cubic (fcc) solid solutions. γ-iron is an alternative term, but this will be reserved for the pure fcc iron which is the stable form between 908°C and 1388°C (1666°F and 2530°F). The allotropic change of pure iron from the α-form to the γ-form at 908°C (1666°F) causes a reduction in volume, because the fcc structure has a denser atomic packing (Section 1.2.6). The austenitic solid solution can accept up to 1.7% carbon; it is a relatively soft and ductile material which is well suited to most fabrication processes.

2.3.3 δ-Iron

δ-iron is the name given to the bcc pure allotrope of iron which is stable between 1388°C (2530°F) and its melting point of 1535°C (2795°F). The term δ-iron is also used to describe the solid solutions of iron–carbon at these higher temperatures, with a maximum solubility of 0.09% carbon. It is sometimes referred to as δ-ferrite.

2.3.4 Cementite

Cementite is the name given to iron carbide (Fe_3C), which has a crystal lattice containing iron and carbon atoms in a three to one ratio. It has an orthorhombic unit cell with 12 iron atoms and 4 carbon atoms per cell and, therefore, a carbon content of 6.67%. Compared to ferrite or austenite, cementite is a hard and brittle constituent. It is relatively weak as it cannot adjust to stress concentrations, but when combined with ferrite produces a strong steel.

2.3.5 Pearlite

Pearlite has a lamellar structure composed of layers of ferrite and cementite. It is a specific mixture formed by the breakdown of austenite of the eutectoid composition (0.83% carbon). Mixtures of ferrite and carbide can be formed from other reactions, but the microstructures are not lamellar and the properties are different.

2.3.6 Martensite

If austenite is rapidly cooled it does not have time to change to ferrite and cementite; then all the carbon remains in the solid solution. The mixture is ferrite highly supersaturated with dissolved carbon, and it is known as martensite. On cooling, the fcc structure of austenite changes to a body-centered structure, but in order to accommodate all the carbon this is a body-centered tetragonal (bct) structure for martensite, rather than the bcc structure of ferrite solutions. Martensite is hard, strong and brittle.

2.3.7 Bainite

If austenite is cooled at a slower rate than that required to produce martensite, but at a rate too fast to allow pearlite to form, then bainite is the decomposition product. Bainite is also formed by isothermal transformation.

2.3.8 Sorbite and troostite

When martensite or bainite are reheated to temperatures below 700°C (1292°F) (a process called tempering) to reduce their brittleness, then structures known as sorbite and troostite are formed. These structures may simply be referred to as tempered martensite.

2.3.9 Ledeburite

This is the name given to the particular composition (the eutectic) containing 4.3% carbon, which solidifies isothermally at a temperature of 1135°C (2075°F).

2.4 THE IRON–CARBON PHASE DIAGRAM

Phase diagrams and equilibrium relationships have been discussed in Section 1.4 for the general case of two metals combining to form alloys. Different situations that can occur were considered. The iron–carbon phase diagram is used to determine the properties and applications of steels, and this diagram will be discussed in this section. Steels are alloys of iron and carbon and for most practical purposes they contain between 0.1% and 1.7% carbon. The addition of other alloying elements produces steels which possess special properties, and these effects are considered in Section 2.11. Before the individual regions of the phase diagram are described, the theory of nucleation and the growth of new phases within steels is discussed.

2.4.1 Nucleation and growth of new phases

Most solid phase transformations occur due to the *nucleation* and *growth* of a new phase. These are temperature-controlled processes, whereas the rate of growth is a diffusion-controlled process; however, diffusionless transformation, such as the shearing of fcc austenite to form the bct martensite phase, can also occur. The resulting martensite lattice is elastically strained due to the presence of excess carbon atoms. The equilibrium phase diagram is an obvious starting point for the consideration of the way in which phase transformations occur. However, this diagram can be used to predict the stable phases that are present only under equilibrium conditions at a particular temperature. The phase diagram provides no information regarding the rate of a phase transformation.

A nucleation process begins with the formation of small stable regions or grains of the new stable phase at a particular temperature, as predicted by the equilibrium phase diagram. These new phase regions are surrounded by the old (thermodynamically) unstable phase. Consider the formation of a small spherical β nucleus in an α matrix (liquid phase). When β nuclei form, an interface is created. The total free energy change involved in making a spherical nucleus of the β phase of radius r is the sum of the bulk free energy change (which is negative at temperatures below the equilibrium transformation temperature) and the surface energy change (always positive, since energy is always expended in creating an interface). When the radius (r) is less than a critical value (r^*), the surface term dominates and the total free energy change increases with r. For $r > r^*$, the volume term dominates and the total free energy change decreases with increasing r. Therefore, a free energy barrier exists for the formation of a β nucleus. Any β nucleus with $r < r^*$ will redissolve in the α matrix, even at temperatures below the equilibrium transformation temperature. Only when a nucleus exceeds the critical size can it continue to grow with a decrease in free energy. The energy barrier is overcome by the addition of thermal energy. Values of r^* are quite small, typically less than 1 micron.

The rate at which the phase transformation occurs depends upon the diffusion of atoms to and from the new phase, as it grows within the crystal structure. When a metal alloy is cooled below the equilibrium transformation temperature, initially the rate of formation of the new phase is slow due to the dependence of the number of nucleation sites on undercooling, although it increases as the temperature is further decreased. The initial low rate is because of the existence of the critical size of nucleus below which growth is unstable (thermodynamically endothermic). However, if the temperature falls too far, then atomic movement is reduced and the rate of transformation is again reduced.

The nucleation process as described is known as *homogeneous nucleation*, and it depends upon the minimum (or critical) number of atoms combining at the same point within the phase so that the phase formation becomes exothermic. *Heterogeneous nucleation* occurs if impurity particles (inhomogeneities) are present which act as nucleation sites. Examples are grain boundaries or dislocations within a crystal lattice, slag in a molten alloy or sand particles on the surface of a casting mold (see Section 8.1). The process of *heterogeneous nucleation* is often preferred because of the lower surface energy required. *Inoculants* or nucleating agents are often added to a material to provide control over the microstructure.

A complete phase transformation requires the diffusion of atoms over large distances; this occurs by both bulk diffusion and accelerated diffusion along grain boundaries or dislocations. Since crystal diffusion often occurs in certain directions, nucleation may be *anisotropic* in certain directions or planes. Disk or needle-like (acicular or dendritic) crystals may be formed.

The microstructure, and therefore the mechanical properties of a material, can be controlled by careful consideration and selection of both the transformation temperature and the length of time that the process is performed. The rate of phase transformation depends upon the nucleation rate and the growth rate. Since the maximum overall rate is usually proportional to some product of these rates, it is higher than the peak nucleation rate. At temperatures close to the equilibrium transformation temperature (from the phase diagram) only a few nuclei are formed. However, the growth rate is rapid and the microstructure consists of a few large crystals. At lower temperatures the opposite is true and a fine-grained microstructure is obtained. At some intermediate temperature a maximum overall rate of transformation occurs. The diffusion rate is much more temperature-sensitive than the driving force (equivalent to the free energy change and proportional to the degree of undercooling); the maximum growth rate usually occurs at a higher temperature than the maximum for the nucleation rate. Since the transformation time is inversely proportional to the transformation rate, the time–temperature relationship is shown in Fig. 2.7. This curve is known as a C curve because of the general shape, or a TTT curve (Time–Temperature–Transformation). The curves (a) and (b) shown in Fig. 2.7 represent the start (0%) and finish (100%) of the *isothermal* transformation process. There is also a family of intermediate curves representing the time required for the process to proceed to a certain extent, e.g. 10% of the total volume transformed or some other basis. The intermediate curves are not shown on Fig. 2.7.

An understanding of the mechanisms of nucleation and growth assist in the intepretation of the TTT curves, which are discussed in more detail in Section 2.5. This discussion will also be useful when considering the effects produced by various heat treatment processes as described in Sections 2.18 and 2.19, and the metallurgical changes that occur as a result of hot-working fabrication processes (see Section 8.2.7).

An example of the influence of the phase transformation conditions upon the nature of the material obtained, is the age-hardening of aluminum alloys (see Section 3.3.1). The transformation time and temperature are selected to produce precipitated particles of the required size, distribution and shape, so that dislocation motion is severely restricted and high mechanical properties are obtained.

Regions of the iron–carbon phase diagram shown in Fig. 2.2 will now be described. It may appear complex but explanation becomes easier if individual regions are considered. It is important at this stage that the preliminary work of Section 1.4 has been followed.

2.4.2 The eutectic region

Consider the region of the phase diagram with a carbon content greater than 0.9% carbon; this is reproduced in Fig. 2.3. It is the eutectic region of the iron–carbon phase diagram. A liquid alloy containing 4.3% carbon has the eutectic composition; it solidifies instantaneously when a temperature of 1147°C (2097°F) is reached. For compositions other than the eutectic, the eutectic reaction is such that upon slow cooling of the liquid alloy, it commences to freeze at a temperature above 1147°C (2097°F) (the eutectic temperature).

Solidification is complete when a temperature of 1147°C (2097°F) is reached and the solid formed is a mixture of eutectic and a solid solution. If the liquid alloy contains more than 4.3% carbon, then as solidification begins the compound cementite (Fe_3C) which contains 6.67% carbon begins to precipitate out of solution. The remaining liquid becomes depleted in carbon and, if cooling is sufficiently slow, contains 4.3% carbon when the temperature reaches 1147°C (2097°F). A complete solid phase is then obtained which is a mixture of the eutectic (4.3% carbon) and cementite. If the liquid alloy contains less than 4.3% carbon, as solidification commences a solid solution of carbon in fcc iron is formed which is known as austenite. The maximum solubility limit in the solid solution is 1.7% carbon as shown in Fig. 2.2, occurring at a temperature of 1147°C (2097°F). Therefore, as a liquid alloy containing less than 4.3% carbon solidifies, the austenite solid solution precipitates out until the liquid reaches the eutectic composition and complete solidification is obtained.

2.4.3 The eutectoid region

Consider the region of the iron–carbon phase diagram with alloys containing less than 1.5% carbon, at temperatures below 1000°C (1832°F). This region,

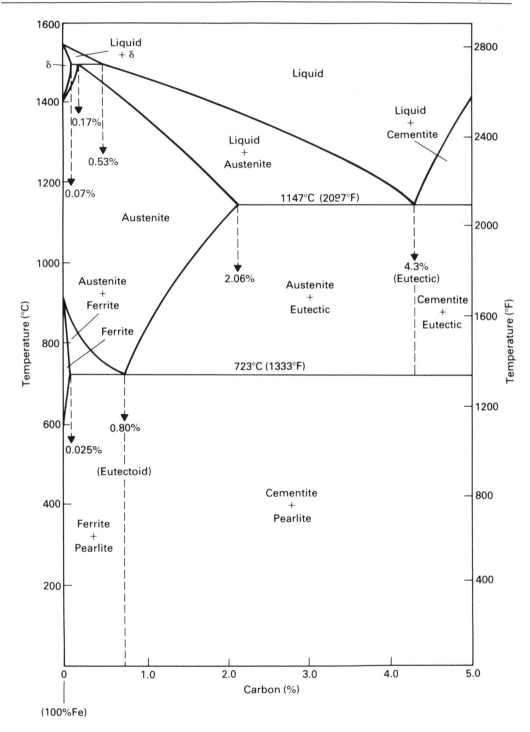

Figure 2.2 The iron–carbon phase diagram.

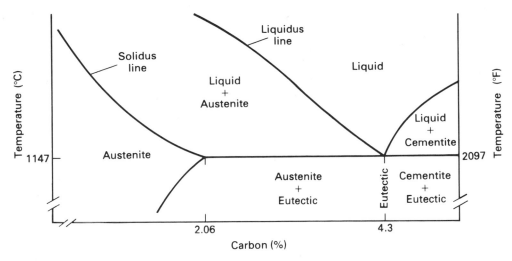

Figure 2.3 The eutectic region.

the eutectoid region, is reproduced in Fig. 2.4. This is the most important region of the iron–carbon phase diagram for understanding the nature of steels, and it is used to predict structural changes that occur in the solid state during cooling. An iron–carbon alloy containing 0.80% carbon is known as the eutectoid. At temperatures above 723°C (1333°F), this alloy is in the form of austenite which is fcc iron containing carbon atoms in solid solution. Upon cooling to 723°C (1333°F), the austenite changes to pearlite (the eutectoid). Pearlite is the name given to the particular composition made up of laminations of ferrite and

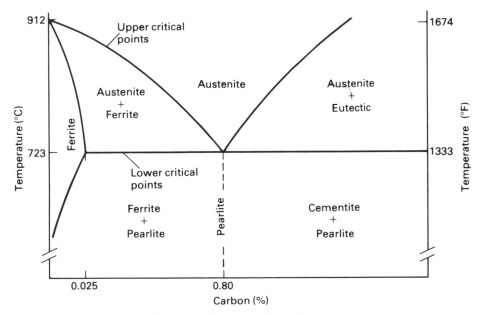

Figure 2.4 The eutectoid region.

cementite. The allotropy of pure iron has already been discussed in Section 1.2.7; from Fig. 2.4 it can be seen that the allotropic change from bcc α-iron to fcc γ-iron occurs at 912°C (1673°F). Figure 2.4 also shows that the alloying effect of carbon on iron is to lower the temperature at which the $\alpha - \gamma$ transformation takes place, from 912°C (1673°F) for pure iron to 723°C (1333°F) for the 0.80% carbon alloy. The maximum interstitial solubility of carbon in the bcc ferrite is 0.03% carbon at 723°C (1333°F); this is very much less than can be dissolved in the fcc austenite (maximum 1.7% carbon at 1147°C; 2097°F).

The terms hypoeutectoid and hypereutectoid are used to denote steels that contain less carbon and more carbon respectively than the eutectoid composition (0.80% carbon). In Fig. 2.4 the temperature–composition points at which austenite begins to transform to ferrite are called the upper critical points; the eutectoid temperature (723°C, 1333°F) is known as the lower critical temperature, with corresponding lower critical points. When a hypereutectoid steel is cooled to its upper critical point, then cementite (Fe_3C) containing 6.67% carbon begins to precipitate out and the relative proportion of carbon in the austenite falls. The carbon content reaches the eutectoid composition (0.80% carbon) at a temperature of 723°C (1333°F); the transformation of austenite to cementite and pearlite is then complete. For a hypoeutectoid steel containing between 0.03% and 0.80% carbon, upon cooling to the upper critical point ferrite begins to precipitate out. The ferrite can only contain a small amount of carbon in solid solution, and therefore the relative proportion of carbon remaining in the austenite is increased. As the temperature falls the carbon dissolved in the ferrite increases, as does the carbon in the austenite. At 723°C (1333°F) the ferrite contains 0.03% carbon, the austenite is of eutectoid composition and transformation to pearlite occurs. An alloy which contains less than 0.03% carbon is fully transformed from austenite to ferrite (all carbon in solid solution) before 723°C (1333°F) is reached. At room temperature the maximum solubility of carbon in ferrite is 0.006%. Therefore, if an alloy contains between 0.006% and 0.03% carbon, a further structural change occurs when it is cooled to some temperature below 723°C (1333°F). This situation occurs when the lower temperature solubility of carbon is reached; the carbon then precipitates out of the ferrite as pearlite.

2.4.4 The peritectic region

Consider the region of the iron–carbon phase diagram at temperatures above 1300°C (2372°F) and compositions of less than 0.8% carbon. This is the peritectic region of the phase diagram and is reproduced in Fig. 2.5. The effect of the carbon present is to reduce the temperature at which a single liquid phase is obtained. Pure iron is molten at 1538°C (2795°F), but this temperature is 1495°C (2723°F) if 0.53% carbon is present (Fig. 2.5) or 1147°C (2097°F) if 4.3% carbon is alloyed (Fig. 2.3). At temperatures between 1388°C and 1535°C (2530°F and 2795°F), the stable allotrope of pure iron is δ-iron, which has a bcc structure. δ-iron can dissolve up to 0.09% carbon (at 1495°C;

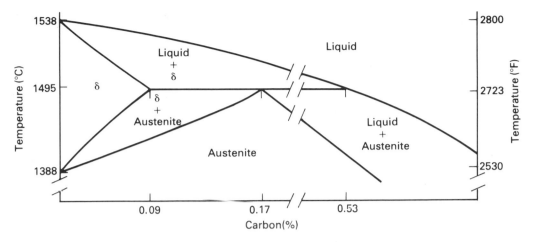

Figure 2.5 The peritectic region.

2723°F) in solid solution (Fig. 2.5). This is less than the fcc austenite (1.7% carbon at 1147°C/2097°F in Fig. 2.3), but is more than the lower temperature bcc ferrite (0.03% carbon at 723°C/1333°F in Fig. 2.4) due to increased atomic spacing caused by the extra thermal energy. From Fig. 2.5, if a liquid alloy containing less than 0.09% carbon is cooled, then the δ-solid solution precipitates out of the liquid, completing the process by the time the temperature is 1495°C (2723°F). With further cooling, at a particular temperature the δ-solid solution begins to transform to austenite and this is complete before 1388°C (2530°F) is reached.

    With carbon contents between 0.09% and 0.53% carbon, the peritectic reaction occurs at 1495°C (2723°F). This has been defined (Section 1.4.6) as the reaction between a solid produced during cooling, and the remaining liquid to form another phase. From Fig. 2.5, the first solid solution produced during cooling is the δ-phase. As cooling proceeds more carbon is accommodated in the δ-phase interstices, up to a maximum of 0.09% carbon at 1495°C (2723°F). The amount of carbon dissolved is small and the relative proportion of carbon in the liquid increases (along the liquidus line), until at 1495°C (2723°F) the liquid contains 0.53% carbon. The new phase which is formed by the peritectic reaction is the fcc austenite solid solution.

    If the original liquid alloy contained 0.17% carbon, then the peritectic reaction (at 1495°C/2723°F) between δ-phase (0.09% carbon) and liquid alloy (0.53% carbon) would produce a single solid solution of austenite (with 0.17% carbon). If the original carbon content was between 0.17% and 0.53%, then the peritectic reaction would produce austenite (containing 0.17% carbon), but because of the excess carbon not all of the liquid is consumed. The alloy must be cooled still further and the solubility of carbon in austenite increases, the relative proportion of carbon in the liquid also increases, and the amount of liquid present decreases. When the solidus curve for the particular alloy is reached, the last drops of liquid are converted into austenite and the alloy is a single phase. If the alloy contains between 0.09% and 0.17% carbon, then when cooled to 1495°C (2723°F) there is more δ-solid solution formed than is necessary to react with the liquid. When the peritectic reaction is complete a mixture of austenite and excess δ solid solution is present as a solid phase (at 1495°C/2723°F). Further cooling results in the δ-phase being transformed to austenite of the same composition as the original liquid alloy.

2.4.5 Uses of the iron–carbon phase diagram

The iron–carbon phase diagram is used to determine the microstructure of an alloy, the constituents that are present as it is *slowly* cooled from the liquid state. However, it must be understood that this diagram applies only to changes that occur under equilibrium conditions, i.e. very slow heating or cooling, with sufficient time allowed at each stage for equilibrium to be attained. It is not usually possible to allow sufficient time for equilibrium conditions to be established, due to the costs involved. Also the phases that are obtained under equilibrium conditions may not possess the properties required by the steel. Ferrite is comparatively soft and ductile with a low tensile strength, whereas pearlite has less ductility but improved strength. Cementite is both hard and brittle. The equilibrium microstructures of some plain carbon steels are shown in Fig. 2.6. If austenite is cooled quickly so that equilibrium is not attained, then the pearlite formed consists of fine lamellae of ferrite and cementite and can have very high strength, depending upon the cooling rate. If a steel is cooled very quickly, then there is insufficient time for the austenite to transform to pearlite. Instead the austenite changes to a body-centered lattice with all the carbon trapped in an interstitial solid solution. This carbon distorts the lattice making it body-centered tetragonal (bct), and the structure is known as martensite. Martensite is hard and brittle but it is a useful constituent of steel for increasing the surface hardness. The austenite can be cooled in either water, oil or an air blast, a process known as quenching. The ways in which the properties and structures of steels can be varied by using heating–cooling cycles are discussed in Section 2.19.

2.4.6 The structure of steels

If large amounts of alloying elements are added, then austenitic alloys can be stable at room temperature, e.g. austenitic 18–8 stainless steels (see Section 2.10.4). Methods are available for the determination of the high-temperature austenite grain size when the alloy is cooled to room temperature and the transformation completed, e.g. ASTM E112.

Ferrite grains occur in a significant quantity only in annealed or normalized steels, and if the carbon content is below 0.4% carbon. The grain size can be increased by heating to 600°C (1112°F); above 723°C (1333°F) any pearlite present transforms to austenite grains. These grains will grow at the expense of the ferrite grains, which are also growing. When the ferrite to austenite

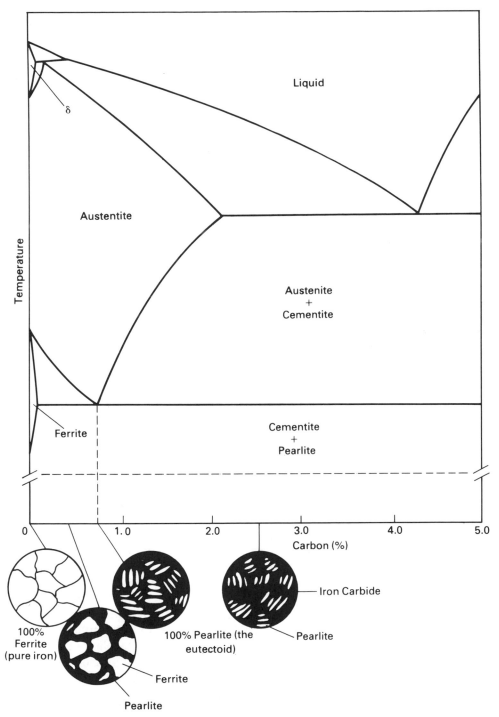

Figure 2.6   Equilibrium microstructures of plain carbon steels (room temperature).

transformation temperature is reached, all the ferrite is converted to austenite if sufficient time is allowed. The size of the ferrite grains depends upon the cooling rate, and coarser ferrite grains are obtained by slow cooling a low-carbon austenite than by rapid cooling. Also a coarse-grained austenite produces a coarser-grained ferrite than would a fine-grained austenite.

If a plain carbon austenitic structure is hot worked, then the grain size depends upon recrystallization and grain growth, both of which are dependent upon the temperature. However, when steel is heated above the critical temperature, austenite grains begin to form within pearlitic regions by nucleation and phase transformation.

Finally, it should be understood that pearlite 'grains' do not exist, since ferrite crystals and cementite crystals are in contact. The apparent grain outline is the old boundary of the austenite grain at which the newly formed ferrite and cementite grains begin to grow.

2.5 TIME–TEMPERATURE–TRANSFORMATION DIAGRAMS

The rapid cooling of steels does not create equilibrium conditions, therefore the metastable phases which may be formed, e.g. martensite, do not appear on the iron–carbon phase diagram. However, the rate of cooling is an important consideration when producing different steels, and an alternative diagram is used to interpret the effects which occur. These diagrams are known as Time–Temperature–Transformation diagrams (TTT diagrams), isothermal transformation curves, 'S' curves, or 'C' curves (due to their general shape). The curves are different for each steel, although their general shape is similar. An example is shown in Fig. 2.7 for a plain carbon steel.

Considering Fig. 2.7 at temperatures above 723°C (1333°F), austenite is the stable form of this plain carbon steel. This temperature depends upon the actual carbon content (see the iron–carbon phase diagram in Fig. 2.2); there is a range of temperatures above 723°C (1333°F) to which the steel can be heated and held as a preliminary heat treatment (see Section 2.19). If the steel is cooled very slowly, then the austenite transforms to pearlite very slowly and a final coarse structure is obtained. With more rapid cooling, as shown by the dotted line (i) in Fig. 2.7, the grain structure is refined and fine pearlite is obtained. If cooling is very rapid, then the cooling rate is shown by dotted line (ii) on Fig. 2.7. The critical cooling rate is such that the 'knee' of the curve is avoided and martensite rather than pearlite is formed. If this rapid cooling curve is followed (curve ii) but is arrested before martensite is formed (and below the knee of the curve), then if the temperature is held constant the product is bainite rather than pearlite. This process is known as *austempering*. Bainite is an intimate mixture of carbide and ferrite (not laminations), it is formed because the transformation curves (a and b) of Fig. 2.7 are really a combination of two sets of curves. This situation is shown in Figs. 2.8 and 2.9 for a plain carbon steel and an alloy steel. Pearlite formation requires sufficient time for diffusion of the carbon and iron

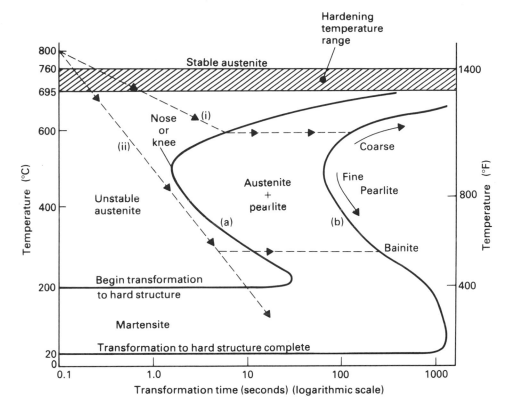

Figure 2.7 Typical Time–Temperature–Transformation diagram for a plain carbon steel. (a) Transformation begins. (b) Transformation ends.

atoms, whereas bainite formation requires sufficient time for the shearing of the fcc structure to a bcc lattice as well as the carbon precipitation.

The critical cooling velocity depends upon the carbon content of the steel and the presence of other alloying elements; an increase in either of these moves the 'knee' of the curve to the right in Fig. 2.7. This reduces the critical cooling rate necessary to form martensite. Typical cooling rates which can be obtained by quenching a steel are approximately $500\,°C/s$, $250\,°C/s$, $50\,°C/s$, ($932\,°F/s$, $482\,°F/s$, $112\,°F/s$) by using water, oil or air as the cooling medium, respectively. The critical cooling rate of a plain carbon steel containing less than 0.3% carbon is about the rate that can be obtained by water quenching. It is therefore impractical to fully harden plain carbon steels containing less than 0.3% carbon by quenching. It is possible to harden the surface layers of a component by the formation of martensite, while the material core transforms more slowly from austenite to pearlite. However, the different cooling rates of the surface (martensite) and the core (pearlite) can result in expansion of the core causing surface cracks. This is a problem in high-carbon steels having low toughness, and in components that have sharp curves and notches, and hence concentrated stresses. To reduce the danger of cracking a process of interrupted quench is used, known as *marquenching* or *martempering*. The steel is quenched to a

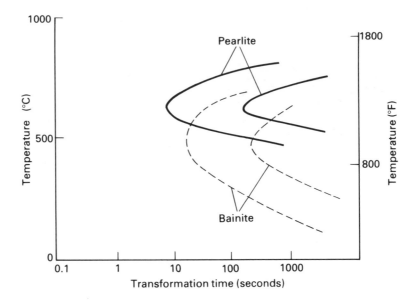

Figure 2.8 Transformation curves for a plain carbon steel.

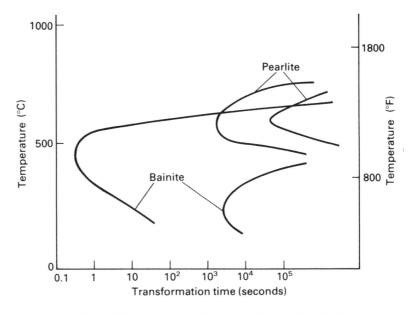

Figure 2.9 Transformation curves for an alloy steel.

temperature where martensite just begins to form; it is allowed to reach equilibrium and is then slowly cooled to form martensite, thus avoiding severe stresses. The steel must then be reheated (tempering; see Section 2.19.4) to improve the ductility. Alloying elements are usually present in the steel to reduce the critical cooling velocity, and to ensure that the required hardness can be achieved in thicker sections. A process known as *ausforming* is used to produce an exceptionally strong product. This is achieved by interrupting the cooling of an alloy steel at approximately 330°C to 440°C (626°F to 824°F), and then performing hot working. This causes strain hardening, and is followed by quench hardening so that martensite is formed. A tempering stage is required and a very strong product is formed.

2.5.1 TTT diagrams and heat treatment processes

TTT diagrams can be used for the selection and specification of several types of heat treatment. The TTT diagram for a steel with a high hardenability (see Section 2.18.2), e.g. AISI 4340, possesses both a pearlite and a bainite 'nose'. The rate of austenite transformation is a maximum in these regions, and has a minimum value between these peaks. The overall transformation rate is greatly reduced and more time is available for martensitic formation; also thicker sections can be hardened. For these reasons it is difficult to produce ferrite and pearlite by full or isothermal annealing in these steels.

Martensite is the hardest steel structure and it is produced by rapid cooling, known as quenching. Rapid cooling, e.g. water quenching, causes internal stresses within the steel, due to non-uniform contraction and crystal distortion, and it is too brittle for many applications. The rate of quenching can be reduced by using an oil quench, or the steel can be reheated (tempering) to preselected temperatures for a short time. Some of the excess dissolved carbon then precipitates in the form of fine carbide particles and the brittleness is greatly reduced, with only a slight reduction in the mechanical properties. There is a critical cooling rate that must be achieved if martensite is the required product; this rate can be determined from the TTT diagram for a particular steel.

If martempering is to be performed, then the TTT diagram can be used to calculate the rapid cooling rate in a quenching bath so that pearlite is not formed. The steel is then cooled at this rate, but is arrested at a temperature just above that where martensite is formed. The temperature is maintained until the core of material reaches the bath temperature (but bainite must not form); the material is then removed and air cooled. Distortion and cracking due to the quenching process are mainly eliminated, although transformation stresses are still present.

Austempering also uses interrupted quenching, but the temperature is held until bainite below the nose of the curve is formed (a finer grain). When the transformation is complete, then the steel is air cooled. The use of TTT diagrams is essential for correct control of both martempering and austempering.

2.5.2 Continuous cooling TTT diagrams

The TTT diagrams described so far are obtained by cooling a steel isothermally at a series of temperatures below 723°C (1333°F). The time for transformation to be completed is then measured and a temperature – log time graph is plotted.

In practice most heat treatment operations are not isothermal, and a series of continuous cooling transformation diagrams are also available for a wide range of steels. These diagrams can be used to determine the microstructure and hardness of a steel after continuous cooling from the austenite region. This information cannot be obtained directly from isothermal TTT curves.

The methods used to obtain continuous cooling curves are not described here. However, the ways in which the data can be presented are considered. A graph of 'temperature against log time' can be presented, although 'zero' time may be taken as the start of cooling or the time when austenite transformation begins. This type of graph is frequently used because of its flexibility; hardness values are usually included for the transformed specimens. A graph of 'temperature vs time to cool over a fixed temperature range', e.g. 800°C to 500°C (1472°F to 932°F) (log scale) can also be presented. Diagrams in common use include a plot of 'temperature or hardness vs (log) bar diameter', and also include an additional axis of cooling rate.

* * *

Self-assessment exercises

1 For the iron–carbon phase diagram, you should be familiar with:
   (a) the general shape of the diagram;
   (b) the different regions that are present;
   (c) important changes that occur, and the approximate temperatures and carbon content;
   (d) the names of the phases that are present in different regions.

2 What information can be obtained from the iron–carbon phase diagram? How can an engineer use this information?

3 What is the main limitation when using a phase diagram?

4 Sketch a typical isothermal transformation curve for a steel. Explain how this diagram can be useful to an engineer.

5 Describe how different phases and microstructures can be present in a steel, depending upon the cooling rate that is used.

6 For welding and some fabrication methods, metals must be heated to high temperatures. Assess the usefulness of phase diagrams and TTT curves for understanding and predicting any changes/effects that may occur.

* * *

2.6 THE PRODUCTION OF STEEL

The basic raw material used in the production of steel is the pig iron which is removed from a blast furnace in the molten state (Section 2.2). The pig iron and sometimes some scrap steel are heated in a furnace to remove carbon and impurities by oxidation or burning; their removal also occurs in the waste products or slag. Other elements (and carbon) can be added to produce a steel of a certain grade with particular properties. Modern steelmaking processes use either a converter or an open-hearth furnace.

One of the earliest processes that used a converter was introduced in 1856 by Bessemer. Different versions of this process have been introduced since 1945. The converter (converts molten pig iron into steel) is a steel shell that is refractory lined. It can be tilted to a horizontal position for charging, sampling and pouring through an open mouth. In the vertical position, air under pressure is blown into the molten metal through a large number of small holes in the refractory. The impurities and some carbon are oxidized, and manganese-rich pig iron is added to control the final carbon content. The capacity is usually between 25 and 60 tonnes and the process takes about 30 minutes.

Modern developments such as the Linz–Donawitz or Kaldo processes employ pure oxygen which is blown onto the molten metal surface. The Linz–Donawitz process is mainly used with pig iron having a low phosphorus content. The Kaldo process uses a rotating converter (up to 30 revolutions per minute) and can be used with irons containing up to 2% phosphorus. These processes typically take 60 minutes to handle a 100 tonne charge.

The open-hearth process uses auxiliary blowing with pure oxygen to reduce the cycle time. The furnace is charged with pig iron and scrap steel, and is heated to between 900°C and 1200°C (1652°F and 2192°F). The molten steel is tapped off and separated from the slag which is formed. The slags produced are acid or basic, and this also refers to the furnace lining. Pig iron produced from ores rich in silicon and low in phosphorus and sulfur produce an acid slag when refined, and a silica refractory is used. A basic slag is produced when pig iron rich in phosphorus is refined, because large amounts of lime must be added. The refractory lining is then made of a basic material. The refined iron has small amounts of carbon, silicon, manganese, sulfur and phosphorus (less than 0.05% of each). Silicon and manganese are added as alloying

Table 2.1 Compositions and typical applications of steels

Carbon content %	Name	Applications
0.05	dead mild steel	sheet and strip for presswork, car bodies, tin plate; wire, rod, and tubing
0.08–0.15	mild steel	sheet and strip for presswork; wire and rod for nails and screws; and concrete reinforcement bar
0.15	mild steel	case carburizing quality
0.10–0.30	mild steel	steel plate and sections; used for structural work
0.25–0.40	medium-carbon steel	bright drawn bar
0.30–0.45	medium-carbon steel	shafts and high-tensile tubing
0.40–0.50	medium-carbon steel	shafts, gears, railroad tires
0.55–0.65	high-carbon steel	forging dies, railroad rails, springs
0.65–0.75	high-carbon steel	hammers, saws, cylinder linings
0.75–0.85	high-carbon steel	cold chisels, forging die blocks
0.85–0.95	high-carbon steel	punches, shear blades, high-tensile wire
0.95–1.10	high-carbon steel	knives, axes, picks, screwing dies and taps, milling cutters
1.10–1.40	high-carbon steel	ball bearings, drills, wood-cutting and metal-cutting tools, razors

elements to deoxidize the metal, and anthracite is added to adjust the final carbon content. The steel at this stage can be made directly into castings, or it can be formed as ingots that can then be fabricated by forging or rolling.

If a high-quality steel is required, then impurities entering the furnace must be reduced, and this is achieved by using an electric furnace. The direct arc type of furnace has between 25 and 100 tonnes capacity; it has carbon electrodes and the lining can be either acid or basic. It is used to produce low-alloy steels, containing nickel and chromium, from the refined steel produced in a converter or open-hearth furnace. The high-frequency induction furnace is not used for refining, only for the production of high-alloy steels and special steels. Capacities are between 350 kg and 6 tonnes.

2.7 CLASSIFICATION OF STEELS

The main types of steel and their applications are summarized in Table 2.1.

2.7.1 Killed steel

The covering of steel in an ingot mold with a deoxidizing material (e.g. silicon) until deoxidation is completed produces what is known as a *killed steel*. If deoxidation is incomplete, then a semi-killed steel is produced. Killed steels have a high carbon content, possess reasonably high strength and are easily forged.

2.7.2 Cast steel

Cast steel is not a special type of steel, but it is a term applied when the molten metal is used directly in castings, usually of complicated shapes.

2.7.3 Plain carbon steel

A *plain carbon steel* contains a maximum of 1.5% carbon, 1.5% manganese and 0.5% silicon, and only traces of other elements. The carbon content affects the hardness and strength of the steel. Plain carbon steels are often classified into the following three main categories.

2.7.4 Mild steel or low-carbon steel

This type of steel contains a maximum of between 0.25 and 0.30% carbon. It is easily machined and welded, but does not respond to heat treatment. It is used for constructional applications and can be forged or rolled to form angle, channel, plate, bar or rod.

2.7.5 Medium-carbon steel

Steel with a carbon content between 0.25% and 0.65% is termed a *medium-carbon steel*. The hardness is often improved by heat treatment. It is a shock resisting steel used for crankshafts and axles.

2.7.6 High-carbon steel

This is a plain carbon steel containing between 0.6% and 1.5% carbon. It is always used in a hardened (or tempered) condition, but it is also brittle. These steels are also known as *tool steels* because of their applications. A *silver steel* has a high carbon content; it contains no silver but has a bright appearance. Silver steel is produced as bar with very close limits of accuracy on the diameter.

2.7.7 Alloy steels

Carbon steels only contain iron and carbon, and sometimes impurities. *Alloy steels* contain one or more elements that are added to produce particular properties in the steel. Alloy steels are more expensive to produce due to the cost of the alloying elements and processing, and they are generally more difficult to fabricate and machine. Special properties are sometimes obtained by heat treatment (see Section 2.18).

2.7.8 Low-alloy steels

Low-alloy steels have a microstructure similar to plain carbon steels, and require the same heat treatment. They have comparable applications and similar carbon contents. Low-alloy steels usually contain up to 3 or 4% of one or more alloying elements to improve the strength, toughness and hardenability. Nickel improves the resistance to fatigue.

2.7.9 High-alloy steels

The structures of *high-alloy steels* and the heat treatment processes that are required are very different from plain carbon steels. Examples of high-alloy steels are stainless steels and high-speed tool steels. These are special steels that have particular characteristics such as corrosion resistance and low temperature applications. The uses of certain high-alloy steels are discussed in Section 2.9.

* * *

Self-assessment exercises

1 Explain what is meant by 'an integrated iron and steel works'.

2 Describe briefly the modern processes used to make steel. Outline the particular features of each process.

3 Consider a country with iron ore deposits but no coal supplies. Assuming there is an abundant alternative energy source, e.g. natural gas, describe a process that could be used to produce iron and steel. What effects could the use of natural gas as a fuel have upon the nature of the steel produced?

4 Can you differentiate between the main types of steel produced, e.g. plain carbon steel, mild steel, high-carbon steel, etc., in terms of the carbon content and typical applications?

* * *

2.8 EFFECTS OF ALLOYING ELEMENTS

The heat treatment of steels is described in detail in Section 2.19. The addition of an alloying element affects the response of the steel to heat treatment, and also influences the properties possessed by that steel. The addition of chromium, manganese and tungsten reduces the critical cooling rate and increases the depth of hardness. Therefore, it will be easier to obtain martensite from the breakdown of austenite, producing thicker sections with a uniform structure when hardened. The addition of nickel or manganese lowers the temperature at which the formation of austenite occurs; less than 5% nickel produces a hard, tough steel. The addition of chromium, tungsten and silicon raises the transformation temperature of ferrite to that of austenite, i.e. the temperature required to improve the hardness. Chromium and molybdenum are carbide-forming elements and produce a greater depth of hardness. The strength of a steel can be increased by increasing the carbon content, which also results in a loss of ductility. However, many alloying elements can be added which enter the solid solution and increase the strength without reducing ductility. Smaller, lighter sections can therefore be used. Nickel and silicon can cause a breakdown of cementite so that graphite is precipitated in the structure. These elements are not therefore added to high-carbon steels. The addition of chromium (which is itself corrosion resistant) in proportions greater than 12% produces a corrosion-resistant (stainless) steel. This is due to the formation of a surface oxide film. The use of nickel and chromium together produces an austenitic structure with superior corrosion resistance, particularly at high temperatures.

The influence of increasing the carbon content on the properties of a steel has already been mentioned. The carbon content also affects the usefulness of the final metal in terms of fabrication, machining and welding operations, or any subsequent heat treatment to be performed. With reference to the phase diagram in Fig. 2.3, the maximum carbon content in iron (as solid solution) is 1.7% carbon at a temperature of 1147°C (2097°F). When this austenitic steel is cooled, the excess carbon forms cementite and crystallizes out at the grain boundaries. At 723°C (1333°F) a stable eutectoid solution (pearlite) is formed containing 0.89% carbon, and this remains stable down to room temperature.

There are also some impurities present in steels that affect the properties of the material. The main impurities are phosphorus and sulfur.

Phosphorus is regarded as a harmful impurity producing hardness and brittleness and lowering the melting point of the steel due to the formation of iron phosphide. The phosphorus content in steels is rigidly controlled, with a maximum of approximately 0.05%. The grading of iron ores is performed according to the phosphorus content. The brittleness produced is more pronounced in high-carbon steels; in low-carbon steels a phosphorus content of up to 0.15% can slightly improve the strength and corrosion resistance.

Sulfur is an impurity that is introduced in the furnace fuel and it forms iron sulfide. This makes the steel unsuitable for cold working because of its brittleness, or for hot working because it has a lower melting point causing steel to crumble at higher temperatures. Sulfur can also cause problems with welding. The sulfur content is usually kept below 0.05%. Manganese added to the molten steel induces the sulfur into the slag; manganese sulfide is then formed instead of iron sulfide, and this is not detrimental to the steel, although the quantity of manganese in the steel must also be controlled. However, in free-cutting steels up to 0.2% sulfur may be present to produce a smooth finish and a short chip when used on automatic cutting machines.

Silicon is sometimes regarded as an impurity because of its graphitizing effect on the steel. It is more commonly referred to as an alloying element and a silicon steel is produced with special properties. The general effects of particular alloying elements and the types of steels produced are summarized in Table 2.2.

Table 2.2   Effects of alloying elements in steels

Alloying element	General effects	Typical steels
Manganese	Increases the strength and hardness and forms a carbide; increases hardenability; lowers the critical temperature range; in sufficient quantity produces an austenitic steel; always present in a steel because it is used as a deoxidizer.	Pearlitic steels (up to 2% Mn) with high hardenability used for shafts, gears, and connecting rods. 13% Mn in Hadfield's steel – a tough austenitic steel.
	Induces sulfur impurity into the slag. Neutralizes effects of impurities in welds. Produces a uniform hardness of greater depth after heat treatment. Makes steel easier to forge due to improved ductility and flow characteristics. Can also produce free-cutting properties.	
Silicon	Strengthens ferrite; raises the critical temperature range; a strong graphitizing tendency; always present because it is used (with manganese) as a deoxidizer.	Silicon steel (0.07% C; 4% Si) used for transformer cores; used with chromium (3.5% Si; 8% Cr) for its high-temperature oxidation resistance in internal combustion engine valves.
	Between 0.5% and 5% silicon imparts magnetic properties to the steel. High silicon content prevents breakdown of a steel due to aging, which occurs at room temperature.	

Table 2.2 (continued)

Alloying element	General effects	Typical steels
Chromium	Increases strength, fatigue resistance and hardness; forms hard and stable carbides; raises the critical temperature range; increases hardenability; amounts in excess of 12% Cr render the steel stainless. After heat treatment it produces a uniform structure and a uniform hardness at greater depths, also improved tensile and elastic properties. Also used for chromium plating steels.	1.0% to 1.5% Cr in medium- and high-carbon steels for gears, axles, shafts and springs, ball bearings and metal-working rolls; 12% to 30% Cr in martensitic and ferritic stainless steels; also used in conjunction with nickel (see below).
Nickel	Marked strengthening effect; lowers the critical temperature range; increases hardenability; improves resistance to fatigue; strong graphite-forming tendency; stabilizes austenite when present in sufficient quantity. Improved ductility and abrasion resistance. Lowers the critical points at which structural changes occur during heat treatment; therefore less heat required and more gradual cooling, also less distortion and cracking. Used as corrosion-resistant plating on steels. Austenitic nickel steels quickly work harden.	0.3% to 0.4% C with up to 5% Ni used for crankshafts and axles, and other parts subject to fatigue.
Nickel and chromium	Frequently used together in the ratio Ni:Cr = 3:1 in pearlitic steels; good effects of each element are additive; each element counteracts disadvantages of the other; also used together for austenitic stainless steels.	0.15% C with Ni and Cr used for case carburizing; 0.3% C with Ni and Cr used for gears, shafts, axles and connecting rods; 18% (or more) of chromium and 8% (or more) of nickel produces austenitic stainless steels.
Tungsten	Forms hard and stable carbides; raises the critical temperature range and tempering temperatures; hardened tungsten steels resist tempering up to 600°C (1112°F).	Major constituent in high-speed tool steels; also used in some permanent magnet steels.
Molybdenum	Strong carbide forming element; improves high-temperature creep resistance; reduces temper-brittleness in Ni–Cr steels.	Not normally used alone; a constituent of high-speed tool steels and creep-resistant steels; up to 0.5% Mo often added to pearlitic Ni–Cr steels to reduce temper-brittleness.
Vanadium	Strong carbide-forming element; scavenging action and produces clean, inclusion-free steels.	Not used alone, but it is added to high-speed steels, and to some pearlitic chromium steels.
Titanium	Strong carbide-forming element.	Not used alone, but added as a carbide-stabilizer to some austenitic stainless steels.
Lead	Improves free-cutting properties, especially for automatic machining.	Used in carbon steels.

2.9 EXAMPLES AND APPLICATIONS OF HIGH-ALLOY STEELS

The specification of steels according to their composition, mechanical properties, suitability for heat treatment, welding and machining, as well as recommendations regarding service conditions, can be found in a range of British and US Standards which are summarized in Section 2.16.

A range of high-alloy steels are produced which are particularly responsive to certain heat-treatment techniques. These are discussed separately in Section 2.20. Other high-alloy steels are produced where the addition of alloying elements imparts to the steel particular characteristics or properties.

2.9.1 High-speed tool steels

The addition of tungsten and chromium to a high-carbon steel causes the formation of very hard and stable carbides, and a subsequent increase in the critical temperature and the softening temperature. These steels can then be used to produce hard-wearing, metal-cutting tools which can be operated up to 600°C (1112°F). A typical composition is 18% tungsten, 4% chromium, 1% vanadium and 0.8% carbon, which is known as 18/4/1 steel.

2.9.2 Wear-resisting steels

An austenitic manganese steel (Hadfield's steel in the UK) contains 12% to 14% manganese, 0.75% silicon and 1% carbon. It may also contain some chromium or vanadium which form carbides and improve the strength. The high manganese content makes the steel austenitic and magnetic at all temperatures. It can be obtained as castings, forgings or rolled sections, but will work harden if machined. It has excellent resistance to abrasion and is used in crushing machinery and pneumatic drill bits. An alternative hard-wearing steel contains 1.4% chromium, 0.45% manganese and 1% carbon.

2.9.3 Corrosion-resistant steels (stainless steels)

The development of stainless steels has been mainly due to their corrosion-resisting properties. However, many of these steels can also be used at high temperatures where alloying additions to the steel can reduce the main problems of loss of strength, creep and oxidation or chemical attack. The classification is the same for both stainless steels and heat-resisting steels:

> *Ferritic*   low carbon content (maximum 0.1%) and between 12% and 25% chromium. The small amount of carbon remains dissolved in the iron up to the melting point, and austenite cannot be formed. Therefore, this steel cannot be hardened by heat treatment but may be strengthened by work hardening.

Martensitic   higher carbon content (between 0.1% and 1.5%) restoring the ferrite-to-austenite transformation, and can be hardened by heat treatment. Contains between 12% and 18% chromium.

Austenitic   contains both chromium and nickel, the carbon content is kept below 0.15% to limit carbide formation which would reduce the corrosion resistance. Cannot be heat treated because the austenitic structure is retained due to the alloying elements. Carbides can form if the steels are slowly cooled from high temperatures, or reheated to 500°C to 700°C (932°F to 1292°F). These carbide particles can cause weld decay; small amounts of stabilizers such as titanium or niobium should be present.

Table 2.3 Standard types of stainless steels

Grade[†]	C(max%)	Cr(%)	Ni(%)	Mo(%)	Ti(%)	Nb(%)
			Austenitic			
18/8	0.2	17–20	7–10			
18/8Ti	0.15	17–20	7–10		$4C_{min}$ ⎫ [‡]	
					0.6_{max} ⎬	
18/10Ti	0.15	17–20	9–12		$4C_{min}$ ⎮	
					0.6_{max} ⎭	
12/12	0.16	11–14	11–14			
18/8 (low carbon)	0.08	17.5–20	8–11			
18/8 (free cutting)	0.15	17–20	7–10	(plus 0.75S and Se, Pb, etc. up to 1%)		
18/8Nb	0.15	17–20	7–10			$8C_{min}$ ⎫ [§]
						1.2_{max} ⎬
18/10Nb	0.15	17–20	9–12			$8C_{min}$ ⎮
						1.2_{max} ⎭
18/8/1½Mo	0.12	17–20	8–12	1.5–2.5		
18/8/3Mo	0.12	17–20	8–12	2.5–3.5		
25/12	0.2	22–26	11–14			
25/20	0.25	24–26	19–22			
254	0.07	18	18	3.75		0.6 (plus 2.4 Cu)
			Martensitic (hardenable)			
13 Chrome	0.12	12–14	1.0 max			
13 Chrome (free cutting)	0.18	12–14	1.0 max		(plus 0.75S and Se, Pb, etc. up to 1%)	
S 80	0.25	15.5–20	1–3			
			Ferritic			
17 Chrome	0.12	15–18				
20 Chrome	0.35	18–22				
27–30 Chrome	0.35	25–30				

† Refer to Table 2.7 for steel classification.

‡ Ti: minimum % ~ 4 × carbon content;
      maximum % ~ 0.6%.

§ Nb: minimum % ~ 8 × carbon content;
      maximum % ~ 1.2%.

Table 2.4   Types of heat-resisting steels

Property required at high temperature	Alloying element
Strength	Nickel or chromium, producing an austenitic structure. Nickel also retards grain growth.
Creep resistance	Small amounts of titanium, aluminum and molybdenum, with carbon.
Resistance to oxidation and chemical attack	Silicon or chromium.

Type	Composition (all %)					Maximum temperature (°C/°F)	Comments
	C	Cr	Ni	Si	Mn		
Ferritic	0.06	21	—	0.4	0.8	900/1652	Work-hardened steel, resistant to
	0.1	29	—	1.2	1.3	1100/2012	sulfur gases.
Martensitic	0.4	11.5	—	0.3	0.5	750/1382	Hardened by heat treatment, easily machined, welding not recommended.
Austenitic	0.12	18.5	8.5	1.0	0.7	350/662	Easily welded, difficult to machine.
	0.2	23	11.5	1.6	0.4	1105/2021	3% W
	0.15	13.5	63.0	0.8	1.3	1100/2012	Very low coefficient of thermal expansion.

Details of certain standard grades of stainless steels are presented in Table 2.3. The effects of the main alloying elements on the heat-resisting properties of steels, and their classifications, are summarized in Table 2.4.

The different types of stainless steels will now be considered in more detail because of their obvious engineering importance, and also the effects of particular elements on the corrosion resisting properties.

2.10 TYPES OF STAINLESS STEELS

2.10.1 Ferritic

Containing 16–18% chromium. A rustless iron with low carbon content. It has a high resistance to corrosion, but low impact strength and cannot be refined by heat treatment alone. Prolonged service at 480°C (894°F) can cause embrittlement.

Containing 25–30% chromium. An iron for furnace parts which is resistant to sulfur compounds. It forms a sigma phase.

2.10.2 Martensitic (hardenable alloys)

Containing 12–14% chromium. Both iron and steels whose mechanical properties are largely dependent on the carbon content. High strength is combined with considerable corrosion resistance.

Stainless iron (12% Cr, 0.1% C) can be welded and easily fabricated, useful for turbine blades.

Stainless steel (12% Cr, 0.2% C) is difficult to weld, corrosion resistant only when hardened and polished. Three main types:

(a) *mild* (0.2% C) – used for steam valves and piston rods, but not in contact with non-ferrous metals or graphite packing owing to galvanic action.

(b) *medium* (0.3% C) – used for table cutlery and tools, and also parts for use at elevated temperatures.

(c) *hard* (0.4%–2.0% C) – used as springs and ball bearings that are subjected to corrosive conditions.

2.10.3 High-chromium steel (18% Cr, 0.1% C, 2% Ni)

It has a higher resistance to corrosion than a martensitic stainless steel, due to the higher chromium content. It is used for pump shafts, valves and fittings that are subjected to high temperature and high pressure.

2.10.4 Austenitic steels

Plain 18–8 austenitic steels
15–20% Cr, 6–11% Ni, 0.05–0.15% C.
May be subject to weld decay after welding.
Soft austenitic steels
12% Cr, 12% Ni.
Used for table wear and ornamental goods; less difficult to cold work.
Decay-proof steels
These are of similar composition to the plain 18–8 austenitic steels, but are specially designed for welding purposes, either low carbon or with small additions of silicon, titanium, molybdenum, and niobium.
Special purpose austenitic steels
Similar to decay-proof steels with the addition of 2% copper, 2–4% molybdenum and 10–18% nickel to improve resistance to ammonium chloride, sulfuric and sulfurous acids. Selenium or sulfur are added to improve machining.
Heat-resisting steels
These are chromium steels with a high nickel content (10–65%) together with tungsten, silicon and other elements; designed for resisting oxidation at elevated temperatures.

The hardenable alloys possess critical ranges comparable with ordinary carbon steels. Therefore, they can be hardened and tempered, and refined by heat treatment that is not dependent on recrystallization after cold working.

2.11 EFFECTS OF ALLOYING ELEMENTS

The principal effects of certain alloying elements are summarized in Table 2.5.

Table 2.5 Effects of alloying elements

Element	Typical ranges in alloy steels (%)	Principal effects
Aluminum	<2.0	Aids nitriding; restricts grain growth; removes oxygen in steel melting.
Sulfur	<0.5	Aids machinability; reduces weldability and ductility.
Chromium	0.3 to 4.0	Increases resistance to corrosion and oxidation; increases hardenability; increases high-temperature strength; can combine with carbon to form hard wear-resistant microconstituents.
Nickel	0.3 to 5.0	Promotes an austenitic structure; Increases hardenability; increases toughness.
Copper	0.2 to 0.5	Promotes a tenacious oxide film to aid atmospheric corrosion resistance.
Manganese	0.3 to 2.0	Increases hardenability; promotes an austenitic structure; combines with sulfur and reduces its adverse effects.
Silicon	0.2 to 2.5	Removes oxygen in steel making; improves toughness increases hardenability.
Molybdenum	0.1 to 0.5	Promotes grain refinement; increases hardenability; improves high-temperature strength.
Vanadium	0.1 to 0.3	Promotes grain refinement; increases hardenability; will combine with carbon to form wear-resistant microconstituents.

2.11.1 Chromium

Refer to Figs. 2.10 and 2.11. Chromium is the principal alloying element found in stainless steels. Corrosion resistance is due to the formation of a thin film (0.02 microns thick) of Cr_2O_3 at the surface. This film is complete when the chromium content exceeds 12%. In pure iron, chromium readily dissolves in the α or γ form. When chromium exceeds 12%, γ disappears and the alloy is ferritic, i.e. chromium is a γ loop forming element. The sigma phase (σ) is a hard and brittle compound of iron and chromium, of varying composition. It reduces corrosion resistance and lowers the mechanical properties, and is an undesirable phase.

The effect of chromium on solid solution hardness is not as marked as with other alloying additions. The maximum hardness of iron–chromium alloys is a Brinell hardness of 350 for a chromium content of 70%. At higher chromium levels the sigma (σ) phase occurs. This is the intermetallic compound FeCr (Vickers pyramid hardness 800) and is formed from the α phase; the contraction of the metal gives rise to fine cracks and reduced impact values.

In the presence of carbon, chromium forms very stable compounds, such

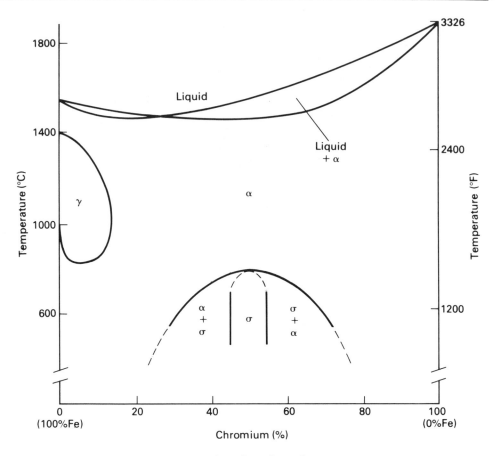

Figure 2.10 Iron–chromium phase diagram.

as cementite $(FeCr)_3C$, $Cr_{23}C_6$ $(CrFe)_3C_2$. These carbides are hard and promote wear resistance; they dissolve much more slowly in the α phase than pure Fe_3C.

Chromium raises the critical points and decreases the amount of carbon necessary to form the eutectoid composition. The effect is that the range of carbon content and the temperature within which the γ phase is stable is restricted by the chromium content. Chromium additions also decrease the rates of reaction by a large amount as shown below:

Chromium (%)	Critical cooling velocity ($^\circ C$/sec)
0	1000
0.5	725
1.1	180
2.0	40
2.6	15–30

This means that chrome steels may air harden, and hardenability is increased by the addition of chromium. As the chromium content of the steel

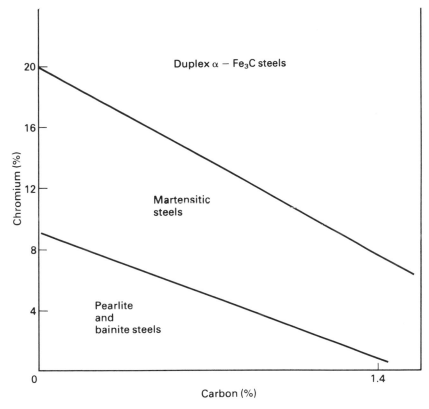

Figure 2.11   Structures present in steels containing chromium.

increases, so the tendency for grain growth increases with an undesirable embrittling effect. This can be countered by additions of a grain growth inhibitor, such as nickel, vanadium and tungsten. Low-chromium steels are particularly suitable for nitriding because the chromium assists in developing a suitable microstructure, and in forming a hard nitrided case (due to the formation of chromium nitride).

2.11.2 Nickel

Refer to Figs. 2.12 and 2.13. Nickel is the most important addition to a high-chromium steel. The effect of nickel on pure iron is the opposite to that produced by chromium. Nickel forms a continuous series of solid solutions and because of the effect of depressing the γ–α transformation temperature, alloys of 30% nickel and higher are fully austenitic at room temperature. With additions of more than 5% nickel, alloys are almost entirely martensitic even if slowly cooled. Alloys containing between 15% and 20% nickel are martensitic and maximum hardness is achieved. With higher percentages of nickel, the structure of stable austenite begins to appear with a corresponding drop in hardness. Alloys above 35% nickel are wholly austenitic.

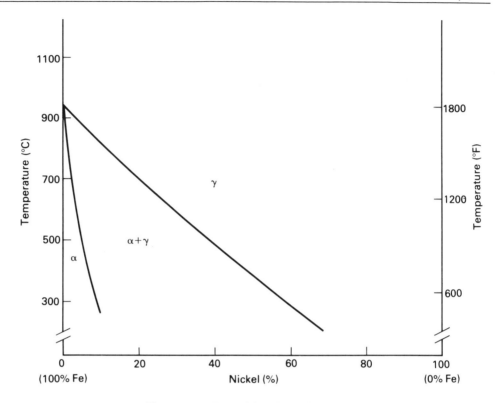

Figure 2.12 Iron–nickel phase diagram.

In the presence of carbon, the effect of nickel on hardness is to superimpose the effects due to nickel on those due to carbon. No carbides are formed, nickel is a graphitizing agent. Nickel lowers the critical temperatures and decreases the eutectoid temperature and composition.

Other effects due to nickel are as follows:

it inhibits grain growth;

it promotes toughness in pearlitic ferrite steels;

in large quantities, nickel imparts special thermal expansion properties and magnetic properties to steels

The effect of nickel on various steels is now discussed in more detail.

12–14% CHROMIUM STEELS

Up to 2% nickel may be present due to its occurrence in the raw material; the effect is to harden the steel slightly and increase its hardening capacity. 12% nickel may be added for cutlery or ornamental work, the effect being to produce a material that is austenitic at all temperatures.

16–20% CHROMIUM STEELS

Small additions may be present in the raw materials. The effects are to increase hardenability and (with 2% nickel) to greatly improve the impact strength of a material tempered at 600°C (1112°F).

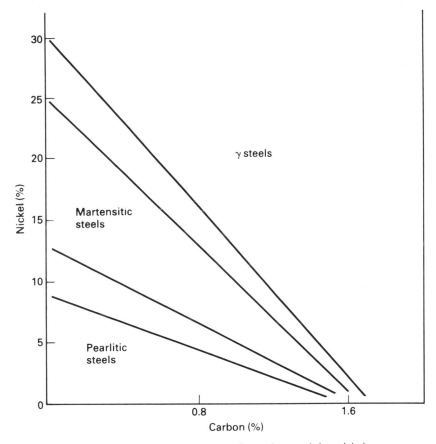

Figure 2.13   Structures present in steels containing nickel.

As the nickel content is increased, the quantities of the γ phase present increase. Above 7% nickel, the alloy is austenitic at all temperatures. Austenitic chromium–nickel steels may be hardened only by cold work or deep-freeze quenching. With the addition of 10–12% nickel, there is no effect on the heat-resisting properties of the steel. If the nickel is increased to more than 25%, then a marked improvement in resistance to oxidation takes place.

STEELS CONTAINING MORE THAN 20% CHROMIUM
To obtain a fully γ phase structure in the higher-chromium steels a larger amount of nickel must be added: for 25% chromium, add 18% nickel; for 30% chromium add 27% nickel. The sigma phase may form below 30% chromium, with nickel in the range 0–25%.

2.11.3 Manganese

Manganese is a carbide-forming element, the effect produced being greater than with iron and less than chromium. It helps to stabilize the γ phase and strengthen the α phase, but impairs ductility. In general, manganese is not added to steel in large quantities (only up to 0.5%). Higher than 0.5% manganese is said to improve scaling resistance, but promotes grain growth and loss of ductility.

The stabilizing effect of manganese is less than that produced by nickel, and since it is cheaper attempts have been made to produce austenitic chromium–manganese steels. 18–8 chromium–manganese steels may be austenitic, but they are not very corrosion resistant. The formation of the σ phase is more likely with chromium–manganese austenitic steels.

2.11.4 Silicon

The effects of silicon on steel are almost entirely the reverse of those produced by nickel. Silicon is a ferrite strengthener and graphitizer, and is a more powerful gamma loop forming element than chromium. The limit of the γ phase in FeSi alloys is 2.2%.

Silicon retards grain growth, but iron–silicon alloys show good corrosion resistance. Iron–silicon alloys are also very brittle. In martensitic stainless steels the effect of silicon is to raise the hardening temperature without improving hardenability, and it is not found in these steels. In stainless irons, the resistance to oxidation is increased by additions of silicon. In austenitic chromium–nickel steels, the effect of silicon is to promote the formation of duplex steels and widen the temperature and composition range over which the σ phase is stable.

2.11.5 Molybdenum

Molybdenum helps to stabilize the α phase and restricts the formation of the γ phase. Like silicon it is more powerful than chromium in forming a γ loop. The carbide-forming tendency is very strong and it contributes greatly to the hardenability in low percentages. The effect of small additions of molybdenum to a martensitic stainless steel is to increase the temperature range for hardening. It is not usually added to these steels.

In austenitic stainless steels up to 3% molybdenum may be added to improve the corrosion resistance to organics. As with silicon the effect is to produce a duplex structure, although the γ state can be regained by increasing the nickel content. These steels are difficult to work and are likely to crack if hot rolled; molybdenum also tends to promote σ formation. Molybdenum increases the strength of an alloy when hot and induces good creep properties.

2.11.6 Tungsten

Tungsten behaves in a similar way to molybdenum. In martensitic stainless steels, tungsten raises the hardening temperature but it is not generally added to these steels, or to ferritic stainless steels. The effect of tungsten on stainless steels is to produce a duplex structure. Tungsten does not enhance resistance to oxidation, but small additions (up to 4%) improve the strength of the alloy when heated. In this respect tungsten is superior to molybdenum.

2.11.7 Titanium

Titanium is another γ loop forming element, the limit of γ being 0.75% titanium in titanium–iron alloys. The carbide-forming tendency is probably stronger than with any other alloying element. When in solid solution, titanium increases the hardness of a steel by more than any other alloying element. Titanium is not normally added to martensitic stainless steels. If titanium is added to an austenitic stainless steel, then it produces a duplex structure. The nickel content must be increased if a fully austenitic steel is required. Titanium is added to inhibit intergranular corrosion, and with the addition of up to 1.5% tungsten, precipitation hardening effects can be produced.

2.11.8 Niobium

In many ways niobium acts similarly to titanium.

2.11.9 Copper

Copper has little or no effect on the stability of austenite. In α-irons, up to 1% copper gives a better hardenability and increases the toughness. With austenitic chromium–nickel alloys, additions of copper reduce the hardness and tensile strength, but improve cold working. However, hot working properties are impaired. Copper has a beneficial effect on the corrosion resistance properties of these steels, but it is rarely used because of its poor hot working properties.

2.11.10 Cobalt and vanadium

These elements improve creep resistance and strength at high temperatures.

2.11.11 Sulfur and selenium

These elements improve the free-cutting properties of steels by forming sulfides and selenides.

2.11.12 Intergranular corrosion

Intergranular corrosion is caused by weld decay due to treatment at 400°C to 600°C (752°F to 1112°F). Chromium precipitates as carbides in the γ grain boundaries, and impoverishes the stainless steel in chromium adjacent to the boundaries. This sets up an electrogalvanic cell which accelerates corrosion.

2.12 LOW TEMPERATURE STEELS

The use of steels at low temperatures usually requires that the metal maintains its properties (especially its toughness) at the low temperature, and when reheated to normal temperatures. Steels of the 18–8 type (Table 2.3) are found to be most suited to these types of applications.

2.13 APPLICATIONS FOR SPRINGS

Depending upon the properties required by the spring, those used in vehicles usually contain up to 0.8% carbon, with silicon (0.4% maximum) and manganese (0.8% maximum). These steels are rapidly cooled. Valve springs may contain 1.5% chromium and 0.17% vanadium, and also nickel.

2.14 VALVE MANUFACTURE

A typical light-duty valve steel would contain 0.3% carbon, 3.5% nickel, 0.4% chromium and 0.4% silicon, with greater amounts of chromium and silicon used for heavy-duty valves. Valves for aero-engines use austenitic steels containing 10% nickel and 15% chromium.

* * *

Revision test

The student should be able to answer the following basic revision questions after studying Sections 2.8–2.14.

1 List the main alloying elements that are either present in steel or are added to steel.

2 Outline the reasons why particular elements are added, and the effects they produce.

3 What influence does increasing the carbon content of a steel have upon its properties?

4 List the main impurities that are present in steel, and describe the effects they produce.

5 List the main types of high-alloy steels that are produced and are usually categorized according to their particular application.

6 Describe the main differences between ferritic, martensitic and austenitic stainless steels.

7 Are stainless steels always 'stainless'? If not, then describe the conditions that should be avoided.

8 List the alloying elements that are added to stainless steels, and describe the effects that they produce.

9 What are the typical composition ranges for the main alloying elements found in stainless steels? Why are these quantities required?

10 Describe what is meant by intergranular corrosion or weld decay. How can this effect be prevented or reduced?

11 What is the main requirement for steels that are used at low temperatures? How is this achieved?

12 What property is important for a valve steel? What alloying elements are used? What effects do they produce, and why?

* * *

2.15 BRITISH STANDARDS AND ASTM STANDARD SPECIFICATIONS FOR IRONS

2.15.1 Introduction

British Standards (BS) are a useful source of information for the scientist and the engineer. Specifications are available for different types of cast irons and steels and these will be described in Sections 2.15 and 2.16. It is important that the most recent information is available, as standards are often withdrawn, superseded, revised or amended. The latest edition of a standard should be obtained, together with the latest amendments. This can be checked by reference to the current BSI yearbook and the current Sales Bulletin and BSI News.

Sections 2.15 and 2.16 are intended to provide a summary of the information relevant to the engineer and the designer, which is available in the BS specifications for irons and steels. Detailed lists of material grades, composition and properties will not be included and if these are required then reference should be made directly to the appropriate specification. The designation system for cast irons and steels will be discussed. Reference should also be made to BS Handbook No. 19: Methods for the sampling and analysis of iron, steel and other ferrous metals.

The American Society for Testing and Materials (ASTM) publishes an annual book of standards, comprising different sections. New volumes are published annually because of new standards being prepared and significant revisions to existing standards. Approximately 30% of each volume is new or revised. A new edition makes previous editions obsolete. Old editions should not be used for design purposes but they are satisfactory for teaching.

Section 1 contains standards relating to irons and steels, and comprises six volumes. These are:

Volume 01.01: Steel piping, tubing, and fittings.
Volume 01.02: Ferrous castings; ferroalloys.
Volume 01.03: Steel plate, sheet, strip and wire.

Volume 01.04: Structural steel; concrete reinforcing steel; pressure vessel plate and forgings; steel rails, wheels, and tires.

Volume 01.05: Steel bars, chain, and springs; bearing steel; steel forgings.

Volume 01.06: Coated steel products.

Each volume contains many standards often related to particular applications of a material. Some examples of ASTM standards related to topics discussed in this book are included here. However, reference should be made to the detailed ASTM index for each section to determine which standards are available for a particular situation or application.

An example of a full ASTM Standard designation is ASTM C564–70 (1982): Title. The standard is issued under the fixed designation C564; the number immediately following indicates the year of original adoption (1970), or of the last revision. A number in parentheses (1982) indicates the year of last reapproval; the letter M, e.g. C564M, indicates a metric standard. Only the standard designation will be used throughout this book, as the latest edition or revision should always be obtained for design purposes.

2.15.2 Gray iron castings

BS1452: Specification for gray iron castings
Contains the requirements for seven grades of gray cast iron based on the tensile strength of the material. The testing and inspection of materials are given, also typical properties and variations in strength which can be expected. The seven grades, designated by a three-digit system, are 150, 180, 220, 260, 300, 350 and 400. This designation number is the minimum tensile strength (N/mm^2) expected on a 30 mm diameter test bar.

ASTM A48: Gray iron castings
This specification covers gray iron castings intended for general engineering use, where tensile strength is a major consideration. Nine classes of materials are included; each class is designated by a number (indicating the minimum tensile strength) followed by a letter denoting the size of the test bar.

ASTM A377: Gray iron and ductile iron pressure pipe

ASTM A126: Gray iron castings for valves, flanges, and pipe fittings

2.15.3 Malleable iron castings

BS309: Whiteheart malleable iron castings
Details are given concerning two grades, designated as W410/4 and W340/3. There must be no primary graphite present in the microstructure, and free cementite should not be present in a form or amount which is detrimental to the physical properties and machinability of the casting. Information is given for mechanical properties, microstructure and freedom from defects; also tensile

testing and permanent-set stress values. Typical properties are given in the appendix. Mechanical properties are more important than chemical composition, although the maximum phosphorus content is given as 0.12% (this applies to most cast irons).

BS310: Blackheart malleable iron castings
The three grades considered are designated B340/12, B310/10 and B290/6. The same considerations apply as for BS309.

ASTM A47: Malleable iron castings
This specification covers ferritic malleable irons used in general engineering castings, for service at both normal and elevated temperatures. Two grades (32510 and 35018) are covered according to their tensile properties. The first three digits indicate the minimum yield strength ($\times 100$ psi) and the last two digits indicate the minimum elongation (% in 2 in.). Physical and tensile requirements are included.

2.15.4 Pearlitic malleable iron castings

BS3333: Pearlitic malleable iron castings
Five grades are considered which are designated by the minimum tensile strength in N/mm^2/minimum % elongation. These are P690/2, P570/3, P450/5, P510/4 and P440/7. Requirements are given for these five grades including mechanical properties, microstructure and freedom from defects. Hardness values are also given for comparison, although these are no longer mandatory in the specification. Typical properties are also listed. The minimum 0.5% proof stress values (N/mm^2) are given for each grade: 540 (P690/2), 420 (P570/3), 340 (P450/5), 310 (P510/4) and 270 (P440/7).

ASTM A220: Pearlitic malleable iron castings
Details as for ASTM A47 (see Section 2.15.3).

2.15.5 Ductile iron castings

BS2789: Iron castings with spheroidal or nodular graphite
All gray cast irons contain graphite; in normal cast iron this is in the form of thin flakes. The flakes are more compact in malleable cast iron, which is obtained by annealing white cast iron. Blackheart and pearlitic malleable cast irons contain temper carbon nodules in the form of flake graphite aggregates. The whiteheart malleable iron has more aggregates, and it also has spheroidal graphite nodules not in aggregates. BS2789 is a specification for iron castings which contain predominantly, or totally, spherical graphite nodules. These materials possess a greater tensile strength, ductility and resistance to impact than flake graphite iron. The specification covers six grades with designations in terms of the minimum tensile strength in N/mm^2/minimum % elongation. The grades are:

370/17 Ferritic matrix with high impact resistance.

420/12 Mainly ferritic matrix, moderately high tensile strength with substantial ductility and toughness.

500/7 Intermediate grade, ferrite–pearlite matrix, possesses strength with reasonable ductility.

600/3 ⎫
700/2 ⎬ Mainly a pearlitic matrix, high tensile strength but ductility and impact
800/2 ⎭ resistance are less important.

The specification sets out requirements for the process of manufacture, freedom from defects, and the mechanical properties, including tensile strength, elongation and 0.2% proof stress. Hardness values are included for guidance only and are not a specific requirement. The impact strength requirements at ambient temperature for the 370/17 grade are given, and this material can be used at temperatures as low as $-40°C$ ($-40°F$). Grades 500, 600, 700 and 800 can all be heat treated using quenching and tempering, induction or flame hardening and normalizing. All grades can be used for pressure purposes up to 350°C (662°F).

ASTM A536: Ductile iron castings
This specification covers castings made of ductile iron, i.e. cast iron with the graphite substantially spheroidal in shape, and essentially free of other forms of graphite. It is also known as spheroidal or nodular iron. Details of the mechanical requirements and heat treatment for five grades are given.

2.15.6 Austenitic cast iron

BS3468: Austenitic cast iron
Austenitic cast irons are high-alloy materials that are austenitic at room temperature due to the presence of alloying elements. Carbon is present as flake or spheroidal graphite and carbides are often present, especially in high chromium grades. The spheroidal graphite grades have superior mechanical and physical properties, heat and corrosion resistance. Martensite may be present in the microstructure by agreement between the supplier and the purchaser. The specification contains requirements for nine grades containing flake graphite, and eleven grades containing spheroidal or nodular graphite in the microstructure. These grades are designated by either the letter L for flake graphite or the letter S for spheroidal graphite, followed by the chemical symbols and their approximate mean percentage levels respectively. The grades covered are listed as follows:

Flake graphite	*Spheroidal graphite*
L-Ni Mn 13 7	S-Ni Mn 13 7
L-Ni Cu Cr 15 6 2	S-Ni Cr 20 2
L-Ni Cu Cr 15 6 3	S-Ni Cr 20 3
L-Ni Cr 20 2	S-Ni Si Cr 20 5 2
L-Ni Cr 20 3	S-Ni 22
L-Ni Si Cr 20 5 3	S-Ni Mn 23 4
L-Ni Cr 30 3	S-Ni Cr 30 1
L-Ni Si Cr 30 5 5	S-Ni Cr 30 3

L-Ni 35   S-Ni Si Cr 30 5 5
S-Ni 35
S-Ni Cr 35 3

Tables are included in the standard of the actual chemical compositions, mechanical and physical properties, and a summary of the main properties and typical applications of each grade.

ASTM A436: Austenitic gray iron castings
Covers castings normally used for resistance to heat, corrosion and wear. They are characterized by the uniformly distributed graphite flakes, some carbides, and the presence of sufficient alloy content to produce an austenitic structure. Many combinations of alloys can be used to obtain an austenitic gray iron; six types (defined by chemical composition limits) are included in this specification.

ASTM A439: Austenitic ductile iron castings
Details as for ASTM A536 (Section 2.15.5) and A436; covers nine general types of ductile iron.

2.15.7 Corrosion-resistant high-silicon iron castings

BS1591: Corrosion resisting high-silicon iron castings
Iron castings containing a high silicon content possess protective properties when brought into contact with strong aggressive liquids. This is due to the formation of a thin surface film of hydrated oxides of silicon. There is a high rate of attack until a complete film is formed, which may take several hours, after which time the corrosion is negligible. These materials have applications in chemical engineering and as cathodic protection anodes. The requirements for four grades, in terms of chemical composition, heat treatment, manufacture and freedom from defects are given:

Grade	Chemical composition (%)		
	Carbon (max)	Silicon	Chromium
Si 10	1.2	10–12	—
Si 14	1.0	14.25–15.25	—
Si Cr 14 4	1.4	14.25–15.25	4.5
Si 16	0.8	16–18	—

Each grade can contain a maximum of 0.5%, 0.25% and 0.1% of manganese, phosphorus and sulfur, respectively.

Si 14 is recommended for general applications requiring corrosion resistance. Si 10 has greater tensile strength than Si 14, but less resistance to corrosion. Si 16 has greater corrosion resistance at the expense of tensile strength. Si Cr 14 4 is used for cathodic protection anodes.

ASTM A518: Corrosion-resistant high-silicon iron castings
Covers two grades containing approximately 14.5% silicon. Both grades are suited for applications in severe corrosive environments, but Grade 2 (higher chromium content) is particularly suited for strong chloride environments.

2.15.8 Abrasion-resistant cast irons

BS4844: Abrasion resisting white cast irons
Part 1: Unalloyed and low alloy grades
Requirements are given for three grades (1A, 1B, 1C) for manufacture, freedom from defects, chemical composition, hardness and inspection. White cast irons contain hard iron carbides in a pearlitic matrix and are generally graphite free, except in heavy slow-cooled sections. The following specifications of chemical composition enable graphite formation to be avoided.

| Grade | Chemical composition (%) | | | | | Brinell |
	Carbon	Silicon	Manganese	Chromium (max)	Phosphorus (max)	hardness (min)
1A	2.4–3.4	0.5–1.5	0.2–0.8	2.0	0.15	400
1B	2.4–3.4	0.5–1.5	0.2–0.8	2.0	0.5	400
1C	2.4–3.0	0.5–1.5	0.2–0.8	2.0	0.15	250

Grade 1C is heat treated for improved machinability but this reduces the abrasion resistance. The low phosphorus content of grade 1A improves the resistance to impact loading. The main difference in the grades is the phosphorus content. Machining of these castings is limited, and heat treatment is not normally beneficial.

ASTM A532: Abrasion–resistant cast irons
This specification covers a group of white cast irons of three classes and various types, according to their chemical composition. The main applications are mining, milling, earth-handling and manufacturing industries.

Plain and low-alloy white cast irons consisting essentially of iron carbides and pearlite are specifically excluded.

BS4844; Part 2: Nickel-chromium grades
The structure of these grades consists of iron-chromium carbides in a hard, predominantly martensitic, matrix that is normally graphite free. High chromium grades have superior resistance to single and repeated impacts, because the carbide is discontinuous and in a less massive form. Both low and high chromium grades can be heat treated. The grades differ mainly in the carbon content, and also between low and high chromium contents. Details of the chemical composition of the grades are given:

| Grade | Chemical composition (%) | | | | Brinell |
	Carbon	Silicon	Nickel	Chromium	hardness (min)
2A	2.7–3.2	0.3–0.8	3.0–5.5	1.5–2.5	500
2B	3.2–3.6	0.3–0.8	3.0–5.5	1.5–2.5	550
2C	2.4–2.8	1.5–2.2	4.0–6.0	8.0–10.0	500
2D	2.8–3.2	1.5–2.2	4.0–6.0	8.0–10.0	550
2E	3.2–3.6	1.5–2.2	4.0–6.0	8.0–10.0	600

Each grade has a manganese content between 0.2% and 0.8% and maximum molybdenum, phosphorus and sulfur contents of 0.5%, 0.3% and 0.15%, respectively. A low carbon content improves impact resistance, but reduces hardness. A high carbon content improves resistance to abrasive wear. The requirements are similar to those outlined for Part 1 of the standard, including heat-treatment specifications and information on section size and chemical composition to develop the maximum hardness.

ASTM A743: Castings, iron–chromium, iron–chromium–nickel, nickel–base, corrosion resistant, for general application

ASTM A744: Castings, iron–chromium–nickel, nickel–base, corrosion resistant, for severe service

ASTM A567: Iron, cobalt, and nickel–base alloy castings for high strength at elevated temperatures

BS4844; Part 3: High-chromium grades
The structure normally consists of discontinuous iron–chromium carbides in a matrix of austenite and martensite. They are generally free from graphite and pearlite. Heat treatment to improve the hardness produces a mainly martensitic matrix. The five grades covered vary according to the chromium content, shown as follows:

Grade	Chemical composition (%)				Brinell hardness (as cast)
	Carbon	Chromium	Molybdenum	Nickel	
3A	2.4–3.0	14–17	0–2.5	0–1	450
3B	3.0–3.6	14–17	1–3	0–1	500
3C	2.2–3.0	17–22	0–3	0–1.5	450
3D	2.4–2.8	22–28	0–1.5	0–1	400
3E	2.8–3.2	22–28	0–1.5	0–1	450

For each grade the range of manganese is 0.5–1.5%, copper is up to 1.2%, the maximum silicon and phosphorus contents are 1.0% and 0.1% respectively.

As the carbon content increases, the resistance to abrasive wear increases, but the toughness and resistance to repeated shock decreases. Details of appropriate heat treatment, annealing and hardening, and the hardness levels acquired are also included.

ASTM A297: Steel castings, iron–chromium and iron–chromium-nickel, heat resistant, for general application

ASTM A447: Steel castings, chromium–nickel–iron alloy (25–12 class), for high-temperature service

2.15.9 Ferroalloys

The following ASTM standard specifications for ferroalloys are contained in Volume 01.02.
ASTM A323: Ferroboron
ASTM A550: Ferrocolumbium
ASTM A482: Ferrochrome–silicon
ASTM A101: Ferrochromium
ASTM A 99: Ferromanganese
ASTM A701: Ferromanganese–silicon
ASTM A132: Ferromolybdenum
ASTM A100: Ferrosilicon
ASTM A324: Ferrotitanium
ASTM A144: Ferrotungsten
ASTM A102: Ferrovanadium

2.16 BRITISH STANDARDS AND ASTM STANDARD SPECIFICATIONS FOR STEELS

2.16.1 BS2094: Glossary of terms relating to iron and steel

Part 1: General metallurgical, heat treatment and testing terms
Part 2: Steelmaking
Part 3: Hot rolled steel products (excluding sheet, strip and tubes)
Part 4: Steel sheet and strip
Part 5: Bright steel bar and steel wire
Part 6: Forgings and drop forgings
Part 7: Wrought iron
Part 8: Steel tubes and pipes
Part 9: Iron and steel founding

2.16.2 Wrought steels

BS970: Specification for wrought steels for mechanical and allied engineering purposes
This specification in six parts uses a steel designation with six digits. This replaces the old En series and allows easier standardization of steels and more flexibility for new steels (which was not possible with the En series). A knowledge of the type of steel from its designation is possible, and the British system is now comparable with that used by the AISI in the United States (see Section 2.17). In referring to a six-digit designation, the first digit is taken to mean that on the far left and the sixth digit on the far right. The fourth digit is generally a letter and the other five are numbers. For example: (first) XXXLXX (sixth digit). In general the first three digits represent a particular series or class of steels. The fourth digit is A or M representing material supplied on the basis of specified analysis or mechanical properties, H represents a hardenability

requirement for the material, or S for stainless steels. The fifth and sixth digits are approximately $100\times$ the mean carbon content; if the average carbon content is 1% or more, then 99 is used for these digits. For stainless steels, the fifth and sixth digits used are: 01 to represent a basic steel composition, or 11 to 99 for specific alloys.

Part 1: General inspection and testing procedures and specific requirements for carbon, carbon manganese and stainless steels
Carbon steels are designated by the first three digits as the 000XXX to 299XXX series. Plain carbon and carbon manganese steels are the 000XXX to 199XXX series (the first three digits representing approximately $100\times$ the mean manganese content). The 200XXX to 240XXX series are free cutting steels (the second and third digits represent $100\times$ the mean or minimum sulfur content). The free cutting steels generally contain at least 0.08% sulfur or lead, or both. Lead-bearing steels generally contain a minimum of 0.12% lead. In carbon steels the maximum sulfur and phosphorus contents are 0.05%. Additions of less than 0.4% nickel, 0.3% chromium and 0.15% molybdenum are usually considered incidental. Heat-resisting and stainless steels are designated by the 300XXX to 499XXX series, some of which correspond to AISI steels. Table 2.6 contains details of the designation of steels in the BS and AISI systems, and typical chemical compositions of some steels.

This standard (Part 1) contains details of specific requirements (as appropriate), covering chemical composition, mechanical properties, hardenability, heat treatment for hot-rolled and normalized steels, bright bar, micro-alloyed steels, through hardening, case hardening, stainless and heat-resisting steels.

Part 2: Direct hardening alloy steels, including alloy steels capable of surface hardening by nitriding
This covers the 500XXX to 999XXX series which are arranged in groups of ten, or multiples of ten. A low residual sulfur content can adversely affect the machinability; its presence depends upon the process used to produce the steel. The range of sulfur content and the maximum phosphorus impurity from each process are given as follows:

Steelmaking process	Sulfur (%)	Phosphorus (max %)
Acid process	0.05	0.04
Basic open-hearth, electric melting, or oxygen process	0.025–0.05	0.04
Electric quality (for nitriding steels)	0.015–0.04	0.025

The groups of steels within the 500 to 999 series, and their main alloying additions are given as follows:

Series	Main alloying additions
503XXX	1% nickel
526	0.75% chromium
530	1% chromium
534 535	1.5% chromium (1% carbon)
605	1.5% manganese; 0.32% (max) molybdenum
640	1.25% nickel; 0.8% (max) chromium
653	3% (total) nickel and chromium
708	1% chromium; 0.25% (max) molybdenum
816–835	Nickel; chromium; molybdenum
945XXX	1.5% manganese; 0.9% nickel; 0.6% chromium; 0.25% molybdenum (all max%)

Part 3: Steels for case hardening

Series	Main alloying additions
523XXX	0.5% chromium
527	0.75% chromium
635	0.75% nickel; 0.75% (max) chromium
637	1.00% nickel; 1.0% (max) chromium
655	3.25% nickel; 1.0% (max) chromium
659	4.00% nickel; 1.3% (max) chromium
665	1.75% nickel; 0.3% (max) molybdenum
805	0.50% nickel; 0.6% chromium; 0.25% molybdenum
815	1.50% nickel; 1.2% chromium; 0.20% molybdenum
820	1.75% nickel; 1.2% chromium; 0.20% molybdenum
822	2.00% nickel; 1.7% chromium; 0.25% molybdenum
835	4.00% nickel; 1.3% chromium; 0.25% molybdenum

Supplement No. 1 to BS970: Part 3
The requirements for carbon and carbon manganese steels for case hardening, including free cutting steels.

Part 4: Valve steels
Includes details of product analysis and permitted variations, En steels replaced, chemical composition and heat treatment.

Part 5: Carbon and alloy spring steels for the manufacture of hot formed springs
Steels for these applications are either carbon steels (060XXX, 070XXX, 080XXX series), silicon–manganese steels (250XXX series) or particular alloy steels. The steels are listed in Table 2.7 with the En steels replaced and the actual chemical composition.

Part 6: SI metric values (for use with BS970: Parts 4 and 5)

Table 2.6 BS steel designations, En steels replaced and the nearest AISI designations for stainless and heat resisting steels

BS designation	En steel (replaced)	AISI steel (nearest)	Type of steel and approximate composition (%)
		Ferritic steels	
403S17	–	403	13 Cr; C 0.08 (max)
430S15	60	430	17 Cr; C 0.10
		Martensitic steels	
410S21	56A	410	13 Cr; C 0.12
420S29	56B	–	13 Cr; C 0.17
420S37	56C	420	13 Cr; C 0.24
420S45	56D	–	13 Cr; C 0.32
416S21	56AM	416	13 Cr; C 0.12; S bearing; free-machining
416S41	56AM	416 Se	13 Cr; C 0.12; Se bearing; free-machining
416S29	56BM	–	13 Cr; C 0.17; S bearing; free-machining
416S37	56C	–	13 Cr; C 0.24; S bearing; free-machining
431S29	57	431	17 Cr; 2.5 Ni; C 0.15
441S29	–	–	17 Cr; 2.5 Ni; C 0.15; S bearing; free-machining
414S92	–	–	17 Cr; 2.5 Ni; C 0.15; Se bearing; free-machining
		Austenitic steels	
304S12	–	304L	Cr Ni 18/10; C 0.03
304S15	58E	304	Cr Ni 18/9; C 0.06
302S25	58A	302	Cr Ni 18/9; C 0.12
321S12	58B/58C	321	Cr Ni 18/9/Ti; C 0.08
321S20	58B/58C	321	Cr Ni 18/9/Ti; C 0.12
347S17	58F/58G	347	Cr Ni 18/9/Nb; C 0.08
315S16	58H	–	Cr Ni Mo 17/10/1.5; C 0.07
316S12	–	316L	Cr Ni Mo 17/12/2.5; C 0.03
316S16	58J	316	Cr Ni Mo 17/11/2.5; C 0.06
320S17	58J	–	Cr Ni Mo 17/12/2.5/Ti; C 0.08
317S12	–	–	Cr Ni Mo 18/15/3.5; C 0.03
317S16	–	–	Cr Ni Mo 18/13/3.5; C 0.06
303S21	58M	303	Cr Ni 18/9; S bearing; free-machining
303S41	58M	303 Se	Cr Ni 18/9; Se bearing; free-machining
325S21	58M	–	Cr Ni 18/9/Ti; S bearing; free-machining
326S36	–	–	Cr Ni Mo 17/11/2.5; Se bearing; free-machining
310S24	–	310	Cr Ni 25/20

ASTM standard specifications for steels (Examples)

   Volume 01.03: Steel plate, sheet, strip and wire

   ASTM A414/A414M: Carbon steel sheet for pressure vessels

   ASTM A568/A568M: General requirements for steel, carbon and high-strength low-alloy hot-rolled sheet and cold-rolled sheet

   ASTM A682/A682M: General requirements for steel, high-carbon, strip, cold-rolled, spring quality

   ASTM A505: General requirements for steel sheet and strip, alloy, hot-rolled and cold-rolled

   ASTM A407/A407M: Steel wire, cold-drawn, for coiled-type springs

ASTM A227/A227M: Steel wire, cold-drawn for mechanical springs

Volume 01.05: Steel bars, chain, and springs; bearing steel; steel forgings
  ASTM A29/A29M: General requirements for steel bars, carbon and alloy, hot-wrought and cold-finished
  ASTM A434: Steel bars, alloy, hot-wrought or cold-finished, quenched and tempered
  ASTM A311: Stress relief annealed cold-drawn carbon steel bars
  ASTM A125: Steel springs, helical, heat-treated
  ASTM A679: Steel wire, high tensile strength, hard-drawn, for mechanical springs
  ASTM A229/A229M: Steel wire, oil-tempered for mechanical springs

2.16.3 High-carbon steels

BS1407: High carbon bright steel (silver steel)
The steel is fully killed and is made in a high frequency electric arc furnace or by the acid open-hearth processes. The chemical composition of the steel is 0.95% to 1.25% carbon, 0.25% to 0.45% manganese, 0.4% (max.) silicon, 0.045% (max.) sulfur, 0.045% (max.) phosphorus and 0.5% (max.) chromium (optional). It is available as rounds with a ground surface (that should be free from decarburization) or a polished surface, or as squares with a drawn surface. The product should be straight and free of seams.

ASTM standard specifications, volume 01.03 (Examples)
  ASTM A620/A620M: Steel sheet, carbon, cold-rolled, drawing quality, special killed
  ASTM A622/A622M: Steel sheet and strip, carbon, hot-rolled, drawing quality, special killed

2.16.4 Tool steels

BS4659: Tool steels
There are six types of tool steel broadly classified according to their application. These are high speed, hot working, cold working, shock resisting, special purpose and water hardening. They are designated by a three-, four- or five-digit system; the first digit is B, the second digit is a letter to designate the application, the third and fourth digits are numbers (the fifth digit is a letter, if applicable). A summary of the application and number designation is given in Table 2.8. The chemical compositions of a range of tool steels are given in Table 2.9, together with the heat-treatment temperatures and hardness requirements. The standard includes case thicknesses for water hardening tool steels, for example BW1B has a maximum case thickness of 3.5 mm for a shallow component and up to 6.5 mm (maximum) for a deep component. Unspecified elements may be included to the following maximum contents: 0.4% nickel, 0.05% tin, 0.2% copper, 0.035% sulfur, 0.035% phosphorus. Sulfur to produce free-machining properties can be present in the range 0.09% to 0.15%, or selenium or lead may be present.

Table 2.7 Chemical compositions and designations of spring steels

BS Designation	En Steel (replaced)	General steel type	Carbon % min	Carbon % max	Silicon % min	Silicon % max	Manganese % min	Manganese % max	Chromium % min	Chromium % max
		Carbon steels								
080A52	43	'52' carbon	0.50	0.55	0.10	0.35	0.70	0.90		
080A67	43E	'67' carbon	0.65	0.70	0.10	0.35	0.70	0.90		
070A72	42	'72' carbon	0.70	0.75	0.10	0.35	0.60	0.80		
070A78	42	'78' carbon	0.75	0.82	0.10	0.35	0.60	0.80		
060A96	44	'96' carbon	0.93	1.00	0.10	0.35	0.50	0.70		
		Silico-manganese steels								
250A53	45	Silico-manganese, '53' carbon	0.50	0.57	1.70	2.10	0.70	1.00		
250A58	45A	Silico-manganese, '58' carbon	0.55	0.62	1.70	2.10	0.70	1.00		
250A61	45A	Silico-manganese, '61' carbon	0.58	0.65	1.70	2.10	0.70	1.00		
		Alloy steels								
527A60	48	0.75% chromium	0.55	0.65	0.10	0.35	0.70	1.00	0.60	0.90
*735A50	47	1% chromium vanadium	0.46	0.54	0.10	0.35	0.60	0.90	0.80	1.10
†805A60	—	0.5% nickel chromium molybdenum	0.55	0.65	0.10	0.35	0.70	1.00	0.40	0.60
‡925A60	—	Silico-manganese chromium molybdenum	0.55	0.65	1.70	2.10	0.70	1.00	0.20	0.40

* Also contains vanadium (0.15% min).
† Also contains nickel (0.40% min. – 0.70% max.), molybdenum (0.15% min. – 0.25% max.).
‡ Also contains molybdenum (0.20% min. – 0.30% max.).

ASTM standard specifications, volume 01.05 (Examples)
  ASTM A681: Alloy tool steels
  ASTM A686: Carbon tool steels
  ASTM A600: High-speed tool steels

Table 2.8 British Standard Tool Steel Classifications

Application	Designation
(1) *High speed*	
Molybdenum grades	BMx
Tungsten grades	BTx
(2) *Hot working*	
Chromium grades	BH10, BH11, BH19
Tungsten grades	BH21, BH21A, BH26
(3) *Cold working*	
High-carbon high-chromium grades	BD2, BD2A, BD3
Medium-alloy air-hardening grades	BA2, BA6
Oil-hardening grades	BO1, BO2
(4) *Shock resisting*	BSx
(5) *Special purpose*	
Low-alloy grade	BL3
Carbon–tungsten grade	BF1
(6) *Water hardening*	BW1A, BW1B, BW1C, BW2

2.16.5 Steel castings for general applications

BS3100: Specification for steel castings for general engineering purposes
This specification uses two designation systems. For carbon and alloy steels, the designation is one or two letters followed by an arbitrary number:

1st letter:	A	– for carbon and carbon manganese steels
	B	– for low-alloy steels
2nd letter:	no letter	– general purpose use
	L	– low-temperature toughness
	W	– wear resistant
	T	– higher tensile strength (low alloy only)
	M	– specified magnetic properties (carbon steel only)

For corrosion-resisting, heat-resisting and high-alloy steel castings, a six-digit system is used, comprising three numbers (corresponding to stainless steels), then C (for castings), then two arbitrary numbers.

ASTM standard specifications for steel castings, volume 01.02 (Examples)
  ASTM A128: Steel castings, austenitic manganese.
  ASTM A148: Steel castings, high-strength for structural purposes
  ASTM A27: Steel castings, carbon for general application
  ASTM A747: Steel castings, stainless, precipitation hardening
  ASTM A781: Castings, steel and alloy, common requirements, for general industrial use

Table 2.9 Chemical composition, heat treatment temperature ranges and hardness requirements for tool steels

Chemical Composition (minimum and maximum %)	Tool-steel designation					
	BM1	BH26	BA6	BS2	BF1	BW1B
Carbon	0.75–0.85	0.5–0.6	0.65–0.75	0.45–0.55	1.15–1.35	0.95–1.10
Silicon	0.4	0.4	0.4	0.9–1.2	0.4	0.3
Manganese	0.4	0.4	1.8–2.1	0.3–0.5	0.4	0.35
Chromium	3.75–4.50	3.75–4.50	0.85–1.15	–	0.25–0.50	0.15
Molybdenum	8.0–9.0	0.6	1.2–1.6	0.3–0.6	–	0.10
Tungsten	1.0–2.0	17.5–18.5	–	–	1.3–1.6	–
Vanadium	1.0–1.25	1.0–1.5	–	0.1–0.3	0.3	–
Cobalt	0.60	0.6	–	–	–	–
Nickel	–	–	–	–	–	0.2
Heat-Treatment Temperature Ranges °C(°F)						
Annealing	850–870 (1562–1598)	870–890 (1598–1634)	730–750 (1346–1382)	790–820 (1454–1508)	780–800 (1436–1472)	740–790 (1364–1454)
Pre-heating	850 (1562)	850 (1562)	650 (1202)	–	–	–
Hardening	1200–1220[a] (2192–2280)	1180–1260 (2156–2300)	830–850 (1526–1562)	870–900[c] (1598–1652)	780–800[d] (1436–1472)	770–790[e] (1418–1454)
Tempering	530–550[b] (986–1022)	550–570 (1022–1058)	150–250 (302–482)	175–425 (347–797)	200–250 (392–482)	180–350 (356–662)
Maximum annealed hardness (HB*)	241	241	241	229	207	207
Minimum hardness after tempering (HB*)	823	763	735	600	760	790

Notes: (a) Hardening temperatures recommended (for full hardening) are for heating in a salt bath, followed by cooling in oil, air or salt bath to 500–560°C (932–1040°F), then air cooling.

(b) Double tempering recommended – for cobalt steels triple tempering. Cool to room temperature after each treatment.

(c) Oil or water quench.

(d) Oil or water quench.

(e) Water or brine quench.

*HB ~ Brinell hardness number.

2.16.6 ASTM standard specifications for castings for high-temperature service, volume 01.02 (Examples)

ASTM A351: Steel castings, austenitic, for high-temperature service

ASTM A216: Steel castings, carbon, suitable for fusion welding, for high-temperature service

ASTM A217: steel castings, martensitic stainless and alloy, for pressure-containing parts, suitable for high-temperature service

2.16.7 BS and ASTM standards relating to pressure vessels and the chemical industries

BS1501: Steels for fired and unfired pressure vessels: Plates

Part 1: Specification for carbon and carbon manganese steels
Part 2: Alloy steels: Imperial units. Including Addendum No. 1 and Addendum No 2.
Part 3: Corrosion and heat resisting steel: Imperial units

Covering steels for plates with requirements of chemical composition, mechanical properties at ambient and elevated temperatures, low-temperature impact properties. Part 3 also includes high-proof stress steels.

ASTM standard specifications for steel plate, sheet, strip, and wire, volume 01.03 (Examples)
ASTM A240: Heat-resisting chromium and chromium-nickel stainless steel plate, sheet, and strip for pressure vessels
ASTM A414/A414M: Carbon steel sheet for pressure vessels

BS1502: Specification for steels for fired and unfired pressure vessels: sections and bars
Requirements for chemical composition, mechanical properties at room temperature and low temperature impact properties for carbon, carbon-manganese low and medium alloy and austenitic steels.

ASTM standard specifications for steel bars, volume 01.05 (Examples)
ASTM A479: Stainless and heat-resisting steel wire, bars and shapes for use in boilers and other pressure vessels
ASTM A484: General requirements for stainless and heat resisting wrought steel products (except wire)
ASTM A739: Steel bars, alloy, hot-wrought, for elevated temperature or pressurized parts, or both

BS1503: Specification for steel forgings (including semi-finished forged products) for pressure purposes
Specifies the requirements for carbon-manganese, low-alloy ferritic steel, martensitic and austenitic stainless steel forgings for pressure purposes. Also semi-finished products for further processing.

ASTM standard specifications for steel forgings, volume 01.01 (Examples)
  ASTM A336: Steel forgings, alloy, for pressure and high-temperature parts
  ASTM A266: Forgings, carbon steel, for pressure vessel components

BS1504: Specification for steel castings for pressure purposes
  Requirements as in BS1502 for carbon, low-alloy, corrosion-resisting, heat-resisting and high-alloy steel castings. Appendices contain information of inspection categories and verification procedure for testing at elevated temperatures.

ASTM standard specifications for ferrous castings, volume 01.02 (Example)
  ASTM A487: Steel castings, suitable for pressure service

BS1501–1506; BS1510. Steels for use in the chemical, petroleum and allied industries

BS1501–1504 replaced by separate standards as detailed above

BS 1505: Unallocated number

BS1506: Carbon and alloy steel bars for bolting material
  Requirements for freedom from defects, margins of manufacture, identification, mechanical tests and inspection. Also chemical composition and mechanical properties. The numbering system gives a broad indication of the steel type. Contains a comparison of British and US specifications.

ASTM standard specifications for steel fittings (bolting materials), volume 01.01 (Examples)
  ASTM A437: Alloy-steel turbine-type bolting material specially heat treated for high-temperature service
  ASTM A354: Quenched and tempered alloy steel bolts, studs, and other externally threaded fasteners

BS1510: Low-temperature supplementary requirements to BS1501–6.
  Contains additional requirements for steels covered by BS1501 and 1506 under severe conditions at low temperatures or when a high degree of notch ductility is required.

ASTM standard specifications for ferrous castings (low-temperature service), volume 01.02
  ASTM A757: Ferritic and martensitic steel castings for pressure-containing and other applications for low-temperature service
  ASTM A352: Steel castings, ferritic and martensitic, for pressure-containing parts, suitable for low-temperature service

2.16.8 ASTM standard specifications for steel pipe and tube, volume 01.01 (Examples)

ASTM A790: seamless and welded ferritic/austenitic stainless steel pipe
ASTM A524: Seamless carbon steel pipe for atmospheric and lower temperatures
ASTM A405: Seamless ferritic alloy-steel pipe specially heat treated for high-temperature service
ASTM A209: Seamless carbon–molybdenum alloy-steel boiler and superheater tubes
ASTM A249: Welded austenitic steel boiler, superheater, heat-exchanger, and condenser tubes

2.16.9 ASTM standard specifications for structural steel, volume 01.04 (Examples)

ASTM A441: High-strength low-alloy structural manganese vanadium steel
ASTM A514: High-yield strength, quenched and tempered alloy steel plate, suitable for welding
ASTM A633: Normalized high-strength low-alloy structural steel
ASTM A709: Structural steel for bridges

2.17 STEEL CLASSIFICATION SYSTEMS – UNITED STATES

The most common steel classification system used in the United States is that of the American Iron and Steel Institute (AISI); it is also used by the Society of Automotive Engineers (SAE). The American Society for Testing and Materials (ASTM) and the American Society of Mechanical Engineers (ASME) have also developed widely used systems.

The AISI-SAE system uses a four-digit designation for carbon and alloy steels and a three-digit system for stainless steels. Tool steels are designated by a letter that indicates the application, followed by one or two digits (the ASTM system uses a letter followed by three digits). For carbon and alloy steels using the AISI-SAE system, the first digit from the left (or sometimes the first and second digits) indicates the type of alloying element that has been added to the iron–carbon alloy. The second digit (sometimes) indicates the percentage of the alloying element added to a particular series. This situation will be clarified by reference to Table 2.10 which summarizes the classes of steels in the AISI-SAE system. The classification 10XX is reserved for plain carbon steels containing a minimum of other alloying elements. The last two (sometimes three) digits indicate the normal carbon content in hundredths of a percent, e.g. a 3142 steel has a normal carbon content of 0.42%. Certain letters are also used in this classification system to provide additional information about the steel; these are summarized in Table 2.11.

Table 2.10 Classification of Carbon and Alloy Steels using the AISI–SAE system

Class of steel	AISI–SAE number	Main constituents
Carbon steels	10XX	Plain carbon steel
	11XX	Plain carbon (resulfurized for easy machining)
Alloy steels:		
Manganese	13XX	Mn, 1.5–2.0%
	15XX	Mn, 0.75–1.25%
Nickel	23XX	Ni, 3.25–3.75%
	25XX	Ni, 4.75–5.25%
Nickel-chromium	31XX	Ni 1.10–1.40%, Cr 0.55–0.90%
	33XX	Ni 3.25–3.75%, Cr 1.40–1.75%
Molybdenum	40XX	Mo 0.20–0.30%
	41XX	Cr 0.40–1.20%, Mo 0.08–0.25%
	43XX	Ni 1.65–2.00%, Cr 0.40–0.90%, Mo 0.20–0.30%
	46XX	Ni 1.40–2.00%, Mo 0.15–0.30%
	48XX	Ni 3.25–3.72%, Mo 0.20–0.30%
Chromium	50XX	Cr 0.25–0.65%
	51XX	Cr 0.70–1.20%
	5XXX	C 1%, Cr 0.5–1.5%
Chromium–vanadium	61XX	Cr 0.70–1.10%, V 0.15% (min)
Multiple alloys	81XX	Ni 0.20–0.40%, Cr 0.30–0.55%, Mo 0.08–0.15%
	86XX	Ni 0.30–0.70%, Cr 0.40–0.85%, Mo 0.08–0.25%
	87XX	Ni 0.40–0.70%, Cr 0.40–0.60%, Mo 0.20–0.30%
	92XX	Mn 0.85%, Si 2%
	93XX	Ni 3.25%, Cr 1.20%, Mo 0.12%
	94XX	Mn 1%, Ni 0.45%, Cr 0.40%, Mo 0.12%
	97XX	Ni 0.55%, Cr 0.17%, Mo 0.20%
	98XX	Ni 1%, Cr 0.80%, Mo 0.25%

Table 2.11 Letters used in AISI-SAE system for carbon and alloy steels

Designation	Meaning
XXBXX	Steel with boron addition
XXLXX	Steel with lead addition for easy machining
Suffix H	Can be hardened
Prefix:	
A	Alloy steel from acid open-hearth furnace
B	Carbon steel from Bessemer process
C	Carbon steel from basic open-hearth furnace
D	Carbon steel from acid open-hearth furnace
E	Made in electric furnace
X	Composition varies from normal limits

The classes of steels which are used primarily for machine design are summarized in Table 2.12, with the appropriate AISI or ASTM grades. The high-strength, low-alloy steels are probably best specified by the ASTM system, followed by the strength grade required. These steels are often used for structural applications. The ultra high-strength steels can be classified according to the AISI designation for tool steels, e.g. H11 (see Table 2.13), or for medium carbon alloys, e.g. 4140. A class of maraging steels (see Section

Table 2.12 Types of steels used mainly for machine design

Classification	Designation	
Carbon steels	AISI 1006–1095	General
	AISI 1108–1151	Easy machining
	AISI 1513–1572	High manganese
	ASTM Volume 01.04: A611	
	Volume 01.03: A619, A620	
Alloy steels	AISI 13XX	Manganese
	31XX to 33XX	Nickel
	40XX, 41XX, 43XX, ⎫	Nickel–chromium
	46XX, 47XX, 48XX ⎭	–molybdenum
	50XX, 51XX, 5XXX	Chromium–vanadium
	61XX	
	86XX, 87XX, 92XX,	Multiple
	93XX	
High-strength low alloy	ASTM Volume 01.04: A242, A441, A572, A588	Structural applications
Ultra high-strength	AISI 4140, 4340	Medium-carbon
	AISI H11, H13	Tool steels
	ASTM Volume 01.04: A538	Maraging
Heat-treated	ASTM Volume 01.04: A514	High yield strength (quenched and tempered)
	ASTM Volume: 01.04: A663	Normalized high-strength
	ASTM Volume 01.04: A724	Carbon steel – quenched and tempered
Special applications	ASTM Volume 01.04: A414	Pressure vessels
	ASTM Volume 01.03: A457	High temperatures

2.20) containing 18% nickel are used for structural applications, and are covered by ASTM volume 01.04: A538/A538M; Specification for pressure vessel plates, alloy steel, precipitation hardening (maraging), 18 percent nickel. The heat-treated steels and steels for special purpose applications are best designated using the ASTM system.

Steels used for cutting and shaping other metals are known as tool steels. These can be classified according to the particular properties they are required to possess and their applications. The AISI system uses a letter to designate the required property or application, followed by one or two digits. The broad groupings are: shock-resistant tools, e.g. punches and chisels; high-speed tools, e.g. drills and cutters; steels used to shape metals in hot (moldings) and cold (dies) conditions. A summary of the AISI classification for tool steels is given in Table 2.13.

The AISI classification of stainless steels uses a three-digit system. The first digit indicates the primary alloying elements present in that series of steels. The 200 series of alloys contain chromium, nickel and manganese, the 300 series are chromium–nickel alloys and the 400 series are chromium–iron alloys. The 500 series are not considered to be stainless steels because they have a low chromium content. Stainless steels should contain a minimum of 10.5%

chromium. The newest range of stainless steels are known as the PH alloys and they all have the 600 series identification numbers. The term PH stands for precipitation hardening, due to the presence of small amounts of titanium, aluminium or copper that precipitate out of solution at low temperatures and cause increased hardening and strength. These stainless steels are low-carbon, chromium–nickel alloys with a particular chromium–nickel ratio. They can be martensitic, semi-austenitic or austenitic. The AISI classification of stainless steels is summarized in Table 2.14; it can be seen that these are basic alloy types which have been modified to obtain other alloys. The 200 series are the austenitic stainless steels, although these are less common. The 400 series contains both martensitic and ferritic alloys; care should be taken to select the correct steel because martensitic alloys are hardenable and ferritic alloys are not. It should be sufficient, at this stage, to be familiar with the various series as given by the first digit.

Table 2.13 Designation of tool steels in the AISI system

Type	Prefix	Specific types
Shock resisting	S	S1, S2, S5, S7
	W (water hardening)	W1, W2, W5
	D (High carbon, high chromium)	D2, D3, D4, D5, D7
	A (medium-alloy air hardening)	A2, A3, A4, A6, A7, A8, A9
	O (oil hardening)	O1, O2, O6 O7
Hot work	H	H1–H19 chromium types H20–H39 tungsten types H40–H59 molybdenum types
High speed	M	Molybdenum types: M1, M2, M3–1, M3–2, M4, M6, M7, M10, M30, M33, M34, M36, M41, M42, M43, M44, M46, M47
	T	Tungsten types: T1, T2, T4, T5, T6, T8, T15
Special purpose	P	Mold steels: P2, P3, P4, P5, P5, P20, P21
	L	Special-purpose steels: L2, L3, L6
	F	Carbon–tungsten types: F1, F2

*   *   *

Evaluation of Sections 2.15, 2.16 and 2.17

Any engineering course contains a unit concerned with materials. This book contains the information which provides the basis for such a unit. However, it is not intended to be either a 'Science of Materials' or a 'Metallurgy for Engineers' type of text – hence the title that was chosen. Both of these topics need to be included in the teaching program, but engineers also need to be instructed in other aspects of materials.

The engineer must be able to specify the materials that he wishes to use in his engineering designs. This means not only knowing the requirements of the material in terms of the physical, chemical and mechanical properties, but also where to obtain appropriate information and how to communicate the requirements to a supplier.

For these reasons, reference is made throughout the book to appropriate British and US standards. Hopefully, this book will be used by engineering undergraduates not only for taught courses, but also when they are involved in design projects. This material should also provide a reference source for the engineer in industry. Sections 2.15–17 are intended to provide a reference source for locating information that is available and which should be used by students. It represents a significant proportion of the material contained in this chapter (similarly in Chapter 3) and for this I make no apology.

It is not necessary to memorize the information contained in these sections. However, the student should be aware that standards:

(a) exist (both British, US and in other countries);
(b) are used by engineers as a useful and reliable source of information;
(c) provide a basis for design work;
(d) cover most established situations but not usually new innovations;
(e) define metal and alloy classification and designation systems;
(f) contain recommended practices and standards for the use of metals and alloys;
(g) contain useful data for most materials used by engineers.

The follow-up work by the student should be twofold. First, to go and locate sets of actual standards, to study the index for each set, to find some of the standards quoted in this book and to become familiar with the information that they contain. Second, to use standards to answer questions that are asked in this book and by the lecturer for the course, and to assist with design projects in other parts of the course. Some of the questions/exercises associated with an 'engineering materials' course unit should provide experience of:

(a) material specification;
(b) calculation of material properties;
(c) material selection.

Decide whether you have achieved these goals at the end of your course.

＊　＊　＊

2.18 HEAT TREATMENT OF STEELS

2.18.1 Introduction

The heat treatment of steels involves a process of heating cycles where the metal is subjected to controlled temperature changes in the solid state. These heating cycles cause changes in the structure of the steel and are used to produce a material with certain desired properties. The heating process can change the size and shape of the grains present in the steel, and also the constituents present. The temperature to which the steel is heated and the rate of cooling are important factors to be considered when heat treating steels. Sometimes cold working (Sections 8.2.7–15) is the only practical way to obtain the desired mechanical properties in a specific metal or alloy.

Table 2.14 Stainless steels categorized by structure and chemical composition (AISI system)

Group	General properties	Hardenability	Type	Analysis built up from basic type
Chromium–iron	Martensitic: non-rusting tools and structural parts	Hardenable by heat treatment	403	Cr 12% adjusted for special properties
			410	Basic type, Cr 12%
			414	Ni added to increase corrosion resistance and physical properties
			416	S or Se added for easier machining
			418 Spec	W added to improve high-temperature properties
			420	C higher for cutting purposes
			420F or Se	S or Se added for easier machining
			431	Cr higher and Ni added for better resistance and properties
			440A	C higher for cutting applications
			440B	C higher for cutting applications
			440C	C still higher for wear resistance
			440F or Se	S or Se added for easier machining
	Ferritic: used for elevated temperatures and non-rusting architectural parts	Non-hardenable	405	Al added to Cr 12% to prevent hardening
			430	Basic type, Cr 17%
			430F or Se	S or Se added for easier machining
			430Ti	Ti stabilizer
			442	Cr higher to increase scaling resistance
			446	Cr much higher for improved scaling resistance
Chromium–nickel	Austenitic: used for chemical resistance	Hardenable by cold work	301	Cr and Ni lower for more work hardening
			302	Basic type, Cr 18% Ni 8%
			302B	Si higher for more scaling resistance
			303	S or Se added for easier machining
			304	C lower to avoid carbide precipitation
			304L	C lower to avoid carbide precipitation
			305	C lower for welding application
			308	Ni higher for less work hardening
			309	Cr and Ni higher with low C for more corrosion and scaling resistance
			309C	Cr and Ni still higher for more corrosion and scaling resistance
			309S	Co, Ta added to avoid carbide precipitation
			310	C lower to avoid carbide precipitation
				Cr and Ni highest to increase scaling resistance

continued over

Table 2.14 (continued)

Group	General properties	Hardenability	Type	Analysis built up from basic type
			314	Si higher to increase scaling resistance
			316	Mo added for more corrosion resistance
			316L	C lower for welding application
			317	Mo higher for more corrosion resistance and strength at high temperatures
			318	Co, Ta added to avoid carbide precipitation
			321	Ti added to avoid carbide precipitation
			347	Co, Ta added to avoid carbide precipitation
			347F or Se	S or Se added to improve machinability
			348	Similar to 347, but low tantalum content (0.10%)
Chromium–nickel manganese			201	Cr and Ni lower for more work hardening
			202	Basic type, Cr 18%, Ni 5%, Mn 8%
			204	C lower to avoid carbide precipitation
			204L	C lower for welding application
Precipitation hardening (PH)	Martensitic and semi-austenitic, combination of chemical resistance and high strength	Hardened by precipitation heat treatment	630	17% Cr, 4% Ni (17–4), high-strength alloy
			631	17% Cr, 7% Ni (17–7), higher strength than 17–4
			632	15% Cr, 7% Ni (15–7), Mo added for pitting resistance

The heat-treatment process can produce either equilibrium or non-equilibrium conditions. Equilibrium conditions are produced by very slow heating and cooling, the iron–carbon phase diagram (Section 2.4) can then be used to determine the phases present and their relative proportions. In practice steels are not usually treated under these conditions, but the phase diagram can still be used to determine the upper (critical) temperature to which the steel must be heated to obtain a particular constituent, and the temperatures required to produce softening and hardening. For the discussion of heat-treatment methods it will be necessary to refer to the iron–carbon phase diagram (Section 2.4, Fig. 2.2), although only a portion of Fig. 2.2 will actually be required and this is reproduced as Fig. 2.14. When a steel is in a state of equilibrium it is more ductile (although still possessing hardness and strength) than in the non-equilibrium condition. Particular reasons for using heat treatment will be discussed, involving consideration of the properties which are required and the processes which are normally employed.

2.18.2 Hardness and hardenability

Steels possessing hardness and strength are required for structural uses and for applications to resist wear and distortion. The distinction should be made between hardness and hardenability. Hardness is a measure of resistance to plastic deformation and hardenability is the ease with which hardness can be attained. The maximum hardness of a steel is obtained only when it is heat treated to produce a structure of 100% martensite. A steel which rapidly

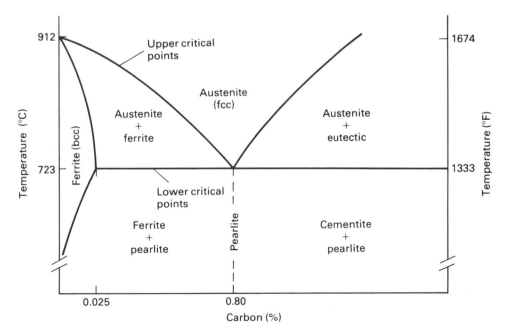

Figure 2.14   Section of the iron–carbon phase diagram.

transforms from austenite into ferrite and carbide, rather than martensite, has a low hardenability. The depth of hardening of a steel is a measure of the hardenability. In plain carbon steels a hard surface layer of martensite can be formed by rapid cooling, but below the surface ferrite and pearlite are formed due to the slower cooling rate. Alloy steels have lower critical cooling velocities and therefore greater hardenabilities than plain carbon steels. Martensite may be present at distances up to 50 mm (2 in.) below the surface. The Jominy test is often used to assess the hardenability of a steel. Standard conditions for the testing procedure are laid down in BS4437: Method for the end-quench hardenability test for steel (Jominy test) and ASTM A255: End-quench test for hardenability of steel. A bar of steel is heated within the austenitic range (Fig. 2.14) and then rapidly cooled at one end by a jet of water. A series of hardness measurements is then made along the length of the bar and a hardenability curve is obtained, as shown in Fig. 2.15. This curve will be obtained for all plain-carbon and low-alloy steels.

2.18.3 Softening

The hardening process is reversible, it is often advantageous to soften a steel before welding or before re-machining a hardened tool.

2.18.4 Recrystallization

When a steel is cold worked, e.g. by rolling or drawing, the original crystal structure is distorted due to slip along the slip planes within the structure, and

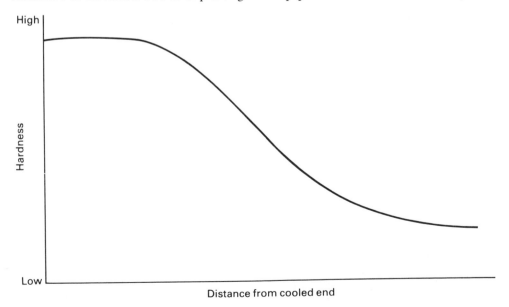

Figure 2.15 Typical results from a hardenability test.

elongated crystals are obtained. This increases the hardness and tensile strength of the metal but reduces the ductility. This effect makes the metal resistant to further cold working and it may become too hard to roll, or it may fracture. If the steel is then heated, recrystallization occurs and the metal returns to its original state of softness and ductility (see Section 1.4.9). The alternative is to use hot working: the metal is shaped at higher temperatures and recrystallization occurs under the action of working alone. The working can cause 'grain refining', producing smaller and more regular crystals. However, if the hot working is performed at too high a temperature (above the upper critical point on Fig. 2.14), then grain growth occurs until the steel is cooled below the lower critical point. This is disadvantageous to the dimensional tolerance of the final product. Further reference can be made to BS4490: Methods for the determination of the austenitic grain size of steel, and ASTM (Volume 03.03) E112: Determining average grain size.

2.18.5 Stress relief

Internal stresses caused by cold working or welding can sometimes be removed by heating to temperatures below the recrystallization point and then allowing to cool in air. If the temperature is kept below 150°C (302°F) then the hardness of the metal is unaffected. The crystal structure is unchanged and dimensional changes are avoided (see Section 1.4.9).

2.19 HEAT TREATMENT PROCESSES

2.19.1 Annealing

Annealing is a process that produces equilibrium conditions (see Section 1.4.9) by heating and maintaining at a suitable temperature, and then cooling very slowly. It is used to induce softness, relieve internal stresses, refine the structure and improve cold-working properties. It is often specified for castings that did not cool uniformly, due to abrupt changes in cross section, and also for forgings or rolled steel parts with nonuniform structures. Recrystallization (see Section 2.18.4) is sometimes known as *process annealing* or *subcritical annealing*; it is generally performed between 500°C and 650°C (932°F and 1202°F). *Full annealing* is a treatment where steel is heated through the recrystallization temperature to remove internal stresses, to a temperature approximately 30–50°C (54–90°F) above the upper critical points on Fig. 2.14. The steel is held at that temperature (soaked) for approximately two hours depending upon the metal thickness and the required properties; this ensures that a uniform austenitic structure is obtained and produces slight grain growth. It is then cooled very slowly, often in a furnace, to produce a pearlitic structure. Process annealing is often used to soften low-carbon alloy steels because full annealing would require heating and holding at temperatures above 900°C (1652°F), with consequent high costs.

Consider the full annealing of a 0.5% carbon steel which is heated above the upper critical region to form austenite. Care should be taken not to overheat the austenite, otherwise crystal growth occurs. As the austenite is cooled below the upper critical point, the ferrite precipitates out at the crystal boundaries making the austenite richer in carbon. As cooling proceeds large new crystals of ferrite are formed; if cooling is sufficiently slow then the carbon diffuses into the center of the remaining austenite crystals. When the temperature passes through the lower critical point, then the remaining austenite changes to pearlite crystals (0.87% carbon). Crystals of ferrite and pearlite are present in fully annealed steels that have a comparatively coarse structure. To obtain softness in the steel the crystals of soft ferrite must be as large as possible; full annealing is therefore usually reserved for low carbon steels.

The main disadvantage of full annealing is the cost of very slow cooling, although this can be reduced by *isothermal annealing*. The steel is again heated and maintained at a temperature that produces a uniform austenitic structure; it is then transferred to a salt bath maintained at a subcritical temperature (approximately 650°C/1202°F). This ensures slow cooling through the lower critical point so that the transformation of austenite to ferrite and pearlite can occur; the required softness is then obtained. When the steel reaches the salt-bath temperature it is cooled freely in air.

If a steel containing more than 0.87% carbon (the eutectoid composition) is slowly cooled from above the upper critical point, then a brittle structure is obtained with cementite at the grain boundaries. To improve the ductility, the steel is heated and kept at a temperature close to the lower critical point for sufficient time for the carbides to agglomerate as large spherical particles. This process is known as *spheroidizing annealing*; it produces a soft and tough steel that is known as *divorced pearlite*. It is used to improve machining, pre-cold working and pre-heat treatment.

2.19.2 Normalizing

Full annealing is an expensive process and normalizing is often used as an alternative where some loss in ductility is acceptable. The product is slightly harder and stronger, with a refined grain size. The final structure is a soft mixture of free ferrite and pearlite. The steel is heated to just above the upper critical point by approximately 40°C (72°F), and allowed to become fully austenitic. It is then removed from the furnace and allowed to cool in still air (drafts may cause it to cool too quickly). The cooling rate for normalizing is very much faster than for the annealing processes and the carbon does not have time to diffuse completely through the ferrite areas and into the austenite. Initially, small crystals of ferrite appear, both at the crystal boundaries and within the bodies of the original crystals. The ferrite regions cannot combine together to form large crystals and the final separation into ferrite and pearlite is incomplete. The final crystal structure is not composed of the large coarse crystals of ferrite and pearlite which are obtained by annealing, but it is an intimately mixed finer structure because most of the ferrite has crystallized with the pearlite. The quantity of pearlite in normalized and annealed steels, with the same carbon content, is identical. However, the refined crystal structure and the effect of spreading the pearlite over the soft ferrite can result in up to 20% increased hardness in a normalized steel. Normalizing is used with low- and medium-plain carbon steels and alloy steels, and steels that have been rolled or cast.

2.19.3 Quenching

Quenching involves cooling a steel very rapidly by immersion in oil or water. If an austenitic steel with more than 0.35% carbon is quenched, excess carbon is not precipitated but is trapped within the bcc ferrite lattice to form a supersaturated solid solution known as martensite. This type of steel has a needle-like (acicular) structure, it is extremely hard and strong but it is also very brittle, and therefore has few uses. It can be used for structural or wear resistant applications, but it is not suitable for shock or impact situations. There is a minimum (or critical) cooling rate for a steel which ensures that all the austenite is transformed into martensite, and it therefore possesses maximum hardness. If quenching produces a cooling rate slightly lower than the critical rate, then *troostitic pearlite* is formed. If this rate is reduced further (but above that at which pearlite is formed), then *sorbitic pearlite* is produced. Both of these structures consist of a very fine form of pearlite, and are stronger and harder than the equilibrium structures.

As steel cools there is an expansion in volume as bcc austenite transforms to fcc ferrite, followed by a contraction with further cooling of the ferrite. This effect progresses towards the core of the material and, due to the brittleness caused by quenching, cracking can occur. This is particularly important for thin sections, abrupt changes in cross section and sharp corners. Cracking can be avoided by correct choice of the quenching medium, e.g. oil (alloy steels), air blast (special alloy steels), or water (plain carbon steel). The thickness of material is an important consideration. It is possible to obtain a hard martensitic surface with a fine pearlite structure at the slower cooling core. There is an interrelationship between the final properties, size and shape of a metal component, due to the fact that the size and shape determine the cooling rates attainable in the core during heat treatment. This is known as the *mass effect*. Other important considerations are the design of uniform section thicknesses within a structure, and the use of supports during heat treatment.

2.19.4 Tempering

Steel that has been quenched is strong and hard, but also very brittle due to the unstable condition of the bcc lattice. Tempering is gentle reheating of the steel to between 180°C and 650°C (356°F to 1202°F), the temperature depending upon the required properties (i.e. below the lower critical point). The carbon can then be precipitated in a more stable form. Internal stresses are removed and the ductility and toughness are improved at the expense of the hardness and strength – a compromise based upon the required properties. The martensite is transformed into either *sorbite* by tempering above 450°C (842°F) or *troostite* by tempering below 450°C (842°F) (these are different from troostitic pearlite and sorbitic pearlite considered for quenching). Troostite and sorbite are fine dispersions of iron carbide in a ferrite structure. These are different from the laminated solid solutions of cementite and pearlite produced by the slow cooling of austenite.

Sometimes quenching and tempering are performed as one process by rapid cooling to the tempering temperature, then holding to allow carbon precipitation. This reduces the risk of cracking due to the volume expansion when austenite transforms to ferrite. The carbide is then present in a different, finely divided form known as *bainite* (see Section 2.5.1).

2.19.5 Surface hardening

A steel can be directly hardened by quenching an austenitic structure. The hardness and properties of the steel produced depend upon the carbon content, the heating time, the temperatures used and the cooling rate. As the hardness and strength of a material increase, then the toughness decreases. Many applications require a tough core of material with a very hard surface, which can be achieved by heat treating a steel to produce the required hardness only at the surface. Surface-hardening methods can involve heating either the whole component (e.g. case hardening and nitriding) or only the surface of the component (e.g. flame or induction hardening). All methods except induction hardening involve diffusion of another material into the metal surface.

2.19.6 Case hardening

This is a process for increasing the surface hardness of a steel, and requires heating of the whole component. This process is usually performed on low-carbon steels (approximately 0.15% carbon) which will not respond to direct hardening.

The first stage involves heating the steel in contact with carbon, to increase the carbon content at the surface. This is known as *carburizing* or *case carburizing*. The steel is heated above its upper critical point (approximately 903°C/1705°F) and is surrounded either by carbon in solid form *(pack carburizing)*, or by liquid sodium carbonate mixed with either sodium or barium chloride. Alternatively *gas carburizing* can be used with an atmosphere of methane or mixed hydrocarbon gases. The temperature must be maintained long enough for the carbon to penetrate to the required case depth, usually 0.5 mm to 0.7 mm. The metal surface is transformed from a low-carbon to a high-carbon steel.

During carburizing, grain growth occurs within the steel. The second stage of the process involves heat treatment to harden the high-carbon skin. The carburizing process causes a variation in carbon content (approximately 0.83–0.15% carbon) from the case to the core over a depth of about 0.5 mm. The difference between these regions is sometimes taken to be the point where the hardness measures Rockwell C50. The heat treatment required by these two regions is different and requires different temperatures. The core structure is first refined by heating above the upper critical point of the core composition (approximately 870°C/1598°F); it is then quenched to prevent grain growth during cooling. This high temperature produces a coarse austenitic structure within the case, and when this is quenched it transforms to extremely brittle martensite. The case is then refined, by heating to just above the lower critical point (approximtely 760°C/1400°F) of the higher carbon layer; this also tempers the outer layers of the core. The steel is then quenched and final tempering of the case is carried out at 200°C (392°F) to relieve quenching stresses. The minimum of hardened surface metal (less than 20%) should be removed after carburizing, as it detracts from the purpose of this heat treatment process.

Carbonitriding is liquid carburizing with diffusion of some nitrogen. High hardness can be achieved with the following advantages: hard and shallow case, lower temperatures and less distortion. The process can be applied to any steel that can be carburized, and also for materials that cannot be nitrided (see Section 2.19.7). It is a useful way of providing a wear-resistant surface on cheap low-carbon steels.

Other carburizing processes have been developed such as *vacuum carburizing*. Parts of the cycle are carried out under vacuum and process times are reduced, although operating costs are higher. Carburizing can also be followed by other heat treatments to provide particular advantages. However, any additional processing costs must be justified.

Cyaniding is similar to carburizing, but the steel component is heated in a bath of molten sodium cyanide and sodium carbonate (at 950°C/1742°F). Both carbon and nitrogen diffuse into the steel surface and the surface hardness is increased by the presence of iron nitrides. The component must still be heat treated as previously described. Components must usually be ground to the final dimensions, due to volume changes during the case-hardening process.

2.19.7 Nitriding

This process can only be used for low-alloy steels (that can be directly hardened) which contain small amounts of chromium, nickel, vanadium or aluminum. The carbon content is usually between 0.2% and 0.5% carbon. Plain-carbon steels cannot be hardened by this process because the nitrides would form throughout the structure causing brittleness. It is also used for some tool steels, gray cast irons, medium carbon steels (400 series) and special alloy steels (high aluminum or chromium). The material is first heat treated and oil quenched to strengthen the core by providing required hardness and toughness, and then tempered at 550°C to 750°C (1022°F to 1382°F) to relieve stresses. It is machined to allow for the small growth of 0.02 mm (0.0008 in.) during nitriding.

The component is then nitrided by heating to 500°C (932°F) in a constant closed circulation of ammonia. A case depth of 0.1 mm (0.004 in.) can be produced in about 10 hours, although a depth of 0.7 mm (0.028 in.) can take up to 100 hours. The core properties are unaffected by nitriding. Finally, the temperature is allowed to fall to 150°C (302°F) before the component is removed from the ammonia atmosphere. The advantages of nitriding are that an extremely hard surface is produced, cracking and distortion are minimized due to the lower temperatures used, and no subsequent heat treatment is required. The component can be produced more cheaply because machining can be performed before hardening, due to the small volume changes which occur.

The disadvantages of nitriding are the costs of the surface pretreatments and the length of the nitriding cycles required. The range of materials for which nitriding is applicable is small and some of these materials are expensive and difficult to machine. The nitrided case produced is relatively thin and only a small amount of material can be subsequently removed by machining. However, nitriding is an accurate process and machining should not be necessary. Masking paints or plating (both subsequently removed) can be applied to enable nitriding to occur only in selected regions. *Soft nitriding* is gas nitriding with carbon added to the furnace atmosphere. Extremely precise case depths can be produced which are shallower than for the conventional process, and the applicable range of alloys is wider.

2.19.8 Flame hardening and induction hardening

In both these processes the steel (containing between 0.4% and 0.7% carbon) is rapidly heated to the hardening temperature, using an oxy-acetylene flame or a high-frequency induction coil. The surface is rapidly quenched to form martensite, the subsequent cooling and outflow of heat from the core tempers the surface layers. The flame temperature must be above the melting point of the steel, surface melting can occur if case hardening depths beyond 0.8 mm (0.032 in.) are attempted. For induction hardening, it takes about 5 seconds for a case depth of 3 mm (0.12 in.) to reach its hardening temperature.

The time that the component is heated is the crucial factor. After the required depth of metal is heated, the part must be quenched immediately. Problems can occur if the cross section changes abruptly, or if there are thin-walled sections that may heat all the way through. Induction hardening is faster, requires less expensive equipment, and can harden thicker cases than carburizing. Cracking is the main problem, especially with high-alloy steels that are more sensitive to heat transfer into the core. A component may be case hardened at selected regions to reduce wear and abrasion, and nonhardened areas are easier to machine. However, stresses will be present in the hardened/ soft metal region, and these should be located away from features which induce stress concentrations (see Section 5.5.5).

2.19.9 Age hardening or precipitation hardening

Age hardening occurs when the solubility of an alloying element decreases as the temperature decreases. At lower temperatures the element in solution becomes supersaturated and starts to precipitate out, causing increased hardness and improved mechanical properties. This age hardening process develops with time. Precipitation hardening has previously been mentioned for stainless steels (Section 2.17). Particular alloying elements are added to the steel and they precipitate out of the structure at lower temperatures, thus improving the strength and hardness. Maintaining the material at the aging temperature for a prolonged time can cause softening following the aging process, known as *overaging*. Precipitation hardening is also specified for some high-strength and heat-resistant alloys (see Sections 2.9 and 2.20), and for some alloys of aluminum, copper and magnesium. At the higher temperature the solute dissolves to form a single-phase solid solution. If the solid solution is suddenly cooled (quenching), then the excess solute has insufficient time to precipitate out and it is trapped in an unstable state as a supersaturated solid solution. The process of treating a raw material by this method is known as *solution treatment* (see Section 3.3.1). The unstable alloy is a soft material that can be cold worked or machined. The finished component can be reheated to precipitate out excess solute and increase the hardness.

Aluminum alloys are hardened either by cold working or by precipitation hardening, due to the addition of copper, manganese, nickel or silicon. The precipitation of the solute phase and the hardness depend upon the temperature used and the time allowed. Precipitation hardening is also used with beryllium copper, aluminum bronze, chromium copper, zirconium copper, cupronickels with silicon and phosphorus, and for magnesium alloys.

2.19.10 Austempering and marquenching

Austempering involves quenching an austenitic structure to a temperature above that at which martensite is formed. The steel is held at that temperature (isothermal transformation) until bainite formation is complete. Bainite is a more ductile phase than martensite. An austempered steel does not require tempering as it is not stressed and does not suffer from distortion.

Marquenching is sometimes called *interrupted quenching* or *martempering*, it is used to harden steel without the cracking sometimes caused by quenching. An austenitic steel is quenched to a temperature above that required for martensite transformation; this temperature is maintained in order to establish thermal equilibrium throughout the entire component. Slow cooling then produces tempered martensite, although this may be followed by further tempering.

Austempering and marquenching are both lengthy and costly operations. However, their advantages are the production of less-stressed parts and less risk of cracking or distortion than when quenching is performed. Both processes can be readily used on high-alloy steels, but plain-carbon steels require more control if a thick section is being treated. Sometimes, with rapid quenching, small quantities of austenite do not have sufficient time to transform to martensite. This austenite is trapped in the microstructure, but it transforms later (several hours or days) causing dimensional changes. This effect can occur with quenching and tempering, or marquenching. Further details of these processes are given in Section 2.5.1.

2.20 HEAT TREATMENT FOR ALLOY STEELS

A relatively new class of steels known as maraging steels has been developed (see Section 2.17). These steels are strong and hard, but less brittle than conventional heat-hardened steels. A steel containing the minimum amount of carbon is alloyed with between 18% and 25% nickel and other alloying elements. A typical composition is 18% nickel, 8% cobalt, 5% molybdenum and 0.4% titanium. This alloy has a type of martensitic structure even when slowly cooled, and it is hard and tough. The other alloying elements improve the strength. The steel is annealed and martensite is formed; this is followed by shaping and machining. Hardening is achieved by heating at 450°C to 500°C (842°F to 932°F) for three hours, followed by air cooling. The surface hardness can be improved at this stage by nitriding. The lower temperature treatment means that cracking and distortion are eliminated. Rapid heating does not cause hardening, which makes these steels suitable for welding. However, maraging steels (named from *mar*tensitic *ag*e harden*ing*) are more costly to produce.

The addition of nickel to steels that have been carburized prevents grain growth, and chromium is added to offset the mass effect (see Section 2.19.3) in thicker components. Typical nickel case-hardened steels contain 0.12% carbon, 3% nickel, 0.45% manganese, and are water quenched. The low carbon content means that the steel does not respond to direct hardening. If the nickel content is increased to 5%, then the steel may be oil quenched. A steel containing 15% carbon, 4% nickel, 0.8% chromium and 0.4% manganese has improved strength and hardness due to oil quenching.

Steels containing sufficient chromium (about 2%) can be quenched in air. Alloying elements are added to steel so that the critical cooling rate and the hardening temperature are reduced. A steel containing 1.5% manganese (0.35% carbon) is relatively cheap and can be oil quenched to give good strength. A steel containing 0.3% carbon, 3% nickel, 0.6% manganese has good strength and hardness when oil quenched. The addition of chromium improves the toughness of a steel and lowers the hardening temperature.

Air quenching can be employed if sufficient elements are added, and this reduces the risk of distortion. Typical steels contain 0.3% carbon, 3% nickel, 0.8% chromium, 0.6% manganese for oil quenching and 0.3% carbon, 4.25% nickel, 1.25% chromium, 0.5% manganese for air cooling. When nickel–chromium steels are tempered between 250°C and 400°C (482°F and 752°F), depending upon the composition, they may become brittle. This can be avoided by the addition of 0.3% molybdenum, which also reduces mass effect by forming carbides. About 0.5% vanadium improves the shock resistance of chromium steels, but can give rise to mass effect if it is used to replace nickel.

2.21 SUMMARY

It seems appropriate to end this chapter with a comparison of some of the important properties of the main types of irons and steels. This is presented in Table 2.15: the compositions quoted are approximate and the values of the

Table 2.15 Properties of ferrous metals and alloys

Material and approximate composition (mass %)	Melting point or range, °C(°F)	Young's modulus GN/m² (psi × 10⁻⁶)	Tensile strength MN/m² (psi × 10⁻³)	Yield strength MN/m² (psi × 10⁻³)	Coefficient of linear thermal expansion (× 10⁶)
Iron; annealed (C < 0.02; Mn < 0.02)	1536 (2797)	200 (29.0)	290 (42.0)	130 (18.0)	11.8
Gray cast iron (C 2.75–3.50; Si ≃ 2.0; Mn ≃ 0.7)	1120–1175 (2048–2147)	100 (14.5)	250 (36.2)	–	10.9
Malleable cast iron; ferritic (C ≃ 2.3; Si ≃ 1.0; Mn 0.55 max.; P and S < 0.20 each)	1120–1175 (2048–2147)	170 (24.6)	365 (52.9)	240 (34.8)	13.6
Low carbon steel; hot rolled (C 0.1–0.25; Si 0.1; Mn ≃ 0.6; P 0.04 max.)	1510–1525 (2750–2777)	205 (29.7)	400 (58.0)	250 (36.2)	15.3
Medium carbon steel; hot rolled (C 0.28–0.50; Si 0.1–0.6; Mn ≃ 0.7; P and S 0.04 max. each)	1480–1510 (2696–2750)	205 (29.7)	600 (87.0)	375 (54.3)	15.1

Table 2.15 (continued)

Material and approximate composition (mass %)	Melting point or range, °C(°F)	Young's modulus GN/m² (psi × 10⁻⁶)	Tensile strength MN/m² (psi × 10⁻³)	Yield strength MN/m² (psi × 10⁻³)	Coefficient of linear thermal expansion (× 10⁶)
High carbon steel; hot rolled (C 0.5–1.05; Si 0.1–0.5; Mn⌢0.6; P and S 0.04 max. each)	1425–1470 (2597–2678)	205 (29.7)	850 (123.2)	500 (72.5)	14.7
Ferritic stainless steel; cold worked (C 0.12 max; Mn 1.00 max; Cr 14–18; Si 1.00 max; P and S 0.04 max. each)	1425–1510 (2597–2750)	200 (29.0)	550 (79.7)	435 (63.0)	10.5
Martensitic stainless steel; heat treated (C 0.15 max.; Cr 11.5–13.5; Mn 1.00 max.; Si 1.00 max.; P 0.04 max.)	1480–1530 (2696–2786)	200 (29.0)	1000 (145.0)	700 (101.4)	10.0
Austenitic stainless steel; cold worked (C 0.08 max.; Cr 16–18; Ni 10–14; Mo 2–3; Mn 2.0 max.; Si 1.0 max.)	1370–1400 (2498–2552)	190 (27.5)	850 (123.2)	600 (87.0)	16.2

Table 2.15 (continued)

Material and approximate composition (mass %)	Melting point or range, °C(°F)	Young's modulus GN/m² (psi × 10⁻⁶)	Tensile strength MN/m² (psi × 10⁻³)	Yield strength MN/m² (psi × 10⁻³)	Coefficient of linear thermal expansion (× 10⁶)
Stainless steel; precipitation hardening (C 0.09 max.; Cr 17.0; Ni 7.1; Al 1.25; Mn and Si 1.0)	—	200 (29.0)	1350 (195.7)	1280 (185.5)	15.5
Maraging steel; oil quenched and aged (C 0.026; Ni 18.5; Co 7.0; Mo 4.5; Mn 0.1; Si 0.11; Ti 0.22; B 0.003)	—	180 (26.1)	1900 (275.4)	1850 (268.1)	10.2

tensile strength and yield strength are average values for the ranges normally encountered. This table is meant to provide a quick and easy comparison of different materials. It is not comprehensive: many other alloys and properties could have been included.

* * *

Revision test

Sections 2.18 and 2.19 describe the heat treatment of steels, and should enable the following questions to be answered:

1 Why is heat treatment performed on steels?

2 What properties and effects are produced by heat treatment?

3 What processes are used, and what particular changes occur?

4 What are the advantages and disadvantages of using heat treatment?

5 What are the alternatives to heat treatment?

6 How does heat treatment differ for alloy steels?

7 How can the iron–carbon phase diagram and appropriate TTT curves be used to determine the changes that occur, and the phases produced during heat treatment of a steel?

8 Which steels or classes of steels can and cannot be heat treated?

* * *

Exercises

1 List the elements present in irons and plain carbon steels. List the microstructures that may be present in these steels.

2 Which irons and steels contain no carbon? In which forms can carbon be present in irons and steels?

3 For the following temperature ranges, state which form of iron is present and its crystal structure:
(a) 900–1300°C (1652–2372°F);
(b) 0–900°C (32–1652°F);
(c) 1300–1600°C (2372–2912°F).

4 Calculate the percentages of the microstructures present in slowly cooled steels containing the following percentages of carbon:
(a) 0.05; (b) 0.15; (c) 0.3; (d) 1.5.

5 Which plain carbon steels and low alloy steels can be hardened by heat treatment, and which perform better?

6 Three wrought carbon steels contain the following percentages of carbon:
(a) 0.2; (b) 0.35; (c) 0.6. Describe the differences in their properties.

7 Which material will carburize more quickly at 875°C (1607°F) – ingot iron or 0.2% carbon steel? Explain the reasons.

8 Describe the main effect on the properties of a steel of adding each of the following elements:
chromium; manganese; nickel; silicon; molybdenum.

9 Sketch the TTT curves for steels containing 0.2%, 0.5% and 1.0% carbon.

10 Which steels have the highest (possible) tensile strength?

11 Why is it difficult to weld malleable cast iron? How can ductility be achieved in a cast iron welded joint?

12 Explain what is meant by weld decay, and the influence of carbon content when welding austenitic stainless steels.

13 Why must the nickel content be increased in an 18–8 stainless steel if molybdenum is present?

14 Why is a 25 mm (1 in.) bar of 4340 steel difficult to machine?

15 State the form and shape of the carbon present in gray, ductile and malleable irons.

16 What are the amounts of ferrite and cementite present in a steel containing 0.75% carbon?

17 Calculate the maximum and minimum hardenabilities that can be obtained using a Jominy test for 4140 steel with the specified composition range.

18 Describe the microstructures that are present in steels containing 0.3% and 1.0% carbon, if they are in equilibrium.

19 For cast and wrought metals with the same compositions, which have the better mechanical properties?

20 What quenching mediums are used for cooling steels from the austenitic region? What are the approximate cooling rates?

21 For a 0.5% carbon steel, describe what happens when it is heated and held at 900°C (1652°F), and then quenched to room temperature.
What will be the effect of reheating the steel to 300°C (572°F)?
What would be the effect of interrupting the quench, holding, and then air cooling to room temperature?

22 Describe the essential properties of tool steels, and how they are obtained.

23 Describe how a large screwdriver could be heat treated to produce a fine pearlite shank and a tempered martensitic tip.

* * *

Complementary activities

1 Examine a metallurgical microscope and determine its principle of operation.
How is the magnification of the microscope obtained?

2 Find out the procedures to be followed for the preparation of metallurgical specimens for examination using:
(a) hand magnifying lenses;
(b) a metallurgical microscope.

3 Obtain some plain carbon steel samples with a known treatment history, prepare them (if not already prepared), examine them under the microscope and using hand lenses.

4 Perform hot and cold working on some plain carbon steel samples, and carry out some types of heat treatment. Then examine the samples under the microscope.

5　Repeat Exercises 3 and 4 for alloy steels.

6　Examine structures containing:
   (a)　pure iron;
   (b)　martensite;
   (c)　ferrite;
   (d)　pearlite;
   (e)　cementite.

7　Is it possible to examine the structures of pure γ iron or an fcc austenite (Fe–C) solid solution?

8　If equipment is available, determine the cooling curves for a plain carbon steel and an alloy steel.

9　For an identifiable steel with a known treatment history, carry out heat treatment and then cool it quickly. Observe the structure obtained under the microscope. Repeat the experiment for different cooling rates, and compare the results with TTT curves published for that steel.

10　Examine different metal parts obtained from a scrap automobile.

* * *

KEYWORDS

iron ore
steel
blast furnace
gangue
slag
hearth
ingot
pig iron
wrought iron
puddling furnace
cast iron
reverberatory furnace
rotary furnace
electric furnace
ordinary cast iron
high duty cast iron
alloy cast iron
gray cast iron
white cast iron
malleable cast iron:
　whiteheart
　blackheart
　pearlitic
spheroidal graphite cast
　iron (SG iron)
ductile ferritic SG iron
pearlitic SG iron
ferrite
α-iron (pure bcc iron)
Curie temperature
austenite

γ-iron (pure fcc iron)
δ-iron (pure bcc iron)
δ-ferrite
cementite
pearlite
martensite
bainite
sorbite
troostite
ledeburite
iron–carbon phase
　diagram
equilibrium diagram
nucleation:
　homogeneous
　heterogeneous
growth
phase transformation
microstructure
grain boundary
inoculant
diffusion
anisotropic
eutectic region
eutectoid region
hypoeutectoid
hypereutectoid
peritectic region
upper critical point
lower critical point
liquidus line

solidus line
TTT diagram:
　C curve
　S curve
　isothermal
　　transformation
critical cooling rate
austempering
marquenching
martempering
ausforming
strain hardening
continuous cooling
　transformation
　diagram
converter
open-hearth furnace
Bessemer converter
Linz–Donawitz process
Kaldo process
acid slag
basic slag
killed steel
cast steel
plain carbon steel
mild steel
low-carbon steel
medium carbon steel
high carbon steel
alloy steel
low-alloy steel

high-alloy steel
high-speed tool steel
wear-resisting steel
stainless steel:
 ferritic
 martensitic
 austenitic
high chromium steel
plain 18–8 austenitic
 stainless steel
soft austenitic
 decay-proof steel
special purpose steel
heat-resisting steel
grain growth
hardenability
toughness
grain refinement
machinability
graphitizing agent
carbide-forming element
free cutting properties
intergranular corrosion
low-temperature steel
spring steel
valve steel
BS

ASTM
Specification
classification
designation
standard
AISI
SAE
ASME
maraging steel
heat treatment
hardening
Jominy test (end
 quench)
softening
recrystallization
stress relief
annealing
process annealing
subcritical annealing
full annealing
soaking
spheroidizing annealing
divorced pearlite
normalizing
quenching (water, oil,
 air)
troostitic pearlite

sorbitic pearlite
mass effect
tempering
surface hardening
case hardening
carburizing
case carburizing
pack carburizing
gas carburizing
carbonitriding
vacuum carburizing
cyaniding
nitriding
soft nitriding
flame hardening
induction hardening
age hardening
precipitation hardening
overaging
solution treatment
interrupted quench
martensitic age-
 hardening

* * *

BIBLIOGRAPHY

Angus, H.T., *Cast Iron: Physical and Engineering Properties,* 2nd Ed., Butterworth and Co. (Publishers) Ltd, London (1976).

Bodsworth, C and Bell, H.B., *Physical Chemistry of Iron and Steel Manufacture*, Longman Group Ltd, London (1972).

Child, H.C., *Surface Hardening of Steel*, Engineering Design Guide No 37, Oxford University Press, Oxford (1980).

Davies, D.J. and Oelmann, L.A., *The Structure, Properties and Heat Treatment of Metals*, Pitman Books Ltd, London (1983).

Honeycombe, R.W.K., *Steels: Microstructure and Properties*, Edward Arnold (Publishers) Ltd, London (1981).

Kern, R.F. and Suess, M.E., *Steel Selection: A Guide for Improving Performance and Profits*, John Wiley and Sons Inc., New York (1979).

Krauss, G., *Principles of Heat Treatment of Steel*, American Society for Metals, Metals Park, Ohio (1980).

Lampman, J.R. and Peters, A.T., (Eds.), *Ferroalloys and Other Additives to Liquid Iron and Steel*, American Society for Testing and Materials, Philadelphia, Pennsylvania (1981).

Leslie, W.C., *The Physical Metallurgy of Steels*, McGraw-Hill Book Co., Inc., New York (1981).

Moore, C. and Marshall, R.I., *Modern Steelmaking Methods*, Institution of Metallurgists, London (1979).

Peacey, J.G. and Davenport, W.G., *The Iron Blast Furnace: Theory and Practice*, Pergamon Press Ltd, Oxford (1970).

Peckner, D. and Bernstein, I.M., *Handbook of Stainless Steels*, McGraw-Hill Book Co., Inc., New York (1977).

Pickering, F.B., *Physical Metallurgy and the Design of Steels*, Applied Science Publishers Ltd, Barking, Essex (1978).

Pickering, F.B., (Ed.), *The Metallurgical Evolution of Stainless Steels*, American Society for Metals, Ohio and the Metals Society, London (1979).

Porter, D.A. and Easterling, K.E., *Phase Transformations in Metals and Alloys*, Van Nostrand Reinhold Co., New York (1981).

Scholes, J.P., *The Selection and Use of Cast Irons*, Engineering Design Guide No. 31, Oxford University Press, Oxford (1979).

Sharp, J.D., *Electric Steelmaking*, Iliffe Books Ltd, London (1966).

Thelning, K.E., *Steel and its Heat Treatment – BOFORS Handbook*, Butterworth and Co. (Publishers) Ltd, London (1975).

Tremlett, H.F., *Welding Manual for Engineering Steel Forgings*, Pentech Press Ltd, London (1976).

Wilson, R., *Metallurgy and Heat Treatment of Tool Steels*, McGraw-Hill Book Co., Inc., New York (1975).

Low Carbon Structural Steels for the Eighties, Institution of Metallurgists, London (1977).

Production and Application of Clean Steels, International Conference at Balatonfüred, Hungary, June 1970; published by The Iron and Steel Institute, London (1972).

3
Nonferrous Metals

CHAPTER OBJECTIVES

To obtain an understanding of:
1 the general properties and classification systems of nonferrous metals and their alloys;
2 selected properties of these materials, and the relationship to their engineering applications;
3 the use and importance of some nontraditional materials for design purposes;
4 the factors influencing the selection of a material for a particular application.

IMPORTANT TERMS

copper   Monel   Nilo alloys
brass   aluminum   Corronel B
bronze   nonheat treatable   Hastelloys
cupro-nickel   age hardening   Langalloys
wrought alloy   magnesium   Inconel
cast alloy   nickel   Nimonic series

Nimocast series	strength: weight ratio	alloying addition
zinc	lead	'new' metals
galvanizing	soft solder	
titanium	tin	

3.1 INTRODUCTION

In texts that describe the properties and applications of engineering materials, it is common practice to include separate chapters for iron and its alloys, and for all other metals. This approach has been adopted in this book and Chapter 2 deals extensively with the properties, applications, standards, specifications, etc., of iron and steels. This chapter aims to provide similar relevant information for other common metals and their alloys, with which the engineer should be familiar. There are approximately 70 known metals, but the importance attached to iron compared with all other metals is not as unbalanced as it first appears. Iron is readily available at a reasonable cost and it is the basis of a large number of alloys that have a wide range of properties and applications. In contrast only a few nonferrous metals are produced in large quantities, some of which are used in the pure state or as alloy base metals.

The most important of these metals are copper and aluminum and they will be considered in detail. Several other metals, namely magnesium, nickel, zinc, titanium, tin and lead will be considered in less detail. A brief summary of the British and US standards relevant to these metals and their alloys is included at the end of each section. Many metals are used in small quantities, mainly as alloying additions that impart special properties to materials. Details of the properties and main uses of these metals are tabulated later in the chapter. There is also another group of rare and expensive metals which are of particular importance; their details are tabulated at the end of the chapter.

Properties of the metals that are produced in large quantities are given in Table 3.1. A brief summary of the properties and uses of some of the metals discussed in this chapter is given in Table 3.2 for reference. ASTM standards quoted in this chapter can be found in the *Annual Book of ASTM Standards*, Section 2: Non Ferrous Metal Products.

* * *

Self-assessment exercises

1 For your particular branch of engineering list the five most important nonferrous metals, and five important nonferrous alloys. (Define the criteria used to assess 'importance'.)

2 Explain why these particular metals and alloys are important. Obtain values of their physical, chemical and mechanical properties from handbooks and

decide which properties influence the applications and importance of each material.

3 Obtain data for the prices of these materials over the past 12 months, compare these data with figures for other engineering materials. Which factors influence the price of each material? Which factors influence the cost of obtaining each material?

4 Decide which materials (if any) could be used as substitutes if these important materials became unavailable 'overnight'. What effects would these substitutions have upon the uses and applications of the materials?

5 Which nonferrous metals and alloys do you think, will assume more importance in the future? What applications and uses could be developed? What problems may be encountered?

* * *

3.2 COPPER

Copper has many applications, both as a pure metal and as an alloy base metal. Copper is a malleable and ductile metal and can be easily rolled, drawn or forged. The tensile strength and hardness of copper can be improved by cold working although the ductility is reduced. Conversely, annealing can improve the ductility at the expense of the tensile strength and the hardness. The importance of copper is due mainly to its very high coefficient of electrical conductivity which is surpassed only by silver. The metal also possesses very good thermal conductivity and corrosion resistance. The impurities present in copper can have a serious effect upon the properties of the metal. The electrical conductivity is reduced by 25% due to the presence of 0.04% phosphorus, although the addition of 1% cadmium, to improve the strength when used in telephone wires, has minimal effect on the electrical conductivity. The oxygen content (pitch) in the form of copper oxide should be kept below 0.1% so that the mechanical and physical properties are not affected. The presence of oxygen (or copper oxide) makes the metal unsuitable for welding due to the formation of blow holes, although the pure metal is easily soldered, brazed or welded.

Copper is available in a variety of grades depending upon the extraction and refining processes that are used. Copper can be extracted from its ores by a smelting process using a reverberatory (blast) furnace. The crude copper produced is approximately 99% pure; it is highly oxidized, brittle and porous and is known as blister copper. Copper can also be extracted by an electrolytic process that produces cathode copper of a purity between 99.2% and 99.7% copper. Both types of crude copper can be refined either in a furnace or by an electrolytic process. The furnace method involves remelting to oxidize impurities, and subsequent slag removal. The electrolytic refining process uses the crude copper as the anode in an electrolytic cell. The metal dissolves and is subsequently redeposited at the cathode to produce a 99.97% pure product. In electrolytic refining, gold and silver are recovered as by-products. 'Tough pitch' copper contains small amounts of copper oxide impurity and is generally the

Table 3.1 Properties of some pure metals

Metal	Melting point °C (°F)	Specific gravity	Specific heat (0–100°C; 32–212°F) J/kg°C (Btu/lb°F)	Coefficient of linear expansion ($\times 10^{+6}$)	Poisson's ratio	Relative electrical conductivity	Relative thermal conductivity at 20°C (70°F)
Aluminum	660 (1220)	2.7	980 (0.234)	24.0	0.35	64	61
Copper	1083 (1982)	8.9	390 (0.093)	16.6	0.34	100	100
Iron (pure)	1535 (2795)	7.9	450 (0.108)	11.9	0.29	17	15
Lead	327 (621)	11.4	130 (0.031)	29.1	0.44	8	9
Magnesium	651 (1204)	1.7	1030 (0.246)	26.1	0.30	39	39
Nickel	1458 (2656)	8.9	450 (0.108)	12.8	0.28	25	17
Tin	232 (450)	7.3	220 (0.053)	21.4	0.33	15	16
Titanium	1660 (3020)	4.5	450 (0.108)	9.0	0.32	3	4
Tungsten (wire)	3410 (6170)	19.3	140 (0.033)	4.5	0.34	34	39
Zinc	420 (788)	7.1	380 (0.091)	33.0	0.25	29	27

Young's modulus E MN/m² × 10⁻³ (psi × 10⁻⁶)	Tensile strength MN/m² (psi × 10⁻³)	Modulus of rigidity MN/m² × 10⁻³ (psi × 10⁻⁶)	Crystal structure*	Properties and uses
70 (10.2)	60 (8.7)	26 (3.8)	fcc	The most widely used of the light metals. Common in the Earth's crust.
125 (18.1)	160 (23.2)	48 (7.0)	fcc	Used mainly in the electrical industries, because of its high conductivity. Also in bronzes, brasses and cupro-nickel.
206 (30.0)	270 (39.2)	82 (11.9)	α bcc < 908°C γ fcc 908–1388°C δ bcc > 1388°C	A fairly soft metal when pure, but hard and strong when alloyed to form steel.
16 (2.3)	15 (2.2)	5 (0.7)	fcc	A soft, heavy metal.
45 (6.5)	100 (14.5)	17 (2.5)	cph	Used with aluminum in the light alloys.
200 (29.0)	370 (53.7)	78 (11.3)	fcc	Used to toughen steels and many non-ferrous alloys; also for plating.
40 (5.8)	13 (1.9)	19 (2.8)	bc tetragonal	A rather expensive metal. 'Tin cans' carry only a very thin coating of tin on mild steel. Very resistant to corrosion.
114 (16.5)	460 (66.7)	43 (6.2)	α cph < 880°C β bcc > 880°C	Structural uses due to excellent strength–weight ratio, good corrosion resistance, and can be used up to 500°C.
400 (58.0)	4500 (652.5)	140 (20.3)	bcc	Its very high melting point makes it useful for electric lamp filaments. Also used in high-speed and heat-resisting steels.
90 (13.1)	155 (22.5)	35 (5.1)	cph	Used widely for galvanizing mild steel and as a basis for a group of die-casting alloys. Brasses are copper–zinc alloys.

Some values are approximate because many properties depend upon past treatment.
*fcc ~ face-centered cubic; bcc ~ body-centered cubic; cph ~ close-packed hexagonal.

Table 3.2 Metals commonly used in engineering

Metal	Chemical symbol	Melting point °C(°F)	Properties and uses
Aluminum	Al	660 (1220)	The most widely used of the light metals. Common in the Earth's crust.
Antimony	Sb	631 (1168)	A very brittle metal, used mainly in bearing alloys and type-metal.
Beryllium	Be	1285 (2345)	A light metal used to harden copper, and for nuclear-power equipment. Not widely used in aircraft and spacecraft due to its scarcity and, hence, the cost.
Cadmium	Cd	321 (610)	Sometimes used to plate other metals and to strengthen copper telephone wires.
Chromium	Cr	1890 (3434)	The high resistance to corrosion makes it useful for plating, and as an addition to stainless steel.
Cobalt	Co	1495 (2723)	Used in permanent magnets and in high-speed steels.
Copper	Cu	1083 (1982)	Used mainly in the electrical industries because of its high conductivity. Also in bronzes, brasses and cupro-nickel.
Iron	Fe	1535 (2795)	A fairly soft metal when pure, but hard and strong when alloyed to form steel.
Lead	Pb	327 (621)	A soft heavy metal, but not the densest.
Magnesium	Mg	651 (1204)	Uses similar to those of aluminum, a light metal.
Molybdenum	Mo	2620 (4748)	A heavy metal used in high-speed and other alloy steels.
Nickel	Ni	1458 (2656)	Used to toughen steels and many nonferrous alloys; also for plating.
Niobium	Nb	1950 (3542)	Small amounts used in some steels and aluminum alloys; also in nuclear-power equipment. Formerly known as 'columbium' in the USA.
Tin	Sn	232 (450)	A relatively expensive metal. 'Tin cans' contain only a very thin coating of tin on mild steel. Very resistant to corrosion.
Tungsten	W	3410 (6170)	Its very high melting-point makes it useful for electric lamp filaments. Also used in high-speed and heat-resisting steels.
Vanadium	V	1710 (3110)	Used as a hardener in some alloy steels; also in nuclear-power equipment.
Zinc	Zn	420 (788)	Used widely for galvanizing mild steel and as a basis for a group of die-casting alloys. Brasses are copper–zinc alloys.

product of fire refining. Oxygen-free, high-conductivity copper (OFHC) is obtained from electrolytic refining, it has the highest conductivity and can be remelted and cast in inert conditions to the required shapes.

Copper can be deoxidized by the addition of small amounts of lithium, phosphorus (with reduced conductivity) or up to 0.5% arsenic. Arsenic reduces the electrical and thermal properties, but improves the tensile strength and oxidation resistance for temperatures up to 300°C (572°F). Copper is available in the form of wire, bars, billets, rod, plate, sheet, strip and foil. Various grades of pure copper are used for electrical windings and wiring (in the form of wire), for cladding and castings (from sheet) and for heat exchangers and domestic installations (as tubing). Despite its abundance, ease of manufacture and excellent electrical conductivity, copper has been replaced in many applications by aluminum despite its inferior properties, due to the cheaper production costs.

With copper pipes there are no threaded joints, for low-temperature duty soldering is used. For high-temperature applications and for medium-sized pipes, then flanged connections are used. These flanges are brass or bronze (with similar thermal expansion) not mild steel or cast iron. Large pipes are welded. For the fabrication of copper vessels the following conditions apply:

(a) Gas or arc welding can be used but the welded metal must be deoxidized, generally using a filler rod containing either phosphorus or silicon and either tin or zinc. This is known as bronze welding and produces a strong corrosion-resistant joint.

(b) Brazing generally uses brass, and therefore a lower temperature is required (approximately 850°C/1562°F), although this can give rise to electrolytic corrosion.

(c) Silver soldering uses copper and silver, with zinc or phosphorus. This produces strong, corrosion-resistant joints.

(d) Soft solder is only used for low temperature, non-corrosive conditions.

(e) Riveting is rarely used.

Copper is rapidly attacked by sulfuric acid, hydrochloric acid, nitric acid, ammonia, sodium hydroxide, potassium hydroxide and amines. Attack occurs under certain conditions with chlorine, fluorine, oxidizing or hydrolyzable organic salts. Copper can be used safely with sulfurous acid (in the paper industry), neutral salts, e.g. sodium chloride, hydrocarbons, alcohols, acetic acid, aldehydes, ketones, ethers, lactic and tannic acid.

Copper is used extensively in chemical plant for evaporators, kettles, stills and heaters, mainly due to its corrosion resistance, ease of cleaning and high thermal conductivity.

* * *

Self-assessment exercises

1 Why is copper an important metal?

2 Identify the advantageous properties of pure copper.

3 List the main applications and uses of copper.

4 What are the main geographical sources of copper?

5 What is the price of copper, and how has this changed over the last 5 years?

6 What are the disadvantages of using pure copper?

* * *

3.2.1   Copper alloys

Many copper-base alloys are produced that have numerous engineering applications. These alloys often possess several of the following advantages: good thermal and electrical conductivities; good mechanical properties; corrosion and wear resistance; ease of fabrication, machining, joining, polishing and plating; lower pressing and forging temperatures than are required by ferrous alloys. However, the use of copper alloys is declining due to the increasing cost of

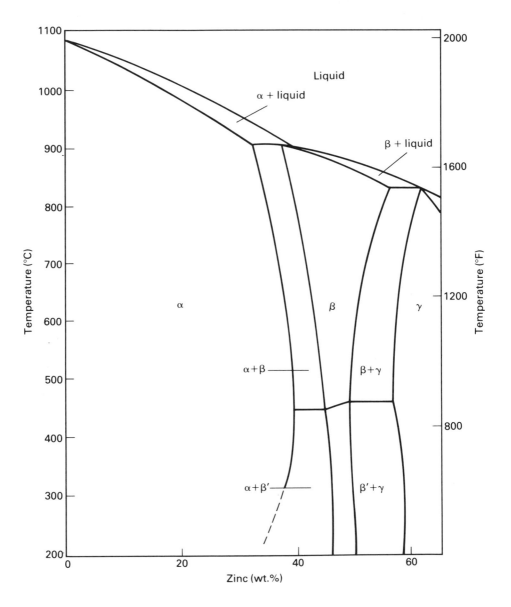

Figure 3.1   Copper–zinc phase diagram.

copper, the availability of cheaper satisfactory materials and improved fabrication techniques. The most important copper alloys are the brasses (copper and zinc), the bronzes (copper and tin or aluminum) and copper–nickel alloys. Copper is also alloyed with small quantities of other elements to produce alloys with particular properties. The phase diagrams for copper and zinc, copper and tin, and copper and aluminum are reproduced in Figs. 3.1, 3.2, and 3.3,

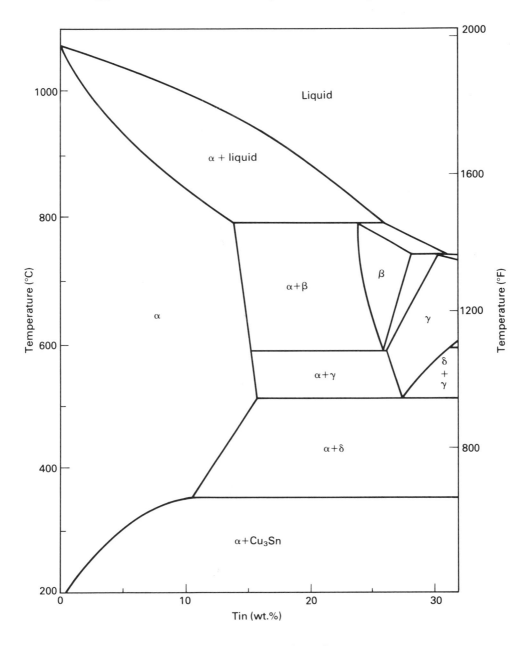

Figure 3.2 Copper–tin phase diagram.

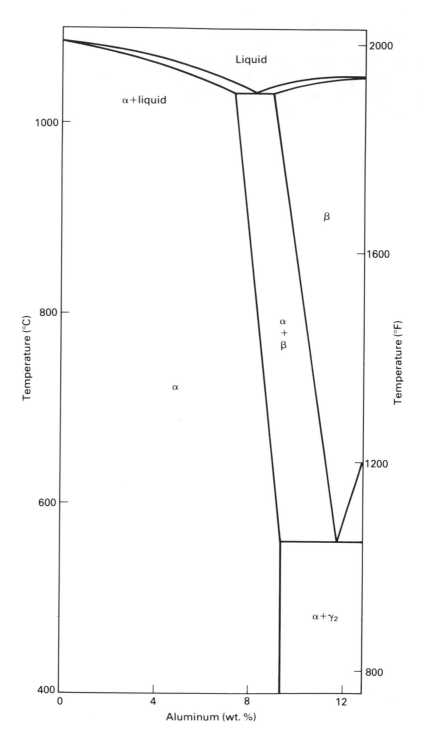

Figure 3.3 Copper–aluminum phase diagram.

respectively. These diagrams are complex and contain several intermediate phases. In each case the copper-rich α phase has a face-centered cubic (fcc) structure, and is ductile and suitable for cold working. The β phases are body-centered cubic (bcc) and are not easily cold worked. The γ and δ phases are very hard and brittle and possess a complex crystal structure.

Brass is a general name used for alloys of copper containing up to 45% zinc. The equilibrium phase diagram of Fig. 3.1 shows that up to 37% zinc can be dissolved in copper, forming a continuous solid solution known as the alpha (α) phase. This solid solution is ductile and increases the strength of the copper, as well as reducing the cost for certain applications. The ductility of copper–zinc alloys actually increases as the zinc content increases, reaching a maximum value when a 30% zinc content is achieved (known as cartridge brass). The ductility of the alloy then decreases as more zinc is dissolved, until a composition of 37% zinc is achieved (known as basic brass) with a ductility comparable to pure copper. The strength of the alloy increases with increasing zinc content due to the lattice distortion. The solid solubility of zinc in copper does not increase with temperature and α brasses are not therefore precipitation treated. Brasses containing less than 37% zinc are suitable for cold working due to their strength and ductility. They are widely used for deep drawing and spinning and are formed into sheet, strip, tube and wire. They are unsuitable for hot working. An α brass containing 35% zinc is often used for general applications; it is slightly α brass containing 35% zinc is often used for general applications; it is slightly cheaper than the 30% zinc brass and has reasonable ductility. Alpha brasses are subject to stress corrosion cracking between the crystals some time after cold working, this is known as *season cracking*. These alloys are therefore annealed before cold working, and are subjected to stress relief annealing at 250°C (482°F) after cold working to prevent this type of failure.

Alloys containing more than 37% zinc are known as the α + β¹ brasses due to the presence of both phases. The ductility of the α phase and the temperature, but with good strength. At temperatures above 454°C (851°F), the β¹ phase becomes a ductile β phase and dissolves the α phase (which is brittle at these higher temperatures). It is very malleable at temperatures between 600°C and 800°C (1112°F and 1472°F), and at these higher temperatures the coarse cast structure becomes transformed to a fine granular structure. Brasses containing more than 37% zinc are suitable only as hot working alloys (between 600°C and 800°C/1112°F and 1472°F) for hot rolling, extruding, stamping and pressing. Brasses containing more than 45% zinc contain entirely β¹ phase crystals; they are very brittle and structurally useless. Brasses containing more than 50% zinc also contain the extremely brittle γ phase. The brasses with high zinc content are used as brazing spelter because of their lower melting point (see Fig. 3.1). Also the joint is not made brittle by loss of volatile zinc during forming, or by diffusion of zinc into the parent metal. During casting, the vaporization of zinc (low boiling point, 907°C/1661°F) causes deoxidation and provides a protective atmosphere over the metal surface. Small amounts of other elements are added to the α + β¹ brasses to improve their properties. Up to 1% tin is added to increase corrosion resistance, particularly in marine environments. The addition of manganese, tin, aluminum, iron or nickel also improves the strength of the brass. The addition of up to 2% lead improves the free-cutting properties or machinability of the brass, due to the presence of tiny insoluble globules. Up to 2% aluminum can be added to improve the corrosion resistance of a brass.

Brasses are easily tinned, they may be joined by soft soldering using tin-based solders with an antimony content below 0.5% (reducing the risk of brittle joints). As the zinc content decreases, the risk of cracking during tinning or soldering is reduced. Silver soldering using alloys of copper, zinc and silver is carried out between 600°C and 800°C (1112°F and 1472°F). Due to the tendency for zinc vaporization, welding operations require an oxidizing oxy-acetylene flame and a borax–boric acid flux. A brass filler rod containing silicon is used, forming a protective oxide layer over the weld. High-temperature arc welding is not recommended, although inert gas-shielded arc welds have been performed. The main types of brasses and their important properties are listed in Table 3.3.

Bronze is the general name given to alloys of copper and tin, although it is also used to describe other copper alloys that are not tin-based. In this case, the term 'tin bronze' avoids confusion. Bronzes have become less popular materials recently because of the cost of the metals used; however, tin bronzes, gunmetals (tin bronze containing zinc), aluminum bronze, silicon bronze and phosphor bronze (tin bronze containing phosphorus) all have particular applications. Bronzes generally have superior mechanical properties and corrosion resistance to brasses. They are high-strength alloys and are often used as bearing materials in the cast form. Bronzes are usually classified as either wrought bronze, cold worked sheet, strip and wire, or as cast bronzes which are often used for bearing materials.

The phase diagram of Fig. 3.2 shows that copper can contain up to 14% tin in the form of the α phase solid solution. However, extremely slow cooling is required to obtain complete equilibrium and this would be industrially prohibitive. Also, a structural change occurs at 350°C (662°F) and alloys containing more than 5% tin contain some brittle δ phase, even when slowly cooled. Wrought alloys do not usually contain more than 7% tin, and therefore any hard and brittle δ phase present (with the ductile α phase) does not impair the required cold working properties. The δ phase present in an alloy containing 10% tin can be made to enter into solid solution by an annealing process, although this is a lengthy treatment. In the phase diagram of Fig. 3.2, the solidus and liquidus lines are far apart and cooling an alloy (containing less than 14% tin) results in the formation of a cored structure. The reason is that the last drops of liquid to solidify have a very high tin content and each grain contains a higher proportion of tin (the lower melting point element) at the surface. This effect can be removed by lengthy annealing. Cast alloys usually contain between 10% and 18% tin, have a lower melting point and combine the toughness of the α phase with the hardness of the δ phase. Bronzes required for bearing materials usually contain between 8% and 12% tin and are composed of hard particles in a soft matrix.

Phosphor bronzes contain residual phosphorus, above that required to deoxidize the alloy. Typically a residual 0.1% phosphorus in the α bronzes, and

Table 3.3 Properties of some typical brasses

Composition %			Type of product	Condition	Typical mechanical properties		Common name and uses
Cu	Zn	Other metals			Tensile strength MN/m² (psi × 10⁻³)	Elongation %	
97	3	—	sheet, strip & wire	A*	230 (33.4)	50	*Cap copper.* Copper deoxidized with zinc (3% remaining), very soft and ductile, good conductivity.
				H*	430 (52.4)	3	
90	10	—	sheet, strip & wire	A	280 (40.6)	55	*Gilding metal.* Gold color, used for imitation jewelry and decorative work. Good ductility, can be brazed and enameled, can be newly worked, less prone to season cracking.
				H	510 (74.0)	4	
70	30	—	sheet and strip	A	320 (46.4)	70	*Cartridge brass.* Maximum ductility, used for deep drawing. Good tensile strength. Used for cartridge and shell cases.
				H	700 (101.5)	5	
65	35	—	sheet, strip & extrusion	A	320 (46.4)	65	*Standard brass.* General-purpose cold-working alloy. Harder and less ductile, less copper, therefore cheaper. Used for press work and limited deep drawing.
				H	700 (101.5)	4	
63	37	—	sheet, strip & extrusion	A	340 (49.3)	55	*Basic brass.* General-purpose alloy, limited cold working. Some β¹ phase present, cheaper, suitable for pressure work.
				H	725 (105.1)	4	
60	40	—	plate, rod & extrusion	hot rolled	370 (53.7)	40	*Yellow or Muntz metal.* Limited cold working, used for castings and a brazing alloy for steels. Used for condenser and heat exchanger plates.
76	22	Al 2	—	—	350 (50.8)	45	*Aluminum brass.* Good corrosion resistance, used for marine condenser tubes.
58	39	Lead 3	rods and sections	hot rolled	450 (65.3)	30	*Free-cutting brass.* Lead improves machinability, impairs ductility and impact strength but not tensile strength. Used for high-speed machining.

Table 3.3 (continued)

| Composition % | | | Type of product | Condition | Typical mechanical properties | | Common name and uses |
Cu	Zn	Other metals			Tensile strength MN/m² (psi × 10⁻³)	Elongation %	
62	37	Tin 1	extruded	hot rolled	420 (60.9)	35	*Naval brass.* Tin improves corrosion resistance and slightly improves the strength. Structural uses, can be forged and cast; 0.5% to 2% lead can be added.
58	Rem*	Up to 7% of Al, Fe Sn, Pb Mn in total	medium strength	hot worked	470 (min.) (68.2)	20 (min.)	*High-tensile brass (manganese bronze).* Stampings, pressings and castings. Marine castings for propellers and rudders.
			high strength		540 (min.) (78.3)	15 (min.)	
70	29	Tin 1	sheet and strip	–	360 (52.2)	50	*Admiralty brass.* Tin improves corrosion resistance, used for condenser tubes cooled by fresh water.
58–60	Rem	Lead 1.5–2.5	sheet strip and extrusions	A	370 (53.7)	33	*Clock brass (engraving brass).* Lead improves machinability. Sometimes called leaded brass or turning brass.
				H	620 (89.9)	3	

*Rem Remainder
A   Annealed (soft)
H   Hard

up to 1.0% residual phosphorus in the (α + β) bronzes. Some of the residual phosphorus is present in the microstructure as copper phosphate (Cu_3P), which is an extremely hard compound used to improve the bearing properties. Phosphor bronzes have good cold-working properties and are resistant to corrosion and abrasion; they are also hard and tough. Phosphorus reduces the coefficient of friction, but also causes loss of ductility. Up to 5% lead is added to bronzes to improve the machinability of the alloy. Some special bearing bronzes contain up to 25% lead; they can be used at high speeds because the high thermal conductivity means that heat is quickly dissipated. Gunmetal is a bronze, usually containing 10% tin, and with the addition of 2% zinc. The zinc acts as a deoxidizer, improves the casting properties and makes the material cheaper. Lead is also added, for the same reasons as with tin bronzes. Bronze in sintered form and impregnated with a lubricant is used in plate, rod, bar or molded form.

Silicon bronzes are alloys of copper, silicon and zinc. Silicon is added to other copper-based alloys to increase the toughness, hardness and tensile strength of the alloy. A maximum of 3% silicon is present in wrought alloys and up to 5% silicon in cast alloys; this combines excellent corrosion resistance and ductility. The malleable wrought alloys are used mainly for cold-working processes, e.g. to produce hydraulic parts, nuts, bolts, rivets and heat exchanger tubes and plates. Silicon improves the fluidity of the casting bronzes and allows intricate shapes to be cast. These alloys quickly work harden but are suitable for hot working. They are rolled, forged and extruded to produce hydraulic pressure pipeline equipment, pressure vessels, filter screens and marine components.

Nickel bronzes are tin bronzes containing small amounts of nickel and zinc; they are more accurately termed 'nickel gunmetals'. Nickel improves the mechanical properties and the corrosion and wear resistance; it also improves the quality of castings, and the alloys retain their strength at elevated temperatures. Approximately 1.5% nickel is required to achieve these properties. Up to 5% nickel is added to a particular alloy containing 10% tin and 2% zinc; the excess nickel facilitates the heat treatment of the alloy to improve its strength and hardness. These alloys are used for valve and pump parts for boiler feed water.

Aluminum bronzes are similar to the tin bronzes; part of the copper–aluminum phase diagram is given in Fig. 3.3. The cold-working wrought alloys usually contain approximately 5% aluminum; they are malleable and ductile and are homogeneous (α) solid solutions. These alloys are used as sheet or tube for condensers and heat exchangers. The hot-working, cast aluminum bronzes contain approximately 10% aluminum at temperatures above 565°C (1049°F), and are composed of a mixture of the α and β phases. Upon slow cooling the β phase is transformed into the brittle γ_2 phase. The temperature at which the transformation occurs depends upon the aluminum content; it is a minimum (565°C/1049°F) at the eutectoid composition of 11.8% aluminum. Slow cooling produces a coarse lamellar structure of alternate layers of the ductile α and the brittle γ_2 phases. More rapid cooling produces a fine-grained structure which is tougher. This effect is similar to the effects on the steel portion of the iron–carbon phase diagram which was discussed in Chapter 2. Rapid quenching of the alloy by immediate removal of the casting from the mold, suppresses the β to γ_2 transformation and a martensitic-type structure is obtained. Tempering this alloy at 500°C (932°F) produces a fine-grained (α + γ_2) structure. The addition of 2% iron to the alloy also retards the β to γ_2 transformation. These alloys are difficult and expensive to cast because aluminum oxidizes readily at the casting temperature of 1000°C (1832°F), and can cause oxide slag to be trapped in the mold. Heat treatment is not often used for aluminum bronzes. They exhibit excellent corrosion resistance due to the surface oxide film; they retain their mechanical properties at elevated temperatures and have good wearing properties. The cast alloys are used for pump casings, valve parts and gears. A summary of the main types of bronzes and their characteristics is given in Table 3.4

Alloys of copper and nickel are known as the cupro-nickels and contain between 15% and 68% nickel. Nickel is soluble in solid copper in all proportions, forming strong, ductile fcc alloys. The strength, ductility and hardness all increase as the nickel content increases. These alloys are suitable for cold working; an alloy with 20% nickel is usually preferred. If impurities are minimal, then the alloys can also be hot worked.

Monel is an important alloy containing 68% nickel; it has the strength of steel and the corrosion resistance of copper. It is stronger than mild steel even when annealed, but the thermal conductivity is only about 15% of the value for copper, although it is still higher than for mild steel. Monel is tough and shock resistant, retains its strength at high temperatures and is difficult to cast due to blow holes. Various fabrication methods can be used; it can be drawn to fine wire and cold working improves the strength. It is expensive, although cheaper than pure nickel, and in most cases (except with alkalis) it is better than pure copper or nickel. Monel is used for valves and heat exchangers (maintains clean surfaces), sometimes in the food industry (but not for storage), for high-temperature applications except in the presence of sulfur dioxide or oxidizing atmospheres.

Monel is rapidly attacked in oxidizing conditions, e.g. nitric acid, chromic acid, moist chlorine, molten sulfur and hydrochloric acid greater than 5% concentration. Under certain conditions attack occurs with sulfuric acid (in the absence of air it is safe up to 80% concentration), with acetic acid (when aerated, or greater than 20% concentration), and with hot phosphoric acid. Monel is generally safe with hydrogen fluoride, sulfurous acid, liquid ammonia, dry chlorine, neutral salts (not cyanides; not iron, tin or mercury salts), fatty acids and organic solvents.

The cupro-nickels are fabricated by rolling, forging, pressing, drawing and spinning and are used for heat exchangers and condenser tubing. Alloys known as nickel silvers are produced by the addition of zinc to copper-nickel alloys; they do not contain silver. They are ductile, fcc homogeneous solid solutions and are suitable for both cold and hot working. The copper content is between 55% and 65%, nickel content between 10% and 30% and the balance is zinc. These alloys are used for the manufacture of tableware and are often stamped 'EPNS' – 'electroplated nickel–silver'. The addition of 2% lead improves the machinability, the suitability for engraving and use for key manufacture. Typical properties of copper–nickel alloys are given in Table 3.5.

Table 3.4 Properties of some typical bronzes

Composition %				Condition	Typical mechanical properties		Common name and uses
Cu	Sn	Al	other		Tensile strength MN/m² (psi × 10⁻³)	Elongation %	
95.5	3	—	Zn 1.5	Wrought, cold rolled: A*	320 (46.4)	65	*British coinage.*
				H*	725 (105.1)	5	
96.0	3.75	—	P 0.25	Wrought: A	340 (49.3)	65	*Low-tin bronze.* Good corrosion resistance and elastic properties. Used for springs and coils.
				H	740 (107.3)	15	
94.0	5.5	—	P 0.1	Wrought: A	350 (50.8)	65	*Drawn phosphor bronze.* Used as work hardened. Used for corrosive and frictional applications, e.g. steam turbine blades.
				H	700 (101.5)	15	
89.0	10.0	—	P 0.05–0.5	Sandcast	280 (40.6)	15	*Cast phosphor bronze.* Hard and tough, good bearing material. With 0.5% P, poor ductility but can withstand larger loads.
81.0	18.0	—	P 0.5	Sandcast	170 (24.7)	2	*High-tin bronze.* Bearings subjected to heavy loads. Bronze containing more than 20% tin is used for bells (*bell metal*), although it is brittle. Bronzes containing 30% to 40% tin are known as *speculum metal*. They are hard and brittle, but when polished are used for optical instruments and diffraction gratings. Also used as an electroplating.

Table 3.4 (continued)

Composition %				Condition	Typical mechanical properties		Common name and uses
Cu	Sn	Al	other		Tensile strength MN/m² (psi × 10⁻³)	Elongation %	
88.0	10.0	—	Zn 2	Sandcast	290 (42.1)	16	*Admiralty gunmetal.* Good corrosion resistance, used for marine applications, also pumps, valves and bearings.
85.0	5.0	—	Zn 5 Pb 5	Sandcast	220 (31.9)	13	*Leaded gunmetal* (or *red brass*). Improved machinability and used for pressure-tight castings.
75.0	5.0	—	Pb 20	Cast	155 (22.5)	6	*Leaded bronze.* A bearing alloy.
Rem*	—	9.5	Fe, Ni Mn up to 1% each	Cast	520 (75.4)	30	*Aluminum bronze.* The most commonly used composition; for pumps, valves and gears.
95.0	—	5	—	Strip and tubing: A	390 (56.6)	70	Good corrosion resistance, and oxidation resistant when heated; used for decorative work and condenser tubes.
				H	770 (111.7)	4	
80.0	—	10	Fe 5 Ni 5	Hot worked: forged	725 (105.1)	20	High-strength alloys, good corrosion resistance.
95.0	—	—	Si 3	Sheet and strip	400 (58.0)	50	*Silicon bronze.* Excellent corrosion resistance.
96.0	—	—	Be 1.8 Ni 0.3	Strip and foil	600 (87.0)	20	*Beryllium copper* (or *beryllium bronze*). Used for non-sparking tools and springs.

*Rem Remainder
A Annealed (soft)
H Hard

Table 3.5   Properties of some typical cupro–nickels and nickel silvers

Composition (%)			Condition	Typical mechanical properties		Common name and uses
Cu	Ni	other		Tensile strength MN/m² (psi × 10⁻³)	Elongation %	
93	5	Fe 1.2	A*	260 (27.7)	50	Improves corrosion resistance and the
			H*	460 (66.7)	5	mechanical properties of pure copper
75	25	Mn 0.25	Strip: A	350 (50.8)	45	British 'silver' coinage
			H	600 (87.0)	5	
70	30	Mn 0.4	Sheet and tube: A	350 (50.8)	45	High resistance to corrosion, used for condenser and heat exchanger tubes
			H	600 (87.0)	5	
29	68	Fe 1.25; Mn 1.25	All forms: A	560 (81.2)	45	Monel metal. Good mechanical properties and excellent corrosion resistance. Used in chemical plant.
			H	720 (104.4)	20	
60	18	Zn 22; Mn 0.44	Sheet and strip: A	375 (54.4)	35	Nickel silver. Decorative purposes and cutlery
			H	700 (101.5)	3	
60	12	Zn 26; Pb 2; Mn 0.25	A	340 (49.3)	45	Leaded nickel silver. Used for keys, etc.
			H	700 (101.5)	3	

*A Annealed (soft)
 H Hard

There are a number of important copper-based alloys which contain small quantities of alloying elements. Manganese is added to aluminum bronzes and cupro-nickels to facilitate metal working, to improve corrosion resistance and to impart specific electrical properties. Small amounts of silicon improve the strength and corrosion resistance of copper, and improve the fluidity of castings. Silicon is used to deoxidize the weld metal during fusion welding or to increase the resistance during resistance welding. The solubility of chromium in copper to form a homogeneous solid solution is small. The solubility increases with temperature and chromium copper usually contains approximately 0.5% chromium, which is fully dissolved at about 1000°C (1832°F). The alloy is quenched (solution treated) to retain the chromium in solid solution and to produce softness and ductility. After work hardening, the alloy is precipitation treated. Copper chromium alloys have good electrical conductivity (approximately 80% of that of pure copper) and good strength due to the work-hardening property.

Cadmium copper contains about 1% cadmium which improves the strength, toughness and fatigue resistance, as well as raising the softening temperature. This quantity of cadmium only slightly reduces the electrical conductivity of the copper, and it is a useful alloy for telephone and overhead wires. Arsenical copper contains about 0.4% arsenic, which raises the softening temperature of cold-worked copper from 200–550°C (392–1022°F). Although arsenic greatly reduces the electrical conductivity of copper, it is still used for high-temperature steam plant. Silver copper contains approximately 0.08% silver which also increases the softening temperature, and consequently the strength, hardness and creep resistance. The conductivity of copper is only slightly reduced by the addition of silver. Tellurium copper contains approximately 0.5% tellurium which is insoluble in copper, and does not therefore affect the conductivity. The presence of the small tellurium globules makes machining easier by breaking up the chips, having a similar effect of that of lead in steels.

Beryllium copper, or beryllium bronze, contains up to 2% beryllium and may contain small amounts of cobalt. Only a small amount of beryllium (less than 0.5% at room temperature) can dissolve in copper to form a homogeneous, soft and ductile α phase solid solution. A slowly cooled alloy contains particles of the hard brittle γ phase (CuBe) which becomes a more ductile β phase above 575°C (1157°F). Solution heat treatment of the alloy produces a more ductile material that can be cold worked, and it can be obtained with very high tensile strengths (up to 1400 MN/m^2; 100 ton/in.^{2}) if subsequently precipitation treated. These alloys are used for the manufacture of nonsparking tools, springs, pressure diaphragms and cells. Beryllium copper is an expensive alloy and this has limited its applications. Alloys of copper, nickel and silicon containing four parts of nickel to one part of silicon, and a total addition of up to 3% nickel and silicon, are particularly useful. These alloys are solution heat treated followed by precipitation treatment; they possess good electrical and thermal conductivity, good resistance to oxidation and scaling at high temperatures, and acceptable mechanical properties at high temperatures.

* * *

Self-assessment exercises

1 The most widely used copper alloys are brasses, bronzes and cupro-nickels. Identify two main uses of each of these alloys within your engineering field.

2 How can the properties of copper alloys be altered?

3 Which in-service conditions should be avoided when using copper alloys?

4 Explain the difference between copper alloy ingots and castings.

5 In what forms are copper and its alloys available, e.g. plate, rod, forgings?

* * *

3.2.2 British Standards and ASTM standards (volume 02.01: copper and copper alloys)

An ASTM standard with the full designation: A123–84 (for example) is issued under the fixed designation A123; the number (84) immediately following indicates the year of original adoption or of last revision. A designation A123M–84 (1986) indicates a metric standard and 1986 is the year of last re-approval. In this book only the number designations will be given for BS and ASTM standards; the most recent editions and ammendments should always be used. Reference should be made to the appropriate standards handbook.

ASTM copper classification systems
  ASTM B224: Standard classification of coppers
  The standard designations for different types of copper are presented in Table 3.6.

  ASTM E527: Practice for numbering metals and alloys (UNS)
  This practice covers a unified numbering system (UNS) for metals and alloys that have a 'commercial standing'. The system comprises 18 series of numbers for metals and alloys as shown in Table 3.7. A secondary division of some of these series is given in Table 3.8.

  ASTM B601: Practice for temper designations for copper and alloys – wrought and cast

BS6017: Specification for copper refinery shapes
This standard covers seventeen grades of copper in various product forms as shown in Table 3.9, which also includes designation cross references.

ASTM standard specifications for copper, refinery products
  ASTM B115: Electrolytic cathode copper
  ASTM B5: Electrolytic tough-pitch copper – refinery shapes
  ASTM B101: Lead-coated copper sheets
  ASTM B170: Oxygen-free electrolytic copper – refinery shapes
  ASTM B379: Phosphorized coppers – refinery shapes
  ASTM B442: Tough-pitch chemically refined copper – refinery shapes
  ASTM B216: Tough-pitch fire-refined copper for wrought products and alloys – refinery shapes
  ASTM B623: Tough-pitch fire-refined high-conductivity copper – refinery shapes

BS1400: Copper alloy ingots and copper and copper alloy castings
The alloys are divided into three groups:

Group A. Alloys in common use (preferred for all general purposes).
Group B. Special–purpose alloys (for applications requiring particular properties).
Group C. Alloys in limited production.

Table 3.6 Classification of coppers in ASTM B224

Designation	Type of copper	UNS numbers
CATH	Electrolytic cathode	
	Tough-pitch coppers	
ETP	Electrolytic tough-pitch	C11000
RHC	Remelted, high-conductivity tough pitch	C11010
ETP	Electrolytic tough-pitch (anneal resist)	C11100
CRTP	Chemically refined tough-pitch	C11030
FRHC	Fire-refined, high-conductivity tough-pitch	C11020
ETP[a]	Silver-bearing, tough-pitch	C11300, C11400, C11500, C11600
FRTP	Fire-refined, tough-pitch	C12500
FRSTP	Fire-refined tough-pitch with silver	C12700, C12800, C12900, C13000
	Oxygen-free coppers (without use of deoxidants)	
OFE	Oxygen-free, electronic	C10100
OF	Oxygen-free	C10200
OFS	Oxygen-free, silver-bearing	C10400, C10500, C10700
OFXLP	Oxygen-free, extra low phosphorus	C10300
OFLP	Oxygen-free, low-phosphorus	C10800
	Deoxidized coppers	
DLP	Phosphorized, low-residual phosphorus	C12000
DLPS[b]	Phosphorized, low-residual phosphorus silver-bearing	C12100
DHP[c]	Phosphorized, high-residual phosphorus	C12200
DHPS[b]	Phosphorized, high-residual phosphorus silver-bearing	C12300
DPA	Phosphorized, arsenic-bearing	C14200
DPTE[d]	Phosphorized, tellurium-bearing	C14500
	Other coppers	
	Sulfur-bearing	C14700
	Zirconium-bearing	C15000

a includes types ETP, CRTP and FRHC coppers to which silver has been added in agreed amounts;
b includes oxygen-free copper to which phosphorus and silver have been added in agreed amounts;
c includes oxygen-free copper to which phosphorus has been added;
d includes oxygen-free tellurium-bearing copper to which phosphorus has been added in agreed amounts.

The alloys are designated as follows:

Sand cast aluminum bronzes	AB1, AB2
Copper-manganese–aluminum alloys	CMA1, CMA2
High-tensile brasses	HTB1, HTB3 (β brass)
Chill cast	CMA1
Gunmetals	G1, G3, G3-TF, LG1, LG2, LG4
Phosphor bronzes	PB1, PB2, PB4, LPB1
Copper–tin alloy	CT1
Leaded bronzes	LB1, LB2, LB4, LB5
Chill cast aluminum bronzes	AB1, AB2
Brasses	SCB1, SCB3, SCB4, SCB6, DCB1, DCB3, PCB1

| Copper | HCC1 |
| Copper–chromium | CC1-TF |

The chemical compositions and mechanical properties of the various alloys are given in tables in the standard.

ASTM standard specifications for copper alloy ingot and castings
ASTM B644: Copper alloy addition agents

ASTM B30: Copper-base alloys in ingot form
Contains details of the nominal compositions, chemical requirements and designations of the following alloys; leaded red brass; leaded semi-red brass; leaded yellow brass; silicon bronze and silicon brass; tin bronze and leaded tin bronze; high-leaded tin bronze; nickel-tin bronze and leaded nickel-tin bronze; aluminum bronze; silicon aluminum bronze; manganese aluminum bronze; nickel aluminum bronze; cupro-nickel; leaded nickel bronze.

ASTM B148: Aluminum–bronze sand castings

ASTM B176: Copper alloy die castings
Covers six copper–zinc alloys; two standard and four special alloys.

ASTM B584: Copper alloy sand castings for general applications
Contains details for a selection of alloys in ASTM B30, also includes mechanical requirements.

Table 3.7   Primary series of numbers – ASTM E527

Nonferrous metals and alloys

A00001 – A99999	aluminum and aluminum alloys
C00001 – C99999	copper and copper alloys
E00001 – E99999	rare earth and rare earth-like metals and alloys (18 items)
L00001 – L99999	low melting metals and alloys (15 items; see Table 3.8)
M00001 – M99999	miscellaneous nonferrous metals and alloys (12 items; see Table 3.8)
N00001 – N99999	nickel and nickel alloys
P00001 – P99999	precious metals and alloys (8 items; see Table 3.8)
R00001 – R99999	reactive and refractory metals and alloys (14 items; see Table 3.8)
Z00001 – Z99999	zinc and zinc alloys

Ferrous metals and alloys

D00001 – D99999	specified mechanical properties – steels
F00001 – F99999	cast irons and cast steels
G00001 – G99999	AISI and SAE carbon and alloy steels
H00001 – H99999	AISI H-steels
J00001 – J99999	cast steels (except tool steels)
K00001 – K99999	miscellaneous steels and ferrous alloys
S00001 – S99999	heat and corrosion resistant (stainless) steels
T00001 – T99999	tool steels

Specialized metals and alloys

W00001 – W99999	welding filler metals, covered and tubular electrodes, classified by weld deposit composition (see Table 3.8).

Table 3.8 Secondary division of some series of numbers – ASTM E527

I *E00001–E99999 rare earth and rare earth-like metals and alloys, e.g. actinium and cerium*

II *F00001–F99999 cast irons*

III *K00001–K99999 miscellaneous steels and ferrous alloys*

IV *L00001–L99999 low-melting metals and alloys*

L00001–L00999	bismuth
L01001–L01999	cadmium
L02001–L02999	cesium
L03001–L03999	gallium
L04001–L04999	indium
L05001–L05999	lead
L06001–L06999	lithium
L07001–L07999	mercury
L08001–L08999	potassium
L09001–L09999	rubidium
L10001–L10999	selenium
L11001–L11999	sodium
L12001–L12999	thallium
L13001–L13999	tin

V *M00001–M99999 miscellaneous nonferrous metals and alloys*

M00001–M00999	antimony
M01001–M01999	arsenic
M02001–M02999	barium
M03001–M03999	calcium
M04001–M04999	germanium
M05001–M05999	plutonium
M06001–M06999	strontium
M07001–M07999	tellurium
M08001–M08999	uranium
M10001–M19999	magnesium
M20001–M29999	manganese
M30001–M39999	silicon

VI *P00001–P99999 precious metals and alloys*

P00001–P00999	gold
P01001–P01999	iridium
P02001–P02999	osmium
P03001–P03999	palladium
P04001–P04999	platinum
P05001–P05999	rhodium
P06001–P06999	ruthenium
P07001–P07999	silver

VII *R00001–R99999 reactive and refractory metals and alloys*

R01001–R01999	boron
R02001–R02999	hafnium
R03001–R03999	molybdenum
R04001–R04999	niobium (columbium)
R05001–R05999	tantalum
R06001–R06999	thorium
R07001–R07999	tungsten
R08001–R08999	vanadium
R10001–R19999	beryllium
R20001–R29999	chromium
R30001–R39999	cobalt
R40001–R49999	rhenium
R50001–R59999	titanium
R60001–R69999	zirconium

VIII *W0001–W99999 welding filler metals classified by weld deposit composition*

W00001–W09999	carbon steel with no significant alloying elements
W10000–W19999	manganese–molybdenum low-alloy steels
W20000–W29999	nickel low-alloy steels
W30000–W39999	austenitic stainless steels
W40000–W49999	ferritic stainless steels
W50000–W59999	chromium low-alloy steels
W60000–W69999	copper-base alloys
W70000–W79999	surfacing alloys
W80000–W89999	nickel-base alloys

IX *Z00001–Z99999 zinc and zinc alloys*

ASTM B271: Copper-base alloy centrifugal castings
ASTM B505: Copper-base alloy continuous castings
ASTM B369: Copper-nickel alloy castings
ASTM B427: Gear bronze alloy castings

BS2870: Specification for rolled copper and copper alloys: sheet, strip and foil
Sheet is flat and of exact length, over 0.15 mm (0.006 in.), up to and including 10 mm (0.4 in.) thick, and over 450 mm (18 in.) in width.

Strip is over 0.15 mm (0.006 in.), up to and including 10 mm (0.4 in.) thick, of any width and generally not cut to length. Often in coil.

Foil is 0.15 mm (0.006 in.) thick and under, of any width.

Table 3.9   Grades of copper, terminology and designation cross references

Terminology	BS6017 designation	BS2870/5 designation
Cathode copper (standard grade)	Cu–CATH–2	C101, C103, C110
Electrolytically-refined tough-pitch copper (standard grade)	Cu–ETP–2	C101
Fire-refined high-conductivity copper	Cu–FRHC	C102
Fire-refined tough-pitch copper	Cu–FRTP	C104
Phosphorus–deoxidized copper, high residual phosphorus	Cu–DHP	C106
Oxygen-free electrolytically refined copper	Cu–OF	C103
Oxygen-free refined copper, electronic grade	Cu–OFE	C110
Oxygen-free copper silver	Cu–Ag–OF–2 Cu–Ag–OF–4	C103
Tough-pitch copper silver	Cu-Ag–1 to Cu–Ag–5	C101

Copper and its alloys are designated by one or two letters, followed by three digits. The letters used are:

C   Copper
CZ   Copper–zinc
CN   Copper–nickel (and sometimes iron)
PB   Phosphor bronze (copper–tin–phosphorus)
CA   Copper–aluminum
NS   Nickel-silver (copper–nickel–zinc)
CC   Copper-chromium (and sometimes zirconium)
CB   Copper–beryllium

The condition of the material is specified by the following letters:

O	Annealed
¼H, ½H, H, EH	Various harder tempers produced by cold rolling (or sometimes partial annealing)
SH, ESH	Spring hard tempers from cold rolling thinner material
M	'As manufactured'
W	Material is solution heat treated and will respond to precipitation treatments
W(¼H), W(½H), W(H)	Material is solution heat treated and subsequently cold worked to various harder tempers
WP	Solution heat treated and precipitation treated
W(¼H)P, W(½H)P, W(H)P	Solution heat treated, cold worked then precipitation treated

The condition of the final rolling, whether hot or cold, for each alloy is also specified. The chemical compositions and mechanical properties of the various alloys are given in tables in the standard.

BS2875: Copper and copper alloys: plate
Plate should be over 10 mm (0.4 in.) thick and over 300 mm (12 in.) wide.

ASTM standard specifications for copper and copper alloy plate, sheet, strip, and rolled bar

> *ASTM B248/B248M: General requirements for wrought copper and copper–alloy plate, sheet, strip, and rolled bar*
> *ASTM B169: Aluminum bronze plate, sheet, strip, and rolled bar*
> *ASTM B 36: Brass plate, sheet, strip, and rolled bar*
> *ASTM B402/402M: Copper–nickel alloy plate and sheet for pressure vessels*
> *ASTM B152/B152M: Copper sheet, strip, plate and rolled bar*
> *ASTM B291: Copper–zinc–manganese alloy (manganese brass) sheet and strip*
> *ASTM B103: Phosphor bronze plate, sheet, strip, and rolled bar*

These standards are a selection of those contained in ASTM Volume 02.01. Many standards refer to particular material applications, e.g. clad steel plate, solar panels, building construction and heat exchanger tubing.

BS2871: Copper and copper alloys: tubes
Part 1: Copper tubes for water, gas and sanitation
Part 2: Tubes for general purpose
Part 3: Tubes for heat exchangers

ASTM standard specifications for copper and copper alloy pipe and tube (some examples)
> *ASTM B251/B251M: General requirements for wrought seamless copper and copper-alloy tube.*
> *ASTM B706: Seamless copper alloy (UNS No. C69100) pipe and tube*
> *ASTM B466/B466M: Seamless copper-nickel pipe and tube*
> *ASTM B543/B543M: Welded copper and copper-alloy heat exchanger tube*
> *ASTM B608: Welded copper-alloy pipe*

BS2872: Copper and copper alloys: forging stock and forgings

BS2874: Copper and copper alloys: rods and sections (other than forging stock)

ASTM standard specifications for copper and copper alloy rod, bar, and shapes, and die forgings (some examples)
> *ASTM B150: Aluminum bronze rod, bar, and shapes*
> *ASTM B151/B151M: Copper–nickel–zinc alloy (nickel silver) and copper–nickel rod and bar*
> *ASTM B140/B140M: Copper–zinc–lead (leaded red brass or hardware bronze) rod, bar, and shapes*
> *ASTM B283: Copper and copper alloy die forgings (hot pressed)*
> *ASTM B124: Copper and copper alloy forging rod, bar, and shapes*

ASTM B249/B249M: General requirements for wrought copper and copper-alloy rod, bar, and shapes
ASTM B139/B139M: Phosphor bronze rod, bar, and shapes

BS2873: Copper and copper alloys: wire

ASTM standard specifications for copper alloy and copper-clad wire (examples)
ASTM B134: Brass wire
ASTM B250/B250M: General requirements for wrought copper-alloy wire
ASTM B159/B159M: Phosphor bronze wire

BS4608: Copper for electrical purposes, rolled sheet, strip and foil

ASTM standard specifications for copper and copper alloys for electron devices and electronic applications and electrical conductors (examples)
ASTM B9: Bronze trolley wire
ASTM B187: Copper bus bar, rod, and shapes
ASTM F96: Electronic grade alloys of copper and nickel in wrought forms.

*   *   *

Self-assessment exercises

1   Outline the basis of the ASTM metal and alloy numbering system. Identify the designations used for important metals and alloys within your engineering field.

2   You wish to purchase a quantity of pure copper and a copper alloy for use in an industrial situation. Examples could be copper strip to be used as a tank lining or brass pipes for a hot water system. How would you specify the material required? What other information should you provide to the supplier?

*   *   *

3.3   ALUMINUM

Aluminum is obtained from the ore bauxite (Al_2O_3) by an electrolytic process. The pure metal is relatively expensive, requiring approximately 22–26 kWh/kg (34 100–40 350 Btu/lb) for extraction from its ore. Aluminum is usually available in a range of purities, containing 99–99.99% of the metal. The pure metal is soft and ductile and has a low tensile strength, approximately 60 MN/m^2 (8.7 × 10^3 psi) when annealed. This value can be doubled by cold working or by the addition of alloying elements. Aluminum-based alloys are generally used for applications requiring improved strength. Aluminum is an extremely useful engineering material because of its high thermal conductivity and low density. It also possesses very good electrical conductivity; although this is only 50% of that of copper, it has the advantage of being a significantly lighter material. Aluminum has a very high coefficient of linear expansion, approximately four times that of other common metals.

Aluminum reacts readily with oxygen forming a thin, dense oxide film on the metal surface that is impermeable to oxygen, and it therefore exhibits excellent corrosion resistance to the atmosphere and to dilute acids. The oxide film is destroyed by alkalis, causing corrosive attack of the base metal. The corrosion resistance can be increased by a process known as *anodizing*, in which the aluminum is made the anode in an electrolytic cell and the oxygen liberated increases the oxide thickness. Dyes can be absorbed onto the surface oxide layer after special anodizing. The metal is nonmagnetic and nonsparking and can be polished to reflect both heat and light. Aluminum and its oxide are nontoxic and are used extensively in the food-processing industries, and also for food containers and packaging.

Aluminum is malleable and ductile and can be easily fabricated, making it a useful material for domestic cooking utensils. The metal can be riveted, brazed and welded, although molten aluminum absorbs carbon dioxide and nitrogen and forms blow holes when cooled. The preferred joining method is argon arc-welding using a tungsten electrode in a stream of argon. Pure aluminum is difficult to machine and aluminum alloys are often used instead. Aluminum is widely used for heat exchangers, aircraft, chemical plant and the electrical industry. High purity aluminum (more than 99.5%) is used as a corrosion-resistant lining on other base metals.

The following is a summary of the effects of various chemicals on aluminum:

(a) Suitable for biochemical applications and for food processing. Only slight attack by biochemical substances, dissolved aluminum salts are nontoxic forming hydroxy acids. Organic solvents and hydrocarbons can be stored in aluminum tanks, although not extremely pure alcohols (less than 0.01% water). Fatty acids react with aluminum if water is present.

(b) Abrasives remove the protective oxide film and expose the metal to attack, so rubber boots should be worn inside aluminum tanks.

(c) Non-oxidizing acids dissolve the oxide film and attack the metal; examples are sulfuric acid, hydrochloric acid and phosphoric acid.

(d) Pure nitric acid and very dilute nitric acid have only very slight attack on aluminum. The rate of attack increases as the concentration of nitric acid increases from 15% to 50%, but then decreases as the concentration is further increased. For stainless steel, the rate of attack increases as the acid strength is increased, with very rapid attack by nearly pure acid.

(e) Aqueous alkalis dissolve aluminum, although this can be inhibited by sodium silicate. However, dry ammonia gas has no effect on aluminum even at high temperatures. Also, ammonia is not catalytically decomposed ('cracked') by aluminum and the metal can be used for ammonia synthesis plant.

(f) Aluminum should not be used with formic acid.

(g) Acetic acid between 80% and 99.9% pure has only a slight attack on aluminum, but with 100% pure acid the attack is violent.

(h) Rapid attack by phenols.

(i) Organic halogens, aldehydes, esters, benzene, and toluene can be stored in aluminum tanks, but aniline can only be stored cold, not hot.

(j) Hydrogen sulfide and sulfur dioxide have no action on aluminum.

(k) Acid chlorides attack when moist.

(l) Hydrogen peroxide has only slight attack on aluminum and catalytic decomposition does not occur, unlike copper and iron.

(m) Aluminum is attacked by halides, phosphates, cyanides, although this can be inhibited by 1% sodium silicate.

(n) Aluminum resists attack by sulfates, sulfides, molten nitrates and alkaline nitrates.

*   *   *

Self-assessment exercises

1 List the advantageous properties of aluminum.

2 Identify the main uses of aluminum.

3 List the disadvantages associated with using aluminum, and describe how these problems can be overcome.

*   *   *

3.3.1 Aluminum alloys

Aluminum is alloyed with a number of elements to produce a range of materials with particular properties. The main alloying elements are copper, silicon, zinc, magnesium, iron, manganese and nickel. The alloys are usually classified as either wrought or cast alloys. Wrought alloys are available either rolled to sheet, strip or plate, drawn to wire, or extruded as rods and tubes. Cast alloys are shaped by sand casting or gravity die casting and in a few cases by pressure die casting (using a cold chamber). The wrought and cast alloys are further classified into alloys that are not heat treated and those which are subsequently subjected to some form of strengthening heat treatment (not annealing).

Wrought alloys that do not respond to heat treatment are those which contain manganese, magnesium, or other small additions. These elements enter into solid solution with aluminum to a limited extent, having a strengthening effect and often improving the corrosion resistance. The final strength is determined by the amount of cold working, and the final dimensions depend upon the cold working after the final annealing (soft working) process. Alloys containing up to 10% magnesium are particularly useful in marine environments; they have good corrosion resistance that increases with the magnesium content. Commercial alloys containing approximately 1.25% manganese and a total of 2.5% alloying elements also possess improved properties. Other alloys are available, containing up to 1% of alloying additions, and their properties depend upon the elements used and the extent of cold working. Increased strength due to work hardening results in loss of ductility; precipitation hardening is not used because of the small strengthening effects achieved.

Heat-treatable wrought aluminum alloys contain up to 4% copper or up to 2% (total) of silicon and magnesium, or various amounts of copper, silicon and magnesium. The increase in strength is due to the formation of an intermetallic compound such as $CuAl_2$ or Mg_2Si. Other elements such as iron and zinc may be added to these alloys to increase their strength, and nickel is added for use at high temperatures. The solubility of copper in aluminum (as solid solution), increases from 0.2% at room temperature to a maximum of 5.7% at 550°C (1022°F). If a molten alloy containing 4% copper is gradually cooled, then the maximum solid solubility of copper in aluminum occurs at approximately 500°C (932°F). Cooling below this temperature results in a solid solution of (less) copper in aluminum, and $CuAl_2$ crystals at the grain boundaries. If the alloy is reheated (to produce only the solid solution) and then rapidly cooled, $CuAl_2$ is not formed and the resulting structure is a supersaturated (or metastable) solid solution. This process is known as *solution treatment*. It produces a soft ductile alloy, due to the absence of the brittle $CuAl_2$ compound. The copper gradually diffuses out of solid solution and $CuAl_2$ precipitates mainly in the grains. This movement causes strain within the crystal lattice, thus increasing the hardness and strength. This process is known as *age hardening*. If this process does not occur immediately, or does so only very slowly, then it can be increased by reheating the alloy to (typically) between 150°C and 200°C (302°F and 392°F). This tempering process is known as *precipitation hardening. Over-aging* occurs if too high a temperature is used or heating for too long, so that stress relieving occurs and the strength and hardness begin to decrease. Air circulating furnaces or salt baths are often used for these types of heat treatment. Age-hardening alloys of aluminum and copper are used in the aircraft industry for structural applications, even though copper reduces the corrosion resistance, and surface treatment or cladding may be necessary.

Commercially pure aluminum containing small amounts of silicon and magnesium can be cast, being both ductile and corrosion resistant. Other elements are added to improve the hardness, strength and rigidity of the casting. The addition of more than 5% silicon improves the fluidity of the alloy. The aluminum–silicon eutectic composition contains 11.6% silicon. The main alloys that are used 'as cast', and derive negligible benefit from heat treatment are those containing either 10–14% silicon, or 10% silicon and 1.5% copper, or 4.5% magnesium and 0.5% manganese. These alloys have medium strength and good corrosion resistance. The addition of 0.01% (of total weight) of metallic sodium to the molten alloy just before casting produces a finer grain structure and a stronger and less brittle product. This is known as *modification* and causes a shift in the eutectic composition.

Cast aluminum alloys that respond to heat treatment usually contain either 4% copper, or 2% nickel and 1.5% magnesium, or 3% nickel (forming $NiAl_3$). Other elements are often present which act as hardening agents. Aluminum and its alloys are widely used in chemical plant for distillation columns for organic solvents, acetic acid plant, the dairy industry and pharmaceuticals.

3.3.2 British Standards and ASTM standards (volume 02.02: die-cast metals; light metals and alloys)

BS1470–1475: Wrought aluminum and aluminum alloys for general engineering purposes

BS1470: Plate, sheet and strip
Plate is of rectangular section, over 6 mm (0.24 in.) thick. Sheet is of rectangular section, over 0.2 mm (0.008 in.) thick but under 6 mm (0.24 in.). Strip is a cold-rolled product in coil form, over 0.2 mm (0.008 in.) thick but not exceeding 3 mm (0.12 in.).
BS1471: Drawn tube
BS1472: Forging stock and forgings
BS1473: Rivet, bolt and screw stock
BS1474: Bars, extruded round tubes and sections
BS1475: Wire

The BS designations of wrought aluminum alloys were changed in June 1980 (see Amendments 3372–6), and a four-digit system is now used. The first digit indicates the alloy group, as follows:

Aluminum, 99.00% minimum and greater   1XXX
Aluminum alloy groups by major alloying elements:

Copper	2XXX
Manganese	3XXX
Silicon	4XXX
Magnesium	5XXX
Magnesium and silicon	6XXX
Zinc	7XXX
Other element	8XXX
Unused series	9XXX

In the 1XXX group, the last two digits indicate the hundredths of a percent of aluminum above 99.00%, e.g. 99.45% aluminum is designated 1X45. The second digit indicates modifications in impurity levels, a zero indicates unalloyed aluminum having natural impurity levels.

For the 2XXX–8XXX groups, the last two digits serve only to identify different aluminum alloys in the group. The second digit indicates alloy modifications, zero indicating the original alloy.

National variations by another country are identified by a letter following the four digits. 'A' represents the first variation registered (I, O, Q are omitted). Table 3.10 lists the new BS and old BS designations, with the nearest equivalent ISO designation for wrought aluminum alloys. The condition of the alloys covered by BS1470 to 1475 is indicated by the following letters:

M	As manufactured
O	Annealed
H1 to H8	Strain hardened
TB	Solution heat treated, and naturally aged, no coldworking
TD	Solution heat treated, cold worked and naturally aged
TE	Cooled from an elevated temperature shaping process, and precipitation treated
TF	Solution heat treated and precipitation treated
TH	Solution heat treated, cold worked and then precipitation treated.

Tables 3.11 (a) and (b) contain information for the chemical composition and mechanical properties of wrought aluminum and aluminum alloys in the form of plate, sheet and strip, or as bar, extruded round tube and sections.

Table 3.10 Designation cross references for wrought aluminum alloys

New BS alloy designation, June 1980 (international)	Previous BS alloy designation		ISO designation (nearest equivalent)
	BS1470	BS1471	
Unalloyed			
1080A	S1A	T1A	Al 99.8
1050A	S1B	T1B	Al 99.5
1200	S1C	T1C	Al 99.0
Non heat-treatable			
3103	NS3	NT3	Al Mn 1
5083	NS8	NT8	Al Mg 4.5 Mn
5154A	NS5	NT5	Al Mg 3.5
5251	NS4	NT4	Al Mg 2
Heat-treatable			
2014A	HS15	HT15	Al Cu 4 Si Mg
CLAD 2014A	HC15	—	Al Cu 4 Si Mg
6061	HS20	HT20	Al Mg 1 Si Cu
6063	HS9	HT9	Al Mg Si
6082	HS30	HT30	Al Si 1 Mg Mn

BS4300: Specification (supplementary series) for wrought aluminum and aluminum alloys for general engineering purposes.
Part 1:   Aluminum alloy longitudinally welded tube
Part 4:   6463 Solid extruded bars and sections suitable for bright trim reflector applications
Part 5:   2011 Free-cutting bar and wire-alloy
Part 6:   3105 Sheet and strip
Part 7:   5005 Sheet and strip
Part 8:   5454 Plate, sheet and strip
Part 10:   5454 Drawn tube
Part 11:   5454 Forging stock and forgings
Part 12:   5454 Bar, extruded round tube and sections
Part 13:   5554 Welding wire
Part 14:   7020 Plate, sheet and strip
Part 15:   7020 Bar, extruded round tube and sections

ASTM B275: Practice for codification of certain nonferrous metals and alloys, cast and wrought
This system was developed for the designation of light metals and alloys, both cast and wrought; it now includes certain heavier, base-metal diecasting alloys. The letters used to represent the alloying elements are given in Table 3.12, and also the expressions of chemical composition limits.

Table 3.11(a)   Chemical composition limits of wrought aluminum and aluminum alloys

Form of material	Plate, sheet and strip			Bars, extruded round tube and sections		
	Unalloyed	Non heat-treatable alloy	Heat-treatable alloy	Unalloyed	Non heat-treatable alloy	Heat-treatable alloy
Material designation	1050A	3103	CLAD 2014A	1050A	5083	6063
Silicon	0.25	0.50	0.50–0.90	0.25	0.40	0.20–0.60
Iron	0.40	0.70	0.50	0.40	0.40	0.35
Copper	0.05	0.10	3.9–5.0	0.05	0.10	0.10
Manganese	0.05	0.90–1.50	0.40–1.20	0.05	0.40–1.00	0.10
Magnesium	0.05	0.30	0.20–0.80	0.05	4.0–4.9	0.45–0.90
Chromium	–	0.10	0.10	–	0.05–0.25	0.10
Nickel	–	–	0.10	–	–	–
Zinc	0.07	0.20	0.25	0.07	0.25	0.10
Notes	–	0.10 (Zr + Ti)	0.20 (Zr + Ti)	–	–	–
Titanium	0.05	–	0.15	0.05	0.15	0.10
Other { Each	0.03	0.05	0.05	0.03	0.05	0.05
Total	–	0.15	0.15	–	0.15	0.15
Aluminum	99.50 (min)	Rem*	Rem	99.50 (min)	Rem	Rem

Chemical composition (%)

*Rem – Remainder
Composition in max.% unless shown as a minimum or range.

Table 3.11(b) Mechanical properties of wrought aluminum and aluminum alloys

Form of material		Material designation	Condition	Thickness (mm)		Tensile strength (MN/m²)	
				over	up to and including	min	max
	Unalloyed	1050A	0	0.2	6.0	55	95
			H4	0.2	12.5	100	135
			H8	0.2	3.0	135	—
	Non heat-treatable alloy	3103	0	0.2	6.0	90	130
			H2	0.2	6.0	120	145
			H4	0.2	12.5	140	175
			H6	0.2	6.0	160	195
			H8	0.2	3.0	175	—
Plate, sheet and strip	Heat-treatable alloy	CLAD 2014A	TB	0.2	12.5	375	—
			TB	12.5	25.0	385	—
			TF	0.2	3.0	400	—
			TF	3.0	12.5	425	—
			TF	12.5	25.0	440	—
Bars, extruded round tube and sections	Unalloyed	1050A	M	—	—	60	—
	Non heat-treatable alloy	5083	O	—	150	275	—
			M	—	150	280	—
	Heat-treatable alloy	6063	O	—	200	—	140
			M	—	200	100	—
			TB	—	150	130	—
			TB	150	200	120	—
			TE	—	25	150	—
			TF	—	150	185	—
			TF	150	200	150	—

An alloy designation consists of not more than two letters, representing the major alloying elements in order of decreasing percentages (or alphabetical if equal percentages). This is followed by the respective percentages (rounded off) and a serial letter.

For example, Alloy aluminum CS104A in specification B179. The base metal 'Aluminum' may be omitted where it is obvious. 'C' represents copper, the alloying element in the greatest amount, i.e. 10% (between 9 and 11%); 'S' represents silicon present between 3.6 and 4.4%, i.e. 4%. The final letter 'A' signifies that this is the first alloy qualified and assigned under this designation.

Reference should also be made to ASTM E527: Standard Practice for Numbering Metals and Alloys (UNS) which can be found in Volume 02.01 of the ASTM Standards. Brief details are given in Section 3.2.2 in this book.

Table 3.12 Designation of nonferrous metals and alloys in ASTM B275

Letters representing alloying elements:

A	Aluminum	M	Manganese
B	Bismuth	N	Nickel
C	Copper	P	Lead
D	Cadmium	Q	Silver
E	Rare Earths	R	Chromium
F	Iron	S	Silicon
G	Magnesium	T	Tin
H	Thorium	Y	Antimony
K	Zirconium	Z	Zinc
L	Lithium		

For aluminum and magnesium alloys (cast and wrought) standard chemical composition limits are expressed to the following places:

Less than 0.0001% (used only for magnesium alloys)	0.0000X
0.0001–0.001%	0.000X
0.001–0.01%	0.00X
0.01–0.1%:	
Unalloyed aluminum made by a refining process	0.0XX
Alloys and unalloyed aluminum or magnesium not made by a refining process	0.0X
0.1–0.5%	0.XX
Over 0.5%	0.X, X.X, XX.X

ASTM standard specifications for aluminum, aluminum alloys, and aluminum-covered steel

The following standards are selected examples for particular categories of products.

Bars, rods, wire and shapes:
  ASTM B211/B211M: Aluminum–alloy bar, rod, and wire
  ASTM B221/B221M: Aluminum–alloy extruded bars, rods, wire, shapes, and tubes

Forgings:
  ASTM B247/B247M: Aluminum-alloy die and hand forgings

Pipes and tubes:
  ASTM B210/B210M: Aluminum–alloy drawn seamless tubes
  ASTM B241/241M: Aluminum–alloy seamless pipe and seamless extruded tube
  ASTM B483/B483M: Aluminum and aluminum–alloy drawn tubes for general purpose applications
  ASTM B491/B491M: Aluminum and aluminum–alloy extruded round tubes for general purpose applications

Sheet, plate, and foil:
  ASTM B209/B209M: Aluminum and aluminum–alloy sheet and plate
    Standards are also published for the heat treatment of aluminum alloys

(B597), fasteners (F467, F468), welding fittings (B361), and products for electrical purposes (B236, B317, B373).

BS1490: Aluminum and Aluminum Alloy Ingots and Castings
The alloy designation is prefixed by the letters LM, followed by one or two digits. The condition of the casting is specified by the use of the following suffix letters:

M   As cast
TS   Stress relieved only
TE   Precipitation treated
TB   Solution treated
TB7 Solution treated and stabilized
TF   Solution treated and precipitation treated
TF7 Full heat treatment plus stabilization

The requirements for all alloys (ingots and castings) are included in one table, although mechanical properties are not relevant for ingots. Selected information is given here in Table 3.13. The alloys are classified either as general-purpose alloys or special-purpose alloys. Information is also given as to the suitability of the alloys for particular casting processes. A comparison of the casting characteristics and other properties of selected alloys is given in Table 3.14.

ASTM Standard specifications for aluminum and aluminum alloy ingots and castings
    ASTM B179: Aluminum alloys in ingot form for sand castings, permanent mold castings, and die castings
    ASTM B686: Aluminum alloy castings, high strength
    ASTM B618: Aluminum alloy investment castings
    ASTM B85: Aluminum alloy die castings
    ASTM B108: Permanent mold castings, aluminum alloy
    ASTM B26: Sand castings, aluminum alloy

* * *

Self-assessment exercises

1 Explain the difference between wrought and cast aluminum alloys.

2 List the forms in which aluminum alloys are available.

3 List the main aluminum alloys used in industry.

4 Explain why some aluminum alloys cannot be heat treated.

5 Describe (briefly) the heat treatment methods used with aluminum alloys. Emphasize the main features of the treatments, the differences between them and the effects which they produce.

6 Describe the designations that are used to identify aluminum and its alloys.

7 Determine the typical chemical composition of several aluminum alloys. Obtain data for the mechanical properties of these alloys. Determine the range of these properties for aluminum alloys to be used in 'general-purpose' engineering situations.

* * *

Table 3.13 Chemical composition of aluminum and aluminum alloy ingots and castings, and mechanical properties of aluminum alloy castings

BS designation	Chemical composition %											Condition	Tensile strength (minimum) MN/m²(psi × 10⁻³)	
	Copper	Magnesium	Silicon	Iron	Manganese	Nickel	Zinc	Lead	Tin	Titanium	Aluminum		Sand cast	Chill cast
General purpose alloys														
LM2	0.7/2.5	0.30	9.0/11.5	1.0	0.5	0.5	2.0	0.3	0.2	0.2	Rem*	M	–	150(22)
LM4	2.0/4.0	0.15	4.0/6.0	0.8	0.2/0.6	0.3	0.5	0.1	0.1	0.2	Rem	M	140(20)	160(23)
												TF	230(33)	280(40)
LM20	0.4	0.2	10.0/13.0	1.0	0.5	0.1	0.2	0.1	0.1	0.2	Rem	M	–	190(28)
Special purpose alloys														
LM5	0.1	3.0/6.0	0.3	0.6	0.3/0.7	0.1	0.1	0.05	0.05	0.2	Rem	M	140(20)	170(25)
LM9	0.1	0.2/0.6	10.0/13.0	0.6	0.3/0.7	0.1	0.1	0.1	0.10	0.2	Rem	M	–	190(28)
LM10	0.1	9.5/11.0	0.25	0.35	0.10	0.1	0.1	0.05	0.05	0.2	Rem	TE	170(25)	230(33)
												TF	240(35)	295(43)
												TB	280(40)	310(45)

* Rem Remainder

Ingots are not subject to mechanical test requirements.
All limits are maxima unless otherwise indicated.

Table 3.14 Comparison of casting characteristics and other properties of aluminum and aluminum alloy ingots and castings

| BS designation | Form of casting | | | | Fluidity | Resistance to hot tearing | Pressure tightness | Machinability | Resistance to corrosion | Strength at room temperature | Strength at elevated temperature | Shock resistance | Electrical conductivity | Decorative anodizing |
| | Sand | Permanent mold | | | | | | | | | | | | |
		Gravity	Low pressure	Die										
General purpose alloys: die casting (pressure) alloys														
LM2	–	–	–	4	4	4	4	3	3	4	3*	2	2	1
LM20	–	–	–	4	4	4	4	3	3	4	1*	2	3	1
General purpose alloys: permanent mold and sand casting alloys														
LM4	4	4	3	–	3	3	4	3	3	2	3	2	2	1
Special purpose alloys														
LM5	2	2	2	n	2	2	1	3	4	2	2	2	2	4
LM9	3	3	4	n	3	4	3	2	3	3	3	2	2	1
LM10	2	2	n	n	2	3	1	3	4	4	n	4	1	2

(4 denotes highest value or suitability, 1 denotes lowest, n indicates not normally recommended in this form or condition)
* The use of die castings is usually restricted to only moderately elevated temperatures.

3.4 MAGNESIUM

Magnesium and aluminum are both light metals and they have similar properties, e.g. melting point of the pure metal and an affinity for oxygen. Although both metals have an affinity for oxygen, the oxide film formed on magnesium is porous and offers less corrosion protection in moist atmospheres, unless suitably coated. Magnesium is the lightest engineering metal and would probably be more widely employed because of its low density, except that the metal is less ductile than aluminum and is more difficult to cold work. This is due to the hexagonal crystal structure that prohibits 'slip', although the metal can be easily hot worked. Castings can also be produced, providing care is taken to prevent spontaneous combustion.

*   *   *

Self-assessment exercises

1 Magnesium and aluminum have similar properties but aluminum is more widely used. Why?

2 List the main industrial uses of magnesium, and the reasons why magnesium is chosen for these applications.

3 Identify possible modifications or improvements of the properties of magnesium that would increase its range of possible applications.

*   *   *

3.4.1 Magnesium alloys

Due to its low tensile strength poor corrosion oxidation resistance, and difficulties in cold working, magnesium is not used in the pure state. It is alloyed with other elements to improve the strength and other properties when used for structural components. Due to the hexagonal crystal structure, only a few elements enter into solid solution with magnesium. Alloying elements include manganese and zinc in small quantities, and aluminum and silver in larger quantities. The solid solubility of aluminum or zinc in magnesium increases with increasing temperature, to a maximum of (approximately) 13% aluminum at 440°C (824°F) and 10% zinc at 350°C (662°F). This means that the alloys will respond to precipitation treatment for increased strength. 'Rare earth' metals are often added to magnesium for similar reasons, and silver is added to speed up the aging process. Manganese improves the corrosion resistance and zirconium acts as a grain refiner. The addition of thorium produces a creep-resistant alloy. Magnesium-based alloys are generally classified as wrought or cast alloys and details of the main types are given as follows:

WROUGHT ALLOYS
    (a) Magnesium and approximately 1.5% manganese.
    (b) Magnesium and aluminum, zinc and manganese.
    (c) Magnesium and zirconium, and up to 3% zinc.
These alloys are work hardening.

    (d) Magnesium and zirconium, and more than 5% zinc.
These alloys can be heat treated.

    (e) Magnesium and zirconium, zinc and thorium.
These are creep resistant alloys and are generally work hardening.

CAST ALLOYS

    (a) Magnesium with zirconium and approximately 2% zinc.
Some of these alloys can be heat treated.

    (b) Magnesium with zirconium and approximately 5% zinc.

    (c) Magnesium with aluminum, zinc and manganese.

    (d) Magnesium with 'rare earth' metals and silver alloys.
These alloys can be heat treated.

    (e) Magnesium with zirconium, thorium and (sometimes) zinc.
These alloys are creep resistant and can be heat treated.

    The strength of magnesium alloys can be improved by suitable heat treatment under controlled conditions. This treatment can be solution treatment followed by air cooling or quenching, then natural aging or precipitation treatment, or it may be merely precipitation treatment.

    Magnesium alloys can be cast by various methods and the wrought alloys can be rolled, forged or extruded. The alloys can be machined and cut by sharp high-speed tools, but care must be taken to remove the inflammable powder. These operations are normally performed dry, but any lubricant must not be water-based. Magnesium alloys can be bolted, riveted and welded. Alloy surfaces are often treated to prevent corrosion; care must be taken if the alloy is in contact with other metals to avoid the possibility of galvanic cell corrosion.

    Cast alloys are used for petrol tanks, engine casings and in the aircraft industry due to the low density. Wrought alloys are used for railings, ladders and brackets. Magnesium alloys are now used as a canning material in nuclear reactors because of the negligible neutron reaction.

3.4.2 British Standards and ASTM standards (volume 02.02: die-cast metals; light metals and alloys)

BS2970: Magnesium alloy ingots and castings
Alloys are classified according to a British Standard identification and a chemical symbol identification, for example:

BS Alloy Designation MAG 7.
Chemical Symbol Designation Mg-Al 8.5 Zn 1 Mn

    This is a magnesium-base alloy containing nominally 8.5% aluminum, 1% zinc and less than 1% manganese.

    Ingot designation has no suffix letters, e.g. MAG 8, whereas castings have suffix letters, e.g. MAG 8 TF. The condition of the casting is indicated by the suffix letters, these are:

M   as cast
TS   stress relieved only
TB   solution treated only
TF   solution and precipitation treated

The chemical composition of magnesium alloy ingots and the chemical composition and mechanical properties of magnesium alloy castings are given in Tables 3.15(a) and (b).

BS3370, 3372 and 3373: Wrought magnesium alloys for general engineering purposes
  BS 3370: Plate, sheet and strip
  BS 3372: Forgings and cast forging stock
  BS 3373: Bar, section and tubes including extruded forging stock

Table 3.15(a) Chemical composition limits of magnesium alloy ingots and castings

Form	Ingots			Castings		
	General purpose		Special purpose	General purpose		Special purpose
Designation:						
BS2970	MAG 1	MAG 3	MAG 2	MAG 1	MAG 3	MAG 2
Chemical	Mg–Al 8	Mg–Al 10	Mg–Al 8	Mg–Al 8	Mg–Al 10	Mg–Al 8
symbol	ZnMn	ZnMn	ZnMn	ZnMn	ZnMn	ZNMn
	Chemical composition limits (%)*					
Aluminum	7.5–8.5	9.0–10.5	7.5–8.5	7.5–9.0	9.0–10.5	7.5–9.0
Zinc	0.3–1.0	0.3–1.0	0.3–1.0	0.3–1.0	0.3–1.0	0.3–1.0
Manganese	0.2–0.4	0.2–0.4	0.2–0.7	0.15–0.4	0.15–0.4	0.15–0.7
Copper	0.15	0.15	0.005	0.15	0.15	0.005
Silicon	0.2	0.2	0.01	0.3	0.3	0.01
Iron	0.03	0.03	0.002	0.05	0.05	0.003
Nickel	0.01	0.01	0.001	0.01	0.01	0.001
Magnesium	Rem†	Rem	Rem	Rem	Rem	Rem
Cu + Si + Fe + Ni max. total	0.35	0.35	–	0.40	0.40	–

† Rem   Remainder
* All limits are maxima unless otherwise indicated.

Table 3.15(b) Mechanical properties of magnesium alloy castings

BS2970 designation	Condition	Tensile strength (minimum) MN/m² (psi × 10⁻³)	
		Sand cast	Chill cast
MAG 1	M*	140 (20)	185 (27)
	TB	200 (29)	230 (33)
MAG 3	M*	125 (18)	170 (25)
	TB	200 (29)	215 (31)
	TF	200 (29)	215 (31)
MAG 2	M*	140 (20)	185 (27)
	TB	200 (29)	230 (33)

* For information only

ASTM standard specifications for magnesium and magnesium alloys
Refer to Section 3.2.2 (ASTM E527) and Section 3.3.2 (ASTM B275) for details of the designation systems for magnesium and magnesium alloys.

Ingots and castings:
    ASTM B92/B92M: Magnesium ingot and stick for remelting
    ASTM B93: Magnesium alloys in ingot form for sand castings, permanent mold castings, and die castings
    ASTM B94: Magnesium–alloy die castings
    ASTM B80: Magnesium–alloy sand castings
    ASTM B199: Magnesium–alloy permanent mold castings
    ASTM B403: Magnesium–alloy investment castings

Sheet, forgings, bars, rods, and shapes:
    ASTM B90: Magnesium–alloy sheet and plate
    ASTM B91: Magnesium–alloy forgings
    ASTM B107: Magnesium–alloy extruded bars, shapes and tubes

* * *

Self-assessment exercises

1 List the common magnesium alloys used in industry.

2 Which components are produced from magnesium alloys for use in the aircraft industry?

3 Use British Standards and ASTM Standards (or any other US standards) to obtain relevant data concerning applications of a particular magnesium alloy.

* * *

3.5 NICKEL

Nickel is an important engineering material. It is similar to iron with a slightly lower melting point and is slightly stronger and harder. Nickel is the strongest and toughest of the pure nonferrous metals, but still has values less than many alloys. The thermal conductivity is good, but the electrical conductivity is poor compared with copper. Nickel loses strength when heated but possesses excellent corrosion resistance to many alkalis and acids, this makes it valuable for use in chemical plants and in the food industry. Nickel is often used as a cladding on mild steel (Niclad), representing 10% to 20% of the total thickness and providing both strength and corrosion resistance. Nickel plating by elec-trodeposition is common on a range of materials and it is used as an intermediate layer in chromium-plated mild steels. Nickel can be easily joined and hot or cold worked; it can also be machined and softened by annealing. Nickel is used for both anodes and cathodes.

    Like copper, nickel is not suitable for use in oxidizing conditions. It is attacked by sulfuric acid, nitric acid, chromic acid, concentrated hydrochloric acid, acetic acid (stronger than 20%), oxidizing salts, sulfur compounds and alkaline hypochlorite. Nickel is resistant to concentrated nitric acid, hydrochloric acid (up to 15% in the absence of air), cold pure phosphoric acid, ammonia and dry chlorinated hydrocarbons. Nickel is the only common material resistant to attack by concentrated sodium hydroxide, which is an important consideration in its selection. Nickel is used in ammonia oxidation plant, dry-cleaning plant, caustic soda evaporators, and in the food, dairy, cosmetic and pharmaceutical industries (nickel salts are generally non-toxic).

* * *

Self-assessment exercises

1 Prepare a table to compare the physical and mechanical properties of nickel, iron, aluminum and copper.

2 Why is nickel often used in chemical process equipment?

3 What are the natural sources of nickel and where are they located?
How is nickel extracted from its ores?
What is the current price of pure nickel?
What is the current price of the common nickel alloys?

* * *

3.5.1 Nickel alloys

Nickel alloys can be conveniently discussed either in terms of their applications or according to the alloying elements used. The applications include corrosion resistance, high-temperature and low-temperature applications and low expansion coefficients. The most important elements that are alloyed with nickel are iron, molybdenum, copper and chromium, although most nickel is used for the production of alloy steels (see Chapter 2). Nickel alloys will be considered here according to the alloying elements used. (Trade names are those used by Messrs. Henry Wiggin.)

Nickel–iron alloys are used in applications that require small to intermediate coefficients of expansion. The coefficient of expansion is a minimum for the alloy containing 36% nickel and 64% iron, which is known as Nilo 36. The coefficient of expansion increases as the nickel content is increased. These alloys are used for thermostats, glass-to-metal seals and for precision equipment which must operate at different temperatures. Details of these alloys and Nilo K are given in Table 3.16.

Nickel–molybdenum alloys are used mainly at room temperature for extremely good corrosion resistance. They are generally nickel (sometimes with small amounts of iron) alloyed with molybdenum and other elements. Alloys such as Corronel B have high strength and hardness, properties produced by cold working, or annealing and heat treatment. Details of some typical alloys are given in Table 3.17. A group of alloys known as the Hastelloys were developed

Table 3.16 Iron–nickel low-expansion alloys

Tradename	Composition (%)			Coefficient of expansion $\times 10^{-6}$ (at 20°C/68°F)	Uses
	Ni	Fe	Co		
*Nilo 36 (Invar, Nivar)	36	64	—	0.9	Pendulum rods, standard lengths, measuring tapes, delicate precision sliding mechanisms, thermostats for low-temperature operation
Nilo 40	40	60	—	6.0	Thermostats for electric and gas cookers, heater elements
Nilo 42	42	58	—	6.2	Thermostats, also the core of copper-clad wire, for glass seals in electric lamps, radio valves and TV tubes
Nilo 50	50	50	—	9.7	Thermostats, also for sealing with soft glasses used in radio and electronic equipment
Nilo K	29	54	17	5.7	Glass/metal seals in medium-hard glasses used in X-ray tubes and various electronic equipment

* The trade names for these alloys are used by Messrs. Henry Wiggin (UK).

for resistance to hydrochloric acid and other nonoxidizing acids. The addition of tungsten and chromium also makes these alloys resistant to nitric acid. The Langalloys were developed for similar reasons, and the properties of some of these alloys will now be discussed in more detail. Table 3.18 contains the chemical compositions of certain Langalloys.

Langalloy 4R
This alloy possesses excellent resistance to corrosion by most mineral and organic acids. It has outstanding resistance to hydrochloric acid under all conditions, and is one of the few alloys which can be generally recommended for handling this acid.

Langalloy 4R can be used with sulfuric acid up to 60% concentration at its boiling point; with stronger acid 4R is suitable for use at temperatures up to 150°C (302°F). With phosphoric acid, Langalloy 4R is particularly recommended for service at temperatures close to the boiling point, for all concentrations of the acid. Corrosion is less with the pure acid, but even in commercial acid the rate of attack at the boiling point would not usually exceed 1 mm (0.04 in.) per year.

Langalloy 4R is also resistant to neutral and alkaline salts, and to caustic solutions. It is particularly resistant to acid chlorides provided they are not oxidizing in nature. This alloy is not recommended for use under strongly oxidizing conditions, and it not therefore suitable for handling nitric acid.

Langalloy 5R
Langalloy 5R possesses remarkable resistance to organic acids; it is outstanding in its resistance to strongly oxidizing salts, hypochlorite liquors, and wet chlorine and bromine. Langalloy 5R is also suitable for handling sulfuric, hydrochloric and phosphoric acids under many conditions.

With cold hydrochloric acid up to about 5% concentration, Langalloy 5R shows better corrosion resistance than the 4R alloy. Above this concentration it

Table 3.17   Corrosion-resistant nickel-based alloys

Tradename	Composition (%)				Uses
	Ni	Mo	Fe	others	
Corronel B	66	28	6	—	Resists attack by mineral acids and acid chloride solutions. Produced as tubes and other wrought sections for use in the chemical and petroleum industries, also for constructing reaction vessels, pumps, filter parts and valves
Ni-O-Nel	40	3	35	Cr 20 Cu 2	A 'Wiggin' alloy with characteristics similar to those of austenitic stainless steels. More resistant to general attack, particularly in chloride solutions
Hastelloy 'A'	58	20	22	—	Transporting and storing hydrochloric acid and phosphoric acid, and other nonoxidizing acids
Hastelloy 'D'	85	—	—	Si 10 Cu 3 Al 1	A casting alloy that is strong, tough, and hard but difficult to machine (finished by grinding). Resists corrosion by hot concentrated sulphuric acid

Table 3.18   Composition of the Langalloys

Element	4R	5R	6R	7R
Nickel	63.00	56.00	85.00	56.00
Copper	—	—	3.00	6.00
Iron	5.00	5.00	—	5.00
Manganese	0.75	0.75	—	0.75
Silicon	0.85	0.75	10.00	0.75
Chromium	—	15.00	—	23.00
Molybdenum	30.00	17.00	—	6.00
Tungsten	—	5.00	—	2.00

Chemical composition (wt. %)

is not normally employed, unless some other conditions such as the presence of chlorine or ferric salts demands its use. In such cases great caution should be exercised in selecting a material, it is advisable to discuss the problem with the metal suppliers. Langalloy 5R is also highly resistant to both pure and commercial grades of phosphoric acid at normal temperatures, under many conditions no attack can be detected.

In handling sulfuric acid, Langalloy 5R is particularly resistant to the more dilute solutions at temperatures close to the boiling point. 5R is well suited to many processes involving these conditions, together with the presence of organic acids. It is not recommended that Langalloy 5R be used with sulfuric acid at temperatures above $70°C$ ($158°F$), if the concentration is higher than about 50%.

Langalloy 5R is resistant to nitric acid in concentrations up to about 25% at room temperature, and to dilute acid at temperatures up to about $60°C$ ($140°F$). It is resistant to many oxidizing acid mixtures, such as nitric and sulfuric, chromic and sulfuric, and is used with many such mixtures where the 18/8 stainless steels are unsuitable.

Langalloy 6R

Langalloy 6R is resistant to many acids but its main field of application is in handling sulfuric acid. It will handle this acid in all concentrations, and at all temperatures up to the boiling point. It has particular advantages for handling boiling solutions containing between 70% and 90% sulfuric acid. 6R possess good resistance to corrosion by phosphoric and acetic acids, many salts and alkaline solutions, and certain moist gases such as hydrogen sulfide.

It should be noted that Langalloy 6R is a very hard alloy that is not easily machined; although this renders the alloy resistant to abrasion and erosion, it makes it unsuitable for the manufacture of certain equipment. At present, valves are not usually made from the 6R alloy but many valves are produced from Langalloy 4R with the valve and seat made of Langalloy 6R when required.

Langalloy 7R

The main field of application of this alloy is with sulfuric acid, and with mixtures containing this acid and nitric or phosphoric acid. For mixtures of sulfuric acid and nonoxidizing sulfates, 7R is highly resistant even in the presence of hydrogen sulfide.

Langalloy 7R is practically free from attack by caustic solutions, organic acids, most organic compounds and salt solutions.

Although for some applications this alloy may have similar properties to those of other materials in the Langalloy 'R' series, it should be noted that 7R is slightly cheaper than the 4R and 5R alloys and is rather easier to machine.

Nickel–copper alloys form a solid solution in all proportions, producing a wrought alloy with good corrosion resistance. Uses are mainly in chemical plant, and for steam turbines. Monel is an alloy containing approximately 68% nickel, 30% copper and up to 2% iron or manganese. If it is required for casting, up to 4% silicon can be added to improve the strength. The addition of between 2 and 4% aluminum allows precipitation and solution treatment to be employed. The alloy known as K Monel has exceptional strength and hardness. These alloys can be easily fabricated, machined and joined, and they resist attack by acids, alkalis and gases. Steel can be clad with Monel for a combination of strength and corrosion resistance. More details are given in Section 3.2.1.

Nickel–chromium alloys generally possess a high electrical resistance and melting point, and are resistant to high-temperature oxidation. These alloys are suitable for use as resistance wires and heater elements at temperatures up to bright red heat. Typical properties of the Brightray series are given in Table 3.19.

Inconel contains approximately 80% nickel, 14% chromium and 6% iron, and is used at high temperatures. It resists attack by organic and inorganic compounds, and by oxidizing atmospheres due to low grain growth caused by the nickel and the protective surface film of chromium oxide. It can be hot or cold worked and it can be joined or cast. Inconel is often used for high-temperature applications because there is very little creep or scaling. It is a good corrosion–resistant alloy, but not as good as nickel or Monel for contact with seawater. It was developed initially for fat hydrolysis plant, and is stronger at

Table 3.19   High-temperature resistance alloys

Composition %			Maximum working temperature °C (°F)	Uses
Ni	Cr	Fe		
80	20	—	1150 (2102)	Heaters for electric furnaces, cookers, kettles, immersion heaters, hair dryers, toasters
65	15	20	950 (1742)	Similar to above, but for goods of lower quality; also for soldering irons, tubular heaters, towel rails, laundry irons, and where operating temperatures are lower
34	4	62	700 (1292)	Cheaper-quality heaters working at low temperatures, but mainly as a resistance wire for motor starter-resistances, etc.

higher temperatures than nickel or Monel. Inconel is attacked by hydrochloric acid, sulfuric acid (except very dilute or oleum), dilute nitric acid, aqueous solutions of sulfur dioxide, moist chlorine or bromine, heavy metal chlorides (e.g. iron, copper, mercury) and hypochlorites. Inconel resists attack by nitrous acid, hydrogen sulfide, alkalis (but not as well as pure nickel), dry chlorine and bromine, salts, hydrogen peroxide, most organic compounds and fatty acids (there is slight corrosion with acetic and formic acids). Inconel is used in the food industry for stills, in the photographic industry, for furnace parts and thermocouple coverings, heat exchangers for fatty acids and phenols, reaction vessels for plastics, dying, tanning and evaporators for sodium sulfite.

The Nimonic series of alloys based upon Inconel were produced for use in gas turbines. They are basically nickel–chromium alloys (approximately 80:20) that are strengthened by adding small amounts of titanium, aluminum, cobalt and molybdenum. The phases formed by these elements also increase the limiting creep stress at high temperatures. The Nimonic series consists of wrought alloys, which are designated by a two-digit number. The Nimonic 75 alloy becomes work hardened, and the Nimonic 80A responds to heat treatment. This involves solution treating, slow cooling, lengthy precipitation treatment and slow cooling. The Nimonic alloys 90, 105, 110, 115, all contain cobalt and all respond to heat treatment. Details of the Nimonic alloys are given in Table 3.20. The Nimocast series of alloys are the casting equivalent of the Nimonic series.

3.5.2 British Standards and ASTM standards (volume 02.04: nonferrous metals)

BS 375: Specification for refined nickel
Table 3.21 (here) specifies the composition of refined nickel for various grades.

BS3071: Nickel–copper alloy castings
These alloys contain copper (28–32%) and small specified additions of Mn, Si, C, Mg, Fe, S and P. The remainder is nickel and cobalt. Three alloys are covered, designated as:

    NA1 (containing 1% Si)
    NA2 (2.75% Si)
    NA3 (4% Si)

Table 3.20 'Nimonic' high-temperature alloys

Nimonic alloy	Approximate composition %						Tensile strength MN/m² (psi × 10⁻³)			Uses
	Cr	Ti	Al	Co	Mo	Ni	$600°C$ $(1112°F)$	$800°C$ $(1472°F)$	$1000°C$ $(1832°F)$	
75	20	0.5	—	—	—	Remainder	590 (86)	250 (36)	90 (13)	Gas turbine flame tubes and furnace parts.
80A	20	2.2	1	—	—		1080 (157)	540 (78)	75 (11)	Gas-turbine stator blades, after-burners, and other stressed parts working at high temperatures.
90	20	2.2	1.3	20	—		1080 (157)	850 (123)	170 (25)	Rotor blades in gas turbines.
115	15	4	5	15	4		1080 (157)	1010 (146)	430 (62)	Excellent creep-resistant properties at high temperatures.

Table 3.21   Composition of refined nickel (BS375)

Residual elements (max. %)*	Grade (and % nickel + cobalt (min))				
	R.99.95A (99.95)	R.99.95B (99.95)	R.99.9 (99.9)	R.99.8 (99.8)	R.99.5 (99.5)
Ag	0.0001	0.0001	—	—	—
Al	0.0005	0.0005	—	—	—
As	0.0003	0.0001	0.005	0.005	0.005
Bi	0.00002	0.00002	0.0002	—	—
C	0.005	0.015	0.015	0.08	0.1
Cd	0.0001	0.0001	—	—	—
Co	0.005	0.0005	0.1	0.15	1.5
Cu	0.002	0.001	0.01	0.02	0.15
Fe	0.001	0.02	0.02	0.02	0.02
P	0.0005	0.0001	0.002	0.005	0.005
Pb	0.0003	0.0001	0.001	0.005	0.005
S	0.0003	0.0015	0.0015	0.04	0.04
Sb	0.0001	0.0001	0.0005	—	—
Se	0.0001	0.0001	0.001	—	—
Si	0.001	0.0005	0.002	—	—
Sn	0.0001	0.0001	0.001	—	—
Te	0.00005	0.00005	0.0001	—	—
Ti	0.00005	0.00002	0.0001	—	—
Zn	0.001	0.0001	0.0015	—	—

* Values for tantalum and manganese are not specified, but are not normally expected to exceed 0.0001% and 0.001%, respectively.

BS 3072 to 3076: Specification for nickel and nickel alloys
   BS 3072: Sheet and plate
   BS 3073: Strip
   BS 3074: Seamless tube
   BS 3075: Wire
   BS 3076: Bar

ASTM standard specifications for nickel and nickel alloys

   ASTM B39: Nickel
This standard specifies the chemical requirements for refined nickel (Ni 99.80% min) as given in Table 3.22. The principal commercial forms available are cathodes, briquettes, and pellets.

ASTM A494: Nickel and nickel alloy castings
This specification covers nickel, nickel–copper, nickel-copper-silicon, nickel–molybdenum, nickel-chromium, and nickel–molybdenum–chromium alloy castings for corrosion-resistant service. Details of the chemical composition and tensile requirements are given for 11 alloys.

ASTM B564: Nickel alloy forgings
This specification covers forgings of nickel–copper, nickel–chromium–iron,

Table 3.22 Chemical requirements for refined nickel (ASTM B39)

Element	Composition (wt. %)
Nickel	99.80 min.
Cobalt	0.15 max.
Copper	0.02 max.
Carbon	0.03 max.
Iron	0.02 max.
Sulfur	0.01 max.
Phosphorus, Manganese, Silicon, Arsenic, Lead, Antimony, Tin, Bismuth, Zinc	All less than 0.005

nickel–chromium–molybdenum–niobium and nickel–iron–chromium alloys (one of each), including details of the chemical requirements (and variations) and mechanical property requirements.

Standards are also available for nickel and nickel-alloy fittings (B366, B462). The following are selected examples of standards for nickel and nickel-alloy pipe and tube; plate, sheet and strip; and rod, bar and wire.

ASTM B167: Nickel–chromium–iron alloys (UNS N06600 and N06690) seamless pipe and tube
ASTM B161: Nickel seamless pipe and tube
ASTM B619: Welded nickel alloy pipe
ASTM B168: Nickel–chromium–iron alloys (UNS N06600 and N06990) plate, sheet and strip
ASTM B162: Nickel plate, sheet and strip
ASTM B670: Precipitation–hardening nickel alloy (UNS N07718) plate, sheet, and strip for high temperature service
ASTM B335: Nickel–molybdenum alloy rod

*   *   *

Self-assessment exercises

1 Nickel alloys are classified in this text according to their alloying elements. Prepare an alternative classification based upon the typical applications of the more common nickel alloys.

2 Prepare a suitability table (similar to Table 3.14) of typical applications for different nickel alloys. Use '4' to denote highest value or suitability, and '1' to denote the least suitable material or application.

3 Discuss the properties of nickel–copper, nickel–iron and nickel–molybdenum alloys with reference to the crystal structures of these elements, and of the solid solutions that are formed. Compare such characteristics as the atomic packing factor (see section 1.2.6) and the atomic sizes of these elements.

4 Discuss how and why the properties of a nickel–copper and a nickel–iron alloy change as the composition of each alloy varies.

5 Inconel contains approximately 80% nickel, 14% chromium and 6% iron. Deduce a possible crystal structure for the solid solution of this alloy based upon the atomic parameters.

6 Estimate the maximum solubilities of alloying elements such as titanium, aluminum and cobalt in a typical Nimonic alloy, based upon crystal structures and atomic sizes.

7 The tensile strength of Nimonic alloys is significantly lowered by operation in the temperature range 600–1000°C (1112–1832°F), as shown in Table 3.20. Explain why this occurs and suggest possible ways of reducing this effect.

Would it be possible/desirable to introduce carbon into the Nimonic alloys as an alloying addition, in order to form stable carbides which would maintain the alloy strength at higher temperatures?

*   *   *

3.6 ZINC

Pure zinc is relatively soft and weak, and it is brittle at room temperature, which is typical of hexagonal structures. It is therefore difficult to cold work, but can be easily rolled into sheets at temperatures between 100°C and 150°C (212°F and 302°F). In this form it is used for battery cases and roofing, and provides corrosion resistance due to the formation of a dense protective surface layer. Zinc is also used for sacrificial anodes to protect ships' hulls and buried pipes (see Chapter 5). A major use of zinc is providing a coating, known as *galvanizing*, for corrosion protection on ferrous materials. Galvanizing can be achieved by various methods, such as hot dip in a bath of molten zinc or spraying molten zinc using air pressure. *Electrogalvanizing* is used when the structure to be coated is made the cathode, and zinc is the anode in an electrolytic cell. *Sheradizing* produces a layer of uniform thickness by heating the component to 370°C (698°F) in the presence of zinc dust.

3.6.1 Zinc-based, die-casting alloys

The low melting point of zinc–based alloys makes them an excellent material for die-casting purposes, and means they can be cast in relatively inexpensive dies. A small amount of aluminum enters into solid solution with zinc and also lowers the melting temperature. The eutectic composition contains 5% aluminum and has a melting point of approximately 375°C (705°F) (pure zinc melts at 419°C/ 788°F). Alloys used for die-casting usually contain 4% aluminum, 1% or 2% copper, and the balance is zinc, producing a reasonably strong and rigid material. Very high-purity zinc (99.99%) is used for these alloys, as very small quantities of impurities such as cadmium, tin, or lead can lead to intercrystalline corrosion causing swelling and brittleness.

Zinc alloys are die-cast in a hot chamber by a very rapid pressure process.

However, Zn-Al and Zn-Al-Cu alloys are subject to aging (causing shrinkage) and if aging occurs the strength increases. This is also caused by rapid solidification in the casting process. The aging is normally complete after five weeks and the properties are then measured; they are known as the 'original properties'. Stabilizing annealing at 100–150°C (212–302°F) for three hours can be used to speed up the aging process if close tolerances are required before machining. The temperature of the metal and the die during casting influences the properties of the casting which is produced. Castings can be machined and slightly worked, but welding is not recommended because of the aluminum content. If welding is necessary, then the filler rod must be of the same composition as the casting, and a slightly reducing oxyacetylene flame is used. Zinc-based alloy castings are used for lightly stressed car parts, scaled toys, domestic appliances and electrical equipment.

3.6.2 British Standards and ASTM standards (volume 02.04: nonferrous metals)

> *BS 1004: Zinc alloys for die casting and zinc alloy die castings*
> *BS 5338: Code of practice for zinc alloy pressure die casting for engineering*
> *BS 3436: Ingot zinc*

ASTM standard specifications for zinc and zinc alloys
> *ASTM B418: Cast and wrought galvanic zinc anodes for use in saline electrolytes*
> *ASTM B69: Rolled zinc*
> *ASTM B669: Zinc alloys in ingot form for foundry castings*
> *ASTM B6: Zinc (slab zinc)*

* * *

Self-assessment exercises

1 Explain how zinc and magnesium are used to protect components by acting as 'sacrificial anodes'.

2 What is galvanizing and why is it used?

3 Explain why galvanizing is preferable to chromium plating for protection from atmospheric attack. (Also see Sections 5.4.1 and 5.7.3)

4 Why are zinc alloys used for die casting? How is this process carried out and what are the associated problems? How can these problems be overcome?

* * *

3.7 TITANIUM

Titanium is often classified as a 'new metal' but it has been included here in its own right because of its increasing usage. Titanium is the fourth most abundant metal to be found in the Earth's crust and is fifty times more plentiful than copper. It would probably be employed to a very much greater extent were it not for the high cost of extraction. This cost is mainly due to the great affinity of titanium for oxygen.

Titanium has a low density, just over half of that of steel, and it has an excellent strength:weight ratio, placing it between aluminum and steel. Its corrosion resistance is equal to that of 18–8 stainless steel, with the added advantage of chloride corrosion resistance. However, titanium is difficult to cold work due to its hexagonal structure, although it can be hot worked in the β body-centered cubic temperature range (above 880°C/1616°F). It is also difficult to cast, and these factors increase the cost of the metal. Titanium loses strength as the temperature is increased and the upper working limit is approximately 500°C (932°F), although this is still higher than is possible with many other light alloys. Oxygen and nitrogen can enter into solid solution with titanium; as the quantities increase the tensile strength also increases, although the metal becomes less ductile.

3.7.1 Titanium alloys

Titanium is alloyed with various amounts (less than 6% of each element) of other metals to produce particular properties. Copper–manganese or aluminum–manganese are added to improve the tensile strength. Tin can be added to make the alloy easier to shape and weld. Vanadium or zirconium–molybdenum–silicon can be added to produce an alloy that has high tensile strength and is also creep resistant. Alloys with tensile strengths up to 1400 MN/m^2 (203×10^3 psi) can be produced, and these strengths are maintained at higher temperatures than is possible with aluminum alloys. The alloys are generally subjected to solution treatment followed by precipitation treatment, and final cooling in air under controlled conditions to produce the desired properties. Titanium-based alloys can be forged, pressed and machined by normal methods, and they can also be welded. Care may be needed to prevent or remove the oxide surface layer formed by heating in air.

Titanium and its alloys have been mainly employed in jet engines, aircraft, spacecraft and nuclear rockets, and also in chemical reactors and heated vats. These applications reflect the low weight, high strength, corrosion resistance and high temperature advantages of these materials.

3.7.2 British Standards and ASTM standards (volume 02.04: nonferrous metals)

British Standards for titanium and its alloys are contained in the BS Aerospace Series (TA), e.g. *BS 2TA 100 (Aerospace Series): Procedure for inspection and testing of wrought titanium and titanium alloys.*
ASTM B338: Seamless and welded titanium and titanium alloy tubes for condensers and heat exchangers

ASTM B348: Titanium and titanium alloy bars and billets
ASTM B367: Titanium and titanium alloy castings
ASTM B381: Titanium and titanium alloy forgings
ASTM B265: Titanium and titanium alloy strip, sheet, and plate

*   *   *

Self-assessment exercises

1 Determine the annual production of titanium and its price.

2 How is titanium extracted from its ores?

3 Estimate the relative costs of using either titanium, a stainless steel or a low-alloy mild steel for the tubes of a simple shell and tube heat exchanger. Consider problems of fabrication, corrosion, etc., and how they may be overcome.
What would be the effects of using sea water to cool the outside of the tubes (shell side)?

4 What are the advantages of using titanium alloys compared with the pure metal?

5 The British Standards relating to titanium are mainly for the aircraft industry. Explain why this situation has occurred and describe specific applications of titanium in aircraft design.

*   *   *

3.8 LEAD AND LEAD-BASED ALLOYS

Lead is soft and malleable with good corrosion resistance. Lead was used for water pipework but has now been replaced by other materials. Lead is now mainly used as an alloy base for solders, printer's type and bearing materials, and as a radiation shielding material, electric cable sheathing and in storage batteries.

In the UK, commercial lead is available either as Type A (99.99% pure), Type B (containing approximately 0.06% copper) or as Type C (antimonial lead). Copper is added to improve the resistance to attack by sulfuric acid. Tellurium is sometimes added (approximately 0.05%) to Type B lead to improve the creep resistance and the mechanical properties, although the corrosion resistance is lowered. Type A is easier to work and has a longer life in nitric acid. Type B has superior mechanical properties and is used for general-purpose work.

The strength and hardness of lead is increased by alloying with antimony (Type C), which is a hard, brittle, crystalline metal. Alloys containing more than 8% antimony can be machined and screw-threaded; details of lead–antimony alloys are given in Table 3.23.

Tin improves the toughness, lowers the melting point, and increases the fluidity of lead–antimony alloys, thus making them suitable for use as printing type. Typical compositions are 12–30% antimony, 5–15% tin, and the balance

Table 3.23   Lead–antimony alloys

Composition %		Uses
Antimony	Lead	
1		Cable sheathing
up to 4		Collapsible tubes (artists' colors)
7	Remainder	Sheet used in chemical plant
7.5		Accumulator plates
11		Chemical plant pipe fittings

Table 3.24   Type metal

Composition %			Uses
Antimony	Tin	Lead	
13–30	3–10		Type metal
12	5	Remainder	'Linotype' metal
12–23	5–17		'Stereotype' metal

is lead, as shown in Table 3.24. Antimony expands upon solidification; an alloy containing 20–30% antimony produces a clear type face due to negligible contraction.

Lead–tin alloys are used as soft solders because of the low melting points. The eutectic composition (approximately 38% lead and 62% tin) solidifies at 183°C (361°F) and is known as 'tinman's solder'. It is used for electrical joining and tinplate sealing. 'Plumber's solder' contains 66% lead and solidifies over a range of temperatures. Alloys of lead, tin and bismuth are used as fusible alloys as shown in Table 3.25.

Bearing materials should be hard and wear-resistant with a low coefficient of friction, combined with the toughness and ductility to withstand mechanical shock. This is usually achieved by using an alloy composed of a hard intermetallic compound distributed in a ductile solid solution matrix. White bearing metals are either tin-based or lead-based. The tin-based alloys, known as 'Babbitt' metals, are of better quality than the cheaper lead-based alloys. Both types contain approximately 10% antimony. Very hard cuboids of tin and antimony are formed as an intermetallic compound in a solid solution. About 3% copper is added to the alloys and solidifies first, as an intermetallic compound of copper and tin. Copper prevents the cuboids from segregating at the metal surface and forming a mixed network. Details of the white-bearing metals are given in Table 3.26. Lead is also added to steels to enhance the free-cutting properties.

Lead can be used with sulfuric acid at acid concentrations between 0.1% and 60%, and temperatures up to 100°C (212°F). For acid concentrations between 60% and 90% the corrosion rate increases, and proceeds rapidly at higher concentrations. With high concentrations, the lead sulfate coating becomes flaky and is removed. Lead can be used with hydrochloric acid between 1% and 5% strength (conditions should be unagitated and unaerated); however, it would be preferable to use another metal. Nitric acid rapidly attacks lead, having a maximum corrosion rate with 29% acid. However, nitric acid with strengths greater than 80% can be safely employed. Nitrous acid has a more

Table 3.25 Fusible alloys

Type of alloy	Composition %				Melting point °C (°F)
	Bismuth	Lead	Tin	Other	
'Cerromatrix'	48	28.5	14.5	Antimony 9	102 – 225 (216 – 437)
Roses's alloy	50	28	22	—	100 (212)
Wood's alloy	50	24	14	Cadmium 12	71 (160)
Dental alloy	53.5	17.5	19	Mercury 10.5	60 (140)

Table 3.26 White bearing metals

Type	BS 3332 alloy specification and ASTM B23 alloy number	Composition %					Characteristics
		Sb	Sn	Cu	As	Pb	
Tin-based	3332/1	7	90	3	—	—	These are generally
Babbitt	3332/3	10	81	5	—	4	heavy-duty bearing
metals	3332/6	10	60	3	—	27	metals.
	B23/1	4.5	91	4.5	—		
	B23/2	7.5	89	3.5	—		
	B23/3	8	84	8.0	—		
Lead-based	3332/7	13	12	0.75	0.2	Rem*	These alloys are generally
bearing	3332/8	15	5	0.5	0.3	Rem	lower-strength,
alloys	—	10	—	—	0.15	Rem	lower-duty materials.
	B23/7	15	10	—	0.45	Rem	
	B23/8	15	5	—	0.45	Rem	
	B23/15	16	1	—	1.0	Rem	

*Rem–Remainder

rapid attack than nitric acid. Lead vessels and pumps can be used with aqueous solutions of sulfur dioxide (up to 200°C/392°F) for concentrations less than 0.1%, chromic acid, neutral salts e.g. sodium chloride and alum, most organic amines and organic acids. Problems may occur when using lead with moist halogens, ammonia and soft water. Lead should not be used with strong alkalis that form plumbates, sodium hydroxide (concentrations above 25% or aerated solutions), or with concentrated phosphoric acid (although a trace of impurity, e.g. sulfuric acid, greatly reduces the corrosion).

3.8.1 British Standards and ASTM standards (volume 02.04: nonferrous metals)

BS334: Specification for compositional limits of chemical lead
Details chemical lead of Types A, B and C. The metal should be free from laminations and oxide inclusions.

Type A lead contains a minimum of 99.99% Pb. Maximum impurities may be:

Copper	0.003%
Antimony	0.002%

Bismuth	0.005%
Iron	0.003%
Nickel	0.001%
Silver	0.002%
Zinc	0.002%
Tin	0.001%
Cadmium Arsenic, Sulfur }	Traces
Total	0.01% (max)

Type B alloy lead (B1 copper lead, B2 copper tellurium lead and B3 copper silver lead) contains specific alloying additions with antimony not exceeding 0.002%, and total maximum impurity of 0.01% (excluding antimony).

Type C antimonal lead contains antimony in the range 2.5% to 11%, and other specified elements.

The Appendices contain guidelines on the selection of chemical leads, typical mechanical properties at 20°C and 60°C (68°F and 140°F) and some preferred alloys.

BS3332: White metal bearing alloy ingots
Contains specifications of chemical compositions for nine white metal bearing alloys in ingot form, and general clauses applicable to all alloys (see Table 3.26).

ASTM B29: Pig lead

ASTM B23: White metal bearing alloys (known commercially as babbitt metal)
(See Table 3.26)

3.9   TIN AND TIN-BASED ALLOYS

Tin is soft and weak, corrosion resistant and has a low melting point (232°C/ 450°F). It is used as a coating on sheet steel known as *tinplate*, this is used in the food industry because it is not attacked by fruit juices. Tin is used in bearing metals (see Section 3.8) and in low melting point alloys such as solders. It is added to copper alloys such as bronzes, gunmetals and brasses, mainly to improve the corrosion resistance. The high cost of thin tin foil has resulted in its replacement by aluminum.

3.9.1 British Standards and ASTM standards

BS2920: Cold-reduced tinplate and cold-reduced backplate
ASTM B32: Solder metal

Table 3.27 Properties and uses of metals used in small quantities

Metal	Melting point °C (°F)	Relative density	Tensile strength MN/m² (psi × 10⁻³)	Crystal structure	Uses and alloys
Antimony (Sb)	631 (1168)	6.6	59 (9)	rhombic	Used in bearings and low-melting point alloys as a lead hardener. A brittle metal.
Bismuth (Bi)	271 (520)	9.8	—	rhombic	Hard and brittle. Difficult to produce pure. Used in low-melting point alloys.
Boron (B)	2300 (4172)	2.3	—	ortho	Difficult to produce. Used as a powder in sintered metal alloys, e.g. boron carbide is hard and chemically inert.
Cadmium (Cd)	321 (610)	8.6	80 (12)	cph	Soft and weak metal. Protective for steel and brass. Added to copper to improve properties and used in low-melting point alloys.
Cerium (Ce)	640 (1184)	—	—	—	Used as an electrode in electric arc lamps, and as an inoculant in cast iron.
Chromium (Cr)	1890 (3434)	7.1	220 (32)	bcc	Added to stainless steels to impart hardness and corrosion resistance. Used as a plating on stainless steels. Alloyed with nickel to produce heat-resisting alloys.
Cobalt (Co)	1495 (2723)	8.9	250 (36)	cph	Used in permanent magnets, in nickel-based heat-resisting alloys and in ultra high-speed steels and other cutting-tool materials.
Iridium (Ir)	2454 (4449)	22.5	—	fcc	Used in platinum alloys to increase the hardness and strength, and raises the melting point. Also used for crucibles.
Manganese (Mn)	1260 (2300)	7.2	500 (73)	cubic-complex	Used as a deoxidant in steel production. Useful impurity in steels to counteract the effects of sulfur. Used as an addition to many alloys, e.g. Al and Mg alloys.
Molybdenum (Mo)	2620 (4748)	10.2	420 (61)	bcc	Used as an alloying addition to cast iron and alloy steels, especially high-speed steels.
Platinum (Pt)	1773 (3223)	21.4	130 (19)	fcc	Soft, malleable and ductile. Corrosion resistant, e.g. not oxidized or tarnished in air. Alloyed with iridium to increase the hardness. Used as resistance pyrometer wire.
Palladium (Pd)	1555 (2831)	12.0	—	fcc	Slightly malleable and ductile, and oxidation resistant. Alloyed with silver to form an electrical contact material.
Rhodium (Rh)	1985 (3605)	12.44	—	fcc	Resists acids, used as a metal plating, alloyed with platinum as thermocouple wire.
Silver (Ag)	960 (1760)	10.5	140 (20)	fcc	Possesses the highest electrical conductivity, used for electrical contacts, silver solder and for plating.
Selenium (Se)	220 (428)	4.81	—	hexagonal	Improves machinability of copper and copper alloys. Used in photocell (electrical conductivity changes when exposed to light).
Tellurium (Te)	452 (846)	6.24	—	hexagonal	Alloying addition to improve the strength of lead and the free-cutting properties of copper.
Thorium (Th)	1827 (3321)	11.25	—	fcc	Soft and similar to lead. Alloyed with tungsten to produce filament wire, and with magnesium for creep resistance.
Tungsten (W)	3410 (6170)	19.3	420 (61)	bcc	Hard and brittle at room temperature, malleable and ductile at high temperatures. Used as a filament wire. Alloyed in high-speed steels and in sintered cutting tools (as carbide).
Vanadium (V)	1710 (3110)	5.7	200 (29)	bcc	Very hard. Alloying addition to steels to produce toughness.

3.10 METALS USED IN SMALL QUANTITIES

Many metals are used in small quantities, mainly as alloying additions to produce particular material properties. Details of the properties, the main uses of these metals and the alloys in which they are used are given in Table 3.27.

3.11 'NEW' OR 'ALTERNATIVE' METALS

There are several metals that have come to prominence in recent years; this group are often referred to as the 'new metals', 'space-age metals' or other similar titles. It is difficult to find a simple, apt description for this group. Their existence has often been known for many years, but their present uses are new and rapidly expanding. Many of these metals have applications in the fields of nuclear engineering and space travel, but they are unlikely to replace the basic engineering materials upon which we now depend, at least not in the immediate future. These metals would indeed be in more common usage were it not for their high costs, caused by either their scarcity or difficulties in extraction or fabrication. It would certainly not be apt to call this group the 20th century metals simply because of their more recent widespread applications, but it would also be unwise to use the 21st century as a designation at this time. Despite its shortcomings 'new metals' conveys (in part) what is intended by this title, at least to those with some knowledge of the subject. However, 'alternative metals' would do just as well, except that in many cases they offer no alternative because of prohibitive costs. The properties and applications of this group of metals are summarized in Table 3.28.

Table 3.28 Properties and applications of the 'new' metals

Metal	Melting point °C (°F)	Relative density	Tensile strength MN/m² (psi × 10⁻³)	Comments and applications
Beryllium (Be)	1285 (2345)	1.84	300 (44)	Scarce and expensive. The dust and vapor are poisonous. Not ductile and difficult to fabricate, less brittle at elevated temperatures. Shaped by powder-metallurgy methods. Small ingots can be hot rolled or extruded. Machined using carbide tools. Can be brazed to other metals; Be–Be joints are welded. Has a higher melting point and is lighter than aluminum, and is used in aircraft and rockets. Used in nuclear engineering applications as moderators and reflectors.
Germanium (Ge)	937 (1719)	5.32	—	Very brittle and crystalline. Known as a metalloid as it possesses some metal and some non-metal properties. Cannot be cold worked. Good electrical properties.

Table 3.28   *continued*

Metal	Melting point °C (°F)	Relative density	Tensile strength MN/m² (psi × 10⁻³)	Comments and applications
Hafnium (Hf)	2222 (4032)	13.09	450 (65)	Similar to zirconium, good corrosion resistance and strength. Heavier, harder and less malleable than zirconium but can be forged, rolled and drawn. Can be machined and welded. Main use as control rods in nuclear reactors because hafnium absorbs neutrons.
Niobium (Nb) (Columbium (Cb))	2468 (4474)	8.57	280 (41)	Scarce and expensive, tantalum is usually used instead. Malleable and ductile, can be rolled and welded (inert gas shield). Resistant to corrosion by most chemicals and molten metals. Used as a nuclear fuel-canning material and for gas turbine blades. Sometimes as an alloying addition to austenitic stainless steels.
Plutonium (Pu)	640 (1184)	19.84	270 (39)	Not a naturally occurring element, formed by atomic transformation of uranium. Stable in dry air but reacts with moist air. Emits powerful radiation and is used as a nuclear fuel.
Tantalum (Ta)	2996 (5425)	16.6	200 (29)	Heavy, soft, ductile metal protected by a surface oxide film. Resistant to most chemicals (cf. glass). High melting point and is classified as a refractory metal, but can be rolled and drawn and is stable at high temperatures. Used for pumps, stills and agitators. Used in human surgery as replacements. The carbide is hard, used in cemented-carbide tools.
Uranium (U)	1132 (2070)	18.7	370 (54)	Heavy, malleable and ductile metal, can be cast, rolled or extruded. Reacts with air. Continually emits small amounts of radiation. Forms hard, stable carbides. Used as a nuclear fuel because uranium has a large atomic nucleus that can disintegrate by fission with another neutron particle. Releases heat and follows a chain reaction, can cause an atomic explosion.
Zirconium (Zr)	1852 (3366)	6.5	210 (30)	Soft, malleable, ductile metal, can be forged, hot and cold rolled, or drawn. Strong when cold worked, brittle when impure. Good corrosion resistance due to the surface film, must not be heated with oxygen, nitrogen or hydrogen. Machined using low-cutting speeds. Used in magnesium alloys as a grain refiner, in nuclear reactors because it does not absorb neutrons, and for fuel containers.

* * *

Exercises

1   What is the composition of a 'tin can'? What are the main properties it must possess?

2   List the present uses of lead and lead alloys.

3   Discuss the statement: 'Metals such as zinc, tin, lead and their alloys, now have few major uses and they could be replaced entirely by other cheaper, acceptable alloys.'

4   Prepare a feasibility study to assess the problems and the possibilities of recycling and reusing the major nonferrous metals.

5   The 'new' metals or 'space-age' metals (with the exception of titanium) are costly, have few uses and are unlikely to be of significant use in the near future. It would be preferable to devote more time and money to developing improved conventional alloys, rather than the production of these elitist metals. Discuss.

6   Are there any ferrous alloys that possess comparable properties to aluminum alloys?

7   If copper became very scarce or unavailable, which metals or alloys could be used as replacements for particular applications? What effects would these substitutions cause?

8   Describe the changes that occur when brass containing 38% zinc is either quenched or slowly cooled from 850°C (1562°F) to room temperature. What are the crystal structures of the α and β phases?

9   What is the difference between brass and bronze?

10   What is dezincification?

11   Which metals are known as: the light metals; refractory metals; rare earth metals; 'space-age' metals; precious metals?

12   Which nonferrous metals melt below 425°C (800°F)?

13   Describe how a Muntz metal β phase alloy could be treated to transform it from a coarse-grained to a fine-grained structure.

14   Explain why the solution heat-treatment times for equivalent aluminum alloys are longer for castings than for wrought forms. Why are the times longer for wrought forgings than for wrought sheet?

15   Describe how the strength and hardness of heat-treatable and non-heat treatable wrought aluminum alloys can be increased.

16   What is the main advantage of cobalt as an alloying element?

17   What problems are encountered when refractory metals are used at high temperatures?

18   What are the 'superalloys'?

19   Why do titanium alloys contain aluminum, tin or molybdenum?

20   Beryllium has many advantageous properties; state the main disadvantage that has severely limited its use.

21   Explain the difference between precipitation hardening and dispersion hardening of alloys.

22 Which of the following metals and alloys can be strengthened, and how? Nickel; beryllium–copper; Monel; titanium; lead.

23 Describe how ingots of titanium, zirconium and tungsten can be produced.

24 Why are aluminum–copper alloys (as cast or heat treated) less corrosion resistant but easier to machine than aluminum–silicon alloys?

25 Why are wrought magnesium alloys (especially sheet) made from high-purity magnesium, but cast alloys are produced from ordinary electrolytic metal?

26 Select nonferrous materials (if at all possible) for use in the following situations, and explain the reasons for your choice:
   (a) a replacement knee joint;
   (b) an automobile exhaust system;
   (c) a ship propeller;
   (d) a bicycle frame;
   (e) a cutting drill;
   (f) a storage vessel for orange juice;
   (g) the impeller of a centrifugal pump.

27 Design a vessel with a cross-sectional area of approximately 1 m² (or 9 ft²) and 1 m high (or 3 ft) for use as an effluent gas adsorber. The vessel will contain an annular region of activated carbon to adsorb the undesirable gases. Gases enter the vessel and pass through the carbon before leaving.
   Decide upon your own problem specification or operating conditions and concentrate upon the materials selection and specification aspects.

* * *

Complementary activities

1 Read through the complementary activities listed for Chapter 2.

2 Prepare some samples of nonferrous metals and their alloys for examination under a metallurgical microscope.

3 Using the microscope, determine the effects of heat treatment, working and rate of cooling upon the structure of the samples.

4 Measure the melting points of some 'pure' nonferrous metals and alloys. Comment upon the purity or composition of these materials based upon the measurements obtained.

5 Determine the composition of some copper wire, a 'tin' can, aluminum cooking foil and a metal soft drink container.
   Assess the possibilities of recycling these material. Try and purify some of these nonferrous metals.

* * *

KEYWORDS

copper	blow holes	annealing
malleable	'tough pitch' copper	cold working
ductile	OFHC copper	stress relief annealing
tensile strength	fire-refined copper	cap copper
hardness	brass	gilding metal
corrosion resistant	bronze	cartridge brass
electrical conductivity	cupro-nickel	standard brass
thermal conductivity	season cracking	basic brass

yellow or Muntz metal
aluminum brass
free-cutting brass
'naval' brass
high-tensile brass
 (manganese bronze)
admiralty brass
clock (engraving) brass
tin–bronze
gunmetal
aluminum bronze
silicon bronze
phosphor bronze
wrought alloy
cast alloy
Monel
electroplated nickel–
 silver (EPNS)
cadmium copper
beryllium copper
beryllium bronze
solution heat treatment
precipitation treatment
BS
ASTM
UNS
ingots
plate
sheet

strip
rolled bar
pipe
tube
forging stock
rod
bar
die forgings
wire
foil
aluminum
bauxite
nonporous oxide film
anodizing
heat treatment
non-heat treatable
aging
age hardening
over-aging
modification
die casting
'light' metals
magnesium
porous oxide film
creep resistance
nickel
cladding
plating
electrodeposition

Nilo alloys
Corronel B
Hastelloys
Langalloys
Inconel
Nimonic series
Nimocast series
zinc
galvanizing
electrogalvanizing
sheradizing
stabilizing annealing
aging shrinkage
titanium
strength: weight ratio
lead
printer's type
bearing metal
lead type A, B and C
soft solder
fusible alloys
white bearing metals
'Babbitt' metals
tin
tin plate
alloying additions
'new' metals

*   *   *

BIBLIOGRAPHY

Betteridge, W., *Nickel and its Alloys*, Macdonald and Evans Ltd, Estover, Plymouth (1977).

Betteridge, W., *Cobalt and its Alloys*, Ellis Horwood Ltd, Chichester, Sussex (1982).

Betteridge, W. and Heslop, J. (Eds.), *The Nimonic Alloys* 2nd Ed., Edward Arnold (Publishers) Ltd, London (1974).

Cairns, J.H. and Gilbert, P.T., *The Technology of Heavy Non-Ferrous Metals and Alloys: Copper, Nickel, Zinc, Tin and Lead*, George Newnes Ltd, London (1967).

Duncan, R.M. and Hanson, B.H., *The Selection and Use of Titanium*, Engineering Design Guide No. 39, Oxford University Press, Oxford (1980).

Everhart, J.L., *Engineering Properties of Nickel and Nickel Alloys*, Plenum Press, New York (1971).

Fishlock, D., *New Materials*, John Murray (Publishers) Ltd, London (1967).

Gordon, J.E., *New Science of Strong Materials*, Penguin Books Ltd, Harmondsworth, Middlesex (1978).

Greenfield, P., *Engineering Applications of Beryllium (1971)* and *Magnesium (1972)*, Mills and Boon Ltd, London.

Harwood, J.J., *The Metal Molybdenum*, Symposium Proceedings of the American Society for Metals, Cleveland, Ohio (1958).

Kleefisch, E.W. (Ed.), *Industrial Applications of Titanium and Zirconium*, Symposium papers of the American Society of Testing and Materials, Philadelphia, Pennsylvania, (1981).

McGachie, R.O. and Bradley, A.G., *Precious Metals*, Pergamon Press Ltd, Oxford (1981).

Miller, G.L., *Zirconium*, Butterworth and Co. (Publishers) Ltd, London (1957).

Mondolfo, L.F., *Aluminium Alloys: Structure and Properties*, Butterworth and Co. (Publishers) Ltd, London (1976).

Morgan, S.W.K., *Zinc and its Alloys*, Macdonald and Evans Ltd, Estover, Plymouth (1977).

Polmear, I.J., *Light Alloys: Metallurgy of the Light Metals*, Edward Arnold (Publishers) Ltd, London (1981).

Raynor, G.V., *The Physical Metallurgy of Magnesium and its Alloys*, Pergamon Press Ltd, Oxford (1955).

Ross, R.B., *Metallic Materials Specification Handbook*, 2nd Ed., John Wiley and Sons, Inc., New York (1972).

Simons, E.N., *Guide to Uncommon Metals*, Frederick Muller Ltd, London (1967).

Sims, C.T. and Hagel, W.C. (Eds.), *The Superalloys*, John Wiley and Sons, Inc., New York (1972).

Smithells, C.J., *Metals Reference Book*, 5th Ed., Butterworth and Co. (Publishers) Ltd, London (1976).

Sully, A.H., and Brandes, E.A., *Chromium*, Butterworth and Co. (Publishers) Ltd, London (1967).

Van Lancker, M., *Metallurgy of Aluminium Alloys*, Chapman and Hall Ltd, London (1967).

Varley, P.C., *The Technology of Aluminium and its Alloys*, Newnes–Butterworths Ltd, London (1970).

West, E.G., *The Selection and Use of Copper-Rich Alloys*, Engineering Design Guide No. 35, Oxford University Press, Oxford (1979).

West, E.G., *Copper and its Alloys*, Ellis Horwood Ltd, Chichester, Sussex (1982).

Winters, R.F. (Ed.), *Newer Engineering Materials*, Macmillan Publishers Ltd, London (1969).

4
Polymers and Ceramics

CHAPTER OUTLINE

Polymers
Molecular structure, bonding
  and classification
Structure and properties
Amorphous and crystalline states
Modification of properties
Polymer reactions
Strength of polymers
Testing of polymers
Rheology and viscoelasticity
Fabrication methods

Design considerations
Typical polymer materials:
  Natural resins
  Thermosetting polymers
  Thermoplastic polymers

Ceramics
Bonding and properties
Ceramic materials
Composites

CHAPTER OBJECTIVES

To obtain an understanding of:
1 the classification of polymer materials and their important properties;
2 how polymers are produced and fabricated and how they can be strengthened;
3 the advantages and limitations of polymers as engineering materials;
4 the nature and uses of ceramic materials;
5 the properties required by the more common nonmetallic materials for use in engineering situations.

IMPORTANT TERMS
---

thermoplastic
thermosetting
polymer chain
repeating molecules
macromolecule

side-branch chains
linear polymer
monomer
degree of
  polymerization

molecular weight
amorphous polymer
crystalline polymer
glass transition
  temperature

degree of crystallinity	cermet	clay
addition polymerization	ionic bond	cemented carbide
copolymerization	covalent bond	cermet
condensation	porosity	cement
polymerization	bend test	concrete
step reaction	Vickers microhardness	glasses
polymerization	alumina	fictive temperature
water absorption	beryllia	thermal tempering
polymer testing	zirconia	chemical strengthening
rheology	silica	crystalline glass
viscoelastic	magnesia	manufactured carbon
Newtonian fluid	refractory:	composite
pseudoplastic	basic	fiberglass
Bingham fluid	acid	laminate
creep	neutral	fiber
fabrication	metal carbide	fabric
abrasive		

POLYMERS

4.1 INTRODUCTION

Polymer is the correct name for that class of materials generally referred to as *plastics*, because not all polymers exhibit plastic deformation. In this text the term polymer will be used (as appropriate), in order to distinguish between the material and a mechanical property or particular behavior.

4.2 MOLECULAR STRUCTURE

Polymer literally means *many parts*: these substances are composed of *long-chain repeating molecules*, built up from shorter molecules and usually based on organic (carbon) materials. A similar system exists based on the silicon–oxygen chain. The continuous polymer chain is a giant *(macro) molecule*, such as high-density polyethylene ($H(C_2H_4)_{2000}H$). If these large molecules could be magnified sufficiently to be seen, then they would resemble a long piece of twisted wire as shown in Fig. 4.1. Carbon is normally present in a tetravalent form, with four equivalent bonds pointing to the corners of a tetrahedron (shown dotted in Fig. 4.1). The difference in structure of a single molecule and a long-chain polymer is illustrated in Figs. 4.1 and 4.2. In practice, the polyethylene chain would be a three-dimensional zig-zag or spiral, as shown in Fig. 4.2. More complex polymers may have side-branch chains or a more twisted structure. Polymers are termed linear despite their twisting chain because the length is at least a thousand times greater than the thickness.

The original molecule is called a *monomer*, e.g. ethylene, C_2H_4, whereas the repeating unit in polyethylene is called a *mer*, e.g. $-CH_2-CH_2-$. Carbon can also be present in the unsaturated double and triple bond forms, i.e. $\overset{\diagup}{C} = C\overset{\diagdown}{}$ or $-C \equiv C-$, although the carbon is still tetravalent. The *degree of polymerization* is the number of repeating units of identical structure within the chain; this value

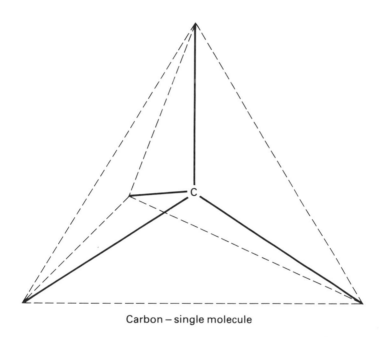

Carbon – single molecule

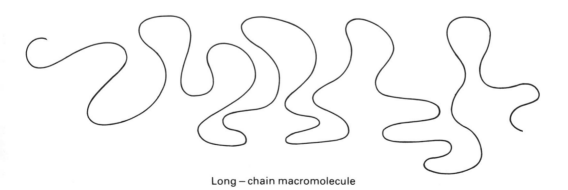

Long – chain macromolecule

Figure 4.1 Comparison of single molecules and macromolecules.

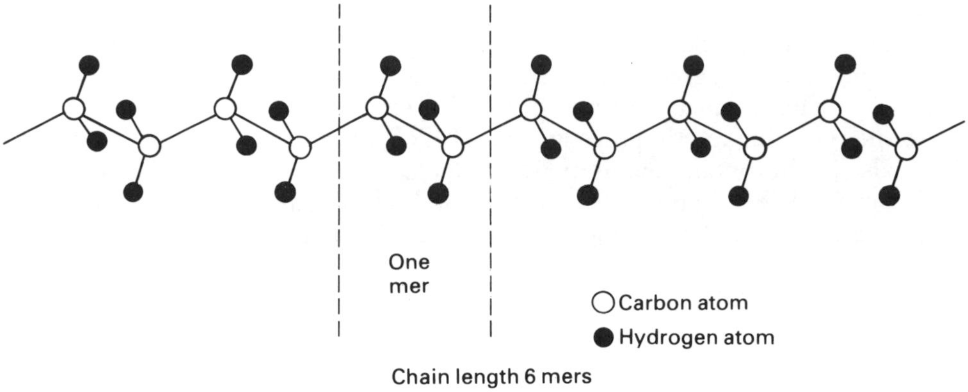

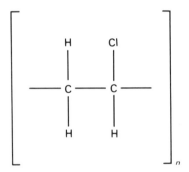

One
mer

○ Carbon atom
● Hydrogen atom

Chain length 6 mers

Figure 4.2   Model of the structure of linear polyethylene.

n is the molecular weight,
i.e. the number of repeating units

Monomer symbol

Figure 4.3   Polyvinyl chloride monomer.

is 6 in the example in Fig. 4.2. The monomer notation used in this text is shown
in Fig. 4.3. For simple polymers based upon ethylene, the value of *n* (in Fig. 4.3)
determines the usual phase of the material, that is:

Value of n	*Phases present*
5–20	gas or liquid
>50	soft wax
200–20000	solid

4.3 BONDING OF POLYMERS

Strong covalent bonds exist between the carbon atoms within a polymer chain, and between the carbon and hydrogen (or other) atoms. The individual molecules interact by intermolecular forces such as *van der Waals* forces; hydrogen bonds or bridges are the main secondary forces. The forces binding the chains to each other are weak electrostatic forces. When a load is applied to a polymer, failure is usually due to separation between chains rather than rupture within particular chains. The linear chains tend to slide over one another; the presence of side chains restricts this motion and increases the strength of the material.

Polymers can also be strengthened, although with some loss of ductility, by creating covalent *cross-linking* between the chains. This usually occurs when double bonds within a chain (usually between carbon atoms) are broken, and individual atoms or molecules link with neighboring chains.

4.4 CLASSIFICATION

Most of the polymers that were originally developed and produced in commercial quantities possessed either thermoplastic or thermosetting properties, and they were classified according to this behavior. These groups of materials will be considered separately in this chapter. Polymers can be classified according to their chemical composition or the group to which they belong; these two classifications are shown in Fig. 4.4. An alternative classification is considered in Section 4.4.3.

4.4.1 Thermoplastic materials

The term plastic as applied to a material means that it can be easily shaped; however, polymers are not plastic at room temperature but only when they are formed or shaped at higher temperatures. The important property of thermoplastic polymers is that they can be repeatedly hardened and resoftened by removing and applying heat and pressure; the temperature must not be too high or decomposition will occur. The linear structure of thermoplastics may be represented by:

$$- A - A - A - A - A - A -$$
$$- A - A - A - A - A - A -$$
$$- A - A - A - A - A - A -$$

'A' represents
the molecule or
monomer

Copolymers contain more than one monomer, which may be randomly distributed or in regular blocks. The chains rapidly become tangled, but with pressure and heat they move apart and slide past each other, and these new

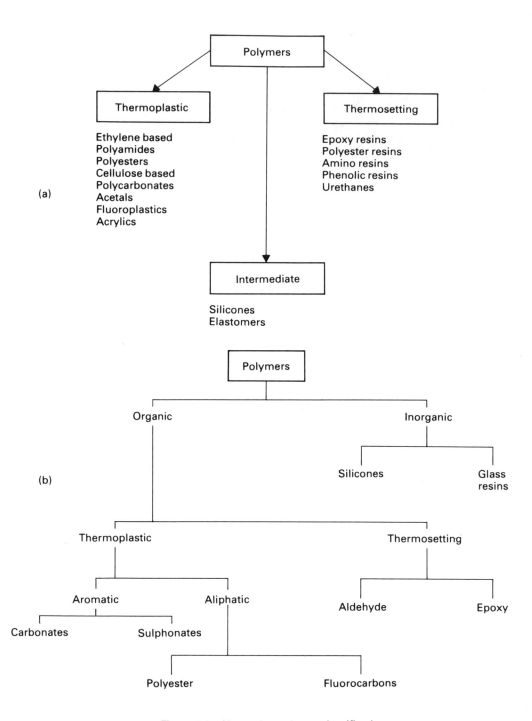

Figure 4.4 Alternative polymer classifications.

positions are retained when the heat and pressure are removed. The properties of these materials depend upon the basic material called the binder, a filler which is added to produce strength, coloring matter, and any plasticizer that acts as an internal lubricant. These polymers may be in the form of films, sheets, rods, tubes or molding materials for fabrication purposes. They can generally be formed or shaped at temperatures of approximately $100°C$ ($212°F$) and require little pressure. They can be joined by either heat and pressure or by the use of a solvent.

4.4.2 Thermosetting materials

Application of heat and pressure to a thermosetting polymer produces a chemical change; the product cannot be altered by further heating or pressure. If further heat is applied the material will char, burn, sublime or decompose.

The network structure of thermosets may be represented by:

$$
\begin{array}{cccc}
| & | & | & | \\
- A & - A & - A & - A - \\
| & | & | & | \\
- A & - A & - A & - A - \\
| & | & | & |
\end{array}
$$

The heating process (curing) initially produces a thermosoftening stage, followed by the cross-linking associated with the thermosetting polymer. The material is rigid and insoluble and the chains are tangled and cross-linked. These materials are available as powders, resins or cloth for subsequent fabrication. Thermosetting polymers contain the basic material, a filler, color, plasticizer, hardening agent (producing cross-links) and an accelerator.

4.4.3 Alternative classification

Although the classification of polymers as either thermoplastic or thermosetting is widely used, the development of a variety of new polymers with particular properties has led to many anomalies in this system. Thermoplastic means heat-flowable, but PTFE and ultra high molecular weight polyethylene do not flow under heat. Many polymers now flow without heat, simply due to mechanical pressure. Thermosetting means heat-set, but silicone, epoxy and polyester can be set without heat. Also phenolic must first be made to flow under heat before it can be heat-set.

A better classification would be between linear and cross-linked polymers. A linear structure means that the polymer chains remain linear and separate after molding. However, cross-linked polymers start from linear chains; these are joined irreversibly during molding into a three-dimensional interconnected molecular network. This explains how polyethylene can now also be available as a thermoset, and polyester and polyurethane can be thermoplastic.

The traditional classification of thermoplastic and thermoset has been used in this chapter, particularly for the materials described in Sections 4.19 and 4.20. However, it would be a useful exercise for the reader to assign the designations linear or cross-linked to polymers as they are mentioned in the chapter.

*   *   *

Self-assessment exercises

Study Sections 4.1 to 4.4 and then answer the following questions.

1 When is the term 'plastic' more appropriate than 'polymer' for that particular class of materials?

2 What is a macromolecule? State some examples.

3 Using specific examples, describe the general molecular structures found in polymers.

4 What type of bonding exists within a polymer? How does polymer bonding affect the properties of the material?

5 Is there any relationship between the molecular structure of a polymer and its bonding?

6 What are the differences between thermoplastic and thermosetting polymers? Give examples of each type and describe their particular applications.

7 Why are thermoplastic materials 'plastic'?

8 Why do thermosetting materials 'set'?

9 After studying Sections 4.1 to 4.4 only, suggest how the properties of polymers can be modified. How can the strength of polymers be improved?

*   *   *

4.5 CONFORMATION AND CONFIGURATION

A polymer macromolecule may be considered to be a very long thin molecule, and these macromolecules are mixed together like a bowl of spaghetti. Flexible plastics are 'squirming' like worms and individual chains move when a force is applied. For thermosetting polymers the strands of spaghetti are tied together and the application of a force causes the entire mass to move. These long chains can assume many coil-like shapes in space, and these are known as *conformations*. If certain fixed positions exist for some of the groups of atoms relative to each other, then these are known as *configurations*. An example is the formation of cis- and trans-polybutadiene as shown in Fig.4.5.

Because polymer chains are coiled and not linear, it is necessary to estimate the end-to-end distance of a polymer. This provides a reasonable estimate of the size of the molecule. Several statistical techniques exist, e.g. the random flight technique; however, several corrections need to be applied. The ratio of the actual average end-to-end distance to the uncorrected value is a measure of the stiffness of the polymer chain.

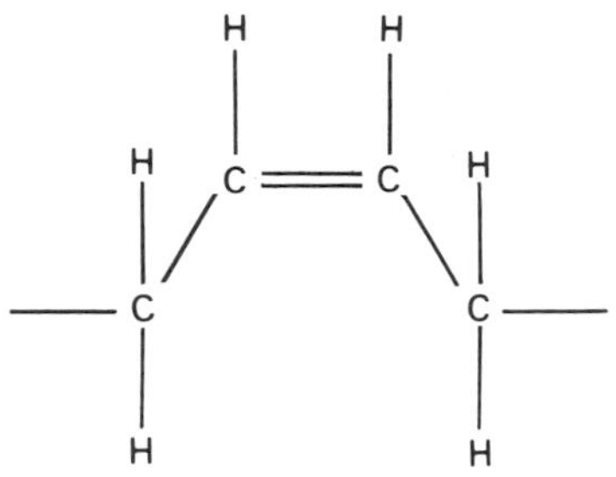

Cis-polybutadiene

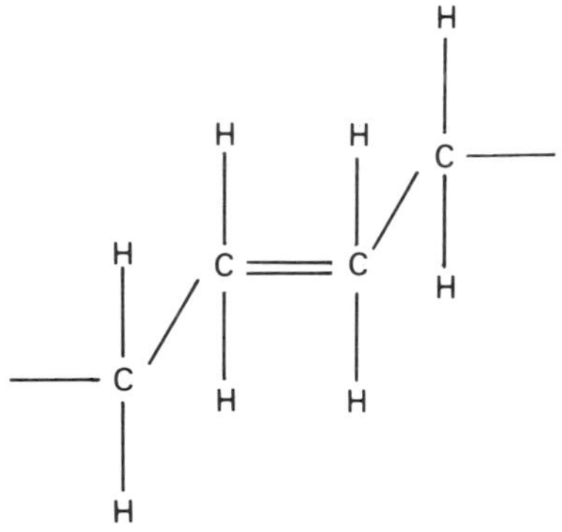

Trans-polybutadiene

Figure 4.5   Example of polymer configurations.

It will be necessary to know the molecular weight of a polymer, and in many cases this will provide an indication of the properties possessed by that material. However, a unique molecular weight is often not obtained and there will exist a molecular weight distribution. This may be a fairly narrow range if controlling catalysts are used, or it may encompass a broad range. Process variables can also be used to control the molecular weight distribution. The molecular weight used to characterize a polymer may be based upon several criteria. The most popular are the viscosity average, number average and weight average molecular weights. For a molecular weight distribution these averages have different values, although they would be identical if the polymer possessed a unique molecular weight.

4.6   THE STRUCTURE AND PROPERTIES OF POLYMERS

It is possible to substitute particular groups, known as *pendant* groups, into the long-chain polymer molecule. An example is the substitution of methyl groups ($-CH_3$) or chloro groups ($-Cl$) for some of the hydrogen atoms in the polyethylene chain. These monomers are shown in Fig.4.6. The properties of a

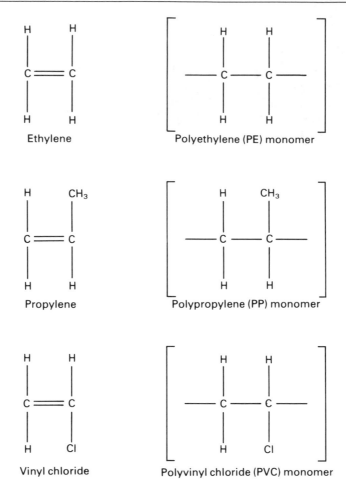

Figure 4.6 Examples of polymer monomers and substitution within the monomer.

polymer depend upon the groups present within the macromolecule and the relative positions of these groups. The flexibility of polyethylene increases as its temperature increases, due to the greater mobility of the polymer chains. The presence of ordered pendant groups in the polymer backbone reduces this flexibility at any particular temperature.

The general term *tacticity* is used to describe the different arrangements of the pendant groups. Several terms are used to describe particular arrangements:

Isotactic   the pendant groups are all on the same side of the polymer chain and are in a regular arrangement.

Syndiotactic alternate pendant groups are on opposite sides of the polymer chain.

Atactic   random arrangement or lack of order for the pendant groups, also called *heterotactic.*

Stereoregular describing isotactic and syndiotactic polymers.

These particular arrangements of the pendant groups are illustrated in Fig.4.7 for the polypropylene macromolecule. The structures shown in Fig.4.7 are all head-to-tail configurations, i.e. the methyl group is present on alternate carbon atoms. This is typical of most polymer chains and it is shown in Fig.4.8 for comparison with a head-to-head configuration.

The chains of an isotactic polymer tend to fit together easily. The structure is more rigid and has a higher melting point than the basic polymer, e.g.

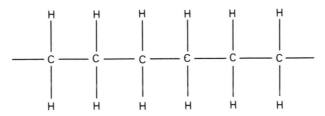

High-density (linear) polyethylene

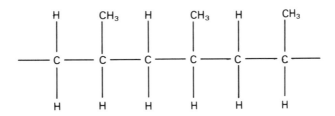

Isotactic polypropylene

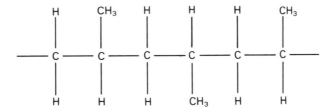

Syndiotactic polypropylene

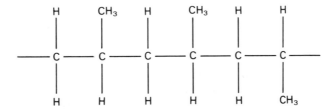

Atactic polypropylene

Figure 4.7 Examples of polymer chains.

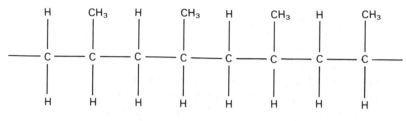

Head-to-tail configuration

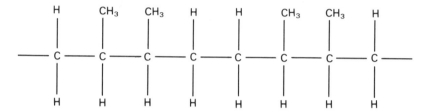

Head-to-head configuration

Figure 4.8   Alternative configurations for polypropylene.

polypropylene compared with polyethylene. The volume occupied by the chain is larger if pendant groups are present, and the density is lower, e.g. 950 kg/m^3 for polyethylene compared to 900 kg/m^3 for polypropylene. However, the presence of a large group, e.g. $-C_3H_7$, instead of the methyl group lowers the melting point and density, because of the bulkier pendant groups and more movement between less closely packed chains. The substitution of long linear pendant groups can increase the melting point because of mutual attraction between the side chains. Materials are often specified because of their volume rather than their weight, and low density is an advantageous property for a polymer to possess.

4.7   THE AMORPHOUS STATE

Most polymers exist in an amorphous or noncrystalline state, unlike metals which have a crystalline structure. The polymer molecules can be considered to be like other small molecules when liquid or solid, except that in polymers the 'small molecules' are joined together and their movement is restricted. There are two main types of macroscopic structure, one is the amorphous or *isotropic* state in which the properties are the same in all directions. The other is *anisotropic*, in which the properties depend upon the direction of measurement.

Some form of order exists even in amorphous polymers; the molecules may be like balls of string or they may align with their ends together. Small regions probably exist where there is microscopic order. This may be due to cross-linking, fillers, high molecular weight, etc., which constrain the movement of the molecules. Movement is then caused by applied pressure, plasticizers, etc. The existence of small ordered regions could explain some of the anomalous properties found for amorphous polymers.

There are definite changes that occur when a polymer is heated or cooled. Consider the cooling of a molten amorphous polymer. A temperature is reached where a change occurs from a viscous liquid to a solid amorphous glass; this is known as the *glass transition temperature* (T_g). There is a rapid increase in viscosity to a very high value and the polymer may be considered to be a solid glass. The change in specific volume as a function of temperature is shown in Fig.4.9; the transition for an amorphous polymer is not sharp but follows a curve. The curvature of the slope depends upon the measurement technique employed. The crystalline polymer curve will be discussed in Section 4.8.

Below T_g, local molecular motion is almost eliminated and significant changes occur in the polymer properties. The polymer becomes hard, brittle and stiff, and is often transparent. Values of T_g cover a wide range, e.g. $-78°C$ ($-112°F$) for amorphous polyethylene to $100°C$ ($212°F$) for polystyrene. For most useful thermoplastics, T_g is between $50°C$ ($122°F$) and $150°C$ ($302°F$). After a material is cooled below T_g, a certain time must be allowed for shrinkage to take place. This time is increased by lowering the temperature at which the material is held.

A free volume must be present within a polymer to allow movement to occur. With rapid heating, T_g is measured at a higher value as more time is available for the polymer regions to reach their equilibrium conditions. This also occurs at higher pressures and with branched or cross-linked polymers. Alternatively, T_g is measured at a lower value with slow cooling, or if plasticizers are present. This is because the free volume is increased, viscosity is reduced and polymer flow is easier.

* * *

Self-assessment exercises

1 Explain, using illustrative examples, the difference between polymer conformation and polymer configuration.

2 Why is it necessary to know the chain length of a polymer? Can this be calculated from a knowledge of the number of repeating units and the length of a monomer unit?

3 What is the molecular weight of polyethylene?

4 How is the molecular weight of a polymer measured?

5 Explain, using illustrations, the meaning of the terms:
   (a) tacticity;   (b) stereoregular;
   (c) atactic;   (d) syndiotactic;
   (e) isotactic;   (f) heterotactic.

6 Distinguish between head-to-tail and head-to-head configurations.

7 Describe, with examples, the effects of pendant groups in a polymer chain upon the properties.

8 What is an amorphous state? Give examples of amorphous polymers, describe their molecular structure, bonding and macrostructure.

9 What is the glass transition temperature of a polymer? What happens when a polymer is cooled below this temperature?

10 What precautions should be taken when measuring the glass transition temperature?

*   *   *

4.8 THE CRYSTALLINE STATE

The crystallization of polymers requires that the polymer molecules have a regular structure, and that there are strong, regularly spaced, attractive forces, e.g. hydrogen bonding, van der Waals forces or dipole–dipole forces. The crystallization process is similar to that for small molecules (see Section 1.2.3), except that the long chains must be accommodated. The *degree of crystallinity* is usually expressed as a percentage, and is a measure of the amount of long range, three-dimensional order in a polymer when compared to its highly crystalline state. Values range from nearly zero for highly cross-linked polymers to over 95% for simple linear polymers. With extensive cross-linking, upon cooling the molecules are restricted and cannot take up a crystalline arrangement. The degree of crystallinity may be estimated by X-ray diffraction, infrared spectroscopy or by NMR techniques. X-ray diffraction is used to compare the areas under the diffraction peaks of the crystalline regions (from specific crystalline planes) and the amorphous region (obtained as a broad peak). The degree of crystallization depends upon not only the method used, but also the model used to predict diffraction behavior for a fully crystalline state.

Crystalline polymers exhibit different behavior on a 'specific volume *vs* temperature' curve, as shown in Fig. 4.9. As a polymer is cooled, if crystallization can occur, at the *crystalline melting point* (T_m) there is a rapid volume decrease. This is similar to the situation for metals (Section 1.4.1; Fig. 1.22) and is due to the close-packed arrangement of crystalline materials. Dilatometry is the classical method of measuring crystallization. The volume change upon crystallization is measured in a dilatometer and the crystallinity is calculated from the known densities of the crystalline and amorphous domains at the experimental temperature. Like T_g, T_m is rarely a specific value but occurs over a temperature range. Many polymers contain both crystalline and glassy regions within their structure. For most polymers the values of T_g and T_m are related. If the chain structure is symmetrical, then T_g is approximately one half to two thirds of T_m (both in kelvins).

Two main theories have been proposed to explain the structure of polymer crystals, or *crystallites*. Early theories were based upon X-ray diffraction studies and were known as *fringed-micelle or oriented molecule theories*. They were mainly applied to solid-phase polymers or to drawn fibers. The polymer was assumed to be composed of small ordered regions where the chains are either aligned or folded over to form the crystallite. These crystalline regions are surrounded by amorphous regions. Possible chain alignments are shown in Fig. 4.10.

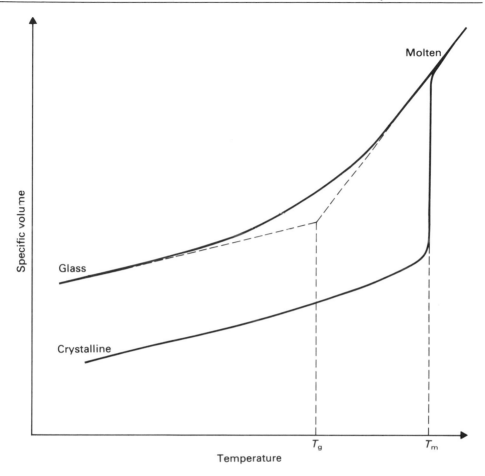

Figure 4.9   Volume changes when cooling a polymer.

Crystallization or nucleation occurs in two ways. Heterogeneous nucleation is due to the presence of impurities or defects that provide sites for the initial orientations of the long chains. The second stage occurs uniformly throughout the mass of the polymer and is homogeneous nucleation. The application of a tensile or shear stress tends to aid this process by causing alignment of the molecules. The growth of the crystallites occurs as more chains attach to the nucleus (secondary nucleation). Some areas remain amorphous or uncrystallized, and these transform to a glassy polymer when cooled to T_g.

Under carefully controlled experimental conditions, a large spherical arrangement of crystallites known as a *spherulite* is formed. It appears to possess a lamellar structure and the individual lamellae each grow from a nucleus. Several types of spherulites have been observed using polarized light; these may be small ribbon structures or large dendritic arrangements. More recent theories suggest that crystalline polymers should be considered analogous to metals, but containing many large defects. Further work is necessary to establish the general applicability of the theory.

Folded polymer chains within polymer crystals

Oriented crystallites

Figure 4.10   Possible alignments of chains in crystalline polymers.

4.9   MODIFICATION OF POLYMER PROPERTIES

Crystallization is the process of crystal formation, and *crystallizability* is the ability to form crystals under favorable conditions. A specimen may contain a large quantity of polymer material that could form crystals, but relatively few crystals may actually be present. Spherulite formation may occur during polymerization, e.g. in polyethylene or polypropylene, if catalysts are present. Polymers are stronger, stiffer, more resistant to failure and more transparent if many small spherulites are formed, rather than a few large crystal structures. If a molten polymer is rapidly cooled (quenched) below T_g, then the crystallinity and the number of spherulites are reduced. Alternatively annealing at a temperature between T_m and T_g increases the number and size of the spherulites. The temperature should be chosen so that crystallization is rapid, however the nucleation of new sites is a slow process.

Mechanical deformation, whether elastic or plastic (see Section 6.2.1), increases the crystallinity and enables the preferred alignment of the polymer chains to be obtained. This method is more effective if it is carried out at temperatures between T_m and T_g. Deformation occurs when a semicrystalline polymer is fabricated, typical processes are cold drawing (see Section 8.2.9) of polymer fibers and rolling (see Section 8.2.8) of sheet material. The crystalline regions are then oriented in the direction of the applied force, as are the chains in the amorphous regions. For drawing, the strength of the fiber increases along its length, i.e. the direction of draw, but it decreases in other directions. For a film or sheet, elongation occurs in two directions simultaneously and the chains are aligned in both directions. The stages through which the semicrystalline polymer passes during drawing are distortion, yielding and disruption. The effects on the polymer chains and the chain arrangements within a drawn or extruded film are shown in Fig. 4.11. The final orientation of the molecules results in a stronger polymer fiber or sheet.

When a fiber is cold drawn, the diameter (or thickness) of the material is reduced. Any heat evolved is reabsorbed during crystallization causing chain orientation in the thinner fiber. The molecules are aligned in the direction of drawing, but they are at right angles to the direction of flow for hot-melt extrusion (see Fig. 4.11). Only subsequent drawing can produce longitudinal strength, or biaxial strength by blow film extrusion (see Section 4.15.6).

The properties of a polymer depend upon, and can be modified by, consideration of several factors. The main properties are strength, stiffness, hardness, melting temperature and melt viscosity. The polymer crystals are composed of lamellae, and some polymer molecules are connected to several lamallae or several crystallites. These molecules are called *tie molecules* and they act to hold the structure together. The number of tie molecules present in a structure determines its strength. Cold drawing can be used to control the number and location of the tie molecules. The polymer properties also depend upon whether a filler is added to the amorphous phase. Sufficient filler is usually added to initiate crystallization and strengthen the polymer. However, the

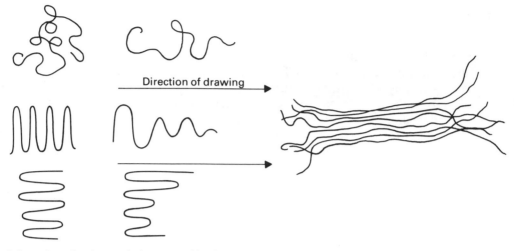

Direction of drawing

Orientation of polymer chains caused by drawing

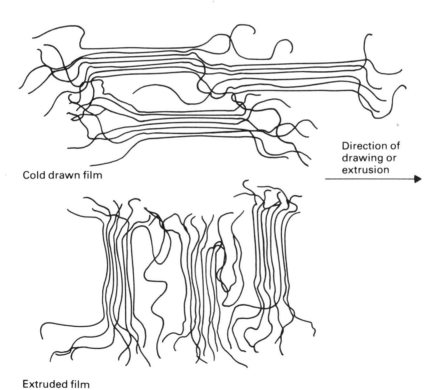

Cold drawn film

Direction of drawing or extrusion

Extruded film

Figure 4.11 Effect of fabrication forces upon polymer chains.

addition of too much filler may restrict cold drawing and molecular orientation, and increase the risk of brittle failure.

There are several modifying agents that can be added to polymers, such as fillers. *Plasticizers* can also be added: these are usually polymers of low-molecular weight that separate the polymer chains and reduce the degree of crystallinity. In glassy polymers, plasticizers increase the ductility and toughness and reduce the stiffness and brittleness. Their main effect is to lower the value of T_g for the polymer.

4.10 POLYMER REACTIONS

Addition polymerization is the joining together of the mers (e.g. $-CH_2-CH_2-$) which make up a polymer chain. An *initiator* such as a radical with either a free electron or an ionized group is added to the basic polymer material, e.g. ethylene C_2H_4. One of the unsaturated bonds, e.g. the carbon double bond in ethylene $H_2C=CH_2$, is broken by attraction to the radical. The other free electron in the mer then attracts an electron in another mer, and so on. When the chain attaches to another radical (the *terminator*) growth is complete. This process is shown in Fig. 4.12 for the growth of polyethylene. The reaction rate can be controlled by the initiators and terminators present. It is possible to form very long chains in just a few seconds, and this type of rapid reaction is known as *chain reaction polymerization*. If a polymer chain is formed by the addition and growth of two different mers, e.g. the formation of polyvinyl chloride, then this is known as *copolymerization*.

Condensation polymerization involves the chemical reaction of two or more chemicals to produce a new molecule. This reaction usually produces a by-product, e.g. water, which must condense; hence the name of the process. The new molecule then becomes the monomer for the formation of a polymer chain or network. This type of reaction can produce either linear thermoplastic polymers, e.g. nylon 6.6 (see Section 4.20.7), or branched-chain thermosetting polymers, e.g. phenol formaldehyde (see Section 4.19.1). This process is slower than addition polymerization because two molecules are required for each step (*step reaction polymerization*). Linear polymers that are formed have a lower degree of polymerization. A catalyst is often required, and it can sometimes be used to control the molecular weight of the polymer.

4.11 STRENGTH OF POLYMERS

Increasing the molecular weight of a polymer may increase the tensile and compressive strength. The mechanical properties may be improved by copoly-merization, or by alloying a weaker polymer to improve its wear characteristics.

Linear polymers are composed of long chains that are entangled and form an amorphous structure, having relatively weak bonding between individual

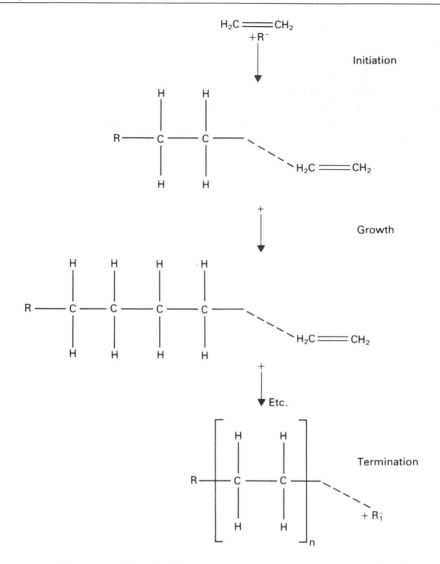

Figure 4.12   Growth of a polyethylene chain by addition polymerization.

chains. *Branched chain polymers* are formed from two chains. This causes strengthening and stiffening due to increased entanglement, as shown:

$$A - A - A - A - A - A - A$$
$$A - A - A - A - A$$

This is typical of elastomers or polymeric rubbers.

Cross linking is the bonding of individual chains to form strong rigid structures. For example,

$$
\begin{array}{c}
A - A - A - A - A \\
| \\
A - A - A - A - A \\
| \\
A - A - A - A - A \\
| \\
A - A - A - A - A
\end{array}
$$

Cross linking usually produces thermosetting materials, some of which can withstand temperatures of $200°C$ ($392°F$) compared to $120°C$ ($248°F$) for many linear polymers. However, several of the newer polymers such as PES (Section 4.20.17) and PEEK (Section 4.20.15) have service temperatures up to $300°C$ ($572°F$). *Chain stiffening* is a method of reducing chain mobility and thereby increasing the rigidity of a polymer. This is achieved by substituting a large molecular group into the monomer; e.g., substituting a benzene ring for a hydrogen atom in the polyethylene monomer produces the polystyrene monomer.

Some crystalline-type polymer structures can be obtained by alignment of the chains, which may be achieved by slow cooling or molding. The effects are to lower the impact strength, raise the melting point, lower the solubility in solvents and generally improve the strength of the material. Plasticizers are added to polymers to act as internal lubricants, and to improve the flexibility due to easier chain movement. They improve the toughness but are often detrimental to the tensile strength of the material. Fillers are added to polymers and their effect is to reduce the movement of the polymer chains. They can increase the strength, or improve the frictional wear effects if the filler is a lubricant. The filler may be powder, fiber, fabric, inert inorganic material, etc. Some fillers can reduce the water absorption of the polymer, but may make fabrication more difficult, depending upon the properties of the filler material.

*   *   *

Self-assessment exercises

1 Explain what is meant by the degree of crystallinity of a polymer.

2 Explain the difference between the glass transition temperature and the crystalline melting point for polymers.

3 What effects occur when a polymer is cooled to the crystalline melting temperature or the glass transition temperature, and then to lower temperatures?

4 What effects are produced when a polymer is *rapidly* heated or cooled through either the glass transition temperature or the crystalline melting temperature?

5 What happens when a *semicrystalline* polymer is cooled through the glass transition temperature?

6 Explain the meaning of the terms:
   (a) crystallite;
   (b) spherulite.

7 Outline the basis of the early fringed-micelle theory for explaining the structure of polymer crystals. Explain how this theory differs from more recent theories, and why none of the theories is generally applicable.

8 Explain the meaning of the terms crystallization and crystallizability.

9 Explain the effects of the following processes upon the properties and structure of polymers:
   (a) annealing;   (d) rolling;
   (b) cold working;   (e) crystallization.
   (c) drawing;

10 What influence do tie molecules have upon the mechanical properties of a polymer? Why is a filler or a plasticizer added to a polymer?

11 Describe the reactions by which selected polymers are formed.

12 Define the terms: stiffness, toughness, rigidity, strength, hardness and brittleness.

13 How can the strength of polymers be improved?

*   *   *

4.12 TESTING OF POLYMERS

Mechanical properties and testing methods are described in Section 6.2. These tests were designed for use with crystalline materials, mainly metals and alloys. Although the same basic principles still apply, other materials possess particular properties or exhibit characteristic behavior. Therefore special tests, or particular features of standard tests, have been developed for materials such as polymers, concrete, ceramics, etc. In this section the mechanical behavior of polymers will be discussed, and also the range of tests which may be performed. This section should be studied in conjunction with Section 6.2.

The specification of a material for a particular application depends upon many factors. One essential consideration is that the material selected possesses the properties required for it to perform satisfactorily when subjected to the in-service conditions.

4.12.1 Standard tests, specimens and procedures

A wide range of standard tests have been established by particular institutions, some of which have gained worldwide recognition and acceptance. British Standards (BS) and the American Society for Testing and Materials (ASTM) Standards provide extensive coverage and information relating to materials testing, and they are widely used. Other standards exist, and some countries have developed their own standards and codes of practice.

The following conditions should be observed when materials testing is performed:

(a) use standard test specimens in respect of size, shape, fabrication, etc.;
(b) store specimens and perform tests under standard conditions, e.g. $23°C \pm 2°C$ ($73.4°F \pm 3.6°F$) and 50% $\pm$ 5% humidity are common conditions for polymers;
(c) follow standard test methods and procedures;
(d) consult the most recently published edition of a standard.

With regards to conditions (a) and (c), the actual part that is being designed and the in-service conditions may be very different from both the standard specimen and standard test conditions specified. However it is important that standard tests are performed; only then can comparisons be made of results for different materials. If nonstandard tests are performed, it is important that the possible effects of changes in the test conditions are considered. It is necessary to repeat the tests several times in order to obtain accurate and reproducible results. Details of the recommended number of tests and the statistical treatment of results are included in BS and ASTM Standards. Reference should be made to the appropriate standards handbook, but the following standards would represent a useful introduction to the subject.

BS2846: Guide to statistical interpretation of data.
Part 1: Routine analysis of quantitative data.
Part 2: Estimation of the mean:confidence limit.
Part 3: Determination of a statistical tolerance interval.
Part 4: Techniques of estimation and tests relating to means and variances.
Part 5: Power of tests relating to means and variances.
Part 6: Comparison of two means in the case of paired observations.

ASTM standard practices for:
ASTM D1898: Sampling of plastics
ASTM E105: Probability sampling of materials
ASTM E122: Choice of sample size to estimate the average quality of a lot or process
ASTM E141: Acceptance of evidence based on the results of probability sampling
ASTM E177: Use of the terms precision and accuracy as applied to measurement of a property of a material
ASTM E178: Dealing with outlying observations
ASTM E691: Conducting an interlaboratory test program to determine the precision of test methods

Reference should also be made to Section 4.16: Design Considerations, and to BS2782 and BS4618, which are quoted there.

4.12.2 Mechanical tests

Standard tensile tests, impact tests and hardness tests are described in Section 6.2. In this section discussion is restricted to conditions that apply particularly to polymers.

Tensile tests are performed using the standard dog-bone or dumbell-shaped specimen, although sometimes tests are performed on sheets of polymer materials of specified dimensions. Polymers are subjected to stresses considerably less than the yield stress and the strength of a material is often specified as *Young's modulus, modulus of elasticity* or *tensile modulus*. This is an important design consideration and is very different from metals where there is no relation between strength and Young's modulus.

Izod or Charpy impact tests with notched specimens are used to determine the *toughness* of a material. Alternative information regarding toughness can be obtained from part of the area under the tensile stress-strain curve. However, this can be misleading, as fracture studies have shown (see Section 6.2.3).

Hardness tests are performed on polymers; the loads required are considerably less than for metals and a round ball indentor is often used. *Scratch tests* are also performed, and *abrasion tests* are used to determine the rate of surface wear by mechanical rubbing.

Flexural testing consists of measuring the force required to cause a simple beam to deflect 5%. Thermoplastic polymers do not usually fail under such conditions, and their quoted flexural strength is the force required to cause a deflection of 5%.

When performing mechanical tests on polymers some caution should be exercised with respect to time effects and time restraints. Effects such as creep (Section 6.2.5) and fatigue (Section 6.2.6) do not become apparent from tensile or impact testing, but may be serious in-service problems. The results of accelerated tests, e.g. for water absorption, may not be representative of the actual conditions to which a final component is subjected. Other basic measurements such as the *secant* and *tangent modulus* may be more useful (see Section 6.2.1).

4.12.3 Mechanical properties and structure

A variety of structural factors influence the mechanical behavior of polymers. However, there are certain basic principles that apply to many polymers. As the degree of crystallinity possessed by a polymer increases, the tensile strength increases, and the yield strength and the stiffness are also affected. If the yield strength and stiffness increase with a corresponding change in the tensile strength, then a brittle polymer is usually obtained. Polymers and metals must possess a certain amount of ductility to be useful for engineering applications. The density of a polymer can be used as a measure of the degree of crystallinity, and the viscosity of the polymer melt as an indication of the molecular weight. If these values are plotted (using a logarithmic scale for the melt viscosity), then it can be shown that:

    (a) an increase in either value produces an increase in the tensile strength;

    (b) in general, an increase in molecular weight has little effect upon stiffness, although for polyethylene there is an increase in modulus with molecular weight;

    (c) an increase in crystallinity produces increased stiffness.

For many polymers the mechanical properties depend upon not only the degree of crystallinity but also the molecular weight. Different criteria that can be used to calculate an average molecular weight value were mentioned in Section 4.5. One such value is the number average molecular weight ($\overline{M}_n$), this can be calculated from the equation:

$$\overline{M}_n = \frac{\Sigma x_i M_i}{\Sigma x_i}$$

where x_i is the number of molecules in the ith size fraction and M_i is the average molecular weight in the ith size fraction.

Polymers that possess a unique molecular weight, e.g. proteins, are called *monodisperse*, whereas polymers consisting of a range of molecular weights are called *polydisperse*. The ratio of the weight average to the number average molecular weights is a measure of the polydispersion within a polymer. Monodisperse polymers have a value of 1, and for polydisperse polymers the value is greater than 1. Several techniques have been developed for determining both weight and number average molecular weights.

4.12.4 Stress–strain curves

Polymers that are viscoelastic materials (see Section 4.14) may not obey Hooke's law (see Section 6.2.1) at temperatures below T_g, but stress–strain curves can still provide useful information. The curves can be used to provide a classification of polymer materials and their properties as shown in Fig. 4.13. A hard tough polymer has a high value of elastic modulus; any deformation is elastic below the yield point and Hooke's law is obeyed. Beyond the yield point the elongation is plastic, and the area under the curve is a measure of the toughness of the material (drawing does not occur for the curves shown in Fig. 4.13).

All stress–strain curves are time and temperature dependent and, as discussed in Section 4.12.1, standard testing conditions must be employed. If the stress rate is reduced, then a lower value of elastic modulus is obtained, this also occurs if the temperature is increased.

Poisson's ratio (see Section 6.3.2) is the ratio of the lateral contracting strain to the elongation strain, due to application of a tensile load. This ratio can be used to provide some indications regarding the mechanical properties of a polymer. The value of this ratio is approximately 0.5 (at constant volume) for elastic materials and it decreases as the stiffness increases. For very stiff materials the value is about 0.25.

4.12.5 Thermal tests

There are several tests available that can be used to measure the thermal properties of polymers. The *Vicat penetration test* records the temperature at

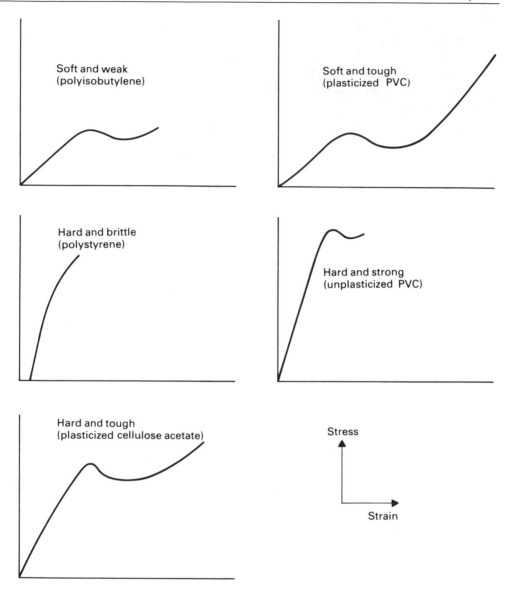

Figure 4.13 Tensile stress–strain curves for different polymers.

which a standard needle penetrates the specimen to a particular depth. The temperature of the specimen is increased at a constant rate.

Tests have been developed to determine the coefficients of linear and cubical thermal expansion. The brittleness of a cantilevered specimen at low temperatures is usually measured from the results of an impact test. Other thermal tests include deflection–temperature tests and flammability tests.

4.12.6 Environmental tests

Standard tests have been devised to measure the corrosive effects of the environments in contact with polymers. The atmosphere is the least severe, but several polymers are affected by the ultraviolet radiation in sunlight, e.g. polyvinyl chloride. Oxidation can occur from the air or oxidizing acids, e.g. for polystyrene, especially at moderate temperatures. The effect of corrosive liquids is usually determined as a weight loss, or by changes in the mechanical properties, e.g. hardness or flexural strength.

Environmental stress cracking tests have been developed to measure the effects of polar solvents and aqueous detergents, particularly upon containers, sheet, pipe and blow-molded vessels. Standards are available for *weathering tests*, although because of the time involved accelerated processes must usually be employed. It is doubtful if there is a reliable correlation between the results obtained from accelerated tests and from natural weathering tests.

The effects of solvents are usually determined by total immersion of polymer specimens. Excessive attack can be observed by changes in the surface or the shape, such as swelling. Alternatively changes can be detected by measurements of the weight, hardness, dimensions or electrical properties.

4.12.7 Other tests

A wide range of other tests is available for the determination of polymer properties. Accurate measurement of the *density* or *specific gravity* of a specimen is obtained using a density gradient column, or the *bulk density* or *apparent density* of granular molding powders in a pycnometer. Because of the extensive use made of polymers in electrical applications, many tests and standards have been established. These include determination of:

(a) *arc resistance*, the ability of a polymer to withstand a high-voltage-low-current discharge across its surface;
(b) *dielectric strength*, the maximum voltage that a thick plastic sheet can withstand for a given time without failure;
(c) *dielectric constant*, the ratio of capacitance of a condenser made from a polymer to that of a condenser using an air gap;
(d) *power factor*, the ratio of the power loss to the total energy passing through a polymer capacitor.

Standards concerned with the determination of optical properties include procedures for measuring the *refractive index, haze* and *luminous reflectance.*

4.12.8 Nondestructive testing

The tests described in Sections 4.12.2–7 often result in the destruction of the specimen. Sometimes it is necessary to preserve a component after testing has been performed, e.g. inspection of a large or intricate fabricated component. Nondestructive testing is also discussed in Section 7.24 for the inspection of welded joints; some of the methods used are similar to those employed with polymers. The main testing methods are:

(a) visual examination, sometimes using reflected or transmitted light;
(b) ultrasonic techniques, useful for detecting porosity, voids and non-uniform regions;
(c) radiographic techniques;
(d) microwave or radar applications;
(e) holographic;
(f) nuclear quadrupole resonance.

* * *

Self-assessment exercises

1 Prepare a table for easy reference and comparison, listing the polymer properties that need to be determined and the testing methods available.

2 Why is it important to measure the coefficient of expansion of a polymer?

3 What properties are important when selecting a polymer for applications involving:
(a) atmospheric exposure;
(b) friction and wear;
(c) constant loading;
(d) fluctuating loading;
(e) exposure to water or solvents;
(f) low temperatures.

4 Is there any value in performing standard tests when the in-service conditions are very different?

5 Why are the Izod and Charpy tests not completely satisfactory for the determination of toughness?

6 Explain how the density and melt viscosity of a polymer can be used as measures of the degree of crystallinity and the molecular weight, respectively.

7 Are the data from a polymer tensile test applicable to real situations? Since many polymers do not obey Hooke's law, is it necessary to plot stress–strain curves?

8 Derive an expression' that includes the elastic modulus of the material, for the change in dimensions of a specimen due to an applied load.

9 Which method(s) would you select to detect any imperfections in a fabricated polymer component, if results are required 'immediately'?

* * *

4.13 RHEOLOGY

Study of the deformation and flow of matter is known as *rheology*. Polymers that exhibit both solid and fluid properties are termed *viscoelastic* materials; their properties will be discussed in Section 4.14. For polymer processing, the rheological properties of melts and elastic solids are of most relevance.

Viscosity is the resistance to flow of a fluid; it produces a frictional energy loss which appears as heat. Many polymers below T_g obey Hooke's law, where the applied shear stress is proportional to the elastic deformation or strain. The *shear modulus of elasticity* is given by the ratio of the shear stress to the shear strain. This behavior is independent of time and occurs when an amorphous polymer is subjected to a small shear stress. The fluids described so far are *ideal* or *Newtonian*, and for situations of *dynamic shearing* (i.e. fluid flow) the viscosity is constant with respect to time. The shear stress is proportional to the velocity gradient within the flowing fluid; the proportionality constant is the coefficient of viscosity. This is a statement of Newton's law for viscous flow.

For most practical situations, fluids are not Newtonian and the viscosity changes with time. If viscosity increases it is called *rheopexy*; if it decreases it is called *thixotropy*. Since the shear rate is proportional to the applied stress for ideal fluids, if the shear rate does not increase as rapidly as the applied stress the system is *dilatant*. For most polymers the shear rate increases more rapidly and it is *pseudoplastic*. As the shear rate increases, a fluid may first exhibit Newtonian behavior which becomes non-Newtonian, and then ideal again at high shear rates. If a certain stress level must be applied before flow commences, the fluid system is termed *Bingham*.

There are numerous methods for the measurement of viscosity, and instruments that can be used. However there are three main categories:

(a) *capillary viscometer* – the polymer solutions flow through standard capillaries under gravity;
(b) *rotating disk* – a disk or cylinder is suspended on a calibrated wire and the disk is rotated in a molten polymer;
(c) *ram extruder*.

Newtonian flow is encountered in the processing of polymers by extrusion, milling, molding, etc. (Section 4.15). The viscosity of a polymer increases with pressure and decreases as the temperature is lowered. The viscosity increases with the molecular weight until the chains become entangled; the rate of increase of viscosity then rises. Successful polymer processing depends upon both the structure and the molecular weight distribution.

4.14 VISCOELASTICITY

A molten polymer subjected to low shearing rates behaves like a viscous Newtonian fluid; with high shear rates the behavior is characteristic of an elastic solid. The actual behavior of most polymers is complex and falls between the two extremes described by Hooke's law and Newton's law. Even when a polymer solidifies, viscous forces are still present.

The application of a load to a linear amorphous polymer (above T_g) produces an immediate elastic response. This is followed by time-dependent viscous flow which decreases until equilibrium is attained. When the load is removed there is an immediate elastic recovery, followed by time-dependent recovery. If the load is only applied for a short time, the behavior approaches that of an elastic solid and the material is said to possess a good *polymer memory*. The longer the load is applied then the greater the remaining *permanent strain*, and the polymer possesses little memory. This viscoelastic behavior of a polymer is shown in Fig. 4.14. If a polymer is stressed for a long time then the chains become disentangled and slip, and the flow is called *creep*.

The viscoelastic behavior of a polymer can be represented by models that use a spring (Hooke's law) and a dashpot (Newtonian). The *Maxwell model* uses these devices in series and the resulting creep curve is shown in Fig. 4.15. This is composed of initial elastic response by the spring followed by slow viscous flow in the dashpot. The *Voigt model* combines these devices in parallel and exhibits retarded elastic (creep) deformation and retarded elastic recovery. In this model the total stress is absorbed by the dashpot and is transferred to the spring for equilibrium. The deformation curve for the Voigt model is shown in Fig. 4.16. These models can be combined, but they still represent only an approximation of the actual behavior. It is possible to investigate viscoelastic flow and elastic behavior independently.

If a tensile test is performed under constant strain conditions, then the stress decreases with time. This is known as *stress relaxation*, it is important for certain in-service conditions particularly at elevated temperatures. Stress relaxation can cause the removal of internal stresses within amorphous thermoplastic polymers near the glass transition temperature.

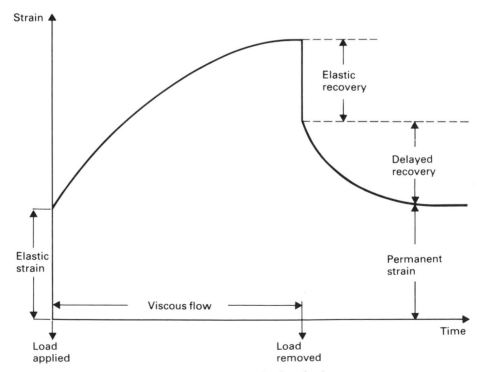

Figure 4.14 Viscoelastic behavior of polymers.

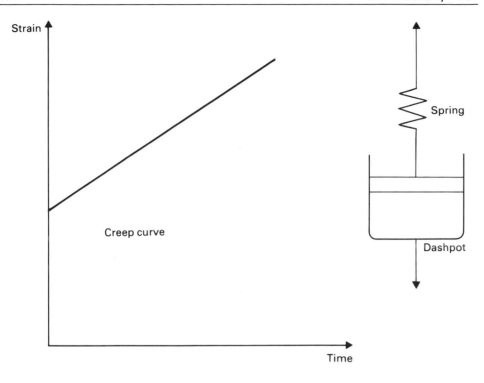

Figure 4.15 The Maxwell model for viscoelastic systems.

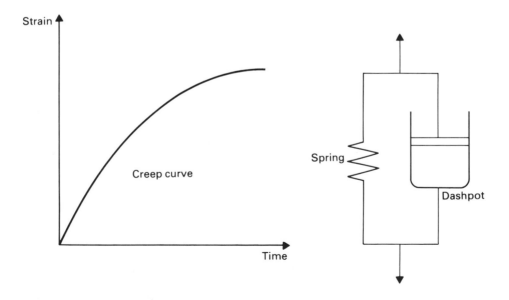

Figure 4.16 The Voigt model for viscoelastic systems.

* * *

Self-assessment exercises

1 Explain the meaning of the term viscosity.

2 Define the terms:
   (a) rheology;   (d) rheopexy;
   (b) viscoelastic;   (e) thixotropy;
   (c) Newtonian fluid;   (f) pseudoplastic.

3 Describe the type of behavior that obeys Hooke's law and Newton's law.

4 Describe two methods for measuring the viscosity of an ideal fluid and a viscous polymer.

5 Explain how the viscosity of a polymer influences the choice of a fabrication process.

6 Explain the meaning of the following terms:
   (a) recovery;   (d) creep;
   (b) polymer memory;   (e) stress relaxation.
   (c) permanent strain;

7 Explain how the Maxwell and Voigt models can be used to describe the viscoelastic behavior of polymers. Why does neither model provide a true representation of the actual behavior of most polymers?

* * *

4.15 FABRICATION METHODS

Refer to Figs. 4.17–26 for illustrations of some of these processes.

4.15.1 Compression molding (Fig. 4.17)

This is the most common method for shaping thermosetting materials. It is usually used with thermoplastics only if the part required is too large to be produced by injection molding. The equipment consists of a press and a punch and die unit; the cavity chamber is heated. This cavity is the same size and shape as the required product, including shrinkage allowance.

With a positive mold an exact amount of polymer powder is placed in the die before it is closed; this determines the thickness of the finished product.

Using a flash mold (or semipositive mold) a slight excess of material is used which is forced into a flash gutter, leaving a fin or flash to be removed from the finished product.

Transfer molding is used for delicate or intricate shapes, or when a flash is not acceptablé. The material is heated in a separate chamber and forced into the closed mold through small channels.

In all cases, when the mold has been filled the temperature and pressure are held for about six minutes to cure the material, then the mold is opened and the product is ejected.

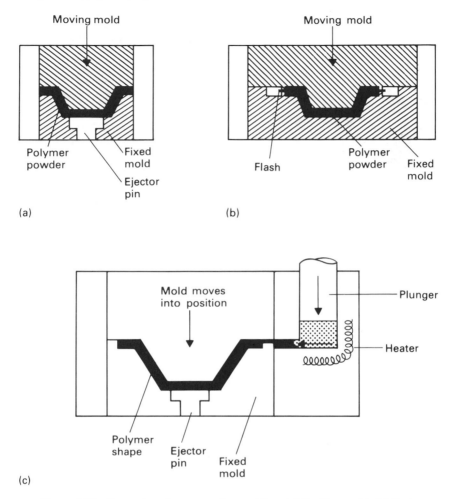

Figure 4.17 Examples of compression molding. (a) Positive mold; (b) flash mold; (c) transfer molding.

Cold molding involves compressing the material in a cold mold, and then heating in an oven to achieve hardening. It is used for rapid production of thick shapes, although the surfaces tend to be inferior to those produced in a heated mold and warping can also occur.

4.15.2 Injection molding (Fig. 4.18)

Injection molding is an extremely quick fabrication method and is suitable for most thermoplastic materials. The material must be softened by heat and flows under pressure to assume the required shape. The required quantity of material is fed to the heated injection chamber, and then forced by a plunger into the closed chamber. After filling, the chamber is rapidly cooled and when the material has solidified, the shape is ejected.

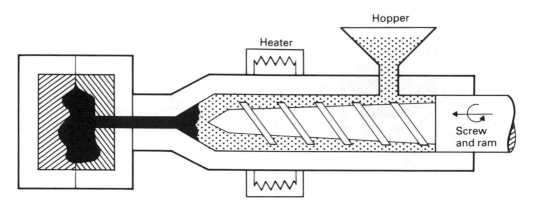

Figure 4.18 Injection molding.

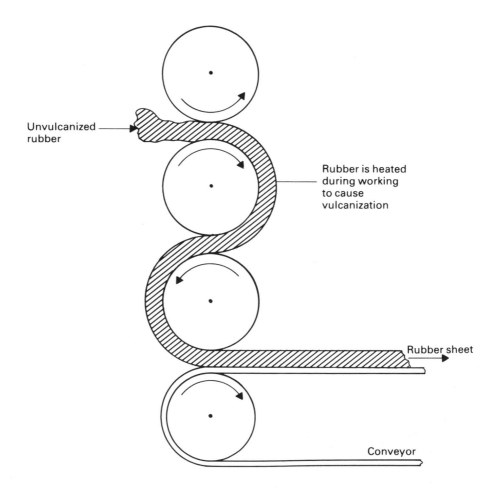

Figure 4.19 Calendering.

4.15.3 Extrusion molding

This process is similar to injection molding, the difference being that it is a continuous rather than a batch process. The material is fed continuously through the heating zone using a rotating screw, and is forced through a metal die of the required cross section.

4.15.4 Casting

This is similar to the process used for metals (see Chapter 8) and involves pouring molten polymer or catalyzed resins into a mold (which may be open at the top), without the use of pressure. This material is then solidified or cured. The molds can be made of metal, lead, rubber or PVC. Polymer film can be produced by pouring liquid polymer onto a rotating drum and continuously removing the solid product.

4.15.5 Calendering (Fig. 4.19)

The polymer material is fed as a warm doughy mass and passed between heated rollers. It is thoroughly worked and the last roll determines the sheet thickness. Elastomer sheets are often produced by this process and it can be used to produce a coating for paper or fabric.

4.15.6 Extrusion blow molding (Fig. 4.20)

A tube of heated polymer is extruded into the mold, air is injected and the material expands to assume the desired shape. The extruded material can have its end closed by the mold movement. It is a quick process and can produce hollow shapes with a reasonably uniform thickness from thermoplastic materials.

4.15.7 Blow forming (Fig. 4.21)

Blow forming produces part-spherical shapes without a mold by using compressed air. For other shapes the sheet material is blown into a female mold. However, local thinning can occur with this process.

4.15.8 Simple pressing (Fig. 4.22)

This method involves shaping a heated sheet clamped at the edges, by using a movable former.

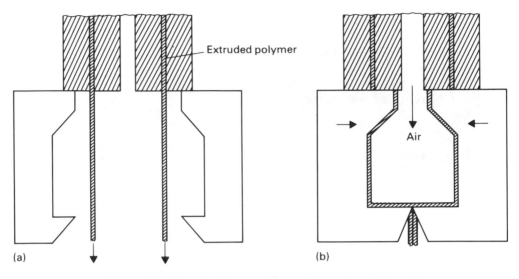

Figure 4.20 Extrusion blow molding. (a) Extrusion; (b) blowing.

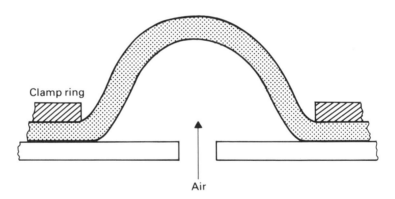

Figure 4.21 Blow forming.

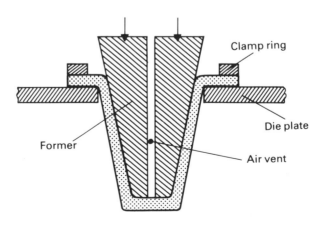

Figure 4.22 Simple pressing.

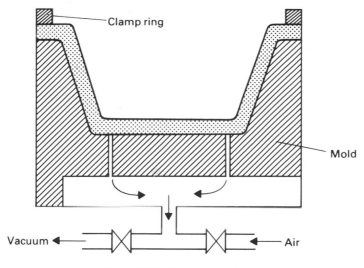

Figure 4.23   Vacuum forming.

4.15.9 Vacuum forming (Fig. 4.23)

A heated sheet of material is clamped at the top of the mold, and as it sags it is sucked into shape using vacuum. It is cooled while in the mold using compressed air, which is also used to eject the finished shape. It is suitable for low volume production but the vacuum also causes localized thinning.

4.15.10 Vacuum-assisted pressing

This is simple pressing using vacuum forming during the early stages.

4.15.11 Plug-assisted vacuum forming (Fig. 4.24)

This is used when the molding has a deep recess in an otherwise flat section.

4.15.12 Drape forming (Fig. 4.25)

The sheet material is clamped in position and the mold moves and stretches the material; the final shape is produced by vacuum. It produces a uniform product thickness.

4.15.13 Bubble forming (Fig. 4.26)

This method is similar to drape forming. Sheet material is clamped in position, the bubble is formed by compressed air, the mold moves into position and the final shape is produced by vacuum.

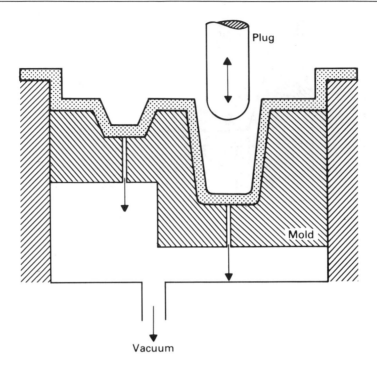

Figure 4.24 Plug-assisted vacuum forming. The plug is only used for pre-forming the large recess, before vacuum is applied.

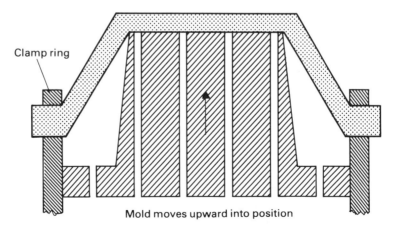

Figure 4.25 Drape forming. The final position is obtained and then vacuum is applied (downwards) to produce an accurate shape.

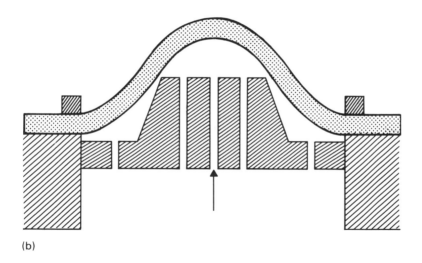

Figure 4.26   Bubble forming. (a) Bubble is formed; (b) mold moves into position; (c) vacuum is applied around mold (not shown).

4.15.14 Laminates

Laminated products are sheets of paper, fabric or glass fiber (as reinforcement) which are impregnated with a thermosetting resin material. The process used can be high-pressure (with heating of several layers), or a low-pressure technique involving curing of the built-up product.

*   *   *

Self-assessment exercises

Study Sections 4.15.1 to 4.15.14. Describe how the following component shapes can be fabricated.

1 A plastic milk bottle.

2 A toothbrush.

3 An automobile steering wheel.

4 A garden fish pond.

5 A 'biro' body and cap.

6 Pieces of a child's construction toy.

7 Small, lightweight polymer packaging pieces.

8 A computer keyboard casing.

9 Packing pieces for transporting a television or a home computer.

10 A garden hose.

11 Stacking plastic chairs.

12 A motorcycle helmet.

13 A plastic identity disk with a company logo on one side and a number on the reverse.

14 Rubber sheets to be used to absorb machine vibration.

*   *   *

4.16 DESIGN CONSIDERATIONS

The shaping and fabrication methods generally employed with polymer materials have been described in Sections 4.15.1–14. Polymer materials have many engineering applications mainly because of their relative cheapness, corrosion resistance and good electrical properties. However, water absorption and generally lower elastic moduli than metals can cause problems when using some materials in certain situations. Polymer materials can be joined by using solvents, by heating or by welding. Thin sheets of material can be joined by heating and applying pressure along the edges, thus forming a low-strength seal. Polymers are generally machined by conventional methods (see Section 8.4). Care must be taken to observe the correct operating conditions because of the lower melting points of thermoplastics and the brittle nature of some thermosetting materials. Polymers are subjected to mechanical tests (see Sections 4.12–4.12.8) similar to those performed on metals; details are given in *BS2782: Methods of testing plastics*.

Reference should also be made to *BS4618: Recommendations for the presentation of plastics design data.*

The appropriate US standards can be found in ASTM Standards, volumes 08.01, 08.02 and 08.03. Tests are also made for the softening temperature, thermoplasticity and water vapor absorption. Polymers are often poor conductors of heat, and temperature gradients within a specimen can cause internal stresses and associated problems of fatigue in the material.

Polymers are frequently employed in situations which make them act as load-bearing materials. Care must be taken to select a polymer possessing the required properties for the in-service situation. Material selection generally requires consideration of the coefficient of thermal expansion, the stiffness, effect of temperature on the mechanical properties, creep and fatigue, flammability and water absorption. *Anisotropy* is the variation in the properties of a reinforced polymer depending upon the direction of loading; it should be considered for component design. Polymers are used in systems which experience friction or wear. These can be sliding systems, or plain bearings where a shaft rotates in a fixed cylinder sometimes called a bushing. Friction is the tendency for two moving surfaces to bond together but because of the dissimilar nature of polymer and metal surfaces and their molecules, frictional effects can be reduced. Wear can be adhesive between two smooth surfaces, or abrasive between a hard and a soft material. Perspex-type materials (see Section 4.20.8) are used as substitutes for glass because of their nonsplintering property, and also for curved surfaces. Molded materials are used for small castings, and built-up laminates increase the strength of large castings. Plastic adhesives, especially metal-to-metal joints, have important applications in engineering design and construction. Polymers are used in the food industry where odor and taste, and the ability of the material to hold its color, are important considerations.

* * *

Exercises

1 What are the advantageous properties of polymers when considered for engineering applications?

2 What are the disadvantages associated with using polymer materials, and how are they avoided or overcome?

3 Describe typical test procedures that can be carried out with polymers.

4 Identify particular polymer materials that could be used under extreme conditions, e.g. low temperature, moderate pressure, exposure to polluted atmospheric conditions.

5 Compare the mechanical and physical properties of common polymer materials with those of steels and nonferrous alloys.

6 Compare the prices of polymers and metals.

7 Obtain data to compare the production and usage of polymers and common alloys over the last ten years.

8 Identify design problems and situations where polymers could be substituted for metals and alloys.

* * *

4.17 TYPICAL POLYMER MATERIALS

There are thousands of polymer materials available for use in engineering situations. Some of these have general applications while others are more specific, some have no practical application (at the present) and others are being superseded for technical or economic reasons. Sections 4.18–20 describe briefly some of the more common and important polymers under headings of natural resins, thermosetting materials and thermoplastic materials. Reference should be made to the classification in Section 4.4. Typical properties of selected polymers and elastomers are given in Table 4.1, and the effect of temperature upon the modulus of elasticity for selected polymers is shown in Fig. 4.27.

4.18 NATURAL RESINS

These materials are either exuded directly by certain trees or obtained from insects which feed on the sap juices. They are used in the manufacture of varnishes, paints and enamels, soaps, inks, glues, and as plasticizers or modifiers with other polymer materials.

4.18.1 Shellac

Shellac is a thermoplastic material in its molded form. It has resistance to a wide

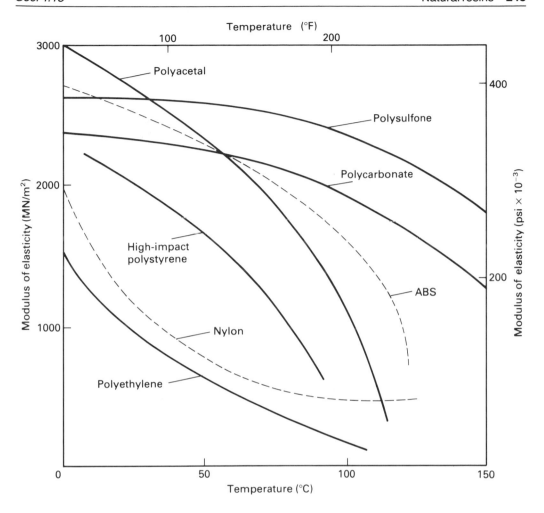

Figure 4.27   Effect of temperature on modulus of elasticity for selected polymers.

range of solvents and can be used as a hard gloss coating, or as a bonding agent with a low coefficient of expansion. It has been superseded in certain applications by less brittle and more flexible materials but is still used in grinding wheels (Section 8.4.10), where some elasticity is required (which is provided by the heat generated). Some mica products are bonded with shellac, and it is also used in polish, wax, ink and lacquer.

4.18.2 Rosin

Rosin is produced by the distillation of pine tree sap, turpentine is produced as a by-product. Rosin is used in paints and varnishes and in the manufacture of soap, paper and ink.

Table 4.1 Properties of polymers and elastomers

Name, structure and state	T_g °C (°F)	T_m °C (°F)	Specific gravity	Tensile strength MN/m² (psi)	Elastic modulus MN/m² (psi $\times 10^{-3}$)	Rockwell hardness	Izod impact kg–m (lb–ft)	Coefficient of expansion °C $\times 10^6$ (°F $\times 10^6$)
			Thermoplastic polymers					
Polyethylene [–CH₂–CH₂–]ₙ High density, linear (95% crystalline):	−120 (−184)	115 (239)	0.95	27.6 (4000)	828.0 (120)	40	0.83 (6)	216 (120)
Low density, branched (60% crystalline):	−120 (−184)	138 (280)	0.92	13.8 (2000)	172.5 (25)	10	2.21 (16)	180 (100)
Polypropylene $\left[-CH_2-\overset{\displaystyle CH_3}{\underset{\displaystyle \;}{CH}}- \right]_n$ isotactic, 60% crystalline	−10 (14)	176 (349)	0.91	34.5 (5000)	1380.0 (200)	90	0.69 (5)	308 (170)
Polystyrene [–CH₂–CH–]ₙ atactic, amorphous	100 (212)	—	1.05	48.3 (7000)	3105.0 (450)	75	0.04 (0.3)	68.5 (38)
Polytetrafluoroethylene (PTFE) (Teflon) [–CF₂–CF₂–]ₙ 85% crystalline	−120 (−184)	327 (621)	2.15	17.25 (2500)	414.0 (60)	70	0.55 (4)	99 (55)
Polyamides (6.6 Nylon) $\left[-NH-\overset{\displaystyle \;}{\underset{\displaystyle O}{\overset{\| }{C}}}- \right]_n$ crystallizable	57 (135)	265 (509)	1.10	81.42 (11800)	2829.0 (410)	118	0.14 (1)	99 (55)

Table 4.1 continued

Name, structure and state	T_g °C (°F)	T_m °C (°F)	Specific gravity	Tensile strength MN/m² (psi)	Elastic modulus MN/m² (psi × 10⁻³)	Rockwell hardness	Izod impact kg−m (lb−ft)	Coefficient of expansion °C × 10⁶ (°F × 10⁶)
Thermosetting polymers								
Polycarbonates $\left[-O-C-O-R- \right]_n$ $\overset{\parallel}{O}$	150 (302)	265 (509)	1.2	62.10 (9000)	2415.0 (350)	118	1.93 (14)	45 (25)
mainly amorphous								
Thermoplastic polymers								
Phenolics (phenol formaldehyde)	—	—	1.4	51.75 (7500)	6900.0 (1000)	125	0.04 (0.3)	81 (45)
Epoxies	—	—	1.1	69.00 (10000)	6900.0 (1000)	90	0.11 (0.8)	72 (40)
Silicones $\left[-Si- \right]_n$	—	—	1.75	24.15 (3500)	8280.0 (1200)	89	0.04 (0.3)	36 (20)

Table 4.1 continued

Elastomers (rubbers)	Maximum service temperature °C (°F)	Minimum service temperature °C (°F)	Specific gravity	Tensile strength MN/m² (psi)	Percentage elongation	Resistance to oil or gas	Oxidation resistance	Crystallization on stretching
Natural rubber $\left[-CH_2-C=CH-CH_2-\right]_n$ $\quad\quad\quad CH_3$ Cis-1,4 polyisoprene	82 (180)	−54 (−65)	0.93	20.7 (3000)	800	Poor	Poor	Good
Nitrile or Buna N $\left[-(C_4H_6)_x-CH-CH_2-\right]_n$ $\quad\quad\quad\quad CN$ Butadiene acrylonitrile copolymer	121 (250)	−54 (−65)	1.00	4.83 (700)	400	Excellent	Poor	Very poor
Neoprene (GR-M) $\left[-CH_2-C=CH-CH_2-\right]_n$ $\quad\quad\quad Cl$ polychloroprene	100 (212)	−54 (−65)	1.23	24.15 (3500)	800	Good	Good	Fair

Thermoplastic polymers

4.18.3 Copal resins

Copal resins are used in paints, varnishes and lacquers. When they are mixed with celluloid, a hard film lacquer that is resistant to water and abrasives is produced.

4.19 THERMOSETTING MATERIALS

4.19.1 Phenolic materials

These were the first thermosetting materials to be used in large quantities. Phenol formaldehyde is polymerized by a series of condensation reactions, the monomer bonds at the three unsaturated carbon sites in the phenol molecule (the 2, 4 and 6 sites), forming a highly cross-linked rigid polymer.

phenol + formaldehyde + phenol $\longrightarrow$ [monomer] + water

The liquid resin can be poured into molds and then cured to produce castings, or filler materials and color can be added and the product obtained as a powder. The pure polymer is opaque and milky white, but discolors with age and is therefore manufactured dark brown or black. The tradename is 'Bakelite'. The properties of the product depend upon the filler material used, e.g. wood, flour, asbestos, mica, graphite, fibers, etc. The phenolic materials can only be compression molded and they are used mainly for electrical applications and insulation. The material can also be produced in laminate form as sheets, tubes or rods.

4.19.2 Amino-formaldehyde materials

These are formed by condensation reactions involving an aldehyde and a compound containing an amino group. The most important materials are urea formaldehyde and melamine formaldehyde.

urea + formaldehyde + urea $\longrightarrow$ [monomer] + water melamine + formaldehyde + melamine $\longrightarrow$ [monomer] + water

The properties of urea formaldehyde are similar to those of phenol formaldehyde, except that it has a lower resistance to water and heat. Both materials are produced in a similar manner but urea formaldehyde molding powder is slightly more expensive. The main advantages of urea formaldehyde are that it can be produced in a range of stable colors and does not impart taste or odor to foodstuffs.

Melamine formaldehyde is more expensive to produce but its advantages are improved mechanical strength, heat resistance up to 95°C (203°F) and water resistance. A stable color range is available with no taste or odor effects, also good electrical resistance, and it can be used to produce a very hard surface.

4.19.3 Polyester resins

An ester is the reaction product of an acid and an alcohol. The most common engineering application of thermosetting polyesters is the combination of a glass fiber filling (or reinforcement) with a resin of relatively low molecular weight and a catalyst (hardening agent), to produce a rigid structure. The addition of solvents and oils to the polyester (or alkyd resin) enables it to be used as a hard-drying, coating material. Many different polyester resins are available; the condensation reaction of maleic acid and propylene glycol produces an unsaturated polyester which can then be blended with an unsaturated monomer such as styrene.

Alkyd molding compounds are based on thermosetting polyester-type resins; they are combined with cross-linking monomers and other additives to produce the desired properties. They are fabricated by compression, transfer or injection molding. The most significant properties are low moisture absorption, excellent dimensional stability and good electrical properties. Applications of alkyd compounds include high-impact grades (high glass content) for switchgear and electrical terminal strips, and mineral-filled grades for electronic components.

4.19.4 Allyls

Diallyl phthalate (DAP) and diallyl isophthalate (DAIP) are the commercially significant polymers of the allyl family of phthalate esters. They are used as monomers, and as prepolymers that can be converted to thermoset molding compounds and resins. Materials based on allyl prepolymers are reinforced with fibers and particulate fillers. Allyl moldings maintain their electrical properties at high temperature and humidity levels; they also possess excellent dimensional stability and resistance to moisture, strong and weak acids, alkalis, organic solvents and liquid oxygen. The advantages of allyls compared to polyesters are low toxicity, low evaporation losses during fabrication, and freedom from styrene odor. Applications of allyls include electronic equipment for use in extreme environments, pump impellers and sterilizing equipment.

4.19.5 Epoxy resins

These are more expensive than polyester resins but are extremely important materials. There is a highly reactive epoxy group

$$\overset{\displaystyle O}{(CH_2-CH-)}$$

at each end of the low-molecular weight polymer chain. This resin is often called a polyether as the polymer frequently contains this molecule, e.g.

$$\overset{\displaystyle O}{CH_2-CH}-(R-O-R)_n-\overset{\displaystyle O}{CH-CH_2}$$

The resin can be manufactured in a variety of forms, either solid or liquid. A hardener, usually containing amino groups, is added to the resin to produce the cross-linking which ensures a rigid structure. The hardener is incorporated into the polymer and no by-product is produced; therefore only very small shrinkage occurs. A filler powder and/or fiber reinforcement can also be used. Epoxy resins are used as high strength adhesives, when reinforced with graphite fiber they have improved values of Young's modulus.

4.19.6 Polyurethanes

These materials are produced by the condensation reaction of an isocyanate $(R=N=C=O)$ and an alcohol (ROH) to form the monomer:

$$\begin{bmatrix} & N & O \\ & | & || \\ R- & N-C & -OR \end{bmatrix} + H_2O$$

They can be produced either as hard rigid thermosetting materials or as elastomers, i.e. rubbers. Due to the abrasion-resistance properties, industrial applications include coatings either as varnish or as thicker floor coverings. Carbon dioxide may sometimes be evolved as a condensation by-product, and this may be used to produce the polymer as a hard or soft foam.

4.20 THERMOPLASTIC MATERIALS

(Ethylene-based polymers: Sections 4.20.1–4.20.19)

4.20.1 Polyethylene

Polyethylene, commonly called polythene, possesses the monomer:

$$
\begin{bmatrix}
& H & H & \\
& | & | & \\
- & C & - & C & - \\
& | & | & \\
& H & H &
\end{bmatrix}
$$

The material properties depend upon the chain length; the classifications used are generally low density, high density and ultra-high molecular weight. Low-density polyethylene is manufactured by vapor phase polymerization at high temperatures and pressures, and there is some branching in the product. The high-density materials are linear and highly crystalline and are made by low-pressure addition reactions in solution. Polyethylene has good resistance to many solvents, acids and chemicals. In general, it is tough and flexible and its uses include electrical insulation, cold water pipes, domestic containers, film and sheet for packaging. Its low softening temperature makes it suitable for molding and shaping. Ultra-high molecular weight polyethylene has good resistance to abrasion and wear, but it is more difficult to mold and is generally produced and used in strip form.

4.20.2 Polypropylene

Polypropylene possesses the monomer:

$$
\begin{bmatrix}
& H & H & \\
& | & | & \\
- & C & - & C & - \\
& | & | & \\
& H & CH_3 &
\end{bmatrix}
$$

It is a crystalline polymer, stronger and more rigid than polyethylene and slightly more expensive to produce, but it can be used for load-bearing applications. It is easily injection molded, and has a higher softening temperature and chemical resistance than polyethylene. The main uses are sterilized chemical containers and parts that must repeatedly bend without cracking. It is not wear resistant, unlike ultra-high-molecular-weight polyethylene.

4.20.3 Polyvinyl chloride (PVC)

Polyvinyl chloride possess the monomer:

$$
\left[
\begin{array}{cc}
H & H \\
| & | \\
-C & -C- \\
| & | \\
H & Cl
\end{array}
\right]
$$

The material usually has a lead salt added which acts as a stabilizer and prevents decomposition due to sunlight or heat. Rigid PVC is used for low-cost piping, it is fairly weak, requires support and should not be scratched. Plasticized PVC has a plasticizer added (an oil or solvent) that produces a soft, flexible, rubbery material, commonly known as 'vinyl' and is used for upholstery and rainwear. The quantity of plasticizer may be from 5% (low) to 50% (high) of the final product weight; the amount influences the product properties. The addition of inert filler materials can reduce the cost and alter the properties. Both rigid and plasticized PVC have good chemical resistance, but polyethylene has a better solvent resistance.

4.20.4 Polystyrene

Polystyrene possesses the monomer:

$$
\left[
\begin{array}{c}
H \quad\ H \\
| \qquad | \\
-C - C- \\
| \qquad | \\
H \quad\ C \\
\end{array}
\right]
$$

It is rigid and breaks easily due to the chain stiffening effect (Section 4.11) of the large benzene ring. It is used as an electrical insulator and as a foam for thermal insulation, and may be toughened by copolymerization with acrylonitrile. The addition polymer of acrylonitrile, butadiene and styrene (ABS) is of particular importance due to its resistance to acids and alkalis, and it can be molded to form battery casings, car bodies and telephones.

4.20.5 Polymethyl pentene

Polymethyl pentene (PMP) is a moderately crystalline polymer which is transparent in thick sections. It is almost optically clear, having a light transmission value of 90%, which is slightly less than acrylic. The reason for the transparency is that the crystalline and amorphous phases have the same index of refraction. PMP has a lower specific gravity than other polyolefins; for short periods it retains its physical properties and heat resistance up to 200°C (392°F). For prolonged use above 150°C (302°F), antioxidants must be added. PMP has similar properties to polyethylene and polypropylene except that its properties are retained at higher temperatures; it is comparable with PTFE (Section 4.20.6) up to 150°C (302°F). It is used for laboratory ware, molded food containers, hot liquid level indicators and plumbing systems.

4.20.6 Fluorocarbons

These may be classified either as a member of the ethylene-based polymer family or as a separate group of thermoplastic materials. The oldest and most common fluorocarbon is polytetrafluoroethylene (PTFE), with monomer:

$$
\begin{bmatrix}
& F & & F & \\
& | & & | & \\
- & C & - & C & - \\
& | & & | & \\
& F & & F &
\end{bmatrix}
$$

The tradename is 'Teflon'.

This material has a fluorine atom substituted for each hydrogen atom in the polyethylene monomer. PTFE is a crystalline material up to 325°C (617°F); it does not soften appreciably on heating but retains its strength up to 300°C (572°F) and its flexibility down to −200°C (−328°F). The advantages of PTFE are that it is completely chemically inert and has a low coefficient of friction, hence its applications in resistant coatings, non-stick films, pistons, etc. PTFE cannot be molded and is fabricated by hot pressing and powder sintering. The growth of the fluorocarbon family was due to the need to produce materials which could be injection molded, although these polymers generally are less resistant to heat and chemicals. All fluorocarbons are relatively expensive.

The range of fluorocarbon materials now extends beyond PTFE and includes fluorinated ethylene propylene (FEP), perfluoroalkoxyethylene (PFA) and several other fluoroplastics. Unlike PTFE, which must be shaped by press-and-sinter methods or lubricated extrusion and sintering, FEP can be molded by conventional melt processing. FEP resins possess nearly all the advantageous properties of PTFE, but the recommended service temperature is lower by approximately 50°C (90°F).

PFA resins are easier to fabricate than FEP resins and they are also melt extrudable. The service temperature of PFA is comparable with that of PTFE, and PFA has higher mechanical properties at elevated temperatures than FEP. The uses of FEP include wire and cable insulation for computer and electronic applications, linings (extruded sheet and film) for chemical processing tanks and piping. PFA is used for high-temperature wire and cable insulation, chemical-resistant linings and in semiconductor processing equipment.

Polychlorotrifluoroethylene (CTFE) is particularly sensistive to processing conditions, mainly because thermal degradation begins at approximately 275°C (525°F). For this reason, compression molding is a common fabrication method. The properties of partially degraded polymer are not seriously affected, but a high degree of degradation results in crystallinity and reduced physical properties. Long-term usage above 120°C (248°F) also causes crystallinity.

The properties of CTFE plastic materials include thermal stability, good electrical properties, an applications range of −250 to +200°C (−418 to +392°F) and chemical inertness. CTFE also has the lowest permeability to water vapor of any plastic. CTFE resin has moderate strength and resilience, high compressive strength and good abrasion resistance. The resin also possesses some improved properties compared to PTFE, FEP and PFA resins, e.g. creep resistance and hardness.

Recent developments of the fluoropolymers include modified ETFE (a copolymer of ethylene and PTFE) and ECTFE (a copolymer of ethylene and CTFE). These new materials can be easily fabricated by conventional thermoplastic methods, unlike PTFE and CTFE. The maximum service temperature range (no load) for ETFE and CTFE is 150–175°C (302–347°F) compared with 195°C (383°F) for CTFE and 300°C (572°F) for PTFE. Many of the properties of ETFE and ECTFE resins are superior to those of other fluoropolymers, e.g. higher tensile strength and toughness, and 'no-break' in notched Izod tests. These resins are used for wire and cable insulation, laboratory ware and chemical-resistant linings.

4.20.7 Polyamides and polyesters

Polyamides are produced by the condensation reaction between an organic acid and an amine; for example:

$$R-\underset{\underset{O}{\|}}{C}-OH + NH_2R \rightarrow \left[R-\underset{\underset{O}{\|}}{C}-NHR \right] + H_2O$$

The general name for the linear polyamides is nylons. In nylon 6.6 the suffixes refer to the number of carbon atoms in each of the reacting substances forming the polymer. Nylons are very strong, tough and abrasion resistant; they are also flexible and have high impact strengths. Their softening temperatures are above

200°C (392°F) and this makes molding difficult except in powder form. They are often produced and used as fibers. Nylons are resistant to most organic solvents and strong alkali solutions, but they are attacked by concentrated acids and phenols. Their main disadvantage is high water absorption, and the consequent size change and strength reduction. This effect is less marked with nylon 6.10 and nylon 11 than with nylon 6.6.

Polyamide–imide is known by the tradename 'Torlon' (produced by Amoco Chemical Corp.). It is an injection molded-amorphous thermoplastic, possessing high strength and good impact resistance. Parts can be used continuously at 250°C (482°F) without loss of structural integrity. Injection molded or extruded parts require post-curing to develop optimum physical properties. Polyamide–imide has good radiation resistance, but it absorbs moisture and water vapor. It is used in structural and mechanical components, and for bearings, seals and gears in equipment such as valves, pumps and compressors.

Polyesters can be thermosetting (Section 4.19.3) or thermoplastic materials, depending upon the reacting acid and alcohol. Polyethylene teraphthalate (tradename 'Terylene') is a linear thermoplastic polyester produced by the condensation of teraphthalic acid and ethylene glycol. It is mainly produced as a fiber by extrusion in the molten state, which is similar to the production of nylon fiber.

4.20.8 Acrylics

These materials are based on acrylic acid, the most common is polymethyl methacrylate (PMMA) which has the tradenames of 'Perspex' and 'Plexiglas'. This polymer is made by reacting methylacrylic acid with an alcohol to form methyl methacrylate; this then undergoes an addition reaction to form the polymer:

$$
\begin{array}{cc}
\text{H} & \text{CH}_3 \\
| & | \\
\text{C} = \text{C} \\
| & | \\
\text{H} & \text{C}-\text{O}-\text{CH}_3 \\
& || \\
& \text{O}
\end{array}
\longrightarrow
\left[
\begin{array}{cc}
\text{H} & \text{CH}_3 \\
| & | \\
-\text{C}-\text{C}- \\
| & | \\
\text{H} & \text{C}-\text{O}-\text{CH}_3 \\
& || \\
& \text{O}
\end{array}
\right]
$$

PMMA monomer

PMMA is hard, rigid, transparent and easily injection molded. It is used for guards, site glasses, lenses, etc. It is resistant to many chemicals but is attacked by petrol, acetone and cleaning fluids.

Acrylonitrile is also made by addition polymerization, monomer:

$$
\begin{bmatrix}
\overset{\displaystyle H}{\underset{\displaystyle |}{}}\ \overset{\displaystyle H}{\underset{\displaystyle |}{}} \\
-C-C- \\
\underset{\displaystyle H}{|}\ \ \underset{\displaystyle C \equiv N}{|}
\end{bmatrix}
$$

It is available mainly in fiber form which is particularly resistant to acids and grease. These polymers are low density; common tradenames are 'Acrilan', 'Orlon' and 'Courtelle'.

4.20.9 Cellulose polymers

These materials are based on the cellulose molecule, or monomer:

$$
\begin{bmatrix}
 & & H\ \ H & & \\
 & & |\ \ \ | & & \\
 & & C = C\,H & & \\
-O-C & \diagup & \ \ \ \ \backslash\ | & & \\
 & |\ \backslash & \ \ \ C-O- & & \\
 & H\ \backslash & \diagup & & \\
 & C-O & & & \\
 & | & & & \\
 & CH_2OH & & &
\end{bmatrix}
$$

Polymerization is achieved by oxygen bonding.

Cellulose nitrate (or celluloid) is tough and water resistant but it cannot be molded due to its high inflammability; because of this property is has now largely been replaced.

Cellulose acetate is cheap, easily molded, less flammable than nitrocellulose and can retain any shade of color. It is tough and has a high impact strength; it can be made flexible and transparent and is easily solvent-bonded. It is not water resistant but this can be improved by using the derivative cellulose acetate-butyrate.

Ethyl cellulose and benzyl cellulose are other common materials. The tradenames 'Rayon' and 'Cellophane' are used for cellulose polymers. Most of the applications of these materials are in domestic situations.

4.20.10 Polyacetals

These are polymers based on a carbon-to-oxygen bond, actually a formaldehyde monomer with end groups on the chain:

$$
HO-\overset{\displaystyle H}{\underset{\displaystyle H}{C}}-
\begin{bmatrix}
\overset{\displaystyle H}{|} \\
-C-O- \\
\underset{\displaystyle H}{|}
\end{bmatrix}_n
-\overset{\displaystyle H}{\underset{\displaystyle H}{C}}-OH
$$

Acetal is the commonly used name. It is often used instead of nylon, having similar properties but with less moisture absorption, and it can also be injection molded.

4.20.11 Polycarbonates

These materials are really polyesters of carbonic acid and phenol, with monomer:

$$\left[-O-\!\!\left\langle\bigcirc\right\rangle\!\!-\!\overset{\displaystyle CH_3}{\underset{\displaystyle CH_3}{C}}\!\!-\!\!\left\langle\bigcirc\right\rangle\!\!-O-\underset{\displaystyle O}{\overset{\displaystyle |\,|}{C}}- \right]$$

Polycarbonates are transparent and temperature resistant and have similar uses to the acrylics; they have a superior impact strength but are more expensive to produce. For structural situations they are similar to nylons and acetals.

4.20.12 Phenylene oxide

Phenylene oxide-based thermoplastics are produced by a process based on oxidative coupling of phenolic monomers. They are known by the tradename 'Noryl' (General Electric Co.). The resins are fabricated by extrusion, injection molding and conventional thermoforming techniques; a range of grades of material is available. Phenylene oxide polymers have excellent mechanical and thermal properties, exceptional dimensional stability and the lowest water absorption of the engineering polymers. Although resistant to many chemicals, the resins are attacked by halogenated or aromatic hydrocarbons. Noryl resins are used for automobile parts, electrical construction equipment, pump impellers and cable covers.

4.20.13 Polyimides

Polyimide polymers possess excellent heat and fire resistance, and they retain their physical and mechanical properties at high temperatures. They are available as thermosets (moldings and laminates) or thermoplastics, and they can be fabricated by many different techniques, e.g. powder metallurgy methods (see Section 8.3), injection, transfer and compression molding, and extrusion. Service temperatures for intermittent exposure can range from cryogenic up to 500°C (932°F); creep is very low as is deformation under load. In addition, polyimides have good wear resistance and low coefficients of friction. Molded polyimides are used for high-speed high-load bearings, jet-engine vane bushings, and gear-pump gaskets. Film applications include insulation for electric motors and for flexible printed circuits; coatings are used on semiconductor devices.

4.20.14 Polyetherimides

Polyetherimide is an amorphous thermoplastic possessing high heat resistance, high strength and modulus, and excellent electrical properties. It was introduced in 1982 and is available under the tradename 'Ultem' (General Electric Co.). The resin is available as unreinforced grade or in three glass fiber-reinforced grades. Polyetherimide can be fabricated by a very wide range of techniques, e.g. injection molding, blow molding, foam molding and extrusion.

The tensile strength and flexural strength values of polyetherimide are higher at 150°C (302°F) than those of most engineering plastics at room temperature. The polymer has good creep resistance and impact resistance values. The resistance to ultraviolet radiation and gamma radiation is good, and polyetherimide also resists attack by a wide range of chemicals. The main applications of polyetherimide are aerospace, e.g. jet-engine components, electrical and electronics, e.g. printed wiring boards, and high-voltage circuit-breaker housings and automobile engine components.

4.20.15 Polyetheretherketone

Polyetheretherketone (PEEK) is a high-temperature crystalline thermoplastic resin, possessing excellent thermal and combustion characteristics (for a thermoplastic) and resistance to a wide range of solvents and other fluids. It was developed primarily as a coating and insulation material for high-performance wiring, and is available in the USA under the tradename 'Victrex' (ICI Americas Inc.). PEEK resins can be molded or extruded using standard thermoplastic fabrication equipment at 190 to 205°C (342 to 374°F). It is produced as unreinforced and reinforced (glass fiber or carbon fiber) grades. PEEK has excellent abrasion resistance and absorbs much less water than many thermoplastics. The resin is used as wire or cable coating for severe conditions and has aerospace, military, nuclear plant and oil-well applications.

4.20.16 Polyphenylene sulfides

Polyphenylene sulfide (PPS) resins are available in the USA under the tradename 'Ryton' (Phillips Chemical Co.) in various grades, e.g. with glass fiber or glass fiber and mineral filler reinforcement. PPS is a high-performance engineering thermoplastic possessing high-temperature stability and wide chemical resistance. The compounds have inherent flame retardance, dimensional reliability, and they retain their mechanical properties at high temperatures. The crystallinity of molded parts can be controlled by the mold temperature. Hot molded parts are highly crystalline and provide optimum dimensional stability at high temperatures, whereas cold molded parts possess optimum mechanical strength. PPS compounds are used for electronic and electrical applications, e.g. telecommunications and computer components, in the chemical and petroleum-processing industries, e.g. pumps, and for automobile engine parts. PPS is transparent to microwave radiation and particular grades have been developed for ovenware and appliance components.

4.20.17 Polysulfones

The sulfone (sulphone in the UK) polymers of polysulfone and polyarylsulfone (PAS), tradenames 'Udel' and 'Radel', are manufactured by Union Carbide Corp., and polyethersulfone (PES), tradename 'Victrex', by ICI Americas Inc. These are the three commercially important engineering resins in this polymer family. They are strong, rigid thermoplastics with high heat deflection temperatures. They require fabrication temperatures of 350–400°C (662–752°F), and when formed they remain transparent up to 200°C (392°F). Thermal stability and oxidation resistance are excellent at service temperatures in excess of 150°C (302°F)

Heat aging of the polymers considerably increases the tensile strength, the heat deflection temperature and modulus of elasticity. However, prolonged heat aging (over one year) decreases the toughness, tensile strength and elongation. Impact strengths of unnotched specimens are high, creep is very low and the resins are resistant to water absorption. The main disadvantage of the sulfones is that they absorb ultraviolet radiation and are unsuitable for outdoor service unless coated or stabilized. These polymers are used for food, photographic and medical applications, the selection of a particular resin depending upon the service temperature.

4.20.18 Reinforced polymers

The mechanical and thermal properties of thermoplastics can be significantly improved by introducing reinforcing fibers of glass, minerals or carbon. The fabrication methods are the same as for unreinforced resins and the higher cost of the reinforcing material is usually offset by the improvements in performance. Nearly all thermoplastics are available as glass-reinforced grades, the glass fibers are high-strength textile-type fibers coated with a binder and a coupling agent. The mechanical properties of a polymer are often increased by a factor of two or more due to glass-fiber reinforcing. Fiber reinforcement of a resin always changes the impact behavior and notch sensitivity; however, this need not result in improved values. Impact-modified (I-M) compounds have been produced, e.g. nylon 6, nylon 6.6, polypropylene copolymer, such that the impact properties of a reinforced I-M material are always superior to the reinforced unmodified grade.

Carbon-fiber-reinforced compounds are now available for several thermoplastics, e.g. nylon 6.6, polyester, ETFE. The cost of these compounds is two to four times that of comparable glass-reinforced materials. However, carbon-fiber reinforcing produces exceptional values of tensile strength, stiffness and mechanical properties. It also produces a lower coefficient of expansion, improved creep resistance, wear resistance and toughness, higher strength-to-weight ratios than glass-fiber-reinforcing of polymers.

Filler materials are also added to thermoplastics to provide internal lubrication and to improve the wear characteristics. Typical applications include gear and bearing components, gasoline metering, and pump components.

Glass is the most widely used reinforcing material for thermosetting polymers, although carbon, graphite, boron, cotton fibers and other materials are used. The addition of reinforcing compounds to thermosets significantly improves the properties, and results in cost–performance benefits compared to the unreinforced resins. Reinforcing increases the tensile, flexural and impact strengths of thermosets, and improves the dimensional stability. More than 95% of reinforced thermosetting components are based on polyester and epoxy resins; the former represent the largest volume production. Other thermoset resins used in reinforced form are the phenolics and silicones.

Glass-reinforced polyesters are used for car panels, boat hulls and tanks. Epoxy applications include circuit boards and aircraft exterior panels (with graphite fibers). Phenolics are used in printed-circuit boards, for gears and insulators; silicones are chosen for service as electrical and thermal insulation.

4.20.19 Structural foams

Structural foam parts are produced by dispersing a gas into the polymer melt during processing. Several different methods can be used, such as direct gas injection or addition of a chemical blowing agent to the resin. Structural foam processes are classified as low- or high-pressure methods; this relates directly to the size range, surface finish, economics and properties of the molded part.

Low-pressure processes are the simplest; they are suitable for production of large three-dimensional shapes. This process is also called 'short shot' or 'free rise' molding. Resin and gas is introduced into the mold at low pressure (1.38–3.45 MN/m^2; 200–500 psi), the mixture only partly fills the mold and bubbles of gas (having been at a higher pressure) expand and fill the cavity. As the cells collapse against the mold surface, a solid skin of melt is formed over the rigid foamed core. With high-pressure molding, the heated melt and a blowing agent are injected into the mold at 34.5–138 MN/m^2 (5000–20000 psi). The mold is entirely filled, and the pressure prevents the occurrence of foaming while the skin portion solidifies against the mold surfaces. The mold pressure is reduced and space is then provided so that foaming can occur between the solid-skin surfaces.

Structural foam parts have a high stiffness-to-weight ratio and a large size capability; low-density structures are produced. Strength-to-weight ratios can be significantly better than many structural metals. The main disadvantage is the surface finish compared with conventional moldings. A characteristic swirl pattern is produced which may be visually unacceptable and also very rough. Surface finishing may be required, especially for parts produced by the low-pressure process. Although the surface finish of components made by the high-pressure process is comparable to parts produced by injection molding, a parting line is formed at the edges where the mold opening made provision for foaming. The cost of the high-pressure process usually limits production to relatively flat parts weighing less then 5 kg (11 lb).

Common structural foam materials are polystyrene, polypropylene, polyurethane (thermoset), and polyethylene. Applications include wood-replacement furniture, building panels, processing tanks and automobile components.

Other polymer materials

4.20.20 Elastomers

Elastomer is a general term used to describe materials that will return to their original dimensions when a loading is removed. The difference in definition between an elastomer and a rubber is that a rubber should withstand a 200% elongation under tensile testing conditions, and return *quickly* to its unloaded original dimensions. A term often used in this connection is *resilience*, which is the ability of a material to recover from elastic deformation. Some of the polymers already discussed have elastomer properties, this property is mainly due to the branching and cross linking of the molecules.

4.20.21 Thermoplastic elastomers

These materials are produced by copolymerization of two or more monomers, using either block or graft polymerization techniques. One monomer develops a hard crystalline segment, and the other develops a soft, amorphous segment (the rubbery characteristic). Thermoplastic elastomers (also called elastoplastics) possess the processing advantages of thermoplastics combined with properties similar to those of vulcanized (thermoset) elastomers and rubbers. The materials are molded or extruded.

Thermoplastic urethanes were the first materials to be produced; although their heat resistance is lower than the cross-linked types, other properties are similar. Abrasion resistance is excellent and other properties are comparable with the 'best' elastomers. They are used for gaskets, diaphragms, gears, and fuel lines.

Copolyester thermoplastic elastomers are generally tougher over a wider temperature range than the urethanes. They are expensive elastoplastics but possess a high modulus, good elongation and tear strength, and the brittle temperature is below $-65°C$ ($-85°F$). Uses include hydraulic hose, couplings and cable jacketing.

Styrene copolymers are the lowest priced elastoplastics, possessing a lower tensile strength and higher elongation than SBR or natural rubber. Styrene-butadiene (SB) block copolymers are used for disposable medical products, tubing and food packaging, and in sealants and adhesives.

Thermoplastic olefin (TPO) elastomers have the lowest specific gravities of all elastomers based on polyolefins, and are available in several grades. Service temperatures as high as $130°C$ ($266°F$) in air are possible; the brittle temperature is below $-65°C$ ($-85°F$) and the materials can be autoclaved. Olefin elastomers are used for medical tubing, seals, gaskets and electrical components.

4.20.22 Natural rubber

Polyisoprene is the most common natural rubber, monomer:

$$
\left[
\begin{array}{ccccccc}
 & & CH_3 & & & & \\
H & & | & & H & H & \\
| & & | & & | & | & \\
-C & - & C & = C & - C & - \\
| & & & & | & & \\
H & & & & H & &
\end{array}
\right]
$$

This material is obtained from the sap or latex (an emulsion containing 40% water) exuded by certain trees found mainly in South East Asia. The rubber particles are coagulated using formic or acetic acid, and then dried and rolled. In this form natural rubber has few uses because of its low tensile strength at room temperature, and it becomes tacky at higher temperatures. To improve the usefulness of the material it is subjected to severe rolling or mixing, and additives such as plasticizers and antioxidants are mixed with the material and an inert filler is also included. Sulfur is also added and the plastic mass is molded or extruded, and then cured at approximately 150°C (302°F). The sulfur allows vulcanization to occur, which is the breaking of some of the double covalent bonds to allow sulfur atoms to form cross-links between molecules, thus creating the elastomer properties. Ordinary soft rubber contains about 4% by weight of sulfur, which causes cross linking between approximately 10% of the double covalent bonds in the molecules. Full vulcanization requires approximately 45% by weight of sulfur to be added, which produces a hard rigid material known as 'Ebonite'.

It has good electrical insulating properties and can be machined or ground to a molding powder, it is used to produce such items as battery casings. Common filler materials are carbon black and silica which may constitute up to 50% of the material volume. These additives may make the product cheaper, reinforce the material, improve resistance to abrasion and tearing, and improve the aging characteristics caused by heat and light. Natural processed rubber is not resistant to oils or solvents, and is attacked by oxygen or ozone, causing embrittlement or perishing – hence the use of antioxidizing agents. A large amount of rubber is recycled in a similar manner to the thermoplastic polymers. Rubber is a useful material for absorbing vibration, e.g. motor mountings.

4.20.23 Synthetic rubbers

These materials constitute approximately 60% of the world rubber consumption; some of them can be mixed with natural or recycled rubber and may be used to impart special characteristics to the material.

Buna S or SBR rubber is a cheap material formed by the copolymerization of butadiene and styrene. An unsaturated copolymer is produced which can be vulcanized; it has the monomer:

$$\begin{bmatrix} \begin{array}{ccccccc} H & H & H & H & H & H \\ | & | & | & | & | & | \\ -C & - C & = C & - C & - C & - C - \\ | & & & | & | \\ H & & & H & H \end{array} \\ \text{butadiene} \\ \text{styrene} \end{bmatrix}$$

Vulcanization prevents attack by organic solvents, although they are still absorbed, causing considerable volume changes.

Nitrile rubber, or *Buna N* or *NBR,* is a copolymer of butadiene and acrylonitrile. It has good resistance to oils and solvents and improved resistance to heat, abrasion and aging. It can be vulcanized but is more expensive to produce.

Polychloroprene (or *neoprene*) has the monomer:

$$\begin{bmatrix} \begin{array}{cccc} H & Cl & H & H \\ | & | & | & | \\ -C & - C & = C & - C- \\ | & & & | \\ H & & & H \end{array} \end{bmatrix}$$

This polymer is oil resistant and possesses good tensile properties, resilience and aging resistance. It can be vulcanized using chemicals other than sulfur, but cannot form a hard rubber. It is used for fuel and hydraulic hoses and gaskets, but it is a more expensive material.

Butyl rubber is a fully saturated addition polymer of iso-butylene:

$$\begin{array}{c} CH_3 \\ | \\ C = CH_2 \\ | \\ CH_3 \end{array} \longrightarrow \begin{bmatrix} \begin{array}{c} CH_3 \\ | \\ -C - CH_2- \\ | \\ CH_3 \end{array} \end{bmatrix}$$

Mixing with isoprene provides some covalent double bonds for vulcanization, and the material then has excellent resistance to oxidation and solvents. It is used for tubing, hoses and tires because of the very low absorption of gases.

Fluoroelastomers, e.g. Viton, have excellent resistance to temperature, chemicals, oxidation and weather, but they are also more expensive.

4.20.24 Vulcanized fiber

This material is formed from paper or cotton (cellulose) and is reacted with sulfuric acid or zinc chloride. It is then washed and steam processed to form a hard, tough product which is often used in laminated sheet form. The color range is limited, generally red. It is a tough, dense material but can be easily machined or drilled, it has good impact strength and abrasion resistance and, if protected by varnish, will not absorb moisture. Filler materials can also be added.

4.20.25 Silicones

These polymers are sometimes classified as inorganic materials. They are composed of a repeating unit of silicon and oxygen that forms the polymer 'backbone'; hydrogen or organic groups (c.g. CH_3) can be linked to the other bonds:

$$
\begin{array}{ccc}
R & R & R \\
| & | & | \\
-Si-O-Si-O-Si-O- \\
| & | & | \\
R & R & R
\end{array}
$$

Both silicon and carbon are Group IV elements, but silicones are more expensive to produce than carbon-based polymers.

Silicones with low molecular weight are liquids or waxes that are chemically inert and act as a water repellent. They are also available as silicone rubber, or as a resin that can be cured with a catalyst to form a rigid thermosetting material. The main advantage of silicones is the wider temperature range over which they can be used. The rubbers retain their flexibility and elasticity down to $-80°C$ ($-112°F$) and the resins can be used at temperatures up to $300°C$ ($572°F$), or even $800°C$ ($1472°F$) for short periods.

* * *

Self-assessment exercises

The information provided in Sections 4.18–4.20.25 should be seen as a brief summary of the published material related to polymers. Extensive handbooks are available that contain details of (nearly) all manufactured polymer materials, and comprehensive data relating to their properties and uses. The reader does not need to memorize this information. It should be sufficient to know where to find it (see bibliography for this chapter and check your own library), to be familiar with the names of the more common and widely used polymers, to know typical values of important properties, to be aware of their uses and the problems that occur.

1 List 10 polymer materials that are important in your branch of engineering.

2 Identify the main advantages or advantageous properties, and the main disadvantages associated with applications of these polymers.

3 Compile a suitability table to compare typical applications for these materials. (Assign a value 4 representing ideally suited, and 1 representing most unsuitable.)

4 Locate and study recent literature describing new polymer materials and assess their (likely) importance in your field of engineering. Summarize any available data for the important properties such as strength, water absorption, softening temperature, etc., and build up a product-applications 'profile'.

Locating literature is an essential part of the student's training. In this example the sources of information are recent handbooks, journals, patents, and personal requests for product information direct to polymer manufacturing companies. The latter is often the most productive, relevant and informative source of information, and the easiest to locate, e.g. a phone call or a letter, but it is often overlooked completely.

* * *

Quick test

What are PTFE, PVC, Perspex, epoxy, Bakelite, silicone, fiberglass, neoprene, Viton, celluloid, polystyrene? What are their main uses?

* * *

CERAMICS

4.21 INTRODUCTION

The term *ceramics* taken from the original Greek word meaning 'fired material', is used to describe a wide range of materials. These include porcelain, earthenware, brick, glass, cement, refractories and abrasives, as well as specialized materials such as cermets. However, the ceramics that are considered structural engineering materials include only a limited number of types. These ceramics are selected for load-bearing applications because of their resistance to high temperatures, corrosive environments and abrasives, and because they possess high hardness, and thermal and electrical insulating properties.

4.22 CERAMIC BONDING

Most ceramics contain both metallic and nonmetallic elements and are held together by ionic or covalent bonding. In contrast, metals are composed of ions that occupy positions in a space lattice, and the outer (valence) electrons that are released form an 'electron sea' (see Section 1.2.2). This electron movement provides the metallic bonding within the metal.

Ionic bonding occurs in compounds such as sodium chloride, where the positive sodium ions are surrounded by the negatively charged chlorine ions and vice versa (therefore no 'molecules' of sodium chloride exist). The strength of the ionic bond depends upon the attraction of positive ions for negative ions and vice versa. However, ionic bonds are usually stronger than metallic bonds.

Ceramics often possess covalent bonding which consists of sharing the outer electrons between neighboring atoms, in order to complete the outer (valence) shell. Diamond provides an example of the strong, hard, rigid structure provided by covalent bonding. The high strength of the covalent bond is due to the restricted mobility of the electrons and the ions. This structure also produces materials that are good electrical insulators. The bonding in most ceramics is a combination of ionic and covalent forces, and weaker van der Waals forces also occur between adjacent planes of atoms (or ions).

The ductility of metals is due to the occurrence of slip under stress between the regular arrangement of metallic atoms within densely packed planes. However, ceramics possess a different type of bonding and a different arrangement of the atoms. Therefore failure occurs due to brittle fracture rather than slip.

The complex crystal structures of ceramics are more difficult to nucleate and grow, and non-equilibrium structures are common.

Glasses are ceramics made up of identical tetrahedral silicate (SiO_4^{4-}) ions, linked together in a large random network. The structure is completely different from that found in metals, but is similar to certain polymers (see Section 4.7). The structure is rigid at room temperature and has a definite elastic modulus; however, it is similar to the structure of liquids and is sometimes referred to as a 'supercooled liquid'.

Engineering ceramics can be differentiated from other ceramic materials, e.g. porcelain, by the strong bonding which they possess. This is achieved by using an appropriate manufacturing method, and by carefully controlling the material and impurity composition. Non-engineering ceramics are usually weaker due to higher levels of impurities and weaker bonding.

4.23 PROPERTIES OF CERAMICS

Property values of ceramics represent only the data obtained from a particular measurement method on a particular piece of material. The data are scientifically valid, but are of little use for engineering applications unless standard measurement techniques are used (see Section 6.2) and the nature (or structure) of the material is fully understood. For example, the porosity of a ceramic can significantly affect the thermal conductivity data.

Most ceramics are brittle and because energy is not dissipated by plastic deformation, cracks can propagate easily under stress and sudden failure occurs (see Section 6.2.3). The fracture toughness of the ceramic is low, although this can sometimes be improved by obtaining crystalline glass with extremely small grain size, e.g. Pyroceram (see Section 4.24.2). For these reasons the tensile strength of a ceramic is difficult to determine and standard tensile testing methods (Section 6.2.1) are not used. In ductile metals, localized stresses that exceed the yield point are usually relieved by local plastic deformation that redistributes the stress into a wider area and prevents fracture. Ceramics have no such yield point and fracture occurs when localized stresses exceed the material strength. This contrasting behavior is shown in Fig. 4.28 by the stress–strain curves for a typical fused-silica glass ceramic and a mild steel. If the curve for the ceramic material in Fig. 4.28 was plotted on the scale used for the metal, then it would be almost vertical.

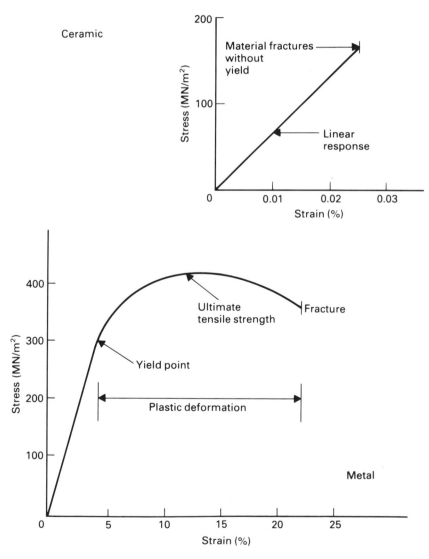

Figure 4.28 Stress–strain curves for typical metal and ceramic materials.

An alternative method, the bend test, is usually employed to measure the strength of ceramics for high-strength applications. This involves applying a bending load to a small beam of the material (approximately 2.5 cm long and 1 cm square cross section; 1 in. by 0.4 in. by 0.4 in.) either at the center (a three-point test) or at two points equidistant from the end supports (a four-point test). The applied load is increased until the sample ruptures; typical test equipment is shown in Fig. 4.29. The modulus of rupture, flexural strength or transverse rupture strength are then calculated from an appropriate simple formula for the bending of beams (see Section 6.3.11). The following formula for calculation of the flexural strength (modulus of rupture) is given in *ASTM Specification C580: Standard Test Method for Flexural Strength and Modulus of Elasticity of Chemical-Resistant Mortars, Grouts, and Monolithic Surfacings*:

$$S = 3PL/2bd^2$$

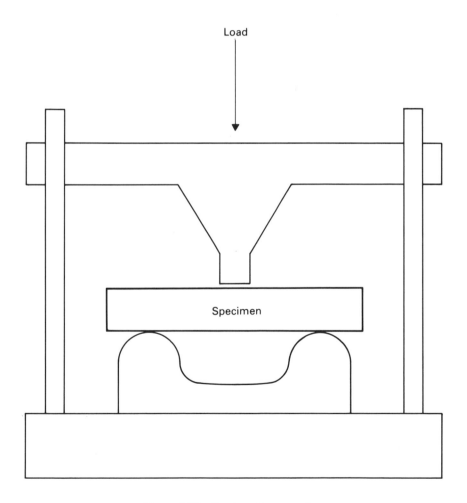

Figure 4.29   Bend test for ceramics.

where:

S = stress in the specimen at midspan (MPa or psi);
P = load at moment of crack or break (N or lb_f);
L = span (mm or in.);
b = width of test beam (mm or in.);
d = depth of test beam (mm or in.).

The flexural strength is equal to the maximum stress at the moment of crack or break. If the specimen does not break, the yield strength is calculated from the above formula by letting P equal the maximum load attained on the stress–strain curve.

For most metals the tensile and compressive strengths are approximately the same; however, this is not usually the case for ceramics, in which the compressive strength is greater than the tensile strength. This lack of strength in tension is due to the presence of impurities and defects, and the associated stress concentrations already discussed. This problem can be overcome by appropriate design, e.g. by prestressing concrete (Section 6.3.16), by using arched load supports, or by heat treatment such as tempering. For limited applications, some ceramics can be strengthened by the addition of other elements or by changes in the crystal structure, but cold working and precipitation hardening are not applicable.

The hardness of a ceramic is an important property, but the standard Brinell or Rockwell tests (Section 6.2.2) would cause material failure. Therefore the Vickers microhardness test is often used, with very light loads and a small indenter. Sometimes the Moh's scratch test is used to determine relative hardness values.

A particular group of ceramic materials is used as refractories, and they are required to possess good resistance to high temperatures. It is important that these materials possess a high melting point or solidus temperature (see Fig. 1.23), and a low linear coefficient of expansion. A refractory material usually possesses either one particular advantageous property or a satisfactory blend of property values.

4.24 CERAMIC MATERIALS

Most metals form at least one oxide but very few oxides are useful as the main constituent of a ceramic. Alumina, beryllia and zirconia are the only ones used in pure form as engineering ceramics. Silica and silicates are common alloying materials for ceramics. When liquid silica is slowly cooled (under equilibrium conditions), three allotropic forms are encountered (cf. allotropes of iron – see Sections 2.3 and 1.27). These are cristobalite between 1710 and 1470°C (3110 and 2678°F), tridymite between 1470 and 870°C (2678 and 1598°F), and quartz below 870°C (1598°F). The phase transformations are more complex than for pure iron, the reactions occur slowly and over a wider temperature range. Silica (SiO_2) is a constituent of refractory bricks that are used at high temperatures (1650°C; 3000°F). However, care must be taken to ensure that cracking does not occur due to thermal shock.

Alumina (Al_2O_3) is used both in the pure form and in ceramic mixtures. Because of its high melting point, alumina is widely used as a refractory. Aluminas are used at temperatures as high as 1925°C (3500°F) provided they are not exposed to thermal shock, impact or highly corrosive atmospheres. Above 2035°C (3700°F), the strength of alumina declines. The creep resistance of aluminas is good below 815°C (1500°F), but other ceramics are superior above this temperature. Aluminas are attacked by strong acids, steam and sodium. Alumina can be alloyed with oxides of silicon, magnesium or calcium, and with chromium (which acts as a strengthening agent). Alumina is also used in lasers, vapor illuminating lamps and as a cutting tool for metal machining.

Magnesia (MgO) is a refractory material, often containing calcium oxide, and is obtained by heating dolomite. Its main advantage is that it does not react with CaO (which is often present in slags) unlike silica refractories.

The term *high-grade refractory* is usually taken to include those materials that are composed of pure or relatively pure oxides. As well as alumina, silica and magnesia, they include zirconia, beryllia and thoria, although these materials are much more expensive. Zirconia is inert to most metals and retains its strength almost to the melting point (2675°C; 4847°F). Beryllia has excellent thermal shock resistance, a low coefficient of thermal expansion and a high thermal conductivity. It is also an excellent electrical insulator and has electrical and electronic applications. Refractories are often classified as being *basic, acidic* or *neutral*. This relates to the behavior of the oxides in an aqueous solution, but for refractories it indicates the tendency for high-temperature, liquid-state attack. For example, magnesia refractories are basic and are attacked by slags rich in SiO_2. Silicate refractories are acidic and are attacked by basic slags.

Metal carbides and carbon are used as refractories, although they do not resist oxidation and are not particularly suitable for high-temperature applications. The most important carbide refractories are those of silicon, zirconium and titanium, and to a limited extent those of boron. Graphite and carbon are used as refractories, although above 800°C (1472°F) protective coatings or reducing atmospheres must be used because of increased oxidation. The advantages of these materials are that at ambient temperatures they are inert and easy to machine.

Many engineering ceramics have multioxide crystalline phases. Mullite is composed of alumina (Al_2O_3) and silica (SiO_2); for mixtures containing less than 70% Al_2O_3 the solidus temperature is 1595°C (2903°F), which limits their use. However, above 70% Al_2O_3, mullite ($3Al_2O_3 \cdot 2SiO_2$) or mullite with corundum (Al_2O_3) are present, having a solidus temperature of 1840°C (3344°F). Therefore, high-alumina refractories can be used in steel-making furnaces, and mullite is an important refractory because of the higher temperatures that can be tolerated.

The phase diagram for the SiO_2–Al_2O_3 system is shown in Fig. 4.30. Clays consist mainly of hydrated aluminosilicates, alkaline-earth (Ba, Ca) oxides or

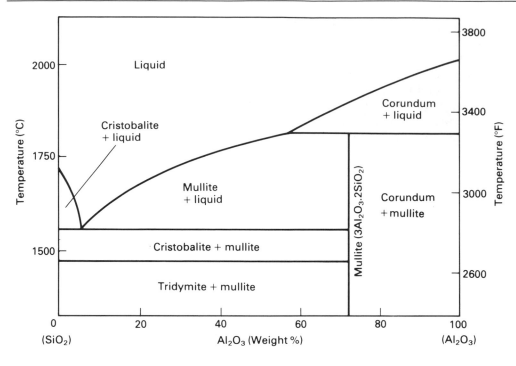

Figure 4.30 The Al$_2$O$_3$-SiO$_2$ phase diagram.

compounds, alkali-metal (Na,K) compounds, and often some iron oxide. The various constituents lower the solidus temperature (Fig. 4.30), and permit fusion and temperatures of approximately 900°C (1652°F). Fireclays are alumino-silicates containing 50–70% SiO$_2$, and 25–40% Al$_2$O$_3$.

The softening temperatures vary considerably with composition and lie in the range 1400–1600°C (2552–2912°F). Porcelain is a relatively pure product with a composition between SiO$_2$ and mullite, i.e. approximately 20–50% Al$_2$O$_3$. Earthenware and stoneware are clay products that are not as pure and are fired at lower temperatures. Brick is similar in composition but usually contains iron oxide and is fired at higher temperatures.

The fabrication of clay products requires the addition of water to make them plastic and easier to shape. A minimum amount of water (the *plastic limit*) is required, but beyond the maximum amount (the *water limit*) the clay becomes wet, weak and sticky. After forming, the clay is dried and fired. Drying removes interstitial water and the rate of drying is the important parameter. This is followed by firing at low temperatures to remove surface water, and then heating up to 600°C (1112°F) to eliminate chemically bound water. Oxidation reactions occur at approximately 900°C (1652°F) and organic materials are removed. Fusion or *vitrification* begins at 900°C (1652°F) and becomes complete at temperatures up to 1400°C (2552°F). Completely vitrified products are strong but they become soft during the fusion process and tend to collapse. For this reason full vitrification is rarely used, except for some porcelain and stoneware. Brick materials and earthenware are used in the construction industries, and porcelain and stoneware products in the chemical process industries. They have good chemical resistance particularly to acids (but not alkalis), but poor shock resistance and only moderate tensile strength (higher compressive strength).

Asbestos is the name given to a group of complex hydrous magnesium silicates, having a fibrous texture and capable of being spun or woven. They are based upon the SiO_4^{4-} tetrahedron with the addition of Mg^{2+} ions, and are terminated by OH^- ions (from water). Chrysotile is the most abundant form and amphilbole produces hard, brittle and short fibers. Asbestos is a light material, noncorrosive, fire resistant, acid resistant and chemically inert. It is used as a filler for polymer materials, and as an insulating material it can be formed into sheets using cement powder. However, there are various health hazards associated with the use of asbestos, particularly in powder form.

Many metals and carbides are used in ceramic materials but only boron carbide, silicon carbide and silicon nitride can be considered as engineering ceramics. Boron carbide has the unusual properties (for a brittle ceramic) of high hardness and low density, it is therefore useful as armour plating. It also possesses excellent abrasion resistance and is used for high-wear applications. However, its use has been limited by the low strength at high temperatures.

Silicon carbide and silicon nitride are the strongest known structural materials for high-temperature oxidation-resistant applications. These materials do not easily self-bond and hot pressing is a common fabrication method. Fully dense shapes are formed using a combination of high temperature and pressure, sometimes with the addition of bond-forming catalysts. However, only simple shapes, bars and billets can be produced and more complex parts must be machined to shape, which is a slow and costly process. Silicon carbide and silicon nitride can be bonded and shaped without pressure by various processes, e.g. reaction sintering, reaction bonding and reaction crystallization (for silicon carbide). For these processes, the parts are injection molded or slip cast (see Sections 4.15.2 and 8.1) and then sintered (see Section 8.3). Complex shapes close to the finished size can be produced, but they have lower strength and less thermal shock resistance than hot pressed parts, and are only about 80% as dense.

Silicon nitride is resistant to most strong acids, and to molten aluminum and other metals with low melting points. It undergoes only small dimensional changes (<1%) during reaction sintering. This means that shapes can be partially sintered (at 400°C/752°F in a nitrogen atmosphere), then machined, and finally fully sintered with only small volume changes. Silicon nitride is stronger than silicon carbide up to about 1500°C (2732°F). At about 1150°C (2102°F), the silicon nitride grain boundaries begin to soften (or 'creep') and the strength declines. Above 1500°C (2732°F), silicon carbide is the stronger material.

Many ceramics that are employed as refractories can also be used for their hardness and wear resistance. Carbon in the form of diamond has the highest hardness known, and both natural and synthetic diamonds are used. Diamond dust is used as an abrasive in a slurry, or as a cutting tool or blade set in a binder of low hardness (grinding and abrasive cutting are discussed in Section 8.4.10). Other hard materials include carbides of boron, silicon, titanium and tungsten, and oxides of aluminum and silicon. The hardnesses of some common abrasive materials are listed in Table 4.2. Most of these materials possess sufficient hardness to act as abrasives, and other characteristics such as toughness, resistance to attrition and friability make a significant difference to their performance and selection.

Cemented carbides are a combination of a ceramic and a metal that are bonded together; they are known as *cermets*. An example is the addition of 10% cobalt to tungsten carbide, which significantly increases the toughness required as a cutting tool. This product has negligible porosity, and high hardness, compressive strength and elastic modulus.

Table 4.2   Microhardness of selected abrasives

Abrasive	Vickers microhardness number
Diamond	8000
B_4C	3700
SiC	3500
TiC	3200
Al_2O_3	2800
WC	2400
Quartz	1250
Hardened tool steel	600

4.24.1 Cement and concrete

Portland cement powder is made by firing a mixture of limestone with shale or clay in a rotating kiln, at a maximum temperature of 950°C (1742°F). The mixture fuses to form a clinker which is ground to a powder, and a small amount of gypsum ($CaSO_4$) is added to produce the setting characteristics of the dry cement. The cement powder is mixed with water, the ideal ratio is 2:5 for water:cement, and it forms hydrated silicates and aluminates of calcium which harden and set as a rigid material. Important factors are the composition of the powder, the amount of water, the time for hardening, and the temperature.

Concrete is a mixture of cement powder, sand and small stones; when water is added the cement paste should coat all the sand and stones and fill any voids. Factors that affect the properties of the concrete are the proportions of water, cement, sand and stones, also the type, average size and surface area of the stones. The compressive strength of hardened cement is approximately 100 MN/m^2 (14.5 × 10^3 psi) compared to 65 MN/m^2 (9.6 × 10^3 psi) for plain concrete. The tensile strength of concrete is only about 5 MN/m^2 (725 psi), although this can be improved by using reinforced steel rods or tubes in regions where tensile stresses will occur, e.g. in the lower layers of a simply supported concrete beam (see Section 6.3.16).

4.24.2 Glasses

Glass is a noncrystalline or amorphous solid, sometimes referred to as a supercooled liquid, or very viscous liquid in the solid state. This is because when molten glass is cooled it does not exhibit temperature changes due to the heat of solidification (see Fig. 1.22 for metals), or volume changes associated with the transition to the solid state (see Fig. 4.9 for polymers).

Glasses are characterized by short-range atomic order rather than the long-range order of most crystalline solids, although special glasses are now made which have a more compact crystalline structure. The properties of glasses depend on composition, but in general:

(a) glasses are harder than many metals;
(b) they are less ductile, but brittle;
(c) their coefficient of thermal expansion is low compared with many metals and plastics;
(d) they can be good electrical insulators;
(e) they are resistant to many acids, solvents and chemicals;
(f) they have elevated temperature applications;
(g) they are slowly attacked by water and some alkaline solutions.

In contrast to other ceramics, the glasses cover a wide range of compositions. Most glass is based on the silicate system and is made from three major constituents: silica (SiO_2), lime ($CaCO_3$) and sodium carbonate (Na_2CO_3). Various oxides are added to produce particular properties for specific requirements. The additions to glass include:

(a) network formers such as SiO_2, GeO_2, B_2O_3, which form triangular or tetrahedral units for the glass framework;
(b) network modifiers, e.g. Na_2O, which lower the melting point of pure silica glass;
(c) intermediates that are added in large quantities to replace silica, e.g. lead oxide.

Nearly all glasses can be categorized into one of six types based upon the chemical composition. Within each type (except fused silica) are several distinct compositions. These types are:

(i) *Soda-lime glass* is the common glass used for bottles, windows, etc. It is a mixture of the oxides of silicon, calcium and sodium. It can be fabricated in a wide variety of shapes but it has a low resistance to high temperatures and temperature changes, and it has only moderate resistance to chemicals.
(ii) *Borosilicate glass* has some of the SiO_2 replaced by boric oxide. It has a low coefficient of thermal expansion and can be used for precision parts. It has good resistance to thermal shock and is used for laboratory ware. Most borosilicate glasses have a better resistance to acids than soda-lime glasses, but inferior resistance to alkalis.
(iii) *Lead-alkali glass* contains some lead monoxide (PbO) to increase the

index of refraction. It is a good electrical insulator and is used for optical applications. It is easy to work but has a low resistance to high temperatures and to thermal shock.

(iv) *Aluminosilicate glass* has some of the SiO_2 replaced by Al_2O_3. These glasses resist chemical attack and are good electrical insulators, they have good thermal shock resistance (cf. borosilicate glass) and are able to withstand high temperatures. However, they are expensive and difficult to fabricate.

(v) *96% silica glass* is highly heat resistant, it is made from borosilicate glass. It can be easily shaped and has properties similar to fused silica (type vi). It is used for laboratory ware, optical components, spacecraft windows and heat-resisting coatings.

(vi) *Fused silica* is composed only of SiO_2 in a noncrystalline state. It is the most expensive of all glasses and is difficult to fabricate. It has the maximum resistance to thermal shock and the highest permissible service temperature, e.g. 900°C (1652°F) for extended periods and 1200°C (2192°F) for shorter periods. It has excellent chemical resistance and the maximum transmission in the ultraviolet range. It is used for severe conditions and applications such as astronomical telescopes and crucibles.

These six types of glass can also be grouped in three pairs. Soda-lime and lead-alkali glasses are *soft glasses* because they soften or fuse at relatively low temperatures. Borosilicates and aluminosilicates are *hard glasses*, and 96% silica and fused silica are the *hardest glasses* of all.

There are also three light-sensitive grades of glass:

(a) *Photochromic glass* darkens when exposed to ultraviolet radiation, and fades when the light is removed or if heated. It is used for spectacle lenses.

(b) *Photosensitive glass* changes from clear to opal when exposed to ultraviolet radiation or heat. This change is permanent but the exposed area is more soluble in hydrofluidic acid, and can be used to etch partially exposed shapes.

(c) *Polychromatic glass* is a full-color photosensitive glass, and is used for windows and information storage.

When glasses cool from the melt to the crystalline melting point, they may either become crystalline or remain in the liquid state. If they are rapidly cooled or the liquid is viscous and cannot assume a crystalline arrangement, they continue to cool to a glass transition temperature range called the *fictive temperature* (T_f). (cf. Section 4.7, T_g for polymers). This temperature depends upon the cooling rate. If glasses are heat treated or reheated in this temperature range, some localized density variations and crystallization occurs known as *devitrification*. Most glasses resist this process.

Glass can be strengthened by *thermal tempering*, which produces residual compressive stresses on the glass surface, thus increasing the static and impact strength. The surface layers of a glass are cooled by an air blast and the center is allowed to cool more slowly. The same result can be achieved by *chemical strengthening* by changing the glass composition at the surface, e.g. by immersion in a molten salt bath. This may allow ions with a larger atomic volume to substitute for surface ions, and may increase the density at the surface and the thermal-expansion coefficient. Chemical strengthening is more expressive, but it produces much larger compressive stresses than tempering.

Materials known as *crystalline glasses* have also been produced. The glass is fabricated in the amorphous state and is then made crystalline by two-stage heat treatment. Nucleating agents such as TiO_2 and ZrO_2 are added to the glass and the first-stage heat treatment involves heating to the nucleation temperature. Small crystalline regions are formed. The second stage is to raise the temperature (5°C per minute; 9°F per minute) to the crystal-growth temperature (100°C/180°F below the crystalline melting point). A high proportion of network modifiers, e.g. Li_2O, MgO or ZnO, are present and a microcrystalline structure is obtained. This consists of small (0.01–1 μm diameter) crystallites dispersed in a glassy matrix. These glasses are stronger and have greater impact resistance than noncrystalline glasses. Pyroceram is a commercial crystalline glass.

4.24.3 Carbon

Carbon can be produced in many forms; the following categories will be used here:

(a) *Carbon*. A general term including any form of elemental carbon, i.e. diamond, graphite or amorphous carbon.

(b) *Graphite*. A crystalline form of carbon found naturally and obtained by heating amorphous carbon.

(c) *Carbon-graphite*. A composite material containing both amorphous carbon and graphite.

(d) *Manufactured carbon*. A bonded-granular carbon structure, this term distinguishes between man-made carbons and those that occur naturally.

Graphite has a layered structure with strong covalent bonding within the sheets, but semimetallic bonds between the sheets. Cleavage occurs between the sheets and gives rise to excellent lubricant properties, as well as high electrical and thermal conductivity parallel to the sheets (semimetallic properties). However, between the sheets there are barriers to electron motion.

Manufactured carbon parts are usually made from mixtures of coke and graphite powder bonded with carbon (derived from coal-tar pitch). The mixture is shaped by compression molding or extrusion (heated). The shape is then fired at high temperatures (up to 1300°C/2372°F) in an oxygen-free environment. The resulting pores are impregnated with resins, glasses or fused salts. Shrinkage occurs during firing but it can be estimated and the final part produced within 1% of required tolerances. Most parts require machining and in some cases this must be performed with carbide or diamond-tipped tools. Manufactured carbon includes a wide range of materials in a variety of grades and shapes, e.g. plates, rods, tubes.

Manufactured carbon has different characteristics from ceramics, in general it is:

(a) a good conductor of heat and electricity;
(b) self-lubricating (it slides on metals);
(c) corrosion resistant, e.g. to solvents, caustics and most acids.

It does not soften or melt and is stronger at 2800°C (5072°F) than at room temperature. The fatigue resistance exceeds that of most metals and the strength is not sensitive to surface flaws (due to its texture). Manufactured carbon is slowly attacked by oxygen at high temperatures but is unaffected by reducing gases. It is weakened by strong oxidizing acids such as hot concentrated nitric acid. Like ceramics, manufactured carbon is brittle.

Manufactured carbon is used for sliding elements in mechanical devices, e.g. mechanical seals, as electrical brushes, and for pistons in chemical-metering pumps. It is also used for bearings due to the self-lubricating property.

4.25 COMPOSITE MATERIALS

A composite material is composed of two materials bonded together, one material serving as a matrix surrounding the particles or fibers of the other. Fiberglass is an excellent example, combining brittle glass fibers in a plastic polyester matrix. The high strength and high modulus of the glass lead to good overall strength, and the plastic improves the toughness. In general, an advantageous balance of properties is achieved in composite materials (in some cases an outstanding combination), rather than an average of the properties of the individual materials.

Surface coatings or laminated constructions may have either an adhesive-type bond, e.g. paint or polymer coatings, or a diffusion bond, e.g. galvanized lining, welded region or nitrided surface. *True* composites may be divided into two important classes, according to whether the fibers of the strengthener are continuous or discontinuous. For continuous fibers, the modulus of the composite is simply the weighted sum of the moduli of the fiber and the matrix, and the load carried in the fiber depends upon the ratio of the moduli of the fiber and the matrix. In the case of composites containing discontinuous fibers, the critical factors are the length-to-diameter ratio of the fiber, the quantity of fiber and the shear strength of the bond between the fiber and the matrix.

The following terms are related to composite materials and fibers:

Fiber is a single, continuous material whose length is at least 200 times its width or diameter.

Filaments are 'endless' or continuous fibers.

Whiskers are single-crystal metal fibers, containing almost no defects and consisting of short, discontinuous fibers of polygonal cross-sections. Common examples are copper, graphite, silicon carbide, etc. Whiskers are usually grown from vapor or metallic depositions of gases or liquids on a surface. Whiskers of oxide, carbide and nitride are of primary engineering interest.

Multiphase fibers consist of materials such as boron carbide and silicon carbide
which are formed on the surface of a very fine wire substrate of tungsten.

Yarn is a continuous strand or bundle of fibers.

Fabrics are sheets of fibers or yarns that are woven, knitted, or otherwise
physically bonded together.

Denier (abbreviation; den) is the weight (grams) of 9000 m of fiber. This reflects
both the size and density of the fiber.

Tenacity is a measure of fiber strength, it is also called 'breaking strength': its
units are g/den.

There is a clear division between natural fibers (from plant, animal and
mineral sources) and synthetic fibers, the latter having been developed to replace
natural fibers. Synthetics usually behave in a more predictable manner and are
more uniform in size; they are sometimes less costly than their natural
counterparts. For engineering applications, glass, metallic and organically
derived synthetic fibers are most significant.

Wood is a natural composite material containing the polymer cellulose and
held together with a glue. Wood has been used throughout history as a
constructional material and is extremely important. Timbers may be broadly ·
classified as soft woods and hard woods. Soft woods are obtained from trees of
the coniferous type including pine, fir, cedar, spruce and redwood. The
cellulose molecules form fibrous cells that lie parallel to the trunk and limbs of
the tree; there are also some cells at right angles lying in a radial direction. Hard
woods are obtained from trees of the deciduous type, including oak, ash, elm,
walnut, beech, hickory and mahogany. The structure is more complex and
contains more radial cells than the soft woods. The majority of cells are fibrous
and lie parallel to the trunk; this determines the grain. However, balsa wood is
both light and soft but is classified as a hard wood. Timber often needs to be
dried and chemically treated in order to prevent attack by fungus and insects.
Chemicals can also be used to make the timber fireproof. The fibrous nature of
wood means that the tensile and compressive strengths are very much greater
along the grain than at right angles to it. The shearing strength is lower along the
grain than across it, which means that wood is better employed in compression
rather than tension. Timber is used as a structural material, for foundry patterns
and tool handles, and in regions where it is plentiful as a packing for cooling
towers.

Wood can be strengthened by using it in a laminate form, composed of
several layers glued together. Plywood is built up of thin layers bonded with a
water-resistant glue or thermosetting resin, with successive grains at right angles
to each other and formed under pressure. An odd number of layers are used so
that shrinkage stresses are symmetrical about the center, causing a minimum of
warping. Plywood sheets can be metal faced or polymer faced for greater
resistance to attack and for extra strength. Plastic wood or dense wood has the
polymer resin forced into the cellular structure before setting. Wood chips can
be shredded in a steam-heated pressure vessel and then hot pressed to form
woodchip fiber boards. These can have low or high density depending upon the
application, e.g. low-density wood fiber is used for insulation.

Glass fibers are probably the most widely used of all synthetic engineering fibers, and also have the smallest diameter (typically 1–4 μm). The surface condition of the fiber has a significant influence on the strength and behavior of a glass fiber because of the large surface area to volume ratio. Glass fibers are used for heat, sound and electrical insulation, polymer reinforcement, fabrics, and fiber optics.

Metal fibers are used in high-strength, high-temperature, lightweight composite materials for aerospace applications. The tensile strength and stiffness of a composite are directly related to the fiber content and to the strength of the fiber, assuming the fibers are aligned in the direction of loading. Anisotropic or directional properties can be designed into a fiber composite by selectively aligning the fiber-base arrangement. The handling, storage and environmental conditions are important for the successful fabrication of composite structures. Bends, nicks and scratches can seriously reduce the properties of a fiber. Even prolonged exposure to air can be detrimental.

A metal-fiber composite can be fabricated by infiltrating molten metal into bundles of parallel fibers. These mats can then be shaped by conventional forming methods, e.g. hot-rolling, forging, pressing. Fibers can be embedded in powdered metals, and subsequently sintered and hot pressed to full density. Another method is that of plating the fibers to the desired thickness, then stacking them together and sintering to consolidate and densify the deposited matrix metal. Among the strongest materials are metal fibers formed by controlled solidification and cold drawing.

Fiber composites improve the strength:weight ratio of base metals, e.g. alumina or silicon carbide may be added to aluminum, or tungsten fibers may be used in a copper matrix. Metal-fiber composites are used in turbine compressor blades, heavy-duty bearings and pressure vessels. Stainless steel fibers are used to provide conductivity in polymers.

Organic fibers have been developed and are characterized by excellent environmental and thermal stability, static and dynamic fatigue resistance, and impact resistance. Aramid fiber (aromatic polyamide) is commercially available in three grades and property levels for specific applications, and has the tradename 'Kevlar'. Nomex aramid fiber is characterized by excellent high-temperature durability with low shrinkage. It will self-extinguish and does not melt, and retains a high percentage of its initial strength at elevated temperatures. Fluorocarbon fibers have very high resistance to chemicals and heat, as well as exceptionally low levels of friction and adhesion. PTFE (Teflon) fibers possess a higher degree of molecular orientation than their resin counterparts. Consequently, strength properties and resistance to cold flow of the fibers are significantly greater than those of the resins.

Carbon and graphite fibers have been developed and have considerable engineering importance, especially in the aerospace industry. Continuous graphite fibers are produced by oxidizing, carbonizing and graphitizing polyacrylonitrile (Rayon) materials at 2500°C (4532°F). The strength:density and modulus:density ratios of graphite composites offer significant advantages over common metals because of their low density (half of that of aluminum, one-fifth that of steel). Graphite composites possess low coefficients of thermal expansion, excellent wear resistance, long fatigue life, and they conduct electricity.

Fabrics are now available for engineering applications requiring resistance to chemicals, moisture and heat, as well as sufficient pliability and strength. The materials include rubberized or metal-coated natural and synthetic fibers, and glass, quartz and ceramic fibers. They are used for high-temperature aircraft skins, seals, diaphragms, gaskets and furnace linings.

* * *

Self-assessment exercises

1 What are ceramics? What are cermets?

2 List some examples of ceramic materials that have industrial applications. Find details of some of the newer materials that are now produced commercially.

3 What are the advantages of ceramics? Name three disadvantages.

4 How are ceramics fabricated and machined?

5 How can the properties of ceramics be modified by using 'alloying' additions?

6 Refractories are high-temperature materials. How are they shaped? What in-service conditions are not suitable for use with these materials? Which chemicals or environments attack refractories?

7 What are cement and concrete? What is the estimated annual usage (or production) of these materials in your country? What quantity of steel reinforcing is used? What type of steel is used, and is corrosion a problem?

8 Assess the importance of other nonmetallic materials, e.g. water, coal, paper, in engineering.

9 What are composite materials? List some engineering composites and prepare a table comparing their important properties. How are these materials fabricated?

* * *

Exercises

1 Explain the difference between molecular weight, average molecular weight, number-average molecular weight and weight-average molecular weight for polymers.

2 Which will have: (a) the greater volume and (b) the greater tendency to form crystals – LDPE or HDPE?

3 Which polymer would have the stronger intermolecular forces – polyethylene or nylon?

4 Why are thermosetting polymers usually supplied as low-molecular-weight, linear polymers before final curing?

5 Explain the changes that occur in the physical state of a plastic when it is cooled from above the glass transition temperature.

6 Select either nylon 6.6, HDPE or PMMA for contact with 20% hydrochloric acid at 60°C (140°F).

7 Select a solvent to dissolve melamine plastic.

8 Describe how the crystallinity of polymers can be controlled by the selection of appropriate polymerization methods.

9 Will a tougher composite be produced by using a spherical or a filamentous filler material?

10 Which additive, and in what form, would be used to produce a strong composite material?

11 Which polymers provide an alternative to zinc and aluminum die castings?

12 Explain why fluorocarbons have good corrosion resistance properties.

13 Describe the simplest technique for producing a low-density casting.

14 Explain why it is important to measure the coefficient of expansion of a polymer.

15 Explain why structural plastics require a high flexural strength.

16 Which is more rigid – isotactic polypropylene or atactic polypropylene?

17 List various polymers that can be used above 200°C (392°F).

18 Describe the property that distinguishes silicones from other polymers.

19 What is a vinyl plastic?

20 Explain why cyanoacrylic esters must be kept dry.

21 Which is the more flexible material – PMMA or polyethyl acrylate?

22 Why is polyacrylonitrile not used as a molded plastic?

23 Which type of rubber is similar to phenolic polymers?

24 Is there a general relationship between bulk density and strength of wood?

25 State the differences in microstructure between metals and ceramics.

26 Explain why glass components require annealing after fabrication.

27 Explain what is meant by the vitrification of ceramics?

28 State four forms in which fiber composites are produced.

29 What is Portland cement? Why should it be finely ground? What types can be used in large volumes in hot weather?

30 Explain the main difference between the structures of cermets and dispersion-hardened alloys.

*   *   *

Complementary activities

An appreciation of the properties and uses of polymers will be obtained only if some practical work is performed. This should include preparing polymers from the basic raw materials and subjecting polymer samples to appropriate tests.

Some of the necessary testing equipment may not be available within your department or working group (and sometimes none at all). However, a wide range of equipment can usually be located within engineering and science departments of educational institutions. Arrangements can generally be made (through appropriate channels) to use this equipment, with some supervision. If this is not the case, then approaches can be made to government establishments or local industries. It may not be possible to gain 'hands on' experience in these situations but practical demonstrations can often be arranged. Personal contact and perseverance can yield positive results. A less satisfactory alternative, although still useful, is to watch films related to polymer manufacture and testing. Many films can be loaned from appropriate industrial companies and professional institutions. The following activities should be performed or observed:

1 Identify the ten polymers most widely used in your country, state or local area. Why are these polymers so popular? What are their uses? What are their selling prices?

2 From the list (in 1 above) select two thermoplastic and two thermosetting polymers. Prepare detailed lists of properties, uses, methods of manufacture, etc., for each polymer.

3 Determine which tests could or should be performed on the polymers in (2) above.

4 Identify, locate and study the relevant standards (BS, ASTM, etc.) that relate to materials and polymer testing.

5 In the laboratory, carry out the chemical reactions that will produce some polymer materials.

6 Determine which chemical reactions are used to produce these polymers industrially, and the operating conditions, e.g. temperatures, pressures, etc.

7 Obtain samples of some common polymer materials (preferably those in 2 above, and in 1 above).

8 Perform appropriate tests (obtained in 4) on the polymer samples. Compare the results with values obtained (by experiment or from the literature) for common pure metals, alloys and other nonmetallic materials.

9 Attempt to shape/fabricate these polymer materials (see Section 4.15), and change their shape by machining (see Section 8.4). Perform some tests (see (4) and (8) above) on the shaped materials, compare the results with those previously obtained in (7) above. Hence, determine the effects (if any) caused by shaping upon the polymer properties.

10 If possible, visit a polymer production plant. Observe the production process and fabrication equipment in use.

11 Perform appropriate tests on some common ceramic materials, and compare the results with published values for ceramics and other common materials.

* * *

KEYWORDS

polymer	covalent bond	curing
plastic	van der Waals forces	thermosoftening
polymer chain	hydrogen bond	hardening agent
repeating molecules	cross-linking	conformation
macromolecule	thermoplastic	configuration
side-branch chains	thermosetting	end-to-end distance
linear polymer	copolymer	molecular weight:
monomer	binder	distribution
degree of polymerization	plasticizer	viscosity average

number average
weight average
pendant group
tacticity
isotactic
syndiotactic
atactic
heterotactic
stereoregular
head-to-tail
head-to-head
amorphous polymer
crystalline polymer
isotropic
anisotropic
glass transition
 temperature
degree of crystallinity
crystalline melting point
crystallites
fringed-micelle theory
oriented molecule theory
crystallization
nucleation
spherulite
lamellar structure
crystallizability
annealing
mechanical deformation
cold drawing
fiber
hot rolling
polymer sheet
lamellae
tie molecules
filler
addition polymerization
initiator
terminator
chain reaction
 polymerization
copolymerization
condensation
 polymerization
by-product
branched chain
step-reaction
 polymerization
catalyst
chain stiffening
lubricant
water absorption
testing
standard specimen
procedure
standards
tensile test

Young's modulus
modulus of elasticity
tensile modulus
Izod/Charpy impact tests
notched specimen
toughness
hardness test
scratch test
abrasion test
flexural test
deflection
monodisperse
polydisperse
stress–strain curve
Hooke's law
Poisson's ratio
thermal test
Vicat penetration test
coefficient of linear
 expansion
coefficient of cubical
 expansion
cantilevered specimen
flammability test
environmental test
oxidation
environmental stress
 cracking
weathering test
accelerated test
total immersion
swelling
density
specific gravity
bulk density
electrical test
optical property
nondestructive test
rheology
viscoelastic
viscosity
shear modulus of
 elasticity
ideal fluid
Newtonian fluid
dynamic shearing
Newton's law
rheopexy
thixotropy
dilatant
pseudoplastic
Bingham fluid
elastic response
polymer memory
permanent strain
creep
Maxwell model

Voigt model
stress relaxation
compression molding:
 positive mold
 flash mold
 transfer molding
 cold molding
injection molding
extrusion molding
casting
calendering
extrusion blow molding
blow forming
simple pressing
vacuum forming
vacuum-assisted
 pressing
plug-assisted vacuum
 forming
drape forming
bubble forming
laminates
glass fiber
reinforcement
resin
softening temperature
thermoplasticity
water vapor absorption
creep
fatigue
flammability
anisotropy
plastic adhesives
natural resins:
 shellac
 rosin
 copal
phenolic polymers
amino-aldehyde
 polymers
polyester resin
alkyds
allyls
epoxy resin
polyurethanes
ethylene-based
 polymers:
 polyethylene
 polypropylene
 polyvinyl chloride
 polystyrene
 polymethyl pentene
fluorocarbons
polyamides
polyesters
acrylics
cellulose polymers

polyacetals
polycarbonates
phenylene oxide
polyimides
polyetherimides
polyetheretherketone
polyphenylene sufides
polysulfones
reinforced polymers
structural foams
elastomers
thermoplastic
  elastomers
natural rubber
vulcanization
synthetic rubber
vulcanized fiber
silicones
ceramic
porcelain
earthenware
brick
glass
cement
refractory
abrasive
cermet
load bearing
corrosion resistant
hardness
thermal insulator
electrical insulator
ceramic bonding
ionic bond
covalent bond
metallic bond
diamond
brittle fracture
porosity

bend test
modulus of rupture
flexural strength
transverse rupture
  strength
Vickers microhardness
linear coefficient of
  expansion
coefficient of thermal
  conductivity
alumina
beryllia
zirconia
silica:
  crystobalite
  tridymite
  quartz
thermal shock
magnesia
high-grade refractory
refractory:
  basic
  acid
  neutral
metal carbide
carbon
graphite
mullite
aluminosilicates
clay
fireclay
plastic limit
water limit
firing
vitrification
asbestos
boron carbide
silicon carbide
silicon nitride

reaction sintering
cemented carbide
Portland cement
concrete
steel reinforcing
soda-lime glass
borosilicate glass
lead-alkali glass
aluminosilicate glass
96% silica glass
fused silica
soft glass
hard glass
photochromic glass
photosensitive glass
polychromatic glass
fictive temperature
devitrification
thermal tempering
chemical strengthening
crystalline glass
manufactured carbon
composite
fiberglass
surface coating
laminate
fiber:
  natural
  synthetic
wood
cellulose
glass fiber
metal fiber
organic fiber
carbon fiber
graphite fiber
fabric

*   *   *

BIBLIOGRAPHY

(Books marked * are reference texts or handbooks.)

Biesenberger, J.A. and Sebastian, D.H., *Principles of Polymerization Engineering*, John Wiley and Sons, Inc., New York (1983).

Birley, A.W. and Scott, M.J., *Plastics Materials: Properties and Applications*, Blackie and Son Ltd, Glasgow (1982).

Brydson, J.A., *Plastics Materials*, 4th Ed., Butterworth and Co. (Publishers) Ltd, London (1982).

Cagle, C.V. (Ed.), *Handbook of Adhesive Bonding*, McGraw-Hill Book Co., Inc., New York (1973).

Davidge, R.W., *Mechanical Behaviour of Ceramics*, Cambridge University Press, Cambridge (1979).

*DuBois, H.J. and John, F.W., *Plastics*, 5th Ed., Van Nostrand Reinhold Co., New York (1974).

Dym, J.B., *Injection Molds and Molding*, Van Nostrand Reinhold Co., New York (1979).

Fenner, R.T., *Principles of Polymer Processing* Macmillan Publishers Ltd, London (1979).

*Frados, J. (Ed.), *Plastics Engineering Handbook*, 4th Ed., Van Nostrand Reinhold Co., New York (1976).

*Harper, C.A. (Ed.), *Handbook of Plastics and Elastomers*, McGraw-Hill Book Co., Inc., New York (1975).

Hawkins, W.L, *Polymer Degradation and Stabilization*, Springer-Verlag, Berlin (1984).

*Ives, G.C., Mead, J.A., and Riley, M.M., *Handbook of Plastics Test Methods*, Iliffe Books Ltd, London (1971).

Katz, S., *Plastics-Design and Materials*, Studio Vista Publishers, London (1978).

Kingery, W.D., Bowen, H.K. and Uhlmann, D.R., *Introduction to Ceramics*, 2nd Ed., John Wiley and Sons, Inc., New York (1976).

*Levy, S. and DuBois, J.H., *Plastic Product Design Engineering Handbook*, Van Nostrand Reinhold Co., New York (1977).

*Lubin, G. (Ed.), *Handbook of Fiberglass and Advanced Plastics Composites*, Van Nostrand Reinhold Co., New York (1969).

McColm, I.J., *Ceramics Science for Materials Technologists*, Leonard Hill (part of the Blackie Group), Glasgow (1983).

Monk, J.F. (Ed.), *Thermosetting Plastics: Practical Moulding Technology*, George Godwin Ltd, London (1981).

Nass, L.I. (Ed.), *Encyclopedia of PVC*, Volumes 1, 2 and 3, Marcel Dekker, Inc., New York (1976).

Norton, F.H., *Elements of Ceramics*, 2nd Ed., Addison-Wesley Publishing Co., Inc., Reading, Massachusetts (1974).

Parkyn, B., *Glass Reinforced Plastics*, Iliffe Books Ltd, London (1970)

Paul, A., *Chemistry of Glasses*, Chapman and Hall Ltd, London (1982).

Postans, J.H., *Plastic Mouldings*, Engineering Design Guide No. 24, Oxford University Press, Oxford (1978).

Powell, P.C., *Engineering With Polymers*, Chapman and Hall Ltd, London (1983).

Reichert, K.H. and Geiseler, W., *Polymer Reaction Engineering: Influence of Reaction Engineering on Polymer Properties*, Hanser Publishers, Munich (1983).

Rosen, S.L., *Fundamental Principles of Polymeric Materials*, John Wiley and Sons, Inc., New York (1982).

Schnabel, W., *Polymer Degradation: Principles and Practical Applications*, Carl Hanser Verlag, Munich (1981).

Turner, S., *Mechanical Testing of Plastics*, Iliffe Books Ltd, London (1973).

*Van Krevelen, D.W., *Properties of Polymers: Their Estimation and Correlation with Chemical Structure*, 2nd Ed., Elsevier Scientific Publishing Co., New York (1976).

*Walker, B.M. (Ed.), *Handbook of Thermoplastic Elastomers*, Van Nostrand Reinhold Co., New York (1979).

Webber, T.G. (Ed.), *Coloring of Plastics*, John Wiley and Sons, Inc., New York (1979).

Wyatt, O.H. and Dew-Hughes, D., *Metals, Ceramics and Polymers*, Cambridge University Press, Cambridge (1974).

Young, R.J., *Introduction to Polymers*, Chapman and Hall Ltd, London (1981).

5
Corrosion

─── CHAPTER OBJECTIVES ───

To obtain an understanding of:
1 the ways in which corrosion can occur;
2 corrosion testing and data presentation;
3 how corrosion can be prevented or controlled.

─── IMPORTANT TERMS ───

electrochemical
    corrosion
standard electrode
    potential
sacrificial anode
protective coating
galvanic series
Pourbaix diagram
filliform corrosion
deposit attack
stress corrosion
    cracking

hydrogen embrittlement
erosion
intergranular corrosion
selective leaching
microbiological corrosion
radiation damage
polymer degradation
corrosion:
    tests
    data
    monitoring
    measurement

corrosion prevention
corrosion control
materials selection
design factors
surface preparation
surface finish
cathodic protection
anodic protection

5.1 INTRODUCTION

A simple definition of corrosion is the deterioration or destruction of a material by its environment. This may involve a weight gain or a weight reduction, or a change in the mechanical properties of the material. In the design of a component the physical, mechanical, thermal properties, etc., must be determined for the conditions which will occur in service. However, the engineer is also required to select a material for the manufacture of the component, and this choice is affected by the environment in which it is required to function.

Corrosion problems are most commonly encountered for metallic materials. Nonmetallic materials are attacked only by specific chemicals under particular conditions, and are generally inert to water and air, which are considered to be mild corrosion environments. Most metals are obtained by reduction of their naturally occurring ores, where they exist in the combined states as oxides, sulfates, etc. The extraction process requires considerable energy and the pure metal is in a higher energy state than the naturally occurring ore. The corrosion process can be considered as the reverse of the extraction process, and an attempt by the pure metal to return to its natural (and lower) energy state.

Metals are good conductors of electricity and many corrosion problems occur because of the flow of electrons, and the resulting electrochemical reactions. Corrosion in nonmetals occurs in a different manner, and usually involves ceramics or polymers. Ceramics are often used in the combined state, e.g. as oxides, and there is little tendency to revert to the natural form. Corrosion then occurs in particular situations and generally at a slower rate. Corrosion of polymers is often due to attack on the polymer bonds, and a subsequent deterioration in the physical and mechanical properties. Temperature effects are more pronounced in polymer materials than in metals; low temperatures can cause brittleness and loss of flexibility and higher temperatures can adversely affect the material properties. Melting points are lower for polymers than metals, and the maximum safe working temperature is usually below the softening temperature. Plastics and ceramics are generally poor conductors of electricity and are used as insulating materials, thus reducing the possibility of electrochemical attack.

The effects of corrosion can be measured quantitatively by the weight loss occurring as a result of the deterioration of the material over a given period of time. This is often expressed as either mass loss per unit area per day or corrosion depth per year, and it should be measured under the conditions that will be encountered in the actual application. If corrosion tests are conducted on relatively small samples compared to the actual component, then the effects of possible surface variations within the component should be considered. Corrosion problems are often considered either as wet or dry processes. The wet process is any situation in which corrosion occurs either as a result of direct liquid attack or where a liquid provides the reaction medium. Dry processes involve direct oxidation, erosion, cavitation, etc. The majority of metallic corrosion problems are wet processes and usually involve electrochemical or galvanic action. However, this broad division will not be used as a basis for the discussion of corrosion in this text, but individual mechanisms that can cause material deterioration will be discussed separately. Galvanic corrosion will be considered first because of its importance and common occurrence. For an adequate appreciation of the process some knowledge of electrochemistry will be required, and this will be outlined here. For more details or specific information, reference should be made to the many comprehensive texts which deal with both electrochemical corrosion and corrosion in general (see Bibliography).

5.2 ELECTROCHEMISTRY

For electrochemical corrosion to occur an electrolyte and an external circuit must be present. The electrolyte is simply a fluid that conducts electricity, and the bolting of two components together is the external circuit. The most obvious example of an electrolyte is water, which is a good conductor due to the presence of hydrogen ions (H^+), hydroxyl ions (OH^-) and various mineral ions. Water is present in the atmosphere as humidity and causes atmospheric corrosion. Even when corrosion occurs in the absence of moisture such as the oxidation of metals, the metal oxide layers conduct ions. Corrosion is often defined as the conversion of metal atoms to ions. When different metals are in contact with an aqueous solution, then corrosion occurs at different rates; gold is unaffected but zinc corrodes rapidly. These differences are due to different thermodynamic driving forces existing between the metal and the electrolyte. If the driving force is large, then corrosion occurs (zinc), if the driving force is low (or zero) then corrosion is negligible. Corrosion rates can be altered by changing the energy difference between the metal and the electrolyte. When zinc is initially immersed in water, the metal is oxidized (loss of electrons) and enters the solution as metal ions according to the reaction:

$$Zn \rightleftharpoons Zn^{2+} + 2e^- \tag{5.1}$$

Free electrons are left behind in the metal. The initial absence of zinc ions in solution means there is a large free energy driving force and corrosion is rapid. As the concentration of zinc ions builds up, the driving force is reduced, and when equilibrium is attained corrosion ceases. The reaction given by Eq. (5.1) is reversible, and any change in the concentration of products or reactants causes the reaction to go the opposite way. When equilibrium is attained the free energy driving force is zero, and the metal then possesses a net negative charge due to the electrons remaining. This charge is known as the *negative electric potential* or the *electromotive force* (E) measured in volts, and is related to the concentrations of products and reactants by the *Nernst equation*:

$$E = E^0 - \frac{RT}{nF} \ln (a_{prod}/a_{react}) \tag{5.2}$$

where: R is the gas constant (8.314 J/mol/K),
T is the absolute temperature (K),
n is the number of electrons transferred in the reaction,
F is Faraday's constant (96 490 coulombs/g equivalent),

a prod and *a* react are respectively the activities of the products and reactants involved in the reaction.

E^0 is the standard electrode potential, obtained when $a_{prod} = a_{react}$.

a is related to the concentration (*c*) in solution by the equation:

$$a = \gamma c \tag{5.3}$$

where γ is an activity coefficient which is usually taken as unity for most dilute solutions, so that $a \simeq c$ (*c* is the molarity).

The electrical energy or potential (*E*) is related to the free energy change (ΔG) of the electrochemical reaction by:

$$\Delta G = -nFE \tag{5.4}$$

Similarly:

$$\Delta G^0 = -nFE^0 \tag{5.5}$$

where ΔG^0 is the standard free energy change at 25°C (298 K) and 101 kN/m² pressure.

Substituting:

$$\Delta G = \Delta G^0 + RT \ln (a_{prod}/a_{react}) \tag{5.6}$$

The ratio (a_{prod}/a_{react}) is known as the *equilibrium constant* (*K*) for the reaction, and Eq. (5.6) can be written:

$$\Delta G = \Delta G^0 + RT \ln K \tag{5.7}$$

For the corrosion of zinc in an aqueous solution:

$$Zn \rightleftarrows Zn^{2+} + 2e^- \tag{5.1}$$

This is an oxidation reaction, and substitution in Eq. (5.2) yields:

$$E_{Zn} = E_{Zn}^0 - \frac{RT}{nF} \ln (a_{zn^{2+}}/a_{zn}) \tag{5.8}$$

E_{Zn} is the reversible oxidation potential of a zinc electrode and E_{Zn}^0 is the standard electrode potential. The activity of solid zinc is assumed to be unity and the activity of zinc ions is taken as equal to the concentration. Therefore:

$$E_{Zn} = E_{Zn}^0 - \frac{RT}{nF} \ln [c_{zn^{2+}}] \tag{5.9}$$

Substituting for:
$$R = 8.314 \, \text{J/mol/K}$$
$$T = 298 \, \text{K}$$
$$F = 96\,490 \, \text{coulombs/g equivalent}$$
$$\ln = 2.303 \log_{10}$$

Then:
$$E_{Zn} = E_{Zn}^0 - \frac{0.0592}{n} \log_{10} [c_{Zn^{2+}}] \tag{5.10}$$

For the oxidation reaction, two electrons are liberated, therefore:

$$E_{Zn} = E_{Zn}^0 - 0.0296 \log_{10} [c_{Zn^{2+}}] \tag{5.11}$$

However, it is possible to arrange conditions such that electrons are taken up by zinc ions and solid zinc is deposited. This is a reduction reaction and can be written as:

$$Zn^{2+} + 2e^- \rightleftarrows Zn \tag{5.12}$$

Substitution in Eq. (5.2) yields:

$$E_{Zn} = E_{Zn}^0 - \frac{RT}{nF} \ln (a_{Zn}/a_{Zn^{2+}}) \tag{5.13}$$

Using substitutions made previously, Eq. (5.13) becomes:

$$E_{Zn} = E_{Zn}^0 + 0.0296 \log_{10} [c_{Zn^{2+}}] \tag{5.14}$$

(Note the change of sign.)
E is then the reduction electrode potential, measured in volts.

The standard oxidation and reduction potentials are numerically equal but have opposite signs. The standard reduction potentials will be used here.

It is not possible to measure absolute values of E^0; values are determined relative to an arbitrarily chosen standard. This standard is the hydrogen electrode consisting of a platinum wire covered with platinum black, immersed in acid of unit hydrogen-ion activity (1N HCl solution) saturated with hydrogen gas at 101 kN/m² pressure and 25°C. The metal is immersed in a solution of its own ions of unit activity. The standard electrode potential is then taken as the measured potential difference (in volts) between the standard hydrogen electrode (assumed zero) and the metal electrode. A typical electrochemical cell is shown in Fig. 5.1. In practice, a saturated calomel electrode ($Hg–Hg_2Cl_2$ in a saturated solution of KCl) is often used, having an electrode potential of 0.280 V at 25°C when compared to the standard hydrogen electrode. Corrections are then made when referred to the hydrogen electrode.

The *electrochemical series* (or *electromotive series*) is obtained when metals are arranged in order of their standard electrode potentials. This series is shown in Table 5.1 for oxidation potentials; the entries in the table are equilibrium values obtained when no current is flowing. These values are sometimes called *Redox potentials* (reduction–oxidation). A high negative potential means a greater tendency to form positive ions; metals with a high negative potential are very active chemically. Using the standard electrode potential and Eqs. (5.5) and (5.6), the free energy (ΔG) can be calculated. If this value is negative, then the corrosion reaction occurs spontaneously; if the value is positive then the reaction does not occur unless energy is added. This information helps determine whether corrosion will occur, merely from thermodynamic considerations.

For electrochemical corrosion of a metal to occur an electrolyte must be present, the two electrodes must be joined and there must be a potential difference existing between them. The electrode which is oxidized by supplying electrons to the wire and positive metal ions to the electrolyte is called the *anode*. The other electrode where ions are reduced is known as the *cathode*. The two most common cathodic reactions are the reduction of dissolved oxygen and the

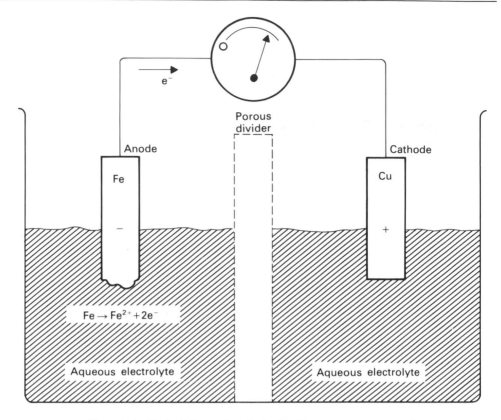

Figure 5.1 Typical electrochemical cell with an iron-copper couple.

formation of hydrogen gas according to the following reactions:

$$O_2 + 4e^- + 2H_2O \rightarrow 4OH^- \tag{5.15}$$

$$2H^+ + 2e^- \rightarrow H_2 \tag{5.16}$$

In this text the anode is defined as the material which allows metal ions to pass into solution, and the electrons generated are transferred directly to the cathode. There are two conventions which affect the sign of Eq. (5.6). The sign must be such that ΔG is negative for ionization in the forward direction. This will not be discussed further because it is the difference in standard electrode potentials which is most important here.

For electrochemical corrosion to occur there must be a complete electrical circuit between the anode and cathode, formed by metallic material and allowing the flow of electrons; the other conducting path is through the electrolyte. The corrosion process in which hydrogen is evolved at the cathode generally occurs in solutions of non-oxidizing acids where hydrogen ions are present. The electrons from the negatively charged anode flow to the cathode, and there combine with hydrogen ions in solution (evolving hydrogen gas), leaving the cathode positively charged. The reaction is:

$$2H^+ + 2e^- \rightleftarrows H_2 \tag{5.16}$$

Table 5.1   Standard electrode potentials* (reduction–oxidation†)

Element	Metal reaction	E^0 (volts)
Lithium	$Li = Li^+ + e^-$	−2.959 (anodic)
Rubidium	$Rb = Rb^+ + e^-$	−2.925
Potassium	$K = K^+ + e^-$	−2.925
Sodium	$Na = Na^+ + e^-$	−2.714
Magnesium	$Mg = Mg^{2+} + 2e^-$	−2.363
Aluminum	$Al = Al^{3+} + 3e^-$	−1.662
Zinc	$Zn = Zn^{2+} + 2e^-$	−0.763
Chromium	$Cr = Cr^{3+} + 3e^-$	−0.744
Iron	$Fe = Fe^{2+} + 2e^-$	−0.440
Cadmium	$Cd = Cd^{2+} + 2e^-$	−0.403
Cobalt	$Co = Co^{2+} + 2e^-$	−0.277
Nickel	$Ni = Ni^{2+} + 2e^-$	−0.250
Tin	$Sn = Sn^{2+} + 2e^-$	−0.136
Lead	$Pb = Pb^{2+} + 2e^-$	−0.126
Hydrogen	$2H^+ + 2e^- = H_2$	0 (arbitrary reference)
Copper	$Cu = Cu^{2+} + 2e^-$	+0.337
Oxygen	$O_2 + 2H_2O + 4e^- = 4OH^-$	+0.401
Iron	$Fe^{3+} + e^- = Fe^{2+}$	+0.771
Mercury	$Hg = Hg^{2+} + 2e^-$	+0.788
Silver	$Ag = Ag^+ + e^-$	+0.799
Platinum	$Pt = Pt^{2+} + 2e^-$	+1.200
Oxygen	$O_2 + 4H^+ + 4e^- = 2H_2O$	+1.229
Gold	$Au = Au^{3+} + 3e^-$	+1.498 (cathodic)

* Values are at 25°C and produce normal ionic activity when referenced to a normal
  hydrogen electrode.
† The choice between oxidation potentials and reduction potentials is arbitrary; the
  values above are oxidation potentials. For reduction reactions, the reaction would
  proceed in the opposite direction and the value of the electrode potential would
  be reversed (same numerical value).

The reduction electrode potential is given by Eq. (5.2) as:

$$E_{H_2} = E^0_{H_2} + \frac{RT}{nF} \ln (a^2_{H^+}/a_{H_2}) \tag{5.17}$$

From Eq. (5.10):

$$E_{H_2} = E^0_{H_2} + 0.0592 \log_{10} [c_{H^+}] \tag{5.18}$$

Since the pH of a solution is given by:

$$pH = - \log_{10} [c_{H^+}] \tag{5.19}$$

and $E^0_{H_2}$ by definition is zero, then:

$$E_{H_2} = - 0.0592 \, pH \tag{5.20}$$

The potential of the hydrogen electrode is zero when pH=0; it is a function of
pH and becomes more negative as the pH increases. The overall reaction when
hydrogen is evolved is obtained by combining Eqs. (5.1) and (5.16) to obtain:

$$Zn + 2H^+ \rightleftarrows Zn^{2+} + H_2 \tag{5.21}$$

This is simply the displacement of hydrogen ions from solution by metal ions, all metals above hydrogen in the electrochemical series (Table 5.1) having a tendency to dissolve in acid solutions. In this type of corrosion process the anode is usually large compared to the cathode; the anodic region becomes more positive as metal ions dissolve and cathodic solutions become more negative as hydrogen ions are used up.

For the oxygen absorption reaction at the cathode [Eq.(5.15)], it can be seen from Table 5.1 that the standard oxygen electrode potential is positive. There is a larger potential difference between a particular metallic anode and the cathode than occurred when hydrogen was evolved. The metal ions may combine with the hydroxyl ions and corrosion then proceeds as long as oxygen is available. In aqueous alkaline solutions the reaction is:

$$O_2 + 2H_2O + 4e^- \rightleftarrows 4OH^- \tag{5.15}$$

The reduction electrode potential is given by:

$$E_{O_2} = E^0_{O_2} + \frac{RT}{nF} \ln (a_{O_2}/a^4_{OH^-}) \tag{5.22}$$

Making previous substitutions, this becomes:

$$E_{O_2} = E^0_{O_2} - 0.0592 \log_{10} [c_{OH^-}] \tag{5.23}$$

In acid solutions the reaction is:

$$O_2 + 4H^+ + 4e^- \rightleftarrows 2H_2O \tag{5.24}$$

The electrode potential is:

$$E_{O_2} = E^0_{O_2} + \frac{RT}{nF} \ln (a_{O_2} a^4_{H^+}/a^2_{H_2O}) \tag{5.25}$$

Upon substitution this becomes:

$$E_{O_2} = E^0_{O_2} + 0.0592 \log_{10} [c_{H^+}] \tag{5.26}$$

or
$$E_{O_2} = E^0_{O_2} - 0.0592\text{pH} \tag{5.27}$$

5.3 POLARIZATION EFFECTS

When an anode and a cathode form an electric circuit, at the instant when current begins to flow the potential difference can be calculated from the equilibrium values given in Table 5.1. This voltage is the driving force for corrosion, but as current flows various effects occur which decrease the potential difference and the electrode potentials at the cathode and anode approach each other. This effect is known as *electrode polarization* and is the difference between the standard electrode potential and the actual value as current flows; the magnitude of this effect is known as the *overpotential* or *overvoltage* (η). The resulting potential is known as the corrosion potential, and the net current flowing is the corrosion current. The larger the polarization effects, at either electrode or at both, then the smaller is the corrosion current. If only one electrode is polarized, then this is the controlling mechanism. The corrosion current cannot be completely eliminated because some current must flow so that polarization can occur. Two types of polarization can occur: these are *activation* and *concentration polarization.*

Activation polarization occurs when one or more of the steps in the sequence of reactions occurring at an electrode are slower than the others. The slow process then becomes the rate-controlling mechanism. This type of polarization can occur at the cathode, with the reduction of hydrogen ions to evolve hydrogen gas. The sequence of reactions involves hydrogen ions travelling from the bulk electrolyte solution to the electrode. The ions are absorbed on the surface and combine with electrons to form neutral hydrogen atoms. These atoms then diffuse along the electrode surface until they combine with other neutral hydrogen atoms to form hydrogen-gas molecules. These molecules must then detach from the surface and escape. If any of these steps is much slower than the others, then this is the step that controls the reaction. In this case the hydrogen reduction slows down, the current flow is reduced and the potential of the cathode becomes more anodic. This decrease in the equilibrium value is known as the *hydrogen overvoltage.* Table 5.2 shows values of the hydrogen overvoltage for selected metals, compared with their standard electrode potential. The final (combined) column in Table 5.2 shows that certain metals are changed from anodic to cathodic due to the hydrogen overvoltage. This means that they would not react with acid solutions as would normally be expected.

Concentration polarization occurs when a concentration gradient exists between the bulk electrolyte and the region immediately surrounding the electrode. If these concentration differences exist, then there is only a small driving force to cause metal ions leaving the anode to diffuse into the bulk solution. This build-up of metal ions around the anode slows down the ionization reaction and the electrode potential of the anode decreases, i.e. becomes more cathodic. This is polarization of the anode. The rapid evolution of hydrogen or depletion of dissolved oxygen at the cathode can result in the cathode polarizing because ions are not available. This is more pronounced if

Table 5.2 Hydrogen overvoltage for selected metals

Metal	η_{H_2} (volts)*	E^0 (volts)	$E^0 + \eta_{H_2}$ (volts)
$Pt = Pt^{++} + 2e^-$	0.12	1.20	1.32
$Ag = Ag^+ + e^-$	0.29	0.80	1.09
$Cu = Cu^{++} + 2e^-$	0.25	0.34	0.59
$2H^+ + 2e^- = H_2$	—	—	—
$Pb = Pb^{++} + 2e^-$	0.60	−0.13	+0.47
$Sn = Sn^{++} + 2e^-$	0.50	−0.14	+0.36
$Ni = Ni^{++} + 2e^-$	0.25	−0.25	0
$Cd = Cd^{++} + 2e^-$	0.50	−0.40	+0.10
$Fe = Fe^{++} + 2e^-$	0.27	−0.44	−0.17
$Zn = Zn^{++} + 2e^-$	0.70	−0.76	−0.06

* η_{H_2} data correspond to metals being immersed in normal acid solutions at a current density of 0.1 A/cm^2.

the bulk concentration of H^+ ions or dissolved oxygen is low. Agitation of the liquid at a polarized anode can help diffuse the ions into the solution. Depletion of the metal ions in the bulk electrolyte increases the concentration gradient, and the driving force for ions to diffuse away from the electrode. A high concentration of H^+ ions or oxygen in solution helps prevent the cathodic region becoming polarized.

Generally both activation and concentration polarization occur in electrochemical corrosion situations. The total polarization overvoltage of an electrode is the sum of the activation and concentration overvoltages. At low reaction rates activation polarization is the controlling mechanism; at higher rates concentration overvoltage becomes more influential.

* * *

Self-assessment exercises

1 What is the standard electrode potential? How can values be measured? Explain the significance of a table of standard electrode potentials for different elements and its practical use.

2 Explain how values of the electrode potential for different reactions can be measured experimentally. How can values be calculated from theoretical considerations? What conditions affect the values that are obtained? What effects are produced by changing the reaction conditions?

3 For a corrosion reaction in which the anodic metal dissolves, describe what may happen at the cathode.
 What effects do acid or alkaline conditions have upon the metal dissolution (corrosion) process? From thermodynamic considerations, describe which metals dissolve in acid environments and which in alkaline conditions.

4 Explain clearly (but briefly), the meaning of electrode polarization and overpotential (or overvoltage). How does this phenomenon influence the corrosion process?
 What are activation polarization and concentration polarization?

* * *

5.4   EXAMPLES OF ELECTROCHEMICAL CORROSION

5.4.1   Electrode corrosion

The reactions occurring at an anode and cathode have already been discussed. In acid solutions the anode corrodes because the metal is oxidized to metal ions, which then displace H^+ ions from solution as hydrogen gas. The corrosion of iron with hydrogen evolution is shown in Fig. 5.2.

This type of corrosion requires only small cathodic areas and these can be in the form of impurity in a large metal anode, e.g. the hull of a ship below the water-line. One method of combating this type of corrosion is to use a *sacrificial anode*. A metal possessing a more negative standard electrode potential than the metal to be protected, i.e. more anodic, is attached to the main structure. The more anodic metal then preferentially corrodes and the metal to be protected becomes the cathode. The sacrificial anode can be replaced when necessary and this is a relatively cheap method of protection. Examples are the use of zinc or magnesium sacrificial anodes on structural steel ships' hulls or pipelines.

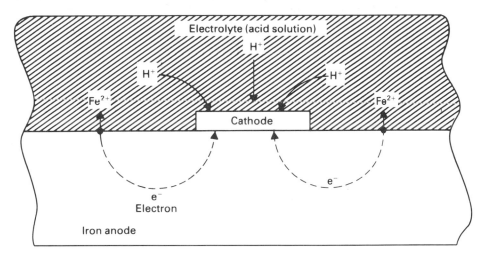

Figure 5.2   Corrosion with hydrogen evolution.

Protective metal coatings are used on several metals to prevent corrosion. However, if the protective layer is subject to cracks or imperfections, care should be taken in the choice of metal coating, depending upon the underlying metal. A protective coating of tin is cathodic with reference to an underlying steel anode, therefore any crack in the surface layer results in corrosion of the steel anode (if an electrolyte is present). However, if zinc is used as a coating on steel, then the zinc is anodic with reference to the cathodic steel and the presence of any cracks in the surface means that the protective layer corrodes. The steel is protected until nearly all the zinc layer has been removed. This type of corrosion also occurs when the reaction mechanism at the cathode involves oxygen absorption.

Consider the case of iron with a surface layer of iron oxide in contact with sodium chloride electrolyte, as shown in Fig. 5.3, where the cathodic area is large compared to the anode. The anode is due to cracks in the surface and leads to localized corrosion. The corrosion current is concentrated at a very small area and attack is strong. Sodium hydroxide is formed at the cathode and ferrous chloride at the anode, and the reaction of these products causes ferrous hydroxide to be precipitated. If sufficient oxygen is present, then ferric hydroxide is precipitated by oxidation (yellow rust), although rust in the form of ferric oxide (Fe_3O_4) may be precipitated if oxygen is limited. Polarization at the electrodes is absent if the ferrous ions diffuse away from the crack, and remove hydroxyl ions from the cathode by precipitation. The corrosion continues as long as oxygen is available. However, high oxygen concentrations can cause immediate precipitation of ferric hydroxide at the anode, which can act as a protective layer and reduce the corrosion rate.

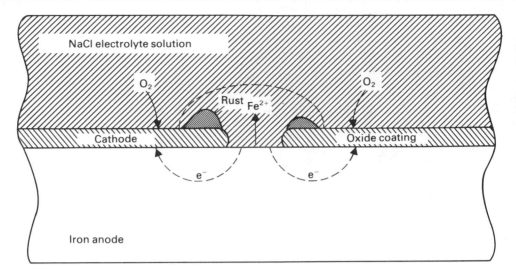

Figure 5.3 Corrosion with oxygen absorption.

5.4.2 Concentration cell corrosion

This type of corrosion occurs when two regions of a metal surface are exposed to different concentrations of the same electrolyte. These variations in concentration result in different potentials between the two regions, one of which is anodic and the other cathodic, and therefore a corrosion current flows. The most common mechanism of this type is known as *differential aeration*, in which corrosion occurs due to different air, or rather oxygen, concentrations. Metal surfaces that are subjected to low oxygen concentrations are anodic and those with high concentrations are cathodic, and a current flows between these two regions. This type of corrosion is important when a metal is partially immersed in an electrolyte. Areas close to and above the electrolyte level have high oxygen concentrations, whereas those at greater depths are anodic due to lack of oxygen, and these will corrode. Oxygen takes up electrons at the cathode to form hydroxyl ions (Eq. 5.15). The corrosion depends upon the magnitude of the oxygen concentration difference and maintenance of this situation. Regions covered by droplets of water or other electrolytes are shielded from oxygen, and so become anodic compared to cathodic areas which have free access to oxygen. This situation is shown in Fig. 5.4.

Differential aeration corrosion occurs in localized areas such as cavities and crevices which are shielded from oxygen; this causes characteristic *pitting* corrosion. It is typical of corrosion at points of contact of wire screens, improperly fitting gaskets and under scale deposits. Corrosion can actually increase with time as corrosion products form oxygen–restricting layers over the anodic region. The problem can be alleviated by restricting the oxygen concentration at the cathodic region, thus reducing the potential difference.

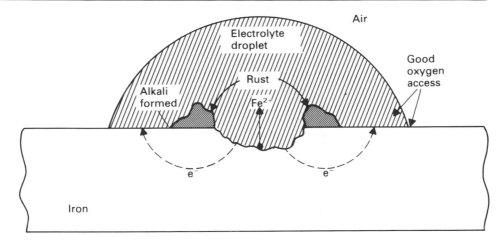

Figure 5.4   Differential aeration corrosion.

Polished surfaces have less cavities and irregularities than roughened surfaces, and are less prone to this type of corrosion.

5.4.3 Passivity

Sometimes metals or alloys exhibit improved corrosion resistance, which would not be predicted from the electrochemical series (Table 5.1). This occurs in strongly oxidizing solutions or air, where the metal becomes inert or passive. This situation is due to the reaction of the oxygen with the metal, creating a protective surface layer. Two alternative theories have been proposed to account for passivity. One theory proposes the formation of a very thin, insoluble, nonporous, self-healing oxide film which acts as a highly protective surface layer. The alternative theory assumes the presence of a surface layer of chemically adsorbed oxygen molecules (one molecule thick) which provides the surface protection. Whichever theory is preferred, the effect of the surface layer is to reduce the corrosion of the metal and make the anode more noble (or cathodic).

For some metals passivity is only temporary and can easily be destroyed, but others exhibit relatively permanent or stable corrosion resistance. Examples of metals exhibiting passivity are iron and aluminum with concentrated nitric acid, although with a dilute nitric acid solution rapid corrosive attack occurs. Stainless steels and titanium have good corrosion resistance to nitric acid over a range of concentrations due to the protective oxide film. Austenitic stainless steels are attacked by dilute sulfuric acid solutions, but exhibit passivity if oxygen is bubbled through the solution. Metals such as tantalum, molybdenum, titanium and zirconium are corrosion resistant due to the presence of a strong oxide film which is also rapidly self-healing in oxidizing environments. However, metals which are passive under oxidizing conditions are rapidly attacked in reducing environments.

5.4.4 Galvanic series

Galvanic corrosion occurs when two metals are in electrical contact in the presence of an electrolyte. This is the general name for the 'wet' corrosion processes which have been considered previously in this chapter. The galvanic cell that comprises the two metals may have two physically separate electrodes, or it may be due to metallic impurities within a metal. The galvanic corrosion involves the flow of electrons from the anodic electrode to the cathode. The anode is the less noble metal and corrodes in preference to the cathode, which is protected. It is important to determine which of the metals forming a galvanic cell will corrode. The standard electrode potentials for metals are given in the electrochemical series (Table 5.1). These values were obtained for definite ion concentrations, and for metals completely free of oxide films. In practice surface oxides are often present on metals and can result in passivity; this shifts the electrode potential to a more positive value than would be obtained from the electrochemical series. For this reason a galvanic series (Table 5.3) is used to predict which metal of a pair will corrode. This series is obtained from observations of corrosion in particular environments and takes into account the effect of passivity. In a galvanic cell, the more anodic (or less noble) electrode will corrode and the cathode will be protected. The further apart two metals are in the galvanic series, the greater are the effects of corrosion. Two metals that are close together in the series can usually be joined with relative safety. However, small changes in environmental conditions can seriously affect the position of a metal in the galvanic series and the predicted corrosion effects.

* * *

Self-assessment exercises

1 Describe practical examples of the use of sacrificial anodes (not ships' hulls or pipelines).

2 What precautions must be taken when using protective coatings? Explain how they work.

3 Describe actual situations where concentration cell corrosion occurs.

4 What is passivity? Why does it occur, and with which metals?
   Calculate, or obtain published data, to demonstrate the effect of passivity on the corrosion of selected metals under particular conditions, e.g. austenitic stainless steel in oxygenated dilute sulfuric acid compared with a nonoxygenated solution.

5 What is the galvanic series? How is this series compiled? Is the series applicable under all conditions, or only in specific situations?
   Why not use a table of standard electrode potentials rather than a galvanic series?

* * *

5.4.5 Pourbaix diagrams

Corrosion products may be soluble ions or precipitates, depending upon the pH of the electrolyte. A 'pH vs. potential' diagram is an equilibrium diagram (but *not* a phase diagram) that indicates the ions or precipitates with the lowest free energy. Such a diagram is known as a Pourbaix diagram and it provides a large amount of valuable data in a concise form. The diagrams (for different metals) are presented as curves that represent various chemical and electrochemical equilibria that should exist between metal and liquid. The curves also define the conditions under which *corrosion, passivation* and *immunity* would be expected.

On a Pourbaix diagram each line represents a balanced reaction. A horizontal line represents an equilibrium involving electrons but not H^+ or OH^- ions, for example:

$$Fe \rightleftharpoons Fe^{2+} + 2e^-$$

A vertical line represents an equilibrium involving H^+ or OH^- ions but not electrons; for example:

$$Fe^{3+} + H_2O \rightleftharpoons Fe(OH)^{2+} + H^+$$

A sloping line represents an equilibrium involving H^+ or OH^- ions and electrons, for example:

$$2Fe^{2+} + 3H_2O \rightleftharpoons Fe_2O_3 + 6H^+ + 2e^-$$

Lines representing the three reactions considered here, and other equilibria, are shown on the Pourbaix diagram for iron in Fig. 5.5.

* * *

Self-assessment exercises

1 Write down the reactions associated with particular lines on the Pourbaix diagram for iron in Fig. 5.5.

2 Why are there families of lines shown on Fig. 5.5? What do the numbers associated with each of these lines represent?

3 Each of the horizontal lines for the equilibria: $Fe \rightleftharpoons Fe^{2+} + 2e^-$ are spaced at intervals of 58 mV. Explain (with calculations) why this spacing is used.

* * *

The families of lines on Fig. 5.5 represent the potentials of the electrode equilibria at ferrous ion 'activities' equal to 10^0, 10^{-2}, 10^{-4}, 10^{-6} times normal, and designated by the numbers $0, -2, -4, -6$ respectively. A simplified form of the Pourbaix diagram for iron is shown in Fig. 5.6, where single lines are shown for equilibrium at normal ionic activity. Also included on Fig. 5.6 are the diagonal lines (a) and (b); the area between these lines represents the conditions under which water is stable. Below line (a), water decomposes to form hydrogen according to the reaction:

$$2H^+ + 2e^- \rightleftharpoons H_2$$

Above line (b), water decomposes to form oxygen according to the reaction:

$$2H_2O \rightleftharpoons O_2 + 4H^+ + 4e^-$$

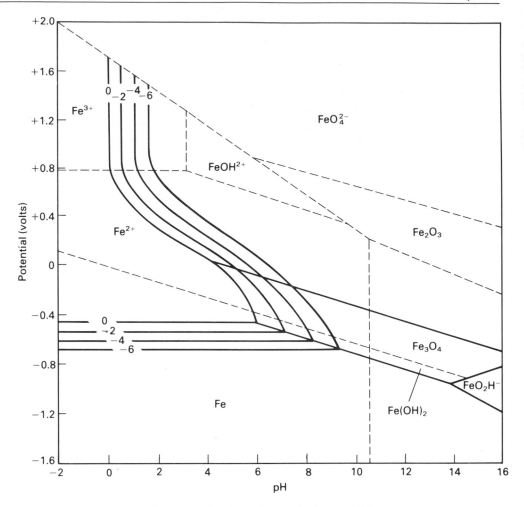

Figure 5.5 Pourbaix diagram for iron at 25°C.

These are both cathode reactions.

An alternative simplified form of Fig. 5.6 is given in Fig. 5.7, showing regions of corrosion, passivation and immunity. Corrosion occurs when metallic iron dissolves into solution, according to the reaction:

$$Fe \rightleftarrows Fe^{2+} + 2e^-$$

The smaller triangular corrosion region represents the formation of ferroates in a strongly alkaline solution. The immunity region occurs because corrosion becomes impossible when a certain concentration of Fe^{2+} ions are present in solution, as given by the horizontal line. Higher concentrations of Fe^{2+} ions would cause metallic iron to be deposited.

In the passive region (Fig. 5.7), the formation of a solid corrosion product (Fe_2O_3) is possible. The formation of Fe^{2+} ions in the liquid is also possible (thermodynamically). When a solid film has been produced, the dissolution of iron is restricted and a condition of passivation has been achieved. Similar

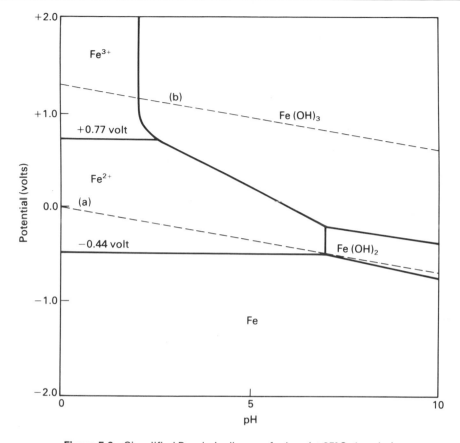

Figure 5.6   Simplified Pourbaix diagram for iron (at 25°C; 1 molar).

simplified Pourbaix diagrams for copper, zinc and chromium are shown in Figs. 5.8, 5.9 and 5.10, respectively.

Referring to Figs. 5.7 and 5.8, it can be seen that ferrous ions (Fe^{2+}) and metallic copper are stable in the neutral or low pH region for potentials between −0.44 volt and +0.34 volt. Therefore, if copper ions and metallic iron are present in an electrolyte, the iron oxidizes to ions and any copper ions (Cu^{2+}) are reduced to metal. A potential of at least +0.78 volt, i.e. 0.34−(−0.44) volt, needs to be applied to the iron before it becomes cathodic to the copper in a 1 molar solution at 25°C. The same deduction could be made from the data of electrode potentials presented in Table 5.1.

In the immunity region, corrosion is impossible based upon energy considerations and is independent of component size. However, passivation occurs due to the geometrical obstruction presented by the metallic film. In this case there is always the possibility of a structural defect occurring in the protective film, or a localized region of internal stress capable of causing continued cracking. The risk becomes greater as the area to be protected is increased; it is important that these effects are considered when interpreting or using the results of experimental tests.

Corrosion, immunity and passivation may be better appreciated if they are

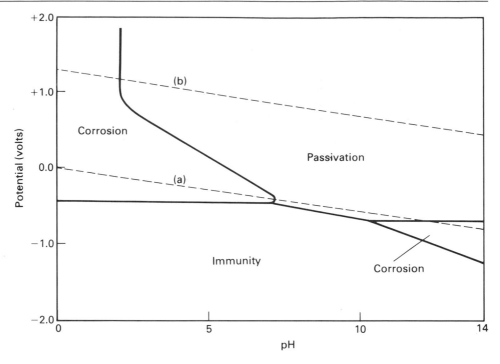

Figure 5.7 Simplified Pourbaix diagram showing reaction condition.

considered as situations in which corrosion does, cannot and does not occur, respectively. The Pourbaix diagram indicates only the lowest-energy species and provides no information about how quickly the corrosion reactions will occur. For a better understanding of corrosion phenomena, the information provided by the Pourbaix diagram should be considered with kinetic and crystallographic data. Determination of the protective nature of different corrosion products requires some knowledge of their properties, structure and behavior.

A comprehensive atlas of Pourbaix diagrams has been published (see Bibliography) and should be available in most libraries of higher educational institutes. Most diagrams in the atlas (covering all important metals) refer to a temperature of 25°C (77°F), although some diagrams at higher temperatures are also included. Other information is required in order to understand the corrosion reactions that can occur in a system, e.g. influence of pH on the solubility of various oxides and hydroxides, composition of the solutions in equilibrium with the metal at different values of pH and potential.

* * *

Self-assessment exercises

1 Locate a copy of the atlas of Pourbaix diagrams (see Bibliography) and obtain a copy of the diagrams for some common metals, e.g. nickel, aluminum, magnesium.

2 For the Pourbaix diagrams obtained in Exercise 1, check the positions of

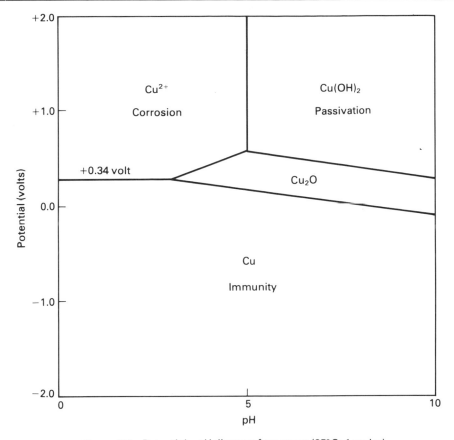

Figure 5.8   Potential – pH diagram for copper (25°C; 1 molar).

some of the lines shown on the diagram, i.e. calculate selected values of potential and pH.

3 Discuss the relevance and usefulness of Pourbaix diagrams for obtaining an understanding of some practical corrosion problems, e.g. atmospheric attack of unprotected metal, corrosion of underground pipelines.

How can Pourbaix diagrams be used by an engineer performing corrosion calculations?

Discuss the limitations of Pourbaix diagrams and the additional information that may be required.

To answer this exercise it may be necessary to refer to a specific corrosion textbook, e.g. Fontana and Greene (1967), Uhlig (1948, 1963), Scully (1975).

✳   ✳   ✳

5.5   TYPES OF CORROSION

There are many situations in which corrosion occurs, the majority of which involve 'wet' environments and some form of electrochemical process. The

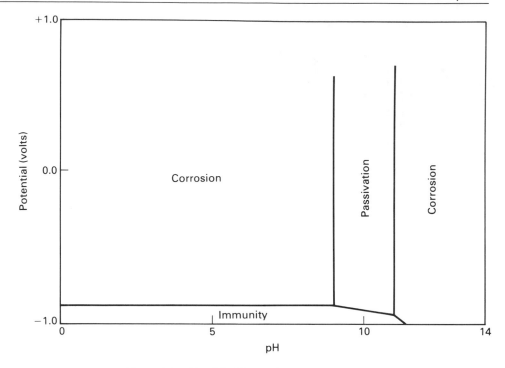

Figure 5.9   Simplified Pourbaix diagram for zinc.

background to electrochemistry and galvanic corrosion has already been discussed, as well as certain specific examples. Other specific types of corrosion problems will now be discussed, and many of these processes involve the electrochemical mechanisms already outlined. However, it is necessary for the engineer to be familiar with the alternative ways in which corrosion can occur, in order to present satisfactory design specifications.

5.5.1 Uniform attack

Corrosion can occur either at certain specific points (localized attack), or over the whole surface of the material (uniform attack). Although localized corrosion may only occur at one point, if the attack is severe this may be sufficient to render a structure unsafe for its intended purpose. Examples of localized corrosion problems have already been considered. Uniform attack may occur as a result of electrochemical reactions, or direct attack as a result of chemical reaction with the environment. Most plastics and ceramics corrode by direct attack because they are poor conductors of electrons and are generally unaffected by electrochemical reactions. Some metals corrode without the formation of a surface film, e.g. molybdenum at $450°C$ ($842°F$) disappears due to sublimation, chlorine attacks certain metals directly, liquid metals and molten salts cause direct attack on metals.

However, uniform corrosion frequently involves electrochemical reactions,

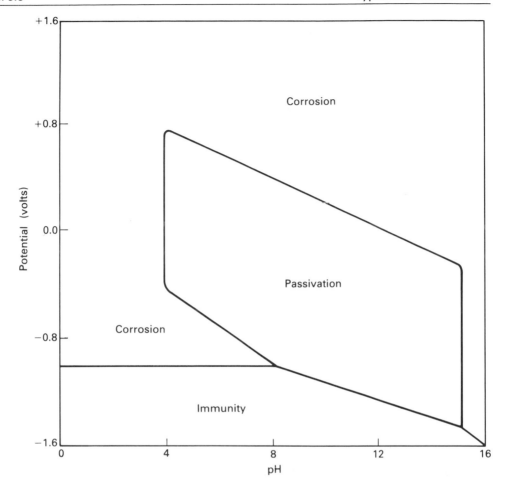

Figure 5.10 Simplified Pourbaix diagram for chromium.

either in the presence of an electrolyte or due to the formation of an oxide film. Examples of electrochemical corrosion involving an electrolyte have already been considered, the most common example being the uniform rusting of iron by moist atmospheres.

The formation of an oxide film on a metal surface can occur in dry atmospheres without an aqueous electrolyte, but still involves electrochemical reactions. The oxide layers formed on metals at room temperature are often very thin, although at high temperatures a thick layer known as scale can be formed. Some surface layers are protective and slow down or eliminate further corrosion, others do not. Pilling and Bedworth (1923) found that a porous oxide film was formed if the volume of oxide produced was greater than the volume of metal reacted. However, stresses may be set up in the film during its formation, and this results in cracks. At higher temperatures, different thermal expansion coefficients between the metal and the oxide can also cause rupture of the protective layer. The corrosion process involves more factors to be considered

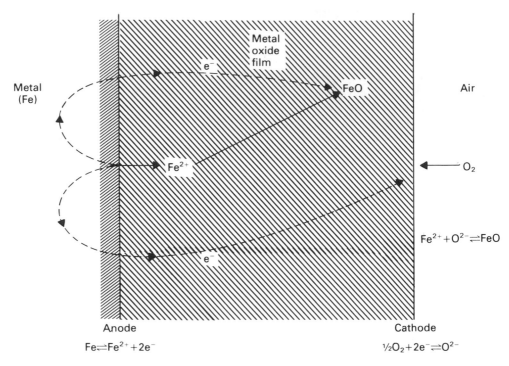

Figure 5.11 Dry oxidation corrosion involving electrochemical reaction.

than merely the porous nature of the film. The formation of the oxide layer is an electrochemical process, as shown in Fig. 5.11. The metal–oxide interface acts as an anode and the oxide–air interface becomes the cathode. The reactions which occur are:

Anode:	$Fe \rightleftarrows Fe^{2+} + 2e^-$	(5.28)
Cathode:	$O_2 + 4e^- \rightleftarrows 2O^{2-}$	(5.29)
Overall reaction:	$2Fe + O_2 \rightleftarrows 2FeO$	(5.30)

The oxide film functions as both the electrolyte and the coupling between anode and cathode for electron transfer. The situation shown in Fig. 5.11 is such that metal ions and electrons diffuse through the oxide layer, and film growth occurs at the oxide–air interface. An alternative mechanism can occur where electrons diffuse to the oxide–air interface, but metal ions do not migrate. The controlling mechanism is the diffusion of oxygen ions and dissolved oxygen through the oxide layer to the metal–oxide interface. The growth of the oxide film then occurs at the metal surface.

For electrochemical corrosion in aqueous solutions, the anode corrodes and this is generally presented in terms of the weight loss per unit area per unit time. When a metal oxide is the product of the corrosion process, this usually results in a weight gain per unit area per unit time. The rate of growth of the oxide film usually follows one of five basic laws. These are:

Linear law:	$W = K_1 t$	(5.31)
Parabolic law:	$W^2 = K_2 t + C_1$	(5.32)

Cubic law:	$W^3 = K_3 t + C_2$	(5.33)
Logarithmic law:	$W = K_4 \log (C_3 t + C_4)$	(5.34)
Inverse logarithmic law:	$\dfrac{1}{W} = C_5 - K_5 \log t$	(5.35)

In each case: W is the weight gain per unit area per unit time; t is the time; C_1 to C_5 are constants; K_1 to K_5 are the rate constants for particular metals.

The linear law is observed when the chemical reaction occurs more slowly than the diffusion of ions, and this becomes the controlling reaction. The oxidation rate is constant at a particular temperature and is not affected by diffusion of ions. This is usually true of group I metals, e.g. sodium, potassium, tantalum, which form porous, permeable films. The oxide–air surface is usually characterized by cracks which permit oxygen diffusion.

The parabolic law is obeyed most frequently in practice, particularly with group II metals, e.g. copper, cobalt, iron and nickel. The oxide film is nonporous, dense and adherent. The controlling mechanism is the diffusion of ions, either metal or oxygen ions. The constant C_1 is often taken as zero and K_2 is a diffusion coefficient.

The cubic law given by Eq. (5.33) (C_2 often zero) is not often observed. However, the oxidation of zirconium for small time exposures is an example. The cubic law applies for a limited temperature range between the logarithmic and parabolic laws and is thought to result from porosity and structural defects in the film, combined with limiting ionic diffusion.

The logarithmic and inverse logarithmic relationships of Eqs. (5.34) and (5.35) are less common, although aluminum, zinc, beryllium and chromium form oxide films according to these relationships. Oxides formed on these metals are extremely protective and usually approach a limiting thickness (approximately 0.01 μm) after which the film does not grow. This effect is thought to be due to a build-up of electrical charge in the oxide layer which restricts the flow of metal ions and eventually causes growth to cease. Metals which form oxide layers according to the logarithmic relationships are most suitable for use at elevated temperatures. This is because of their thin films and excellent resistance to oxidation. For this reason, chromium is an alloying element in heat resistant materials to be used at high temperatures.

Metals can oxidize according to more than one law depending upon the temperature. The logarithmic relationships are obeyed at low temperatures, and as the temperature is increased the growth becomes parabolic; at even higher temperatures the linear law becomes more common. A process known as *catastrophic oxidation* can occur when the linear growth law is obeyed. Rapid exothermic oxidation occurs at the metal–oxide surface, causing the rate constant to increase; the reaction rate increases with further heat release, and so on. This can result in ignition of the metal–oxide surface, which is catastrophic oxidation. This occurs with niobium, molybdenum and vanadium at high temperatures, although the addition of chromium and nickel can retard this effect. It is possible for metals to form different oxides depending on the temperature, e.g. CuO or Cu_2O and FeO, Fe_2O_3 and Fe_3O_4. The oxide closest to the metal–oxide interface would be expected to contain a higher proportion of metal ions, and that at the oxide–air interface more oxygen ions.

5.5.2 Wet electrochemical corrosion

Particular examples of corrosion occurring in the presence of an aqueous electrolyte due to electrochemical reactions have been discussed in Sections 5.2 and 5.4. Specific situations can be identified where this type of corrosion may occur.

Galvanic corrosion occurs when two dissimilar metals are electrically connected in the presence of an electrolyte, and a current flows. The anode corrodes in preference to the cathode by dissolving into solution. It may be possible to prevent galvanic corrosion by using an insulating material, e.g. nonmetallic gaskets or washers between dissimilar metals to eliminate electrical contact. Different situations and problems have already been discussed in Section 5.4.4, and the galvanic series for metals in seawater is presented in Table 5.3.

Pitting corrosion has been discussed in Section 5.4.2, and the mechanism

Table 5.3 Galvanic series of metals and alloys in seawater.

Noble or cathodic	Platinum
	Gold
	Graphite
	Titanium
	Silver
	⌈ Chlorimet 3 (62 Ni, 18 Cr, 18 Mo)
	⌊ Hastelloy C (62 Ni, 17 Cr, 15 Mo)
	⌈ 18–8 Mo stainless steel (passive)
	│ 18–8 stainless steel (passive)
	⌊ Chromium stainless steel 11–30% Cr (passive)
	⌈ Inconel (passive) (80 Ni, 13 Cr, 7 Fe)
	⌊ Nickel (passive)
	Silver solder
	⌈ Monel (70 Ni, 30 Cu)
	│ Cupronickels (60–90 Cu, 40–10 Ni)
	│ Bronzes (Cu–Sn)
	│ Copper
	⌊ Brasses (Cu–Zn)
	⌈ Chlorimet 2 (66 Ni, 32 Mo, 1 Fe)
	⌊ Hastelloy B (60 Ni, 30 Mo, 6 Fe, 1 Mn)
	⌈ Inconel (active)
	⌊ Nickel (active)
	Tin
	Lead
	Lead–tin solders
	⌈ 18–8 Mo stainless steel (active)
	⌊ 18–8 stainless steel (active)
	Ni-resist (high Ni cast iron)
	Chromium stainless steel, 13% Cr (active)
	⌈ Cast iron
	⌊ Steel or iron
	2024 aluminum (4.5 Cu, 1.5 Mg, 0.6 Mn)
	Cadmium
	Commercially pure aluminum (1100)
	Zinc
Active or anodic	Magnesium and magnesium alloys

which occurs. It is an extremely severe form of localized corrosion and tends to occur with certain metals in particular environments. One possible solution to the problem is to avoid the use of materials or environments in which pitting is known to occur. The corrosion pits are often covered by corrosion products and are difficult to detect before failure occurs. This type of corrosion generally occurs in the direction of gravity on the lower surface of a structure, rarely on an upper surface in a vertical direction.

Crevice corrosion occurs in the small space between two surfaces that are joined together. The two types which are generally considered are gasket corrosion and filiform corrosion, and both are severe localized electrochemical forms of corrosion. Gasket corrosion often occurs in crevices under bolt heads or rivets, or under gaskets. The important feature is that the liquid in the crevice remains stagnant compared to the bulk environment, and chemical changes occur in this stagnant layer. It has been proposed that the corrosion is due to a self-catalytic mechanism involving acid formation in the crevice, although anodic metal dissolution and cathodic oxygen absorption are important contributing reactions. Assuming neutral aerated seawater is present (pH = 7), then the initial reactions which occur in the stagnant liquid are the dissolution of metal (assumed iron here) and oxygen absorption according to the equations:

$$Fe \rightleftarrows Fe^{2+} + 2e^- \tag{5.28}$$

$$\frac{1}{2}O_2 + H_2O + 2e^- \rightleftarrows 2(OH^-) \tag{5.15}$$

The oxygen in the stagnant layer is quickly used up, but the transfer of electrons continues within the metal surfaces in contact with the bulk liquid. The oxygen absorption reaction for the bulk liquid continues, and Fe^{2+} ions continue to dissolve into the stagnant layer, causing an increase in concentration and positive charge. This situation is shown in Figs. 5.12 and 5.13. Negative ions diffuse into

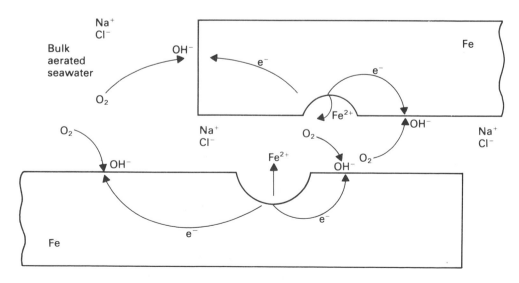

Figure 5.12 Primary stage of crevice corrosion.

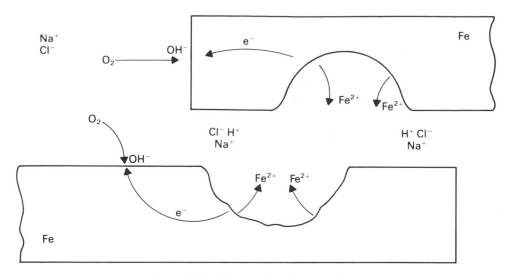

Figure 5.13   Advanced crevice corrosion.

the crevice to neutralize the charge, and these consist of the more mobile Cl^- ions rather than OH^- ions. In the crevice ferrous chloride is formed, but this hydrolyzes in water according to the equation:

$$FeCl_2 + 2H_2O \rightleftarrows Fe(OH)_2 + 2HCl \qquad (5.36)$$

The ferrous oxide is precipitated from solution, thus removing ferrous ions and accelerating the metal dissolution. The presence of hydrochloric acid increases the acidity in the crevice, and the crevice corrosion accelerates relative to corrosion at other sites. Areas of the metal exposed to the bulk solution become cathodic (receive electrons), and are therefore protected in preference to the crevice region. Stainless steels and aluminum are particularly susceptible to crevice corrosion, their passive film being destroyed by high chloride ion concentrations. Changes in the pH of the crevice liquid from 7 to 2 have been measured. This type of corrosion can be avoided by using gaskets that are 100% efficient at eliminating liquids. If any liquid is present, then gasket corrosion occurs, and materials which absorb liquids should obviously be avoided.

Filiform corrosion occurs under coatings on metals and appears as wavy hairlines. This is particularly important on metallic food containers. It is called filiform corrosion because the corrosion pattern appears as a network of tiny filaments. Most metal coatings are partially porous, allowing some diffusion of oxygen and water vapor to the metal below. This type of corrosion rarely occurs if atmospheric humidity is less than 65%. If the humidity is greater than 90%, then corrosion is rapid and large blisters appear on the surface. Reference will be made to Fig. 5.14 to describe the mechanism which probably causes filiform corrosion. Filiform corrosion occurs where there is a crevice region between the metal and the coating, e.g. sharp edges or joins, or due to poor coating techniques. Water and oxygen must be able to enter the crevice, either by diffusion through the film coating or at a crack in the film. The metal ions (Fe^{2+} here) dissolve in the condensed water vapor (Fig. 5.14, region A), oxygen is

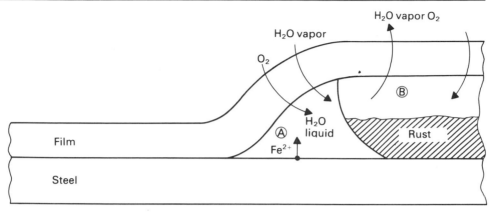

Figure 5.14   Proposed mechanism of filiform corrosion.

absorbed to form hydroxyl ions and $Fe(OH)_2$ is formed. Water and oxygen are depleted and this provides the driving force for further diffusion to the corrosion site. The Fe^{2+} ions are depleted and more ions dissolve in solution. Ferric hydroxide (red rust) is formed by the reaction:

$$2Fe(OH)_2 + H_2O + \tfrac{1}{2}O_2 \rightleftarrows 2Fe(OH)_3 \qquad (5.37)$$

This reaction increases the concentration of dissolved salts (Fig. 5.14) in region B, and water vapor then diffuses out of the film. Region B has a high pH due to oxygen reduction, whereas region A is acidic (low pH) due to the hydrolysis reaction and low oxygen concentration. The filament grows in the direction B to A, having an active head (A) and inactive tail (B).

    Deposit attack is a form of pitting corrosion, where the different chemical environment present under a scale or deposit gives rise to pits. This often occurs at sharp corners, and regions where stagnant or slow-moving liquids allow build-up of sediment.

    Waterline attack has been described (Section 5.4.2, Concentration cell corrosion) where pitting corrosion occurs at the waterline of a container. This is due to the higher concentration of oxygen in the surface layers of the solution. It is sometimes known as differential aeration corrosion.

5.5.3 Stress corrosion

Stress corrosion cracking is a term used to describe the cracking or embrittlement of a metal due to the combined effects of tensile stress and a corrosive environment. Stresses near to the yield stress of the material may be required, although there are many examples where failure has occurred with stress values below the yield stress. The stress may be applied or residual, e.g. due to quenching or cold working, but only tensile stress is damaging. Stress corrosion cracking occurs only if the combined effects of tensile stress and a particular corrosive environment are present. Pure metals are generally immune to this form of corrosion; alloys susceptible to precipitation tend to crack along grain boundaries. Homogeneous alloys may be subject to intergranular and transgranular cracking. Environments that give rise to stress corrosion cracking are generally those which produce a low overall deterioration of the material. This type of corrosion is unique in that corrosion is due to cracking, and not as a result of significant material removal. The corrosion is time dependent and may take months to occur.

The mechanism by which stress corrosion cracking occurs is not adequately understood. Several theories have been proposed but none of them can satisfactorily account for all the situations which have been observed. The theories generally fall into two categories. These are either an electrochemical process where the crack is propagated in a direction normal to the tensile stress, or as a process involving interaction of mechanical and corrosion factors. Most theories assume that localized electrochemical corrosion initiates or generates the crack, thus creating local anodic areas compared to the more cathodic metal surface. It is thought that the stress produces areas of high energy which can then be attacked by certain environments. The corrosion causes fissures to be formed which give rise to cracking. The tip of the crack, or the base of a corrosion pit, act as extremely efficient stress concentration raisers. It has been shown that removal of either the tensile stress or the corrosive environment eliminates the stress corrosion cracking, showing that both features are required simultaneously and must interact in some way. It is sometimes possible to avoid stress corrosion cracking by either eliminating the tensile stress, e.g. by annealing heat treatment, or by avoiding the use of metals and environments which are known to be susceptible to this type of corrosion. Some typical examples of stress corrosion cracking are given in Table 5.4.

5.5.4 Corrosion fatigue

Corrosion fatigue is the term used to describe the reduction of fatigue strength of a material that occurs in corrosive environments. The presence of the corrosive environment significantly reduces the fatigue strength below the typical limiting fatigue stress (approximately half the tensile strength), as shown in Fig. 5.15 for steels. The absence of a limiting fatigue stress means that even low stress levels will eventually cause failure. For corrosion fatigue to occur the environment does not have to be specific, and cracks are propagated in a transgranular manner. The position of the corrosion fatigue curve in Fig. 5.15 is affected by the corrosion rate, but not by the tensile strength of the material. Increasing the material strength will not extend its effective life, and may even be detrimental if the metal becomes more brittle and crack propagation becomes easier.

5.5.5 Corrosion cracking

Metals or alloys that are in contact with liquid metals can produce a crack similar to stress corrosion cracking. This depends upon the alloy, the environment and the type of stress. This type of corrosion is more physical than chemical in nature, and several reaction mechanisms may occur. These include deteriora-

Table 5.4 Some cases of stress corrosion cracking.

Alloys	Corrosive environment
Aluminum-based alloys (high-strength alloys)	Solution of NaCl Solution of NaCl + H_2O_2 Seawater Air Water vapor
Magnesium-based alloys	Solution of NaCl and K_2CrO_4 Marine atmosphere Distilled water
Copper-based alloys (brass) (season cracking)	Ammonia vapors and ammoniacal solutions Mercury salt solutions Amines Water
Low-carbon steel	Solution of NaOH (caustic embrittlement) Solution of NaOH–Na_2SiO_3 Solution of nitrates ($Ca(NO_3)_2$, $NaNO_3$) Acidic H_2S Seawater
High-chromium stainless steel (400 series)	Solution of NaCl + H_2O_2 Brackish water and seawater Solution of H_2S + NaOH Caustic solutions Mixed nitric and sulfuric acid
Austenitic stainless steel (300 series)	Solutions of metal chlorides ($MgCl_2$, $ZnCl_2$, LiCl, NaCl, etc.) Caustic solutions Brackish water and seawater Solution of NaCl + H_2O_2
Nickel	Strong solutions of NaOH, KOH Fused caustic
Monel (Ni–Cu alloy)	Solution of HF Solution of H_2SiF_6 Mercury salts solution
Lead	Solution of lead acetate
Titanium alloys	High-temperature chlorides (fused salt mixtures) Solution of HCl and methanol Red-fuming HNO_3 Chlorinated hydrocarbons
Stainless steels and other Fe–Cr–Ni alloys	Corrosion cracking may be caused by repeated wetting and drying of the metal

tion in the containing metal or alloy, liquid metal entering the solid metal, intermetallic compound formation or metal ion movement due to concentration or temperature differences. The liquid metal often causes brittle cracking of the solid metal by forming brittle compounds. This is a particular problem in nuclear applications when liquid metals are often used in tubular heat exchangers as high-temperature heat-transfer media. Typical examples and the problems associated with particular liquid metals are given in Table 5.5.

5.5.6 Hydrogen embrittlement

Corrosion damage can be caused by hydrogen due to either blistering or

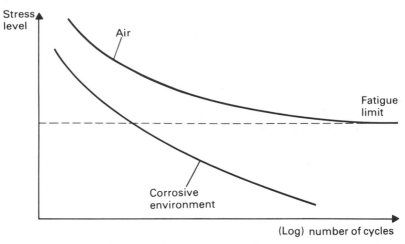

Figure 5.15 Fatigue curves for steel.

embrittlement of the metal. The hydrogen can be produced by either a chemical or an electrochemical reaction at the metal surface, at ordinary temperatures. Atomic hydrogen penetrates the metal (molecular hydrogen is too large) and if the metal is thin it may emerge from the other side. However, if atomic hydrogen enters vacancies, dislocations or large voids within the metal, it can combine to form molecular hydrogen which is then trapped. The high equilibrium pressure of molecular hydrogen under these conditions causes blistering, fissures or complete failure. This situation can also occur at a cathode when hydrogen is evolved. The cathode can become saturated with hydrogen atoms, causing embrittlement. This type of corrosion differs from stress corrosion cracking (Section 5.5.3) in that the crack does not originate at the tip of a local anode under stress. Instead cracking occurs due to internal stresses caused by trapped molecular hydrogen, combined with metallic embrittlement due to the hydrogen.

Table 5.5 Materials to be used with molten metals.

Molten metal	Recommended containing metal	Problems
Na, K, NaK alloys	18–8 stainless steels	Oxidation corrosion due to oxygen; use inert atmospheres
Hg	Carbon steel or 5% Cr steel	
Mg	Plain carbon steel or cast iron	Austenitic stainless steels severely attacked
Pb, Bi, Sn	Ni, Monel, Cu–Ni alloys	Can cause stress corrosion cracking
Fused NaOH	Ni or Ni alloys	
Al		Attacks *all* metals

5.5.7 Erosion

Erosion corrosion or *corrosive wear* are terms used to describe the deterioration of a material due to the combined effects of corrosion and mechanical abrasive action. Abrasive conditions can be caused by fluid impingement, turbulence, solid particles or gas bubbles. The abrasive action may accelerate a corrosion process which is occurring simultaneously, or it may provide appropriate conditions on the metal surface for corrosion to begin. Alternatively, failure may occur entirely due to removal of metal, either uniformly or at local regions. Erosion corrosion is frequently characterized by the appearance of holes, ridges, grooves, etc., on the metal surface; these are also indicative of the fluid direction. The fluid motion can prevent deposition of protective, insoluble corrosion products or it may remove protective layers after they are formed. Corrosion increases due to the removal of protective surface layers such as on stainless steels and aluminum, and the prevention of surface repair to these regions. Corrosion may be due to galvanic cell formation or localized anode pitting. Erosion effects are commonly encountered in equipment where a fluid stream changes direction or encounters a change in cross-sectional area, or at an obstruction to flow. Soft metals are obviously more vulnerable to physical attack by hard particles than are hard metals. Various common situations involving erosion corrosion will now be considered. It should be remembered that failure may be unexpected if the corrosion data and tests were conducted under static conditions.

Corrosion may occur as a direct result of fluid movement. This may be due to direct impingement of a liquid or gas on a metal surface or high-velocity flow parallel to a surface. The result in both cases is to remove the protective metal surface layer, thus allowing chemical or electrochemical corrosion to proceed. Impingement effects can be reduced by using baffles, and liquid erosion can be reduced by using lower velocities or metals with higher fluid velocity tolerances. The presence of gas bubbles or solid particles in the fluid greatly increases the erosion process. With very hard particles such as alumina or silica, ceramics or elastomers should be used instead of metals.

Cavitation corrosion is caused by the presence of gas or vapor bubbles in flowing liquids at low pressures. These bubbles are due to either gas entrainment or the fluid pressure being below the equilibrium vapor pressure. If the liquid pressure is increased, e.g. by pumping or a contraction in the flow area, then the bubbles collapse and the local pressure is extremely high; this can cause severe damage if the collapse occurs at a metal surface. This damage is generally in the form of pitting. Cavitation effects are observed in pump impellers, curved channels and turbines, and are usually accompanied by vibration, noise and loss of pumping efficiency. A metal must possess high hardness and strength and extremely tough surface films to withstand cavitation corrosion. Only titanium and cobalt-based alloys can be recommended for general applications where cavitation may occur.

Fretting is wear that occurs between two contacting, loaded metal surfaces and is due to vibration and slipping between the parts. This causes removal of protective layers from the surfaces, interaction of any surface irregularities, seizing of parts, abrasion due to oxide debris and fatigue in cavities or pits. If this wear occurs in the presence of a corrosive environment, then it is known as fretting corrosion and is characterized by the presence of a red 'oxide' debris. This type of corrosion occurs on bearings, shafts and couplings and can be alleviated by using a soft plating such as cadmium or copper on one metal surface. This acts as a lubricant and minimizes contact between irregularities. If the vibration causes large oscillations of the surfaces, then metal-to-metal abrasive wear occurs and not fretting.

5.5.8 Intergranular corrosion

Intergranular corrosion can occur if the grain boundaries within an alloy are chemically dissimilar to the grains. Grain boundaries have higher energy levels than the grains because they possess less atomic order. These regions are anodic compared to the grains, but corrosion occurs to a significant extent only when alloy precipitation or segregation occurs at the boundary causing galvanic action. High-strength aluminum alloys, copper alloys, and brass (grain boundaries rich in zinc) can be weakened by intergranular attack. It is also an important consideration when using stainless steels at high temperatures. If an 18–8 stainless steel (containing 18% Cr, 8% Ni and typically 0.06–0.08% carbon) is heated within the temperature range 400–850°C (752–1562°F), then carbon diffuses to the grain boundaries. The chromium is less mobile and when chromium carbide ($Cr_{23} C_6$) is precipitated at the grain boundaries the regions nearby are depleted in chromium compared to the bulk grains. The boundary is more anodic than the grain and a galvanic couple is established. This depletion in chromium is known as *depassivation* or *sensitization* and causes the grain material to become essentially a low-alloy steel. Chromium carbides are generally immune to attack; it is the regions around the carbides and the grains which are corroded. This causes grain separation and loss of strength, and may lead to failure. This process can be rapid if the metal itself is also exposed to corrosive environments.

Intergranular weld delay occurs in stainless steels due to the corrosion mechanism already described. When metals are welded, high temperatures are used and there are regions on either side of the weld where the temperatures are in the sensitization range (400–850°C/752–1562°F). Intergranular corrosion occurs in these regions if precautions are not taken. Care should be taken not to maintain any region at the sensitization temperature for long periods (soaking). If possible the metal should be heat treated by quench-annealing (Section 2.19) at approximately 1000°C (1832°F), followed by rapid water cooling (quenching). This will restrict the time available for carbon diffusion and carbide formation, but it is not practical for large or heavy structures. Alternatively, a low-carbon stainless steel can be used (carbon content below 0.02%) so that insufficient carbon is available for carbide formation. The use of stainless steels containing alloying elements that form carbides in preference to chromium is also a possibility. Niobium, tantalum and titanium are often added, although titanium is generally lost at the welding temperatures. A niobium content of approximately ten times the carbon content would be required to prevent intergranular attack.

5.5.9 Dealloying

Dealloying corrosion is also known as *selective leaching* and involves the preferential removal of one or more of the constituents from an alloy. The process is partially similar to intergranular corrosion, the difference being that some of the alloy is actually removed. The most common example is the removal of zinc from brasses containing large proportions of zinc, e.g. 70% Zn and 30% Cu. This process is then known as *dezincification*. The stages include the brass dissolving, the zinc ions go into solution and are removed, and the redeposition of a weak, spongy copper matrix on the underlying brass. The removal of zinc from brass improves its corrosion resistance, but it has undesirable effects on the material strength. If vacuum annealing is used with stainless steels, then chromium can be removed from the surface as metal vapor if the vapor pressure of chromium is exceeded. Gray cast iron is composed of an iron matrix in a network of carbon (as graphite). In the presence of a corrosive environment the iron matrix is anodic and corrodes to form rust, while the graphite is protected. (The process is incorrectly called graphitization because of the appearance of the end product.) Dimensional changes do not occur and it is therefore difficult to detect prior to failure. The solution to the problem is to avoid using alloys and environments where dealloying can occur.

5.5.10 Microbiological corrosion

Corrosion can be caused, either directly or indirectly, by the metabolic activity of various micro-organisms. The metabolic process usually involves the ingestion of certain chemical substances, and the production of corrosive by-products in otherwise noncorrosive environments. Micro-organisms are classified as aerobic or anaerobic bacteria, depending upon whether oxygen is required for their growth.

The most common anaerobic bacteria reduce sulfate ions to sulfide ions by the reaction:

$$SO_4^{2-} + 4H_2 \rightleftarrows S^{2-} + 4H_2O \qquad (5.38)$$

The hydrogen is produced by organic reactions or an electrochemical reaction at a cathode. The corrosion products are precipitates of ferrous sulfide and ferrous hydroxide, which remove ferrous ions and intensify anodic dissolution. This occurs at localized areas, particularly with iron and steel. The presence of certain nutrients, with temperatures between 25°C and 30°C (77°F and 86°F) and a pH between 5 and 9, provides conditions favorable for anaerobic corrosion.

Aerobic bacteria produce sulfuric acid by the oxidation of sulfur or sulfur compounds in the presence of dissolved oxygen, according to the reaction:

$$2S + 3O_2 + 2H_2O \rightleftarrows 2H_2SO_4 \qquad (5.39)$$

Thus the acidity and corrosiveness of the environment increases, and bacterial growth improves in acidic conditions.

Certain aerobic micro-organisms grow by ingesting iron and manganese ions, which are discharged as insoluble hydrates of iron and manganese dioxide. The bacteria grow in stagnant or running water at temperatures between 5°C and 40°C (41°F and 104°F) for a pH range of 4 to 10. Some dissolved oxygen must also be present. Corrosion occurs as a result of differential aeration between regions covered by the bacteria and other free metal regions.

Corrosion can also result from the presence of macro-organisms such as barnacles, mold and fungi. These organisms form crevices on the surface and create oxygen concentration cells, due to differences in the concentrations of dissolved salts and acids. Organic acids may be produced which will increase the pH and the corrosive nature of the environment.

Biological corrosion can usually be controlled by using paints or coatings which are toxic to biological organisms, or chemical additions to solutions where growths may occur. Externally applied cathodic potentials can also be used. Biological growths can significantly impair the heat transfer ability of equipment, and macro-organisms such as barnacles can reduce the effectiveness of streamlined designs.

5.5.11 Radiation

Materials may be damaged by nuclear radiation which can cause changes in the chemical, physical or mechanical properties. The damage depends upon the characteristics of the material and the type of radiation to which it is exposed. The charged particles in gamma or beta radiation can cause ionization in a material. These effects usually dissipate in metals due to the movement of free electrons, and result only in localized heating. In ceramics which are often ionic, this radiation can cause significant permanent changes in the material properties. These changes are usually increases in hardness, strength and brittleness, but also decreases in thermal conductivity.

The radiation can also create point defects within ceramics that possess ionic lattice structures. Radiation due to uncharged particles, such as fast neutrons, can cause more extensive damage. These particles lose their energy by collisions with atoms of the material; these collisions may be inelastic or elastic. The neutrons can travel through the material, and may be involved in many collisions until the energy is dissipated or until neutron capture occurs. This process causes heating within the material and may be sufficient to cause melting. Atoms may be displaced from their lattice positions creating point defects such as vacancies, interstitial atoms and even resulting in dislocations.

In metals these effects can cause increases in hardness, yield strength and tensile strength, and decreases in ductility, impact strength and electrical and thermal conductivity. The effects are similar to those produced by cold working or precipitation hardening and can be alleviated by annealing, although at lower temperatures than normal. This type of radiation can have serious effects on metals and alloys, and other types of materials such as ceramics which show even less resistance. Organic materials such as polymers, rubbers, etc., are particularly sensitive to radiation exposure, although the effects produced depend upon the particular material. The effect on polymers is to cause degradation or cross linking; the latter effect causes increased hardness and a high softening point, although prolonged exposure results in disintegration. Other polymers may actually become soft, as happens with butyl rubber, although natural rubber hardens. Reinforced polymers and those containing an aromatic group generally show improved resistance to radiation.

5.5.12 Polymer degradation

Polymer degradation is any process by which failure occurs as a result of a physical or chemical change in the polymer *structure*. It does not include abrasive wear, tensile stress, viscoelastic phenomena, etc., that can cause failure. Water absorption, heat effects, etc., that change the properties of a polymer and thereby cause brittle failure, etc., are also not classified as degradation processes. There is some overlap between the ultimate failure of polymers and degradation processes, but the latter always involve a change in the molecular structure, e.g. molecular weight, composition, etc. Degradation often results from a combination of factors, such as heat, radiation, abrasion, chemical effects, solvent attack, oxidation, etc. Particular causes and effects of degradation will now be discussed.

The application of heat can result in *thermal depolymerization*, i.e. the reverse of polymerization. This process produces monomer units by breaking down polymer chains, starting at some weak point. If the heat of polymerization is low, then depolymerization is easier, e.g. polymethyl methacrylate (PMMA, see Section 4.20.8) is more than 90% depolymerized. At the other extreme, polyethylene suffers less than 1% degradation. Also the lower the degree of polymerization, the greater the thermal depolymerization. Various techniques and equipment have been developed to measure or monitor depolymerization processes, e.g. mass spectrometry, thermogravimetric analysis, etc. The *ceiling temperature* of a polymer is the temperature at which the rate of the forward reaction (polymerization) and the reverse reaction (depolymerization) are equal. Knowledge of this property has important implications for both the storage and processing of polymers.

Degradation can occur by chemical conversion. An example is the dehydrohalogenation of polyvinyl chloride by removal of hydrogen chloride; this gas catalyzes further degradation. The reaction may be retarded if hydrogen chloride acceptors are present, e.g. lead salts or amines. Slow oxidation occurs in the presence of oxygen, this can result in cross-linking (hardening) or chain breaking (softening), or both. This is a chain reaction comprising initiation, propagation and termination, and any of these stages can provide the rate-controlling step. Ozone attack is different from slow oxidation and is most relevant in elastomers. Ozonides are formed which are relatively stable, but large localized stress concentrations occur. The application of a load causes cleavage, the effect is rapid and specific and elastomers can be used to test for the presence of ozone.

Polymer degradation can also occur due to combustion. When selecting a polymer for a particular application, the resistance to combustion and the products of combustion are important factors. Combustion may yield hazardous gases such as hydrogen chloride (from polyvinyl chloride), carbon monoxide, fluorine, etc. Alternatively a char may form that slows down further combustion.

Plasticizers tend to promote burning, although inert and incombustible fillers can be added to a polymer. Numerous tests have been devised to determine the effects of combustion on polymers; many are only applicable to very mild conditions. A common test is to determine the *self-extinguishing (SE) rating* of a polymer. However, the test results are only relative, and under appropriate conditions polymers will burn.

* * *

Self-assessment exercises

1 After studying Sections 5.5.1–5.5.12, *list* the alternative mechanisms by which corrosion can occur, and also the distinguishing feature of each mechanism.

2 Which corrosion processes are caused by the mechanical properties (see also Section 6.2), fabrication methods (see also Chapter 8) or the in-service loading?
    Which corrosion processes are caused by welding (or joining using heat), chemical effects or actual operating conditions?

3 Describe the actual theoretical equations that can be used to predict the corrosion rate of a metal in aqueous conditions, based upon an understanding of electrochemical principles.
    Choose an example such as the dissolution of zinc acting as a sacrificial anode, or the walls of a mild steel vessel used as a liquid storage tank. Try to obtain the data that are required, check the dimensional consistency of the equations and perform the calculations.

4 Select a particular example of corrosion caused by surface attack and oxide formation, e.g. aluminum or magnesium exposed to air. Determine which rate law will apply to this situation. Obtain data for the constants required and calculate the corrosion rate.

5 Compare the values obtained from Exercises 3 and 4 with published data of actual measured corrosion rates.

6 Can the calculations performed in Exercises 3 and 4 be performed for more complex situations, e.g. a high-alloy iron or steel containing several metallic elements?

* * *

5.6 CORROSION TESTS AND DATA

In the preparation of a design specification it is necessary to consider the materials to be used, and the environments in which they are to be situated. There are considerable data available for the corrosion of all kinds of materials under different conditions; examples of this type of information are given in Tables 5.6–8.

The designer must establish all the pertinent information such as proposed materials, environments, temperature, pressure, abrasive conditions, etc. Corrosion data are usually presented as a change in either material thickness per unit time or mass of material per unit surface area per unit time. Typical results are presented in Table 5.6. In cases of localized corrosion, the data are presented as units of penetration per unit time. This would enable calculations to be

Table 5.6   Corrosion test results on austenitic cast iron compared with phosphor bronze or ordinary cast iron.*

Corrosive medium	Austenitic cast iron (ACI)	Ordinary cast iron (OCI)	Phosphor bronze	Relative resistance to attack	
				ACI to OCI	ACI to phosphor bronze
Acetic acid, 33%	17.0	840.0	18.6	49.0	1.1
Boracic acid, 10%	7.7	57.4	4.6	7.5	0.6
Citric acid, 5%	9.3	1 492.0	4.6	160.0	0.5
Formic acid	13.9	138.0	13.9	10.0	1.0
Hydrochloric acid, 1%	32.5	1 007.0	41.8	31.0	1.3
Hydrochloric acid, 5%	54.2	3 360.0	57.3	62.0	1.1
Hydrochloric acid, 20%	62.0	11 180.0	60.5	180.0	1.0
Nitric acid, 1%	620.0	697.0	2 446.0	1.1	4.0
Nitric acid, 5%	4 060.0	4 680.0	12 420.0	1.2	3.1
Nitric acid, 20%	7 830.0	10 092.0	Dissolved	1.3	—
Oxalic acid, 5%	6.2	55.8	12.4	9.0	2.0
Phosphoric acid, 50%	26.4	4 650.0	7.7	176.0	0.3
Sulfuric acid, 1%	26.4	1 642.0	18.6	62.0	0.7
Sulfuric acid, 5%	37.2	6 880.0	37.2	185.0	1.0
Sulfuric acid, 20%	41.8	13 720.0	38.8	328.0	0.9
Sulfurous acid	240.0	1 032.0	9.3	4.3	0.04
Tartaric acid, 5%	10.8	1 040.0	10.8	96.0	1.0
Vinegar	4.6	104.0	4.6	23.0	1.0
Acetone	1.5	4.6	1.5	3.1	1.0
Aluminum sulfate, 5%	20.0	96.0	10.8	4.8	0.5
Ammonium chloride, 5%	10.8	35.6	57.4	3.3	5.3
Ammonium nitrate, 5%	51.2	163.0	57.4	3.2	1.1
Ammonium sulfate, 10%	9.3	32.6	13.9	3.5	1.5
Sulfuric acid	26.3	11 160.0	21.7	424.0	0.8
Carbon tetrachloride	1.5	3.0	3.0	2.0	2.0
Copper chloride, 10%	1 394.0	8 030.0	543.0	5.8	0.4
Ferric chloride, 5%	667.0	1 038.0	347.0	1.6	0.5
Fuel oil	1.5	1.5	1.5	1.0	1.0
Hydrogen peroxide 20 vols	6.2	9.3	1.5	1.5	0.2
Magnesium chloride 10%	7.7	18.6	6.2	2.4	0.8
Magnesium sulfate, 10%	3.1	14.0	3.1	4.5	1.0
Potassium alum, 10%	15.5	372.0	20.2	24.0	1.3
Seawater	6.2	23.2	6.2	3.7	1.0
Sodium chloride, 3%	7.7	12.4	3.1	1.6	0.4
Sodium hypochlorite	223.0	688.0	80.6	3.1	0.4
Sodium sulfate, 5%	9.3	7.7	9.3	0.8	1.0
Sodium sulfite, 10%	3.1	6.2	1.5	2.0	0.5
Sodium sulfite, 5%	1.5	1.5	17.0	1.0	11.0

* Figures represent the loss in milligrams per square decimeter per day under static conditions at 20°C (68°F).
1g/m^2/day = 1.422 × 10^{-6} lb/in^2/day

performed to establish the life of a component before penetration causes failure.

If the required corrosion data are not available, then laboratory or field tests should be carried out. Small-scale tests can be undertaken to determine the effect of uniform or pitting corrosion by simple immersion, and crevice

Table 5.7 Effects of environments on stainless steels.

Environment	Effects on stainless steels
Atmospheric	All classes show excellent resistance in the absence of corrosive gases, e.g. chlorine, sulfur dioxide.
High-temperature oxidation	Some grades can be used as high as 1100°C (2012°F), generally safe at 400–600°C (752–1112°F), higher for intermittent use.
Neutral water	Most grades are inert for velocities up to 30 m s^{-1} (100 ft/s) in austenitic pumps and piping.
Salt water	Chlorides can cause stress corrosion cracking above 65°C (149°F), pitting may occur. With stagnant liquids avoid ferritic and precipitation-hardening alloys.
Petrol	Many grades have excellent resistance.
Foods	Most grades are resistant to dairy produce; austenitic grades are resistant to fruit and vegetable juices.
Bleaches	Not recommended.
Alkalis	All grades are resistant at room temperature. For higher temperatures choose high-nickel (austenitic) grades.
Sulfuric acid	At room temperature, certain austenitic grades have low corrosion rates for dilute (<10%) or concentrated (>95%) solutions. At higher temperatures they are only resistant to dilute acid.
Nitric acid	Certain grades are available for all concentrations at room temperature. Other grades at temperatures up to the boiling point, if the concentration is below 70%.
Hydrochloric acid	Generally rapid attack.
Phosphoric acid	Most grades are resistant to dilute solutions at room temperature.
Hydrofluoric acid	Not recommended.
Acetic acid	Certain types are resistant at temperatures below the boiling point, and concentrations up to 98%.
Organic solvents	All grades are resistant for pure solvents.

Table 5.8 Effects of environmental conditions on particular metals.

Metal	Suitable environment	Conditions to be avoided
Stainless steels	Oxidizing acids	Reducing conditions
Nickel and its alloys	Basic solutions, e.g. NaOH	
Monel	HF	
Hastelloys	Hot HCl	
Lead	Dilute H_2SO_4 in stagnant conditions	Erosion in flowing conditions
Aluminum	Atmospheric oxidation	
Tin	Very pure distilled water	
Titanium	Very strong, hot oxidizing solutions	
Ordinary steel	Concentrated H_2SO_4	Dilute or aerated H_2SO_4
Tantalum	Most acids, any temperature or concentration	HF and strong NaOH solutions

corrosion by shielding a region with tape or a rubber band. The effects of tensile stress can also be examined in this manner. Simple equipment can be used to determine the effects of abrasive wear, velocity of fluids, or corrosion due to abrasive particles. Electrical equipment is used to monitor electrochemical or galvanic action. The majority of corrosion situations involve electrochemical reactions and require the presence of an aqueous electrolyte; there are many electrical techniques available to determine and monitor corrosive conditions. The main disadvantage of performing corrosion tests is the time required to obtain reliable information. This may be weeks or months, and may be necessary for long-life installations. However, there are now more sophisticated polarization techniques available which can produce the information required in only a few hours.

5.7 CORROSION MONITORING

Before describing ways in which corrosive attack can be detected and measured, it will be useful to consider the justification for employing monitoring techniques. Corrosion is one of the major problems associated with the operation of industrial equipment. Chemical plant in which corrosion could be ignored, or were insignificant, would be 'overdesigned' or completely uneconomic in terms of the materials costs. Although laboratory testing is useful and important, corrosion monitoring provides (or *should* provide) a means of measuring attack in the equipment itself, as it occurs. The aim should be to feed back corrosion measurements to the plant control functions and enable corrective action (if possible) to be taken immediately. Monitoring is concerned with understanding and using appropriate theory for the design of electronic instrumentation, and developing ways of analyzing the data produced and presenting it in an appropriate manner.

There are definite reasons why corrosion monitoring should be performed: these are usually economic rather than technical. It would be possible to design equipment so that corrosion could be ignored; however, the products would be produced at uneconomic costs. There are three main advantages of applying appropriate monitoring procedures:

(a) the ability to operate or manufacture equipment closer to the design limits (safety margins can be reduced);
(b) reduction in the number of breakdowns, or less risk of operating equipment in a dangerous condition;
(c) reducing the time (and cost) of maintenance and inspection ('downtime').

The overall effect should be to improve the reliability and safety of equipment. Corrosion monitoring can be used to assess the suitability of materials for particular applications, particularly when small amounts of impurities are present. It can also provide a warning of unexpected fluctuations in environmental conditions which cause rapid attack.

Monitoring equipment provides measurements of changes in either the process conditions or the wall thickness or structure. It is not usually possible to obtain all this information from one particular measuring technique or item of equipment. The selection of a monitoring method depends upon whether it is to be used to 'inspect' the equipment and assist maintenance, or if it is to determine corrosion due to (changes in) process conditions. The design wall thickness for a vessel or pipe is based upon two considerations:

(a) the structural design thickness to withstand pressures, stresses, etc.;
(b) the corrosion thickness, based upon an estimated corrosion rate.

Measurements of the (remaining) wall thickness for a structure are performed by ultrasonics, radiography, etc., and are intended to ascertain that consideration (a) is still observed or applicable. Corrosion due to the environment (condition b) is more difficult to monitor. Measurements of all the relevant process conditions would be possible but it would require a vast army of equipment duplicated in many locations. Since the detection of corrosion is the immediate requirement, this is usually performed by probes. Thus any deterioration indicates environmental changes which can then be investigated (as soon as possible). Some methods provide an *integrated rate measurement* – typically the average weight loss over the period of operation. Other methods provide *instantaneous readings* which are then used to initiate control action. It is vital that the corrosion monitoring equipment selected provides the information required, rather than selecting the cheapest detection method available.

5.7.1 Monitoring techniques

Monitoring techniques fall into two categories. First, identification of the equipment condition, e.g. wall thickness, on a regular basis. Second, measurement of corrosion due to process or environmental conditions, usually on a continuous basis. Some of the more common methods employed for each category of measurements will now be described.

Monitoring the condition of equipment:
(a) *Visual inspection* is a useful method, and it is easy to initiate. Unfortunately, to be effective it must provide more than an external appraisal of leaks, cracks, distortion, etc. It is necessary to perform detailed internal inspections of equipment, with the corresponding interruption to production and the associated costs.
(b) *Sentry holes* are drilled in the outside of a structure to a depth just beyond the structural design thickness. When internal corrosion reaches this depth, process fluid is released (an indicator) and the hole is plugged until the damage is repaired.
(c) *Ultrasonic methods* use the transmission of high-frequency sound from a piezo-electronic transducer. Access is only required from one side, but a specially prepared surface and couplant fluid are required. These methods can be used at temperatures up to 500°C (932°F), although with some loss of accuracy. This is also true of continuous operation. It is difficult to determine corrosion rates and a skilled operator is required. Methods used include a pulsed echo and a resonant technique.
(d) *Eddy currents* can be induced in a material by a primary solenoid and are then detected by a separate search coil. This method is useful for detecting intergranular attack or pitting and is often used for inspection of heat-exchanger tubes.
(e) *Electrical conduction* method is based upon the change in voltage between two contacts carrying a direct current. The voltage changes with both corrosion and scale formation.

(f) *Magnetic particles and magnetic induction* use the principle that changes in an induced flux are dependent upon material thickness. Magnetic particles can only be used to detect surface cracks, and this can only be done when the equipment is shut down and empty. This method is not really suitable for measuring local corrosion, except localized stress corrosion cracking in vessel walls.

(g) *Radiography* (see Section 7.21.6) is a popular nondestructive testing method that can be used to detect pitting and crack propagation. It is an accurate method, providing a permanent record. Unfortunately it is expensive, time consuming and requires access to both sides of a structure.

Monitoring environmental conditions:

(i) *Coupons* are used to detect the weight loss due to corrosion; sample size, surface preparation and type of support are important. These tests are carried out for long exposure periods and examination should be performed at regular intervals.

(ii) *Process stream (chemical) analysis* is important in high-purity systems, e.g. foods and pharmaceuticals, or circulating systems, e.g. cooling water.

(iii) *Electrochemical methods* include potential measurements of alloy samples or the plant itself, using a high-impedance voltmeter and a suitable reference electrode. The potential measurements are made against a calomel or silver–silver chloride half cell, electrodes of stainless steel, platinum and tungsten are used although they also respond to changes in redox potential of the electrolyte. These measurements do not provide a corrosion rate but an indication of whether the corrosion condition is passive or corrosive. An alternative method is the linear polarization resistance measurement, based upon the relationship of electrode potential to current when close to the corrosion potential.

(iv) *Electrical resistance probes* (wire, tube or strip) are inserted in the process stream to measure changes in the process environment. Corrosion is measured by the increase in resistance due to a reduction in cross section. A reference element is required, and regular probe inspection is necessary to distinguish between uniform and localized corrosion. The probes are often made from different metals than the equipment itself; the hydrodynamics of the process fluid may not be typical at the measuring point and important local variations may not be detected. Access points should be located in high-risk regions, they should be easy to inspect and have suitable space for removal and installation of equipment.

5.7.2 Location of monitoring equipment

Having discussed the techniques available for monitoring corrosion (remembering that new techniques are being developed and existing methods modified and improved), it is necessary to consider the possible location(s) of the detection equipment. The location must satisfy two criteria.

(a) it must be accessible;
(b) it must correspond to a possible area of (high) corrosion activity.

Corrosion may occur because of the materials that are used, the process fluids within the equipment and the operating conditions. Problems are associated with particular effects, such as

(a) temperature gradients;
(b) changes in fluid velocity;
(c) stagnant regions;
(d) erosion;
(e) cavitation;
(f) material residual stresses;
(g) mixed metals;
(h) variations in the operating conditions.

A probe can detect corrosion conditions only at its chosen location; consideration must be given to installing a probe (or several probes) at fluid inlets and exits, fluid levels, within bulk liquids, within vapor phases, at changes in flow directions, in stagnant regions, at heated regions, etc. Ideally probes should be situated at many locations, but the cost of this approach may be prohibitive. A compromise is required between using few probes (least cost) and multiple probes (maximum reliability). An additional problem is the time (and cost) required to process the large amounts of data produced by many monitoring devices. It should be possible to instal and remove monitoring equipment without disruption of the plant operation.

Corrosion monitoring provides the maximum possible benefits only if it is considered at all stages of a project, from design to full operational capacity. There should be adequate consultation between the design engineer, materials/corrosion engineer and the process engineer. Corrosion is a continuous process and monitoring must be carried out at regular intervals, if not continuously. Process conditions may change during the lifetime of an item and minor variations can cause significant increases in plant corrosion.

5.8 CORROSION PREVENTION AND CONTROL

The much quoted phrase 'prevention is better than cure' is not always true when the designer is faced with corrosion problems. After identifying the materials available, the environments to be encountered and the relevant operating conditions, then the designer may produce a specification which is free from any corrosion problems. However, if this design is uneconomic, then the designer has failed in his task. It may be that the materials specified or the nature of the design have become financially prohibitive in order to completely eliminate corrosion. In this situation it may be preferable to allow for the effects of corrosion, and to incorporate various safeguards and periodic checks. Various methods of dealing with corrosion will be considered, although whether these methods result in prevention or control (or merely monitoring of the situation), depends upon the particular application. The economics associated with design problems will not be considered here, although they are always a major consideration for the designer.

5.8.1 Material selection and design considerations

One of the first considerations for minimizing corrosion is in the choice of the materials to be used. Most of the corrosion problems already described in this chapter have been concerned with attacks on metals. However, although nonmetallic materials may possess certain advantageous properties, they are not often used as metal substitutes when corrosion presents a problem. The most obvious alternatives are polymers and ceramics, although few of these materials possess the strength or ductility of steel. Steel is easily fabricated and is probably the most widely used engineering material. Compared to steel, polymers are softer and weaker, more susceptible to swelling and attack by solvents and strong inorganic acids, and have limited applications above ambient temperature. In contrast, ceramics are strong and hard, unaffected by high temperatures and resistant to corrosive environments. However, they are brittle and have low tensile strengths, and can fracture due to impact loads. Nonmetallic materials are generally employed as linings, coatings or gaskets.

There is no single material available which is corrosion free in all environments and for all applications. However, certain materials exhibit superior corrosion resistance in certain situations, e.g. nickel and copper and their alloys are used in reducing or non-oxidizing conditions; alloys containing chromium, especially stainless steels, are used in oxidizing conditions; titanium and its alloys are used in severe oxidizing environments. Often stainless steels are thought to be the cure for all corrosion problems, but this can lead to expensive failures. Stainless steels are exceptionally good corrosion-resistant metals under many conditions, particularly oxidizing environments, due mainly to the presence of chromium, which forms a passivating surface layer. However, they are more liable to pitting crevices, stress corrosion cracking and intergranular attack than steels which do not contain chromium. They are less resistant to attack in chloride solutions and may suffer from galvanic corrosion when two phases are present, e.g. ferrite in an austenitic steel. Despite these limitations, stainless steel is still an extremely useful design material. Table 5.7 contains information about the effects of certain environments on stainless steels. Table 5.8 contains information concerning particular metals that would be preferred or avoided in certain environments. There is often an obvious choice of material for a particular environment and the designer must ensure that all the relevant conditions have been determined, including any unusual effects such as cyclic loads, rapid environmental changes, vibrations, etc.

Corrosion can often be controlled or significantly reduced by the correct choice of materials, combined with the correct design of equipment. There are certain design principles based on an understanding of the basic mechanisms of corrosion previously discussed. If these principles are understood and adhered to, then corrosion rates can be minimized in many common situations. Good design eliminates, as far as possible, situations where particular types of corrosion will occur, e.g. welding rather than riveting reduces available crevices. If materials must be bolted together, then crevices can be avoided by using gasket insulation which is nonporous and provides an exceptionally good fit. The gasket can also be used to insulate dissimilar metals which would be subject to galvanic corrosion. Avoiding high liquid and slurry velocities, regions of stagnant liquid and the deposition of solids should prolong the life of equipment. This may also be achieved by avoiding sharp bends in equipment and including extra protection plates, or conical-bottomed tanks to assist solid and liquid removal. The use of dissimilar metals which are joined together should be avoided, and metals chosen which are close together in the galvanic series. These metals can be insulated for better protection, or sacrificial anodes (see Section 5.8.4) can be used. If a situation is unavoidable, then a large anode and small cathode should be used, e.g. sheet metal anode and cathodic rivets, bolts, etc. In this situation it is usually better not to use anode coatings, because if they crack severe pitting corrosion results. It is preferable to have a uniform rate of corrosion over a large area, and then to include an extra material thickness above that required for mechanicial considerations. Galvanic corrosion can occur in closed circulating systems when two dissimilar metals are used, even though they are not in contact. This can be due to dissolution of metal ions, their transport and subsequent deposition. This is particularly true of copper and aluminum.

Tanks and containers should be designed for complete draining and cleaning. Equipment which may corrode should be installed such that its removal and replacement is easily achieved. Such equipment should also be inspected periodically so that corrosion can be monitored. This is particularly true of moving parts, e.g. stirrers and pump impellers. The entrainment of air in liquid systems, particularly pipes and pumps, should be avoided. It is sometimes advantageous for liquid feed pipes to be situated below the liquid level in a tank to avoid splashing and evaporation. Corrosion may then occur at the liquid surface on the pipe due to differential aeration.

Welds are an obvious source of corrosion problems, either from intergranular attack, cracking due to contraction or poor design of the welded components. Incomplete weld penetration and formation of crevices must be avoided, as must trapped slag or blow holes. Dilution of the parent metal thickness by welding must be avoided and any backing plates must be removed. Systems that use heating coils often use a completely immersed system to avoid splashing and water-line corrosion. The design should ensure elimination of any hot spots which can give rise to localized corrosion.

The location of equipment should be decided after considering the effects of the weather, e.g. prevailing winds, tides, etc., and any changes in the atmosphere which may occur. Vessels and equipment should be situated such that liquids cannot collect in stagnant regions in contact with the outside surfaces. This may be achieved by using supporting legs to provide a free circulation of air.

Examples of particular situations where corrosion can occur, and the ways in which design considerations can be used to avoid or limit the effects are

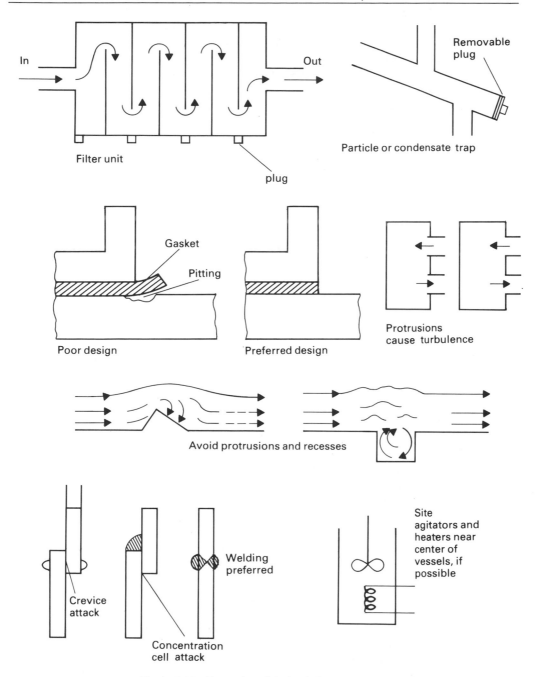

Figure 5.16 Examples of design influencing corrosion.

shown in Fig. 5.16. These are only a few examples: the possibilities and alternatives could fill a book! The following list is intended to provide some ideas regarding the importance of correct design for reducing corrosive attack. These ideas should not be considered exhaustive. If possible avoid or remove:

   (a) stagnant liquids;
   (b) abrasives and solid particles;
   (c) condensation;
   (d) crevices, or access to crevices by corrosive environments;
   (e) excessive velocities or turbulence;
   (f) formation of gas bubbles;
   (g) inaccessible locations;
   (h) adverse effects of winds, tides, weather, etc.;
   (i) adverse effects between items of equipment, e.g. drainage, vibration, etc.;
   (j) lap joints that are not sealed;
   (k) changes in material thickness;
   (l) joining or contact of dissimilar metals;
   (m) extremes of mass or cross section;
   (n) distortion of equipment and structures;
   (o) and so forth.

5.8.2 Surface preparation

Corrosion can sometimes be reduced or eliminated if the surfaces of the material are prepared and maintained in a suitable condition. The structure and properties of a metal surface may be changed by surface hardening processes (see Sections 2.19.5–2.19.8), or the surface may be separated from the corrosive environment by the application of protective coatings (see Section 5.8.3). If these methods are carried out correctly, then in appropriate circumstances the corrosion rate can be reduced.

The surface finish (texture) also has considerable influence upon the corrosion occurring at the surface of a material. Some of the more common surface finishing processes are described in Section 8.4.10. They are performed in order to produce a 'flat' surface, or rather to reduce the imperfections and irregularities present at the surface. Most of these processes remove only small amounts of material; they are not machining processes. However, the smoother the surface required, the greater the associated costs.

A high degree of surface finish is often specified to improve the mechanical performance of a component, i.e. to reduce friction and abrasive wear. This is especially true if two metal surfaces are in contact without the presence of a lubricant, and it becomes more important as relative movement between the surfaces increases. However, the potential savings due to corrosion reduction may be of equal magnitude, whether the surfaces remain uncoated or further surface treatment is applied. The following factors are related to surface preparation (finish) and corrosion control.

   (a) A lubricant film between moving parts must ensure that any surface irregularities are covered.
   (b) The appropriate surface grain should be specified if dry friction occurs.

(c) Fatigue cracks on highly stressed members subjected to load reversals are frequently caused by surface irregularities.

(d) Very rough surfaces retain dust and condensate and increase turbulence and water-scale deposits, all of which increase the tendency for corrosion.

(e) All surfaces must be cleaned prior to surface preparation.

(f) A very rough surface is *not* required for application of a protective coating.

(g) Certain surface cleaning methods are applicable to particular materials to reduce the effects of corrosion, e.g. steel – abrasive blasting; stainless steel and nickel – nonmetallic abrasive blasting; aluminum – fine grade abrasive blasting; copper and zinc – mechanical cleaning followed by appropriate washing.

(h) Oil, grease, salts, and contaminants should be removed by either steam cleaning, vapor degreasing, solvents or alkalis, before or after surface cleaning is performed.

5.8.3 Protective coatings

If there is no metal suitable for a particular application, with respect to the required mechanical properties and corrosion resistance, then it may be necessary to use protective coatings on a metal surface. The coating should, ideally, be chemically inert to the environment under the conditions to be encountered; it should also provide a continuous barrier between the base metal and the corrosive conditions. The coating should be nonporous, impermeable and pinhole-free, which is difficult to achieve or maintain in practice. The protective layer is known as a coating, lining or cladding, depending upon the thickness. It is essential that the surface is correctly cleaned and pretreated before any coating is applied, in order to achieve the maximum protection and useful life. The simplest type of surface protection is a chemical conversion coating which is formed by a chemical or electrochemical reaction at the surface of the base metal. This process is frequently used on magnesium and aluminum and their alloys. The base metal becomes the anode in an electrolytic oxidation process which is known as *anodizing*. The result is the formation of an oxide film of the required thickness. This type of protection is reasonably effective against atmospheric corrosion, but not against chemicals or liquids. The same comment applies to paint films, thick polymer layers and metal platings, e.g. zinc galvanizing, which usually suffer from pinhole cracks. Chromium additions to stainless steels provide surface protection against the atmosphere, as do heat treatment diffusion processes such as nitriding (however, this lowers the chemical resistance). Suitable heat treatment such as annealing and quenching can be used to improve the corrosion resistance, particularly for welds and castings. It can often be worth the cost incurred, due to the savings from reduced corrosion. This applies for both coating applications and direct metal protection. Coatings are generally described as metallic, nonmetallic (inorganic) or organic in nature.

Metallic coatings can be applied by spraying, dipping, electrodeposition, vapor deposition and diffusion. The method chosen depends upon the properties of both the base metal and the metal coating, the thickness required and the protection required. A metal-to-metal coating can give rise to galvanic corrosion if there is a break in the coating and if an electrolyte is present, e.g. atmospheric humidity. If the metal coating is more anodic than the underlying metal, then the coating corrodes in preference to the base metal (cathode) if a galvanic couple is established. Protection is obtained until most of the coating has disappeared, and this is known as a *sacrificial anode*. An example is the layer of zinc on steel known as galvanized iron. A small crack in the surface exposes a small cathodic area (steel) which is protected as the larger anode (zinc) slowly and uniformly disappears. The corrosion rate of the zinc increases as more cathodic areas are exposed. However, if a metal coating is cathodic with respect to the base metal, then any penetration through the layer results in rapid, deep pitting corrosion at the base metal. Applications with large cathodic and small anodic areas should be avoided.

Nonmetallic (inorganic) coatings are usually ceramics, glass linings or porcelain enamels. Ceramic coatings are generally alumina, zirconia or magnesia and are applied by flame spraying. Coatings are 25–75 μm thick, but are often porous and require resin sealing. They are used at high temperatures, prevent oxidation, and reduce abrasive wear. Glass linings are resistant to corrosion by a wide range of chemicals and rarely contaminate food products. Linings are generally 0.6 mm (0.024 in.) thick and have a high silica content. With careful preparation of the base metal to obtain a uniform thickness, the elimination of sharp edges and matching the coefficients of thermal expansion of the coating and the base metal, then the life of a lining can be considerably extended and risk of damage by thermal or mechanical shock reduced. Porcelain enamel has better shock resistance but lower corrosion resistance, and is applied only in thicknesses of less than 0.03 mm (0.0012 in.).

Organic materials are often used to provide corrosion resistance, especially from microbiological growths, and are usually applied as a series of built-up layers. If the final thickness is less than 0.4 mm (0.016 in.) they are known as coatings; if it is greater than 0.4 mm (0.016 in.) layers are called mastics or linings. Paints and polymers are most frequently used, although they are rarely impermeable unless several layers are used. Even then they are only completely effective in environments with a low corrosion rate.

5.8.4 Cathodic and anodic protection

Problems associated with galvanic corrosion and electrochemical reactions have been described in some detail. The protective nature of large anodic coatings when small cathodic regions of the base metal are exposed has also been considered. Since the majority of corrosion problems are due to galvanic attack, it is possible to use an understanding of the electrochemical processes to reduce corrosion in certain cases. Since galvanic corrosion involves reactions at both the anode and the cathode of a couple, a reduction in the reaction activity at either electrode should reduce the overall reaction and the corrosion which occurs at the anode. The techniques that are used fall into two categories, either modifying the cathode or anode metal to increase the corrosion resistance or using particular equipment to create the desired conditions.

Corrosion can be reduced if the number of cathodic sites available to form galvanic cells is reduced. If the anodic areas are far larger than the available cathodic areas, then the reaction at the cathode controls the corrosion process. This is the case with zinc coatings (anode) on steel base metal (small exposed cathode) which has already been discussed. For the reverse situation of small anodes in a large cathodic structure, rapid corrosion occurs at the anodes until these regions have disappeared, and a weakened structure remains. Galvanic corrosion can occur in a metal as a result of impurities which set up a galvanic cell. Minimizing these areas, whether anodic or cathodic, reduces corrosion of the metal. Certain metals are available in very high purity, e.g. 99.99% aluminum and 99.998% lead, but generally the extra cost of obtaining this type of purity is prohibitive.

Most pure metals possess undesirable properties, e.g. low tensile strength, and are therefore unsuitable for many applications, or the extra material thickness which would be required by their use becomes uneconomic. The pure metal must of course be resistant to attack by the environment: otherwise little is gained, e.g. aluminum is attacked by alkaline solutions whether it is pure or coated with an oxide film. Cathodic areas in alloys can generally be reduced by increasing the stability of the solid solution, using heat-treatment techniques or particular alloying additions. A reduction in cathodic activity can be achieved by increasing the overvoltage of the cathodic processes. This is usually the hydrogen overvoltage and it is achieved either by alloying zinc with cadmium or by adding arsenic, antimony or bismuth to steel. Methods of reducing localized anodic activity within a metal have been discussed and these require internal stress relieving by heat treatment, prevention of carbide precipitation and purification of grain boundaries. Increasing the tendency for passivity in a metal and in alloys by the use of alloying additions is another possible technique, e.g. additions of chromium to iron or nickel.

Corrosion can also be controlled by using a knowledge of electrochemical reactions to create conditions where structures are protected. The techniques that are available provide either cathodic or anodic protection. *Cathodic protection* is the only corrosion control method which can completely eliminate attack. This method has been discussed with reference to sacrificial anodes and anodic coatings. The principle behind the method is that the material to be protected is made the cathode in a galvanic cell. This can be achieved by connecting the metal to be protected to a metal which is more anodic, and will therefore be preferentially corroded. The sacrificial anode is generally scrap magnesium or zinc, connected to the protected cathode material by insulated copper wire. This type of protection can also be achieved by connecting the metal to be protected to the negative terminal of a d.c. power source, thus making the metal cathodic (receives electrons). The anodic electrode is then made of steel, graphite or silicon–iron, i.e. a fairly inert material. This is known as an *impressed current method*. Typical applications are protection of boat hulls, propellers and buried steel pipes. Cathodic protection methods cannot be used with metals (acting as protected cathodes) which are susceptible to hydrogen embrittlement, e.g. high yield strength steels in nuclear reactors.

Anodic protection is a more recent development and is a more complicated method of corrosion control. This method is used to protect metals from attack in very corrosive environments, e.g. stainless steel in strong oxidizing concentrated sulfuric acid. The metal is protected by the formation of a passive surface film, due to anode polarization. A high current is applied to the metal to ensure that passivity is quickly achieved, and also to limit the initial dissolution of the metal. When the metal becomes passive, the potential becomes a high positive value, and the current is reduced to the value required to maintain passivity and hence reduce corrosion. The corrosion is not eliminated as with cathodic protection but the rate is very small, also the current required for cathodic protection is larger than for the anodic passivating technique.

5.8.5 Inhibitors and environment control

By alteration of the environment which is responsible for the corrosion, it is sometimes possible to reduce the rate of attack. Obviously lowering the liquid velocity reduces erosion corrosion and cavitation effects, and stagnant liquid regions must be avoided. Lowering the temperature and solution concentration are environmental factors which are generally beneficial, although not in all cases. Traces of impurities can be detrimental to certain metals, e.g. traces of chlorides cause stress corrosion cracking in stainless steels; this is also true of organic solvents which would normally be considered inert. Dissolved gases have been known to lower the pH of distilled water to 2 or 3, and vacuum degassing is often employed. This can of course increase the danger of corrosion if the metal relies on the presence of oxygen to form a protective surface layer. It may be preferable to have a high concentration of oxygen in solution so that water-line corrosion of a partially immersed structure (due to differential aeration) is reduced.

Inhibitors are materials which reduce the corrosive effect of an environment. Their main use is in controlling aqueous electrochemical corrosion by preventing anodic or cathodic reactions. Their use would probably be more commonplace except for the unknown side effects on chemical processes. Some inhibitors, known as scavengers, react with and remove corrosive substances. Others are absorbed onto the metal surface and form a protective film. Absorption inhibitors slow down the rate-controlling reactions in electrochemical corrosion, and thereby reduce the rate of attack. Inhibitors may slow down the rate of anodic dissolution by forming a protective film with the metal ions, resulting in anode polarization. If a complete film is not produced, then severe localized attack may occur. Inhibitors such as arsenic and antimony ions cause cathode polarization by retarding the evolution of hydrogen. Often two or more inhibitors are added to a solution to obtain the best results; sufficient quantity should be added to eliminate corrosion and allow for subsequent depletion. The time required for the inhibitor to become fully effective can be reduced by removing corrosion products from the electrodes before addition of the inhibitor.

The use of inorganic inhibitors, such as arsenic and antimony salts, to increase the hydrogen overvoltage at the cathode has been mentioned. Magnesium, zinc and nickel salts form insoluble hydroxide precipitates at the cathode, and present a diffusion barrier for the reaction ions approaching the cathode. This is also true of calcium bicarbonate which deposits calcium carbonate in hard water. Inorganic anodic inhibitors such as chromates and nitrates are oxidizing agents forming protective oxide films in alkaline solutions. The presence of nonoxidizing ions such as phosphates and silicates induces dissolved oxygen, which provides the passivating effect. Inorganic inhibitors are generally added to neutral or alkaline solutions. In reducing environments and with acid solutions, or where microbiological corrosion is a problem, then polar organic compounds or colloidal organic substances are used as inhibitors. The effect may be increased resistance to current flow at the cathode due to physical adsorption, or anodic polarization due to chemisorption of the inhibitor. The main organic inhibitors are amines, mercaptans and substituted ureas, some of which have a high molecular weight. The mechanism may be anodic, cathodic or both, and they should be adsorbed to produce a film thick enough to significantly reduce corrosion. However, the correct operation of equipment should not be impaired.

5.9 PROTECTION OF PIPELINES

A problem that an engineer often faces is the transport of fluids, slurries, sewage, water, etc., through a pipeline. Such a structure may be buried or exposed to a wide range of environmental corrosion situations. The task for the engineer is either to reduce the rate of corrosion to an acceptable level, or to design the pipeline system such that corrosion is controlled. This section will present a few ideas and suggestions concerning the design aspects of a pipeline, aimed at making this task easier. It should not be considered as a 'blue print' for a 'good' design, but rather as providing the basis for a discussion session or a design project.

    (a) For a buried pipeline, a metallic coating should be selected with as many as possible of the following properties:
        (i)   high insulation resistance;
       (ii)   resistance to deformation stresses;
      (iii)   resistance to temperature deformation forces;
      (iv)   chemical inertness to electrolytes;
       (v)   insolubility;
      (vi)   allowing no microbial attack;
     (vii)   inexpensive;
    (viii)   easy to maintain.

It is not generally possible to satisfy all of these criteria; in any given situation the essential requirements should be defined and the most advantageous blend of optimum properties obtained.

    (b) Plastic or rubber-lined pipes should provide the minimum exposure of unprotected edges. The main problem occurs at joins or changes in direction; a suggestion is illustrated in Fig. 5.17.

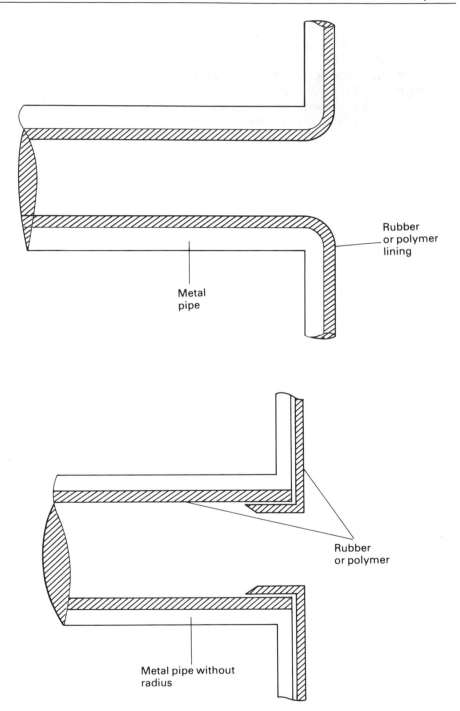

Figure 5.17 Rubber or polymer lining of metal pipes.

(c) Pipes up to 5 cm (2 in.) diameter should be lined with rubber in sections of 3 m (10 ft) or less. Pipes over 5 cm diameter can be lined in lengths of 6 m (20 ft).

(d) High-pressure and low-temperature pipes, e.g. those situated underground, should be plastic-dip coated.

(e) Pipelines encased or coated with concrete should be separated from pipelines buried in soils. This applies whether the pipes are bare, coated with insulating material or uninsulated. An appropriate insulating device can be used as shown in Fig. 5.18, or compensating cathodic protection.

(f) If coatings are not used on the cathodic surfaces of a closed system, the area of optimum protection will cover a 2.5:1 spacing. That is, if the closest anode to cathode spacing is 1 m (3 ft) and a potential of 1000 mV is obtained, then a potential of 850 mV is achieved at 2.5 m (7.5 ft) distance from the anode.

(g) If a fluid flows over a smooth metal surface at a velocity of 0.3 m/s (1 ft/s), a current density of 2.2 A/m^2 (200 mA/ft^2) is sufficient to protect steel, stainless steel or copper-based alloys.

(h) A single sacrificial anode at one end of a pipe provides protection for a distance of only 2 or 3 diameters in length. It may be necessary to use a continuous-strip anode inside the pipe as shown in Fig. 5.19.

(i) Where two pipe flanges are joined together, a sacrificial anode should not be included in the crevice between the flanges. Instead, one of the flanges should be made the sacrificial anode.

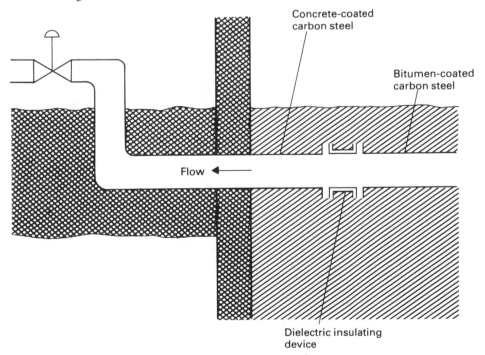

Figure 5.18 Connection of buried pipelines.

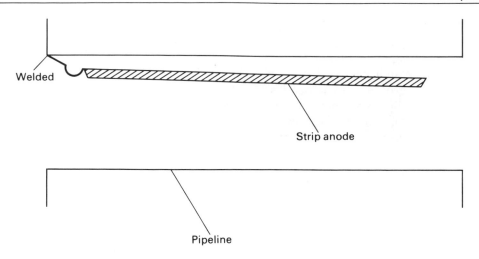

Figure 5.19   Installation of a continuous strip sacrificial anode.

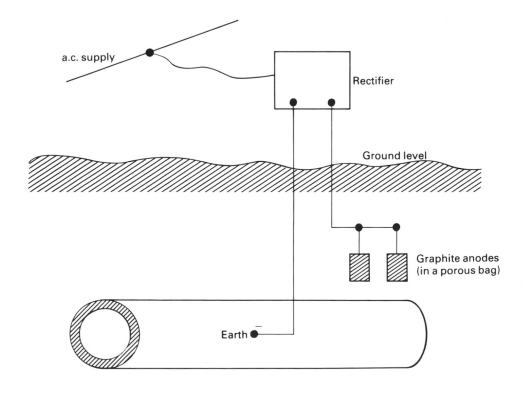

Figure 5.20   Cathodic protection of a buried pipeline by impressed current method.

(j) For boilers, oxygen should be removed from the feed water. The oxygen level should be maintained within the required limits in order to eliminate corrosion and pitting.

(k) The cathodic protection of a buried pipeline by an impressed current (using bonded connections) is shown in Fig. 5.20.

(l) It is important to prevent condensation on the external surfaces of a piping system by appropriate insulation and ventilation.

*   *   *

Self-assessment exercises

1 Consider specific practical situations, e.g. joining pipelines, pump operation, storage vessels, and describe how corrosion of a component can be:
 (a) prevented;
 (b) reduced;
 (c) measured;
 (d) monitored.

2 Explain the differences between cathodic and anodic protection, and the principle upon which each method is based.

3 Select items of equipment relevant to your engineering field. Consider particular design features that could be incorporated to reduce or eliminate corrosion.

    Examples could include: material settling in a tank, corrosion of a pump impeller, liquid trapped in gaskets, erosion at pipe bends due to solid particles in a gas or liquid, corrosion of sewerage pipes by microbiological action, and so on.

4 List the main types of coatings (not just paints) that can be used to provide corrosion protection. What conditions should be observed, and how effective are these coatings?

5 What main types of inhibitors are used and how do they work? With which metals can they be used, and what adverse effects can they create?

*   *   *

Exercises

1 Why does the de-aeration of water reduce corrosion in boilers and water pipes?

2 Explain whether aluminum can be safely coupled with copper, galvanized steel or stainless steel.

3 Recommend gasket materials for use when bolting a stainless-steel vessel containing chemicals.

4 What is the difference (if any) between passivation and polarization in corrosion processes?

5 Explain the effect of changing the anode-to-cathode area upon the rate of galvanic attack.

6 Explain the problems due to pinholes in coatings, where:
 (a) the coating is anodic compared to the base metal;
 (b) the coating is cathodic compared to the base metal.

7 Explain why lead and tin are not attacked by dilute acid solutions, although their electrode potentials are less noble than that of the hydrogen electrode.

8 Describe the action of corrosion inhibitors. Identify different types of inhibitors, the materials with which they can be used and any disadvantages that may occur.

9 Why does corrosion occur in copper tubes when water containing dissolved carbon dioxide flows through the tubes? How can this problem be overcome?

10 Explain why aluminum and stainless steel have a high corrosion resistance in certain corrosive environments.

11 Explain why aluminum is protected when connected to a zinc electrode, but corrodes when coupled to a magnesium electrode.

12 Is pitting a direct chemical attack or an electrochemical process?

13 Explain why cavitation corrosion occurs in pumps and pipelines, and how it can be avoided.

14 Explain whether corrosion is more severe with prolonged steady heating or with alternate heating and cooling.

15 Calculate the standard electrode potentials for the following reactions:
(a) $Cu^+ + e^- \rightarrow Cu^{2+} + 2e^-$
(b) $Fe^{3+} + 3e^- \rightarrow Fe$
Assess the values obtained.

16 What current would be required to deposit 150 g of copper in 12 minutes from an aqueous solution of copper sulfate?

17 What factors affect the rate of corrosion? How can they be altered to reduce or prevent corrosion from occurring?

18 Discuss the corrosion of high-silicon cast irons and austenitic nickel alloys when cast.

19 Explain which of the following metal pairs will corrode when in contact with sea water:
(a) lead–yellow brass;
(b) zinc–cast iron;
(c) copper–zinc;
(d) zinc–steel;
(e) lead–magnesium.

20 For a high-temperature oxidation-resistant alloy of Fe–Ni–Cr, explain why Cr_2O_3 is the first oxide to form and acts to protect the metal surface.

21 Discuss the effect of carbon precipitation into the grain boundaries of steel, when in the presence of chromium.

22 Explain how stress influences corrosion. Describe embrittlement and season cracking.

23 Explain how the corrosion resistance of a metal may be improved by increasing the purity.

24 Explain the difference between the series of standard electrode potentials and the galvanic series.

25 Can a direct-current source be used to prevent corrosion?

26 Discuss the corrosion of metal surfaces in the presence of natural salts, dilute alkalis, nitric acid or hydrochloric acid.

*   *   *

Complementary activities

1 Refer to handbooks, standards, publications of professional bodies, etc., to obtain information regarding the procedures for carrying out corrosion tests. Publications of the National Association of Corrosion Engineers (NACE), Houston, Texas are particularly comprehensive and useful. From similar sources, obtain values of the corrosion rates for some common materials under particular conditions.

2 In what units are corrosion rates measured? Why are different units used and how can they be converted?

3 Using the data obtained in Exercise 1 for selected materials, calculate the expected corrosion over a period of 1 week (say) in some common environments. Perform recommended tests so that measurable and sufficiently accurate corrosion data should be obtained. Compare the experimental data with values published in the literature.

4 Perform visual observations of the materials before and after the corrosion tests are performed. Study prepared samples using a metallurgical microscope.

5 Look for local examples of corrosion, and determine the possible causes. Examine some welded structures (see Section 7.21). Examine some popular cars for signs of corrosion in both visual and hidden regions.

* * *

KEYWORDS

deterioration
destruction
weight loss
  mg/(dm)^2/day
  ipy
  mils/year
electrochemical
  corrosion
electrolyte
electromotive force
Nernst equation
free energy
Faraday's constant
activity
standard electrode
  potential
equilibrium constant
oxidation reaction
reduction reaction
hydrogen electrode
anode
cathode
Redox potentials
hydrogen evolution
electrochemical cell
oxygen reduction
pH of solution
electrode polarization
overpotential
overvoltage

activation polarization
concentration
  polarization
electrode corrosion
sacrificial anode
protective coating
concentration cell
  corrosion
differential aeration
crevice corrosion
pitting corrosion
passivity
self-healing film
galvanic series
noble (cathodic)
active (anodic)
Pourbaix diagrams
uniform attack
localized attack
dry atmospheres
oxide formation
scale
rate laws
filiform corrosion
deposit attack
water-line attack
stress corrosion cracking
corrosion fatigue
brittle cracking
hydrogen embrittlement

blistering
erosion
corrosive wear
cavitation corrosion
fretting
intergranular corrosion
depassivation
sensitization
heat treatment
de-alloying
selective leaching
dezincification
graphitization
microbiological
  corrosion
micro-organisms
anaerobic bacteria
aerobic bacteria
biological growths
radiation damage
polymer degradation
thermal
  depolymerization
ceiling temperature
chemical conversion
combustion
self-extinguishing rating
corrosion tests
corrosion data
corrosion monitoring

feedback	electrochemical method	metallic coating
process conditions	electrical resistance	galvanizing
material structure	probe	nonmetallic coating
integrated rate	corrosion prevention	ceramic
measurement	corrosion control	glass lining
instantaneous reading	materials selection	porcelain enamel
visual inspection	design factors	mastic
sentry hole	surface preparation	paint
ultrasonic technique	surface hardening	polymer coating
eddy current	machining process	cathodic protection
electrical conduction	surface finish	anodic protection
magnetic particle	coating	impressed current
magnetic induction	lining	method
radiography	cladding	inhibitors
coupon	chemical conversion	scavengers
process stream	coating	absorption inhibitors
(chemical) analysis	anodizing	

*   *   *

BIBLIOGRAPHY

(Books marked * are reference texts or handbooks.)

Barer, R.D. and Peters, B.F., *Why Metals Fail*, Gordon and Breach, Science Publishers, Inc., New York (1970).

Bosich, J.F., *Corrosion Prevention for Practicing Engineers*, Barnes and Noble, Inc., New York (1970).

*Brasunas, A. de S. (Ed.), *NACE Basic Corrosion Course*, National Association of Corrosion Engineers (NACE) Houston, Texas (1971).

Butler, G. and Ison, H.C.K., *Corrosion and its Prevention in Waters*, Robert E. Kreiger Publishing Co., Inc., Melbourne, Florida (1978).

Carter, V.E., *Metallic Coatings for Corrosion Control*, Newnes-Butterworths, London (1977).

Diamant, R.M.E., *The Prevention of Corrosion*, Business Books Ltd, London (1971).

Evans, U.R., *The Corrosion and Oxidation of Metals*, Edward Arnold (Publishers) Ltd, London (1976).

Evans, U.R., *An Introduction to Metallic Corrosion*, 3rd Ed., Edward Arnold (Publishers) Ltd, London (1981).

*Fontana, M.G. and Greene, N.D., *Corrosion Engineering*, McGraw-Hill Book Co., Inc., New York (1967).

*Hamner, N.E. (Ed.), *Corrosion Data Survey*, 5th Ed., National Association of Corrosion Engineers, Houston, Texas (1974).

Hanks, R.W., *Materials Engineering Science: An Introduction*, Harcourt Brace Jovanovich, Inc., New York (1970).

Jastrzebski, Z.D., *The Nature and Properties of Engineering Materials*, 2nd Ed., John Wiley and Sons, Inc., New York (1977).

*Mellan, I., *Corrosion Resistant Materials Handbook*, 3rd Ed., Noyes Data Corp., Park Ridge, New Jersey (1976).

Parkins, R.N., *Corrosion Processes*, Applied Science Publishers Ltd, Barking, Essex (1982).

Pilling, N.B. and Bedworth, R.E., *J. Inst. Metals*, **29**, 529 (1923).

Pludek, V.R., *Design and Corrosion Control*, Macmillan Publishers Ltd, London (1977).

*Pourbaix, M.J.N., *Atlas of Electrochemical Equilibria in Aqueous Solutions*, Pergamon Press Ltd, Oxford (1966).

*Rabald, E., *Corrosion Guide*, 2nd Ed., Elsevier Publishing Co., Amsterdam (1968).

*Ranney, M.W., *Corrosion Inhibitors: Manufacture and Technology*, Noyes Data Corp., Park Ridge, New Jersey (1976).

Ross, T.K., *Metal Corrosion*, Engineering Design Guide No. 21, Oxford University Press, Oxford (1977).

Scully. J.C., *The Fundamentals of Corrosion*, 2nd Ed., Pergamon Press Ltd, Oxford (1975).

*Uhlig, H.H., *Corrosion Handbook*, John Willey and Sons, Inc., New York (1948).

*Uhlig, H.H. *Corrosion and Corrosion Control*, John Wiley and Sons, Inc., New York (1963).

Van Fraunhofer, J.A. and Boxall, J., *Protective Paint Coatings for Metals*, Portcullis Press Ltd, Redhill, Surrey (1976).

Controlling Corrosion: Volume 1, Methods; Volume 2, Advisory Services; Volume 3, Economics; Volume 4, Specifications and Standards; Volume 5, Case Studies on Corrosion; Committee on Corrosion, Department of Industry, London (1977).

Corrosion Prevention Directory, Department of Industry, London (1975).

Economics of Corrosion Control, papers from Autumn Review Course, Institution of Metallurgists, London (1974).

6
Applied Mechanics and Materials Testing

CHAPTER OUTLINE

Basic mechanics
Mechanical properties and
   testing of materials
BS and ASTM standards for
   mechanical testing
Systems involving applied
   mechanics

Stress and strain
Thermal stress
Stresses in thin-walled cylinders
Torsion in shafts
Bending of beams

CHAPTER OBJECTIVES

To obtain an understanding of:
1 basic applied mechanics commonly used by the engineer;
2 the mechanical properties that influence the selection of a material, and
   the methods used to determine these properties;
3 particular mechanical systems, and their analyses, that are encountered by
   the engineer when solving design problems.

IMPORTANT TERMS

statics and dynamics
mass
force
acceleration
work, energy and power
momentum
projectile
friction
modulus of elasticity
Young's modulus
impulse

center of mass
mechanical properties
tensile test
stress
strain
elastic limit stress
hardness test
Brinell, Vickers and
   Rockwell tests
impact test
ductility test

creep testing
fatigue testing
shear test
mechanical vibration
Poisson's ratio
thin-walled cylinder
torsion
bending moment of
   a beam
shear force in a beam
composite beam

6.1 BASIC MECHANICS

6.1.1 Definitions

The *mass* of a body is the quantity of matter in a body, or a measure of its tendency to resist a change in motion. Units: kg; lb_m.

The *weight* of a body is the force of gravity on it. Units: newton (N) or kgf i.e. kg $\times g$ or kg-m/s^2 (1 kgf = 9.81 N); lbf.

Force is that which creates motion or tends to create motion, destroys motion (or tends to), or changes the direction of motion (or tends to). Units: newton (N); lbf.

1 newton is the force required to produce an acceleration of 1 m/s^2 on a body of mass 1 kg.

Pressure is force per unit area.
Units: N/m^2; lb/in.^{2} (psi).

The *center of gravity* of a body is the point through which the weight always acts.

The *moment of a force* (or *torque*) about a chosen axis is the product of the force and its momentum.
Units: N-m; ft-lbf.

Hooke's law states that the deformation of an elastic body is directly proportional to the applied force provided that the elastic limit is not exceeded. In a more general case, stress is proportional to strain in the elastic region.

The *resultant* of a system of forces is either the single force that can replace the system and have the same effect, or it is a couple.

A *couple* is a system of two forces in the same plane of equal magnitude and acting in opposite directions, and with different lines of action. The resultant force of a couple is zero, the torque produced by a couple is independent of the position of the axis and is equal to the product of the magnitude of either force and the perpendicular distance between the lines of action.

The *equilibrant* of a system of forces is the single force that when added to a system of forces creates a state of equilibrium, and is equal and opposite to the resultant.

Conditions of equilibrium for a beam are:
    (a) total upward force = total downward force;
    (b) total anticlockwise moment = total clockwise moment.
The general conditions are:
    Σ forces = 0;
    Σ moments = 0.
Energy is a minimum.

Work is done when a force moves along its line of action, and is the product of the force times the distance moved.

Momentum is the product of mass and velocity.
Units: kg m/s; slug$^-$ ft/s.

Speed is rate of change of position (scalar quantity).
Units: m/s; ft/s.

Velocity is rate of change of position in a specified direction (vector quantity).
Units: m/s (ft/s) in a certain direction.

Angular velocity of a rotating body is rate of change of angle by a line that passes through the axis of rotation.

Units: radians/s.
Symbol: ω.
Acceleration is rate of change of velocity (vector quantity).
Units: m/s²; ft/s².
Acceleration due to gravity is usually assumed constant, with a value of 9.81 m/s²;
32.2 ft/s².
Symbol: *g*.

6.1.2 Equations of motion

Speed is a *scalar* quantity. It has *magnitude* only; the direction is unspecified.
Units: m/s.
Velocity is a *vector* quantity. It possesses *magnitude* and *direction*. For example, 8
m/s at a bearing of 145°, or in the direction N26°W, or at an angle of −19° to
the *x*-axis (the direction is often omitted). Consider Fig. 6.1, where $\overrightarrow{AB}$
represents the velocity of a particle at an instant, and $\overrightarrow{AC}$ represents the velocity
of a particle at a later instant. Then $\overrightarrow{BC}$ represents, in magnitude and direction,
the change in velocity during that interval.
    (*Note:* Vectors may be written as $\overrightarrow{AB}$, *AB*, *a*, **a**, $\overrightarrow{a}$, depending upon the
author.)
    If *AB* = *AC*, the speed is unchanged but the velocity has changed, because
the direction of motion has altered.
    For motion in a straight line, changes occur only in the speed (unless the
direction is reversed). For uniform motion in a circle the direction is constantly
changing, although the speed is constant.

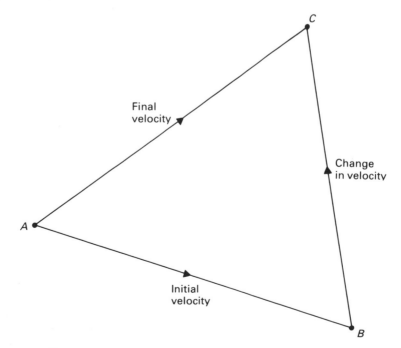

Figure 6.1 Representation of the velocity of a particle.

Acceleration is the rate of change of velocity (vector quantity).
Units: $(m/s)/s$, i.e. m/s^2; $(ft/s)/s$, i.e. ft/s^2.
Uniform acceleration occurs when changes in velocity in equal time intervals are the same in direction and magnitude. The magnitude of acceleration is the 'change in velocity' in unit time.
Notation: dt is the instantaneous increase in time (t);
δt is the small increase in t;
Δt is a larger increase in t.

Motion in a straight line

Conditions: a particle moving in a straight line with *uniform* acceleration.

(*Note:* Acceleration is positive if velocity is increasing; acceleration is negative (retardation) if velocity is decreasing.) A velocity–time graph for the motion of a particle is shown in Fig. 6.2.

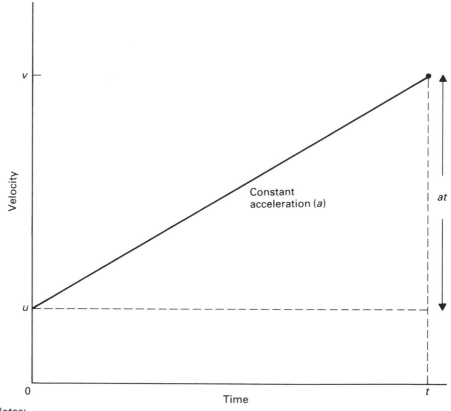

Notes:

(i)   Slope of graph = dv/dt = a

(ii) equation of the line is $v = at + u$
where a is the gradient
and u is the intercept

(iii) distance travelled (s) is equal to the area under the curve

Figure 6.2   Velocity – time graph for the motion of a particle.

Let

u = initial velocity;
v = final velocity;
t = time interval between initial and final velocities;
a = acceleration;
s = distance.

Therefore, $v = u +$ (increase in velocity over time t).
For uniform acceleration (a), the increase in velocity $= at$.
Therefore

$$v = u + at \tag{6.1}$$

Distance, $s = $ (average velocity $\times$ time)

$$s = \frac{(u + v)}{2} t \tag{6.2}$$

Substitute Eq. (6.1) into Eq. (6.2):

$$s = \frac{(u + u + at)\, t}{2}$$

Therefore

$$s = ut + \tfrac{1}{2}at^2 \tag{6.3}$$

From Eqs. (6.1) and (6.3):

$$t = (v-u)/a$$

Hence, $s = u(v-u)/a + \tfrac{1}{2}a\{(v-u)/a\}^2$

Rearranging:

$$v^2 = u^2 + 2as \tag{6.4}$$

Derivations using calculus

$$\frac{d^2s}{dt^2} = a$$

Integrating:

$$\frac{ds}{dt} = at + C_1$$

Since $\dfrac{ds}{dt} = u$ at $t = 0$, $C_1 = u.$

Therefore,

$$\frac{ds}{dt} = at + u$$

or
$$v = u + at \tag{6.1}$$

$$\frac{ds}{dt} = u + at$$

Integrating:

$$s = ut + \tfrac{1}{2}at^2 + C_2 \quad (\text{since } s = 0 \text{ at } t = 0,\ C_2 = 0)$$
$$= ut + \tfrac{1}{2}at^2 \tag{6.3}$$

As before,

$$s = \frac{(u + v)t}{2}$$

Equation (6.4) is obtained from Eqs. (6.1) and (6.3).

Summary Motion in a straight line with uniform acceleration:

$$s = \frac{(u + v)t}{2} \tag{6.2}$$

$$v = u + at \tag{6.1}$$

$$s = ut + \tfrac{1}{2}at^2 \tag{6.3}$$

$$v^2 = u^2 + 2as \tag{6.4}$$

When solving problems, select the relationship(s) which contain only the given quantities and the required unknown.

Motion under gravity
A heavy body falling under gravity experiences an acceleration, i.e. the speed and velocity increase. It can be shown by experiment that if there is no air resistance, then the acceleration is uniform. The acceleration due to gravity (g) is defined as the acceleration of a body falling freely with no air resistance acting. At a given place the value of g is the same for *all* bodies, but varies between places on the earth's surface and with height from the surface.

    For a particle projected upwards, during ascending motion the particle experiences a retardation ($-g$) due to the force exerted by the weight of the body downwards. Assuming g is constant (no variation due to altitude) and no air resistance, then the equations of motion are:

$$v = u - gt$$
$$s = ut - \tfrac{1}{2}gt^2$$
$$v^2 = u^2 - 2gs$$

At the maximum height, $v = 0$.
The time to reach the maximum height is given by (u/g).
When $t > (u/g)$, the body descends and v is negative (and numerically increasing).

The maximum upward distance traveled is given by $(u^2/2g)$.
Upon returning to the position of initial projection,

$$s = 0; \qquad 0 = ut - gt^2/2; \qquad t = 0 \text{ or } (2u/g)$$

This is twice the time taken to reach the greatest height.

If the time is calculated from these formulae for any given height between $s = 0$ and $s = u^2/2g$ (the maximum height), two values of t are obtained, one representing the upward journey and the other representing the downward one. For a height below the point of projection, s is negative and again there are two values of t (one negative, representing a time before the moment of projection). The vertical motion of a particle under gravity is shown in Fig. 6.3.

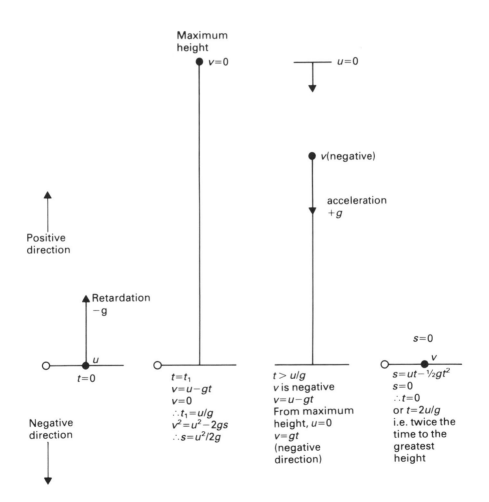

Figure 6.3   Vertical motion of a particle under gravity.

6.1.3 Dynamics of a particle

Consider Fig. 6.4(a) where a particle is at point P at time t. The displacement relative to O is represented by OP in magnitude and direction, and displacement is a vector quantity. Draw axes (usually perpendicular) through O to fix the position of P relative to O. This can be using Cartesian coordinates (x, y) or polar coordinates (r, θ).

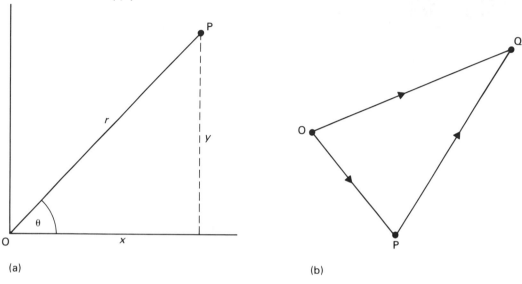

(a)   (b)

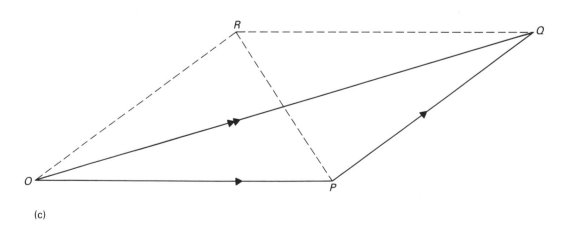

(c)

Figure 6.4 Dynamics of a particle and the laws of vectors. (a) Displacement of a particle. (b) Representation of particle displacement. (c) Parallelogram law of vector addition.

Considering Fig. 6.4(b), the displacement *OP* is denoted by the vector *OP*, and *OP* = − *PO* (equal in length, but opposite in direction). Displacement *OP* followed by *PQ* is geometrically equivalent to *OQ* , i.e. *OQ* = *OP* + *PQ* .

This is the *law of addition of displacements.*

6.1.4 Laws of vectors

The sum of any two vector quantities (of the same kind) represented by $\vec{OP}$ and $\vec{PQ}$ is a vector represented by $\vec{OQ}$. This can be shown graphically. Two vectors are equal if they have the same magnitude and direction.

Completing △OPQ to form a parallelogram (*OPQR*) as shown in Fig. 6.4(c), then $\vec{QR} = \vec{PO}$ and $\vec{RQ} = \vec{OP}$. Also $\vec{OP} + \vec{PQ} = \vec{OQ}$ and therefore *OP* + *OR* = *OQ*.

This leads to an alternative form of the vector law of addition, known as the *parallelogram law*. This states that the sum of two vectors, $\vec{OP}$ and $\vec{OR}$, is represented by the diagonal $\vec{OQ}$ of the parallelogram *OPQR*, having *OP* and *OR* as adjacent sides.
Also, $\vec{OP} + \vec{PR} = \vec{OR}$
$$\vec{PR} = \vec{OR} - \vec{OP}$$
Therefore, $\vec{RP} = \vec{OP} - \vec{OR}$
Therefore, diagonal $\vec{RP}$ represents the difference of the vectors $\vec{OP}$ and $\vec{OR}$. The two diagonals of a parallelogram represent the sum and difference of the vectors comprising adjacent sides.
(*Note:* If *m* is a scalar quantity, then $m.\vec{OP}$ is a vector of magnitude $m \times OP$, and is parallel to $\vec{OP}$.
For two vectors $\vec{AB}$ and $\vec{LM}$, $m.\vec{AB} + m.\vec{LM} = m.(\vec{AB} + \vec{LM})$.

6.1.5 Parallelogram of velocities

A particle or body may have several different velocities simultaneously, e.g. a man walking along a moving train. The several velocities are called the *components*; the equivalent single velocity is called the *resultant*.

Consider a particle which possesses two simultaneous velocities, represented in magnitude and direction by straight lines *OA* and *OB*. It has a resultant velocity represented by the diagonal *OC* of the parallelogram *OABC*, as shown in Fig. 6.5, where:

$$OC^2 = OF^2 + FC^2$$
$$= (OA + AC \cos \alpha)^2 + (AC \sin \alpha)^2$$
$$= OA^2 + 2OA \times AC \cos \alpha + AC^2$$
$$R^2 = u^2 + v^2 + 2uv \cos \alpha.$$

Alternatively, by the cosine rule:

$$R^2 = u^2 + v^2 - 2uv \cos (180° - \alpha)$$
$$R^2 = u^2 + v^2 + 2uv \cos \alpha.$$

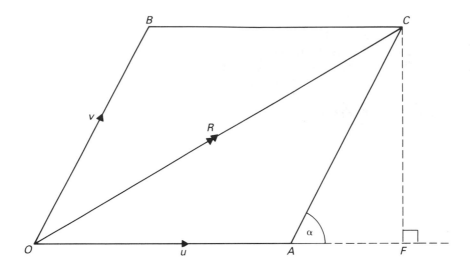

Figure 6.5   Parallelogram of velocities.

6.1.6 Resolution of a velocity into two components

The two components are usually taken at right angles. Considering Fig. 6.6 then the vertical component (*ON*) is *V* sin θ and the horizontal component (*OM*) is *V* cos θ.

Using the sine rule (see Fig. 6.7), for the resolution of *V* into two components at angles α and β to *V*:

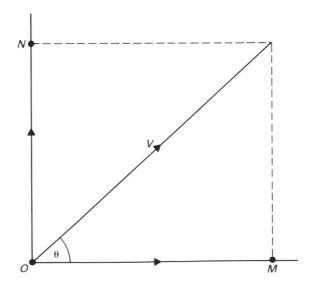

Figure 6.6   Resolution of a velocity into two components at right angles.

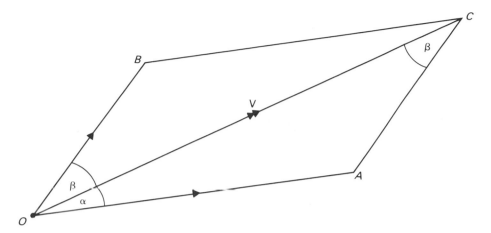

Figure 6.7   Resolution of a velocity into two components (the sine rule).

$$\frac{OA}{\sin\beta}=\frac{OC}{\sin A}=\frac{OC}{\sin(180-(\alpha+\beta))}=\frac{OC}{\sin(\alpha+\beta)}$$

Therefore, $OA = V\dfrac{\sin\beta}{\sin(\alpha+\beta)}$

Similarly, $OB = \dfrac{V\sin\alpha}{\sin(\alpha+\beta)}$

6.1.7 Projectiles

For the projection of a particle (under gravity) in any direction, assume the gravitational acceleration is constant and neglect air resistance.

Terminology

Angle of projection or *angle of elevation* is the angle between the direction in which the particle is projected and a horizontal plane through the point of projection.

Trajectory is the path described by the particle.

Range is the distance from the point of projection to the point where the trajectory meets any plane through the point of projection.

Altitude is the height (usually) above the horizontal plane passing through the point of projection.

The downward acceleration due to gravity causes the path of the particle to be curved.

The following equations are derived, although it is recommended that problems are approached from 'first principles'.

Consider the horizontal and vertical components of the velocity separately. Gravity acts vertically downwards and so has no effect on the horizontal component.

Considering Fig. 6.8:

(a) The horizontal component is constant throughout the motion and is

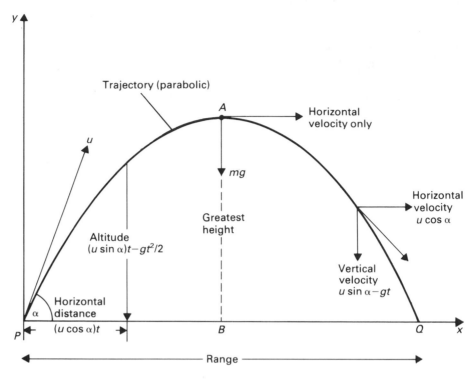

Figure 6.8 Representation of the motion of a projectile.

equal to $u \cos \alpha$.

(b) The vertical component $= u \sin \alpha - gt$ (subject to gravity downwards and time t after projection).

(c) After time t, the horizontal distance $= (u \cos \alpha)t$ and vertical distance $= u(\sin \alpha)t - gt^2/2$.

(d) At the *greatest height* reached (H), all vertical velocity is lost. Considering the vertical direction, initial velocity $= u \sin \alpha$, and $v^2 = u^2 - 2as$
$0 = (u \sin \alpha)^2 - 2gH$
Therefore, $H = u^2 (\sin^2 \alpha)/2g$.

(e) The time (t) to reach the greatest height:
$v = u - gt$
$0 = u \sin \alpha - gt$
Therefore, $t = u (\sin \alpha)/g$.

(f) The *time of flight* (T), i.e. from P to Q:
$s = ut + at^2/2$
($s = 0$ when $t = T$, i.e. horizontal distance only)
$O = u (\sin \alpha) T - gT^2/2$
$T = 2u (\sin \alpha)/g$
Therefore, T is twice the time to reach the greatest height, as expected from symmetry.

(g) The *range* (R) is the distance PQ horizontally·

$R = (u \cos \alpha) \, T$
$= (u \cos \alpha) \, 2u \, (\sin \alpha)/g$
$= u^2 \, (\sin 2\alpha)/g$

R has a maximum value when $2\alpha = 90°$, i.e. $\alpha = 45°$.

For any value of u, the range is always a maximum when $\alpha = 45°$.

(h) To determine the velocity and direction of motion after time t, the horizontal component of velocity $= u \cos \alpha$ (a constant value) and the vertical component of velocity $= (u \sin \alpha)t - gt$.

The resultant velocity is V, where:

$V^2 = u^2 \cos^2 \alpha + [(u \sin \alpha)t - gt]^2$
$V^2 = u^2 - 2ugt \sin \alpha + g^2 t^2$

If θ is the angle that the projectile path makes with the horizontal, then:

$\tan \theta = [(u \sin \alpha)t - gt]/(u \cos \alpha)$

θ is positive for values of t such that $(u \sin \alpha - gt)$ is positive, and θ is negative for values of t such that $(u \sin \alpha - gt)$ is negative.

θ is positive if $t < (u \sin \alpha)/g$, i.e. the time to reach the highest point, and

θ is negative after the highest point.

(i) A given value of u will produce 2 possible values of α for a required range:

$$R = u^2 \sin 2\alpha/g$$

For given values of R and u:

$$\sin 2\alpha = Rg/u^2$$

2α is one value, the other is $(180° - 2\alpha)$ unless $(Rg/u^2) = 1$.

α and $(90° - \alpha)$ are the two possible angles of projection; the directions are equally inclined to the horizontal and vertical axes, and $\alpha = 45°$ (for maximum range) bisects them.

(j) The shape of the path of the projectile after time t is determined by:

$x = (u \cos \alpha)t$
$y = (u \sin \alpha)t - (gt^2/2)$

Eliminating t,

$$y = u \sin \alpha \left(\frac{x}{u \cos \alpha} \right) - \frac{1}{2} g \left(\frac{x}{u \cos \alpha} \right)^2$$

This represents a parabola with a vertical axis, and when $y = 0$,

$$x = 0 \text{ or } 2u^2 \sin \alpha \cos \alpha/g$$

6.1.8 Motion in a circle

Consider a particle moving in a circle with a velocity of constant magnitude. The direction and the velocity are continually changing and the particle must be subject to an acceleration.

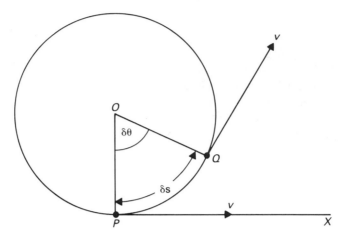

Figure 6.9 Motion in a circle.

A particle moves from P to Q in time δt, travelling along an arc δs through an angle $\delta\theta$, as shown in Fig. 6.9, such that:

$$\delta s = r\,\delta\theta$$

At Q, resolving the velocity: $v \cos \delta\theta$ parallel to PX;

$v \sin \delta\theta$ perpendicular to PX.

The change in velocity parallel to PX is $(v \cos \delta\theta - v)$.

Therefore, average acceleration parallel to PX in time δt is

$$(v \cos \delta\theta - v)/\delta t$$

$$\text{Acceleration along } PX = \lim_{\delta t \to 0} \{v\,(\cos \delta\theta - 1)/\delta t\}$$

$$= \lim_{\delta t \to 0} \left\{ -2v \sin^2\left(\frac{\delta\theta}{2}\right) \delta t \right\}$$

$$= \lim_{\delta t \to 0} \left\{ \left(\frac{-v \sin \frac{1}{2}\delta\theta}{\frac{1}{2}\delta\theta}\right)\left(\frac{\delta\theta}{\delta t}\right)\left(\sin \tfrac{1}{2}\,\delta\theta\right) \right\}$$

$$= 0$$

$$\left(\text{Note: } \left(\frac{\sin \frac{1}{2}\delta\theta}{\frac{1}{2}\delta\theta}\right) \to 1, \text{ but } \sin \tfrac{1}{2}\delta\theta \to 0. \right)$$

$$\text{Acceleration along } PO = \lim_{\delta t \to 0} \left\{ \frac{v \sin \delta\theta}{\delta t} \right\}$$

$$= \lim_{\delta t \to 0} \left\{ v\left(\frac{\sin \delta\theta}{\delta\theta}\right)\left(\frac{\delta\theta}{\delta t}\right) \right\}$$

$$= v(d\theta/dt)$$

The acceleration at any point on the circle is directed towards O, and has magnitude $v(d\theta/dt)$.

$(d\theta/dt)$ is the angular velocity (rate of change of θ) of a particle moving in the circle, written as ω.

Units: radians per second (2π radians = 1 revolution).

Since, acceleration = $v\omega$, and $v = r\omega$;

then, acceleration $= \dfrac{v^2}{r}$ or $r\omega^2$

For a particle of mass m, the force producing the acceleration is given by: (mv^2/r) or $mr\omega^2$

This force always acts towards the center of the circle.

Examples are a train moving round a curved track (inward force provided by the pressure of the outer rail against the wheel flanges), a car on a bend (force provided by friction of tires on the road), a particle attached to an inextensible string and a particle threaded on a smooth circular wire. The string is in a state of tension, but there is no tendency for the particle to move outwards along the radius of the circle. If the string breaks, the particle continues to move along a tangent to the circle, the subsequent path being that of a free projectile.

Conical pendulum

Consider the system shown in Fig. 6.10.

Let length $OP = l$, angular velocity = ω.

Resolving the forces:

Vertically: $T\cos\theta = mg$ (6.5)

Horizontally: $T\sin\theta = m\,(PN\,\omega^2)$

i.e. $T\sin\theta = ml\omega^2\sin\theta$ (6.6)

$$T = ml\omega^2$$ (6.7)

$$= ml4\pi^2 n^2$$

(where n = number of revolutions of P per second)

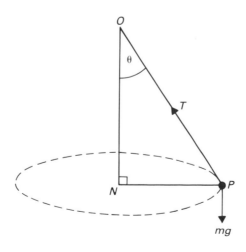

Figure 6.10 Conical pendulum.

Substitute Eq. (6.7) into Eq. (6.5):

$$ml\omega^2 \cos\theta = mg$$
$$\therefore \cos\theta = g/\omega l^2 \tag{6.8}$$
$$ON = l\cos\theta = g/\omega^2 \tag{6.9}$$

ON depends only on *g* and ω.

Using the speed of *P* (i.e. *v*) instead of ω, then:

$$v = PN\omega = l\omega \cos\theta \tag{6.10}$$

Substitute in Eq. (6.6):

$$T\sin\theta = ml\sin\theta \, (v^2/l^2 \sin^2\theta) = mv^2/(l^2 \sin\theta) \tag{6.11}$$

Divide by Eq. (6.5):

$$\tan\theta = v^2/(gl\sin\theta)$$
$$\text{and } v^2 = gl\sin\theta \tan\theta \tag{6.12}$$

6.1.9 Force, momentum and laws of motion

Force – see Section 6.1.1, Definitions.
For a body moving at uniform speed, the force causing the motion must be equal and opposite to any retarding forces, i.e. friction or air resistance. Force has direction and magnitude and is a vector quantity.

Consider a foot kicking a ball as shown in Fig. 6.11. Two bodies are involved and it is said that the bodies are acted upon by mutual forces. There is an action on one body and a reaction on the other. Considering only one body, the force acting on it is *external* or *impressed*. Considering the forces (action and reaction) acting on both bodies, these forces are referred to as *internal forces*.

The *linear momentum* of a particle is the product of the mass of the particle

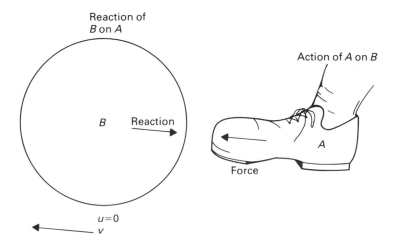

Figure 6.11 Example of bodies acted upon by mutual forces.

and its velocity (usually just called momentum). It is a vector quantity: momentum (kg-m/s; lb-ft/s) in the direction of velocity.

Newton's laws of motion

There is no rigorous proof, but Newton's laws lead to results that are in agreement with observations.

(a) *Every body continues in a state of rest or of uniform motion in a straight line, except in so far as it is compelled to change that state by external impressed forces.* This law is basically a definition of force, i.e. force accelerates a body (unless balanced by resistances). A body at rest on a horizontal table exerts a force downwards, due to its weight. There is an equal and opposite upward force (the thrust of the table) over the surface of contact, and therefore no net resultant force.

(b) *Change of momentum per unit time is proportional to the impressed force, and takes place in the direction of the straight line in which the force acts.*
Force is proportional to the rate of change of momentum.
$F = k$ (rate of change of mv)
$F = k\,m$ (rate of change of v)
$F = k \times$ mass $\times$ acceleration
Choose the unit of force such that $k = 1$:

$$F = ma$$

A force of 1 newton produces an acceleration of 1 m/s^2 when acting on a mass of 1 kg.

(c) *To every action there is an equal and opposite reaction; or the mutual actions of any two bodies are always equal, and oppositely directed.*
This law leads to the principle of linear momentum:
In any system of mutually attracting or impinging particles, the linear momentum in any fixed direction remains unaltered unless there is an external force acting in that direction.

6.1.10 Work, energy and power

When a force moves its point of application it performs work. If the force is constant, then work (W) is the product of force (F) and the distance moved (s):

$$W = Fs$$

The distance moved is the distance through which the point of application moves in the direction of the force, as shown in Fig. 6.12.

If the force is acting perpendicular to the distance moved then no work is done (since cos 90° = 0). Obviously this force will not cause the body to move in the direction shown.

Units: 1 joule is the work performed when 1 newton moves its point of application through 1 m in the direction of the force; that is:

$$\text{Work (joules)} = \text{force (newtons)} \times \text{distance (meters)}$$

Power is the rate of doing work, i.e. the work performed in a unit of time.

$$\text{Power} = \text{work done/time taken} = \text{joules/s} = \text{watts}$$

Alternatively:

$$\text{Power} = (\text{force} \times \text{distance})/\text{time} = \text{force} \times \text{velocity}$$
$$(\text{other units: ft-lb/s; 1 horsepower (h.p.)} = 746 \text{ W})$$

If a force (F newtons) keeps its point of application moving in the direction of the force with uniform velocity (v m/s), the work done per second is (Fv) joules. The power is (Fv) watts.

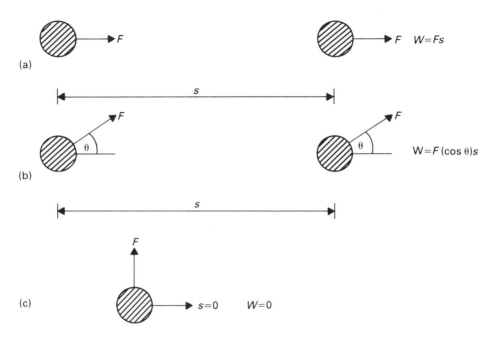

Figure 6.12 Work done when a force moves its point of application.

Resistance

Consider a train moving with uniform velocity (v) exerting a tractive force (F, pull or thrust), experiencing a resistance to motion (R) as shown in Fig. 6.13(a). If the speed is uniform, then there is no acceleration and $F = R$.

$$\text{Power} = Fv \text{ (or } Rv\text{)}$$

If the train is accelerating then $F > R$, and for retardation $R > F$. For problems involving horizontal motion, use $R = \text{mass} \times \text{acceleration}$, and the equations of motion (for uniform acceleration). For vertical motion downwards, as shown in Fig. 6.13(b):

$$\text{Resultant retarding force} = R - mg$$
$$= \text{mass} \times \text{acceleration } (a)$$

Energy

The energy of a body is its capacity for doing work, it can be thought of as a store of work. Units: joules; ft-lb.

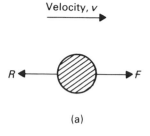

(a)

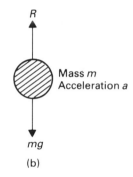

(b)

Figure 6.13   Resistance to motion. (a) Tractive force (F) and resistance to motion (R); (b) Motion under gravity.

Energy may be thermal (heat) or electrical and can be converted into mechanical work. In mechanics we are only normally concerned with mechanical work, which may be *kinetic* (due to motion) or *potential* (due to position).

The *kinetic energy* (KE) of a body is the energy it possesses due to its motion, measured by the amount of work it uses when coming to rest. Consider the system shown in Fig. 6.14, where F is the retarding force and a is the retardation.

$$F = ma$$
$$v^2 = u^2 - 2as$$
$$v^2 = 2ax$$
$$ax = v^2/2$$

Work done in coming to rest $= Fx$

$$= max$$
$$\text{KE} = mv^2/2$$

Units: KE is in joules, if m is in kg and v in m/s.

The *potential energy* (PE) of a body is the work it can do in moving from its actual position to some standard position, e.g. a mass above the ground or an extended spring.

The PE of a particle of mass m at height h above the ground is the work the

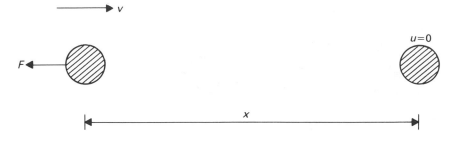

Figure 6.14   Kinetic energy of a body.

particle will do in falling to the ground, and is equal to the work done in raising it to height h.

$$PE = mgh$$

Units: PE in joules, if m is in kg and h in m.
($g = 9.81$ m/s^2, converts mass to force i.e. weight).

Consider a particle at rest ($u = 0$) at height h, falling a distance x to a final velocity v (at point P).

Initially, $KE + PE = 0 + mgh$
At P: $v^2 = 2gx$
              $KE = v^2/2 = mgx$ (from above)
              $PE = mg(h - x)$
  Total $(KE + PE) = mgx + mg(h-x) = mgh$

This is independent of x and constant throughout the motion; that is,

KE gain = PE lost.
At the ground:

$$v^2 = 2gh$$
$$KE = mv^2/2 = mgh \ (=PE \text{ at height } h)$$
$$PE = 0$$

All PE has been converted to KE.

Principle of conservation of energy
The total amount of energy in the universe is constant; energy cannot be created or destroyed, although it may be converted to different forms, e.g. heat, light, electricity.

The principle of energy is a restricted form dealing only with mechanical energy and not the conversion of energy to different forms. This states that with forces such as gravity, the work done in changing position depends only on the initial and final positions and not on the manner in which it is achieved; these are called *conservative forces*. The principle of energy states that 'if a system of bodies in motion are under the action of a conservative system of forces, the sum of the kinetic and potential energies of the bodies is constant.'

The principle cannot be used where there is any friction or a sudden jerk or impact, in which case energy is nearly always converted.

Retarding forces
Consider a mass losing velocity due to a retarding force, e.g. a bullet hitting a
plank or brakes on a train. The retarding force can be found by either:

$$Ft = m(u - v) \tag{6.13}$$

or
$$Fs = m(u^2 - v^2)/2 \tag{6.14}$$

where the retarding force (F) acts for time t, over distance s, on a mass m.

The force F which is calculated is an average value; Eq. (6.13) gives a time
average and Eq. (6.14) gives a distance average. If the force is not constant then
the values are different.

If the force is constant, then

$$\frac{m(u^2 - v^2)}{2s} = \frac{m(u - v)}{t}$$

$$\text{and } \frac{(u + v)}{2} = \frac{s}{t}$$

The average velocity (s/t) is the mean of the initial and final velocities; this is
only true if the acceleration is constant, i.e. if the force is constant.

6.1.11 Tension in an elastic string

This is an example of a variable force. The tension in an elastic string (not an
elastic band which exhibits nonlinear behavior) is proportional to the extension
beyond its natural length (if the extension is small). This statement is known as
Hooke's law and can be written:

$$T = \lambda \left(\frac{l^1 - l}{l} \right)$$

where λ is the *stiffness* of the material, often referred to as the *modulus of
elasticity*,
and ($(l^1 - l)/l$, i.e. extension/original length, is the *strain*.

λ is a constant which depends upon the material and the thickness of the
string. The tension required to extend the string to twice its normal length has a
numerical value equal to λ.

λ has the same units as tension, i.e. N; lbf.
Young's modulus (E) is the value of λ for a string of unit cross sectional area;
$E = \lambda/$(area of cross section).
Units of E: N/m^2; lbf/in^2.

Work done in extending a string
The tension is given by:

$$T = \lambda x/l$$

where $x =$ increase in length.

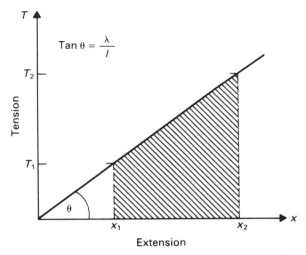

Figure 6.15 Work done in extending an elastic string.

The work done in increasing the extension from x_1 to x_2 is given by the shaded area in Fig. 6.15 and is equal to:

$$[\tfrac{1}{2}(T_1 + T_2)(x_2 - x_1)]$$

Alternatively the work done in increasing the extension from x to $(x + dx)$ is small, so that T is approximately constant:

$$\text{Work done} = T\,dx = (\lambda x/l)\,dx$$

$$\text{Total work} = \int_{x_1}^{x_2} (\lambda x/l)\,dx$$

$$= \frac{\lambda}{l}\left[\frac{x^2}{2}\right]_{x_1}^{x_2}$$

$$= \frac{\lambda}{l}\left(\frac{x_2^2 - x_1^2}{2}\right)$$

$$= \frac{\lambda}{l}\frac{(x_2 - x_1)}{2}(x_2 + x_1) = \tfrac{1}{2}(T_2 + T_1)(x_2 - x_1)$$

Thus the total work performed is the product of the mean of the initial and final tensions and the increase in extension. The PE of a string when extended (x) is $(\lambda x^2/2l)$. This is often referred to as strain energy (see Section 6.3.4).

6.1.12 Impulse

For a constant force (F), the impulse of the force is the product of the force and the time during which it acts.

$$\text{Impulse} = Ft \text{ (but } F = ma)$$
$$= mat$$
$$= m(v - u)$$

Impulse of a force = change of momentum produced.
For a variable force:

$$\text{Impulse} = \int F \, dt$$

Acceleration is variable, and $F = m(dv/dt)$

$$\text{Impulse} = \int_0^t F \, dt$$

$$= \int_0^t m \frac{dv}{dt} \, dt$$

$$= \int_u^v m \, dv$$

$$= m(v - u)$$

Impulsive forces

If a force is very large and acts for a short time, then the body only moves a small distance while the force is acting. The change in position during the action of the force can be neglected and the total effect of the force is measured by its impulse, i.e. the change of momentum it produces. The force should be infinitely large and the time it acts infinitely small. In practice this situation is approximated by the blow of a hammer, a bullet hitting a plank of wood or two billiard balls colliding.

Impact of two bodies

Newton's third law states that if two bodies (A and B) impinge, then during contact the action of A on B is equal and opposite to the action of B on A. Hence, the impulse of A on B is equal and opposite to that of B on A. The sum of the momenta of the two bodies (measured in the same direction) is unaltered by the impact. This is an example of the *principle of conservation of linear momentum*.

Consider the system shown in Fig. 6.16; since there is no net loss of momentum:

$$\text{loss of momentum of } m = \text{gain of momentum of } M$$
$$\text{momentum before} = \text{momentum after}$$

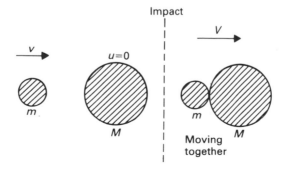

Figure 6.16 Impact of two bodies.

$$mv + Mu = (m + M)V$$
$$V = \frac{mv}{m + M}$$
$$\text{Gain in momentum of } M = MV$$
$$= \frac{MmV}{m + M}$$
$$\text{Loss in momentum of } m = mv - mV$$
$$\text{KE before impact} = mv^2/2$$
$$\text{KE after impact} = (m + M)\,V^2/2$$
$$= \{m^2v^2/(M + m)\}/2$$
$$= \{m/(m + M)\}mv^2/2$$

Since $\{m/(M + m)\} < 1$, KE after impact $<$ KE before. KE is lost in nearly all cases of impact, and the principle of energy should never be used with impulsive forces.

Examples
(1) Bullet striking a block; see Fig. 6.17. The bullet strikes the block perpendicularly and becomes embedded; all momentum is destroyed.
(2) Bullet striking a smooth plane obliquely; see Fig. 6.18. All impulse is perpendicular to the plane (since the surface is smooth), therefore all momentum perpendicular to the plane is destroyed, but there is no change in momentum (i.e. $mu \cos \alpha$) parallel to the surface.
(3) Consider Fig. 6.19 where the component of momentum perpendicular to the plane (i.e. $mu \sin \alpha$) is destroyed. The component parallel to the plane ($mu \cos \alpha$) is shared between the bullet and the block, and may cause the block to move.

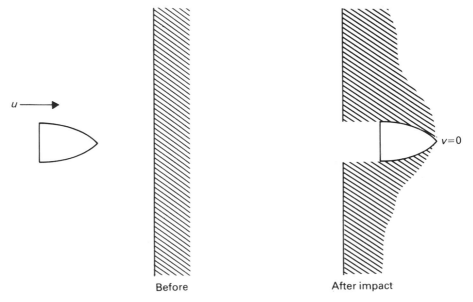

Before   After impact

Figure 6.17 Bullet striking a block.

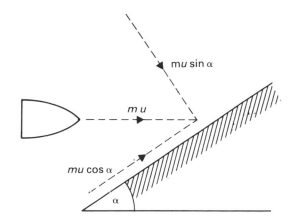

Figure 6.18   Oblique impact of a bullet on a smooth plane.

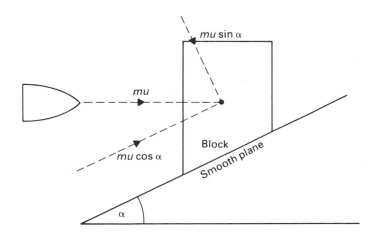

Figure 6.19   Bullet striking a block on an inclined plane.

(4) The forward momentum of the shot is equal to the backward momentum of the gun, as shown in Fig. 6.20.

(5) The horizontal momentum of the shot is equal to the horizontal momentum of the gun, as shown in Fig. 6.21. Vertical momentum of the gun is destroyed by the impulsive pressure of the plane.

(6) A correcting spring as shown in Fig. 6.22. For such problems, ignore the spring and calculate V due to firing; then consider the spring reducing the gun to rest. The spring does not have an effect until compressed. Similarly, gravity is neglected (since it is not an impulsive force) and the impact is over before the effect becomes appreciable.

(7) To find the pressure exerted by a jet of water impinging on a fixed surface, calculate the momentum destroyed per second, i.e. $m(v - u) = Ft$.

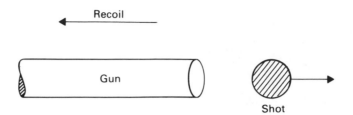

Figure 6.20 Shot from a gun.

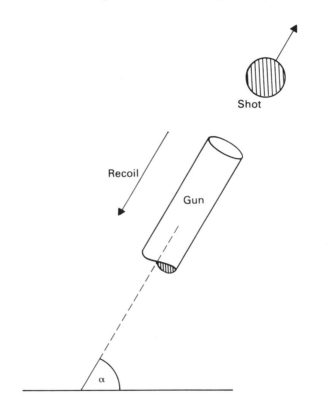

Figure 6.21 Gun fired at an angle.

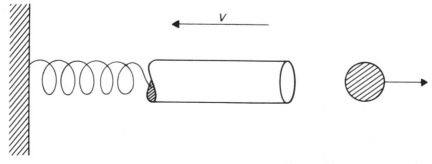

Figure 6.22 Recoil of gun opposed by a spring.

Impulse in strings – connected particles

Consider a light inextensible string as shown in Fig. 6.23 with an impulse (P) on particle B. It is not known in which direction B will move, unless P is perpendicular or parallel to AB. This is because an impulsive tension which affects A and B also occurs in the string.

If A is acted upon only by the impulsive tension in the string, it will move along AB with velocity u as shown in Fig. 6.23. Velocity u is equal to the component velocity of B in this direction, i.e. along AB. Therefore:

horizontal momenta of A and B = horizontal residual momentum of B, that is, $mu + Mu = MV \cos \alpha$.

Perpendicular to AB: $Mv = MV \sin \alpha$.

Figure 6.23   Impulse in connected particles.

6.1.13 Impact of elastic bodies

Consider two hard spheres which collide and separate. The spheres are slightly compressed, and as they tend to return to their original shape they rebound. The time of contact consists of compression and restitution phases. The shape recovery is due to the elasticity of the material. An inelastic body has no force of restitution and will not recover its shape.

If we make the usual assumption that the bodies are smooth, then the only mutual action they have on each other is along the common normal at the point of contact. If the bodies are spheres, then the mutual action is along the line joining the centers.

Direct impact – the direction of motion of both bodies is along the common normal at the point where they touch.

Oblique impact – the direction of motion of one or both bodies is not along a common normal.

Consider the system shown in Fig. 6.24. By the principle of conservation of linear momentum:

$$m_1 u_1 + m_2 u_2 = m_1 v_1 + m_2 v_2$$

If the bodies stay together after impact, then v_1 and v_2 are equal, and only

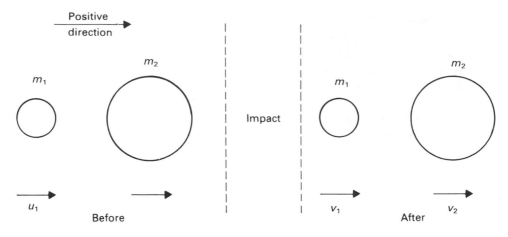

Figure 6.24 Impact of elastic bodies.

one equation is required to solve the problem. The values of v_1 and v_2 depend upon the materials of the bodies.

Newton's experimental law

When two bodies made of given substances impinge directly, the relative velocity after impact is in a constant ratio to the relative velocity before impact, and in the opposite direction. If the bodies impinge obliquely the same result holds for the component velocities along the common normal.

Direct impact

Velocities u_1 and u_2 before impact, v_1 and v_2 after. All velocities are measured in the same direction along the direction of the impact; then:

$$\frac{v_1 - v_2}{u_1 - u_2} = -e$$

or
$$v_1 - v_2 = -e\,(u_1 - u_2)$$

This is Newton's experimental law.

The constant e has a positive value that depends upon the materials of the bodies. It is called the *coefficient of restitution* or *coefficient of elasticity*.

Impact problems can be solved by using Newton's experimental law and the conservation of momentum principle.

Typical values of e: for two glass balls $e \simeq 0.9$;
for two ivory balls $e \simeq 0.8$;
for two lead balls $e \simeq 0.2$.

If $e = 0$ the bodies are inelastic, and if $e = 1$ the bodies are perfectly elastic.

The law is only approximate and values of e alter slightly for very large velocities.

Direct impact of two spheres

Consider the system shown in Fig. 6.24. By the principle of conservation of momentum:

$$m_1u_1 + m_2u_2 = m_1v_1 + m_2v_2 \qquad (6.15)$$

By Newton's experimental law:

$$v_1 - v_2 = -e(u_1 - u_2) \qquad (6.16)$$

Add Eq. (6.15) to Eq. (6.16) $\times$ m_2:

$$(m_1 + m_2)\, v_1 = (m_1 - em_2)\, u_1 + m_2\, (1 + e)\, u_2 \qquad (6.17)$$

Subtract Eq. (6.15) from Eq. (6.16) $\times$ m_1:

$$(m_1 + m_2)v_2 = m_1\, (1 + e)\, u_1 + (m_2 - em_1)\, u_2 \qquad (6.18)$$

Therefore, v_1 and v_2 can be obtained by solving Eqs. (6.17) and (6.18).

Designate the positive direction, usually left to right, then velocities in the opposite direction are negative. Usually assume v_1 and v_2 are positive, if not then the solution will be negative.

Loss of KE due to direct impact

Using Eqs. (6.15) and (6.16) above, square both Eqs. (6.15) and (6.16). Multiply the new Eq. (6.16) by m_1m_2, then add this to the new Eq. (6.15). Then:

$$\tfrac{1}{2}m_1v_1^2 + \tfrac{1}{2}m_2v_2^2 = \tfrac{1}{2}\, m_1u_1^2 + \tfrac{1}{2}\, m_2u_2^2 - \tfrac{1}{2}\left(\frac{m_1m_2}{m_1 + m_2}\right)(u_1 - u_2)^2\, (1 - e^2)$$

KE after impact = KE before − Loss of KE due to impact
Therefore:

$$\text{Loss of KE} = \tfrac{1}{2}\left(\frac{m_1m_2}{m_1 + m_2}\right)(u_1 - u_2)^2\, (1 - e^2)$$

There is always a loss of KE, unless $e = 1$.

It is often easier to calculate v_1 and v_2, and then find the loss of KE from the difference of KE before and after.

Oblique impact of a sphere on a fixed plane

Consider the impact shown in Fig. 6.25. Assume a smooth plane and a smooth sphere; then there is no impulse parallel to the plane. Therefore, the horizontal component of the sphere's velocity ($u \sin \alpha$) is unaltered.

Using Newton's experimental law, the relative velocity along the normal after impact is $-e$ times the relative velocity before impact (measured in the same direction).

If n is the normal velocity after impact, then:

$$n - 0 = -e(u \cos \alpha - 0) = -eu \cos \alpha$$

This is the normal velocity reversed and multiplied by $-e$.

The resultant velocity after impact is $u\sqrt{(\sin^2 \alpha + e^2 \cos^2 \alpha)}$ and

$$\tan \theta = (u \sin \alpha)/(eu \cos \alpha)$$
$$= (\tan \alpha)/e$$

The impulse on the plane due to impact is measured by the change in momentum along the normal:

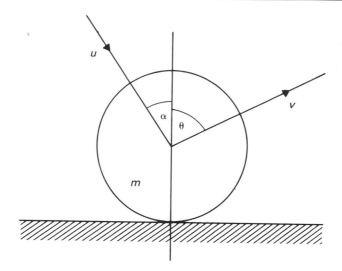

Figure 6.25   Oblique impact of a sphere on a fixed plane.

$mu \cos \alpha + meu \cos \alpha = mu(1+e) \cos \alpha$

If $e = 1$, then $v = u$ and $\theta = \alpha$.

If $e = 0$, there is no velocity along the normal after impact, and the sphere slides along the plane with velocity $u \sin \alpha$.

If impact is direct, there is no component velocity parallel to the plane; the sphere rebounds along the normal with velocity eu.

6.1.14 Friction

Frictional forces occur whenever a tangential force is applied to a body that is pressed normally against another surface. In Fig. 6.26(a), a normal force (P) presses body A against the surface of B, and a tangential force (T) is also acting on body A. Therefore, a friction force is created at the interface of A and B, and acts in a direction such as to oppose the motion of body A, as shown in Fig. 6.26(a). The frictional force and normal reaction acting on a body are shown in Fig. 6.26(b). The equilibrium of a particle on a rough inclined plane is shown in Fig. 6.26(c), and the situation of a particle on a rough horizontal plane acted on by an external force is shown in Fig. 6.26(d).

The following laws are based on experiment and are subject to certain limitations.

(a) The direction of friction is opposite to the direction of motion.
(b) Below a certain value, the magnitude of friction is equal and opposite to the force tending to produce motion.
(c) There is a maximum or limiting amount of friction in any given situation, i.e. the static friction force; this is shown in Fig. 6.27.
(d) A friction force arises from the interaction of the surface layers of two bodies in contact. This interaction is made up of a number of processes, including, in particular, the adhesion of surface atoms.

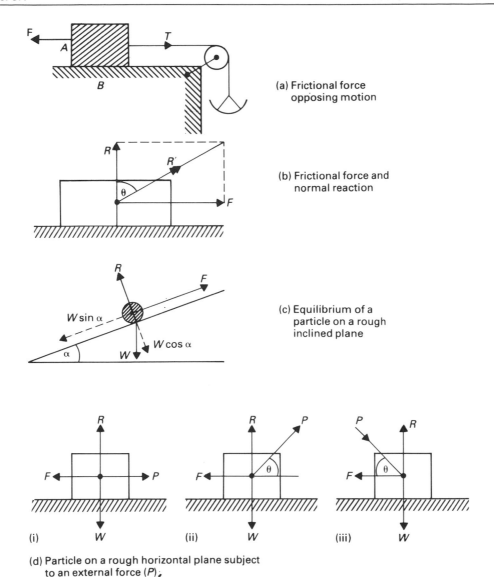

(a) Frictional force
    opposing motion

(b) Frictional force and
    normal reaction

(c) Equilibrium of a
    particle on a rough
    inclined plane

(d) Particle on a rough horizontal plane subject
    to an external force (P).

Figure 6.26 Situations involving frictional forces.

(e) For given surfaces, the ratio of limiting friction to the normal reaction
(R) between the surfaces has a constant value (μ). This value depends
upon the nature of the surfaces and is known as the *coefficient of statical
friction*.

(f) The amount of friction is independent of the areas or shapes of the
surfaces in contact, provided that the normal reaction is unaltered.

(g) If the tangential force (T) exceeds the static friction force, then sliding
occurs. When body A moves over body B, the friction force acting on A
has a direction opposite to the velocity of A relative to B, and its

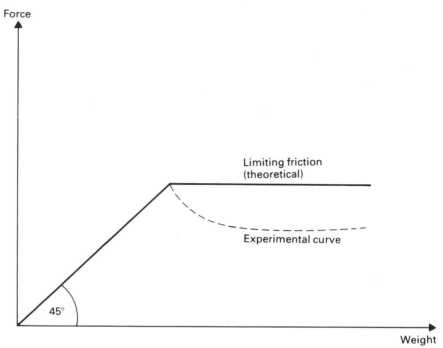

Figure 6.27   Effect of frictional forces.

magnitude is known as the dynamic (or kinetic) friction force.

(h) The dynamic friction force is proportional to the normal force between the surfaces in contact, and the constant of proportionality is known as the *coefficient of dynamical (or kinetic) friction.*

(i) The coefficients of statical and dynamical friction are both (nearly) independent of the area of contact and the shapes of the surfaces in contact, provided that the normal reaction is unaltered. The static coefficient is nearly independent of the time of contact of the surfaces at rest. The dynamic coefficient is nearly independent of the relative velocity of the two surfaces. Typical variations of the static coefficient with time and the dynamic coefficient with the velocity are shown in Fig. 6.28.

(j) When motion occurs, the friction force is slightly less than the value of limiting friction as shown in Fig. 6.27.

From (e) above: $\mu = F/R$ or $F = \mu R$

The frictional force (F) is not always equal to μR; it is so only in the limiting case when motion is about to commence: otherwise it can have any value from 0 to μR. The normal reaction (R) and the frictional force (F) could be replaced by a single force (R') known as the *resultant or total reaction*, as shown in Fig. 6.26(b). It makes an angle $\tan^{-1}(F/R)$ with the normal reaction.

The effect of lubrication and sliding velocity on the dynamic friction coefficient for steel-on-steel surfaces is shown in Fig. 6.29. If the velocity

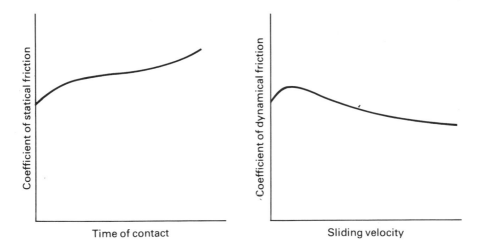

Figure 6.28   Schematic representation of the effects of contact time and sliding velocity upon the friction coefficients.

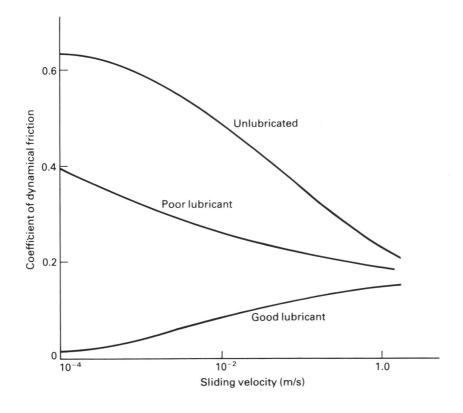

Figure 6.29   Variation of coefficient of dynamical friction with sliding velocity for steel on steel surfaces.

changes by a factor of 10, this only changes the friction by a maximum of 10%. For poorly lubricated surfaces, as the sliding velocity increases the friction decreases. This can lead to frictional oscillations (known as stick-slip), e.g. squeaking brakes or creaking doors. Typical friction coefficients for nonmetal on nonmetal (e.g. leather on wood) and nonmetal on metal are shown in Fig. 6.30; these are the probability range of values. Use of a general purpose friction chart such as Fig. 6.30 is possible because there is only a small difference between the static and dynamic friction values, and effects such as the sliding velocity or stick time are relatively small.

In Fig. 6.30, the ratio of the maximum to minimum friction values for any state of lubrication is approximately 2:1. This approximation is probably the largest source of error in most calculations involving friction. Curves similar to Fig. 6.30 can be obtained for similar and dissimilar metals in contact. Typical values of the friction coefficients are given in Table 6.1.

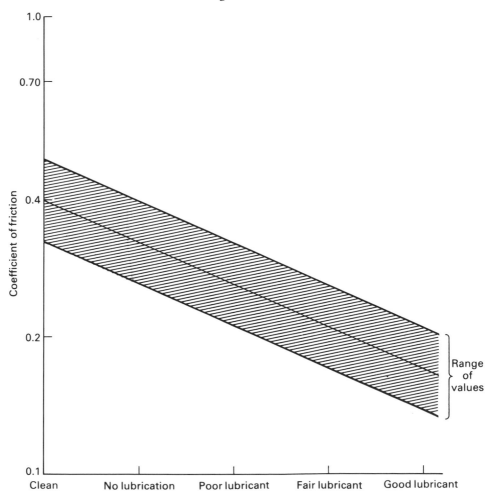

Figure 6.30 General-purpose friction chart for nonmetal on nonmetal or nonmetal on metal.

Table 6.1   Typical average values of friction coefficients for different situations.

Materials	Surface condition	Friction coefficients	
		static	dynamic
Metal on metal	Cleaned	0.7	0.6
	Unlubricated	0.3	0.2
	Good lubrication	0.08	0.08
Nonmetal on nonmetal	Unlubricated	0.7	0.6
	Good lubrication	0.15	0.13
Metal on nonmetal	Unlubricated	0.5	0.4
	Good lubrication	0.08	0.08

In many situations friction has specific functions, e.g. driving or braking a car, locking a nut and bolt. However, in other cases it is desirable to minimize the effects of friction in order to reduce the energy required to overcome it.

The name *tribology* is used to describe the study of interacting surfaces in relative motion, and of related subjects and practices. It is an interdisciplinary subject concerned with mechanics, friction, lubrication, wear, etc., and it is studied by chemists, physicists, engineers and metallurgists.

Friction is often associated with *wear*, and although they both occur at the same place, there is no simple relation between the two effects. The results presented in Table 6.2 show that a high friction coefficient does not necessarily imply high wear. Also the range of the friction values is approximately a factor of 3, whereas the values of the lowest and highest wear rates differ by many thousands. The wear and friction that occur at an interface subjected to a load can sometimes be controlled by consideration of the following factors.

(a) Choice of materials: determination of values such as those given in Table 6.2 and selecting appropriate material combinations, if this is possible. However, friction generates heat and the materials selected must also be capable of withstanding the temperature rises that occur.

(b) Use of surface films: when adhering to the solid surfaces a film

Table 6.2   Coefficient of friction and wear rate for various material combinations.

Materials	Coefficient of friction	Wear rate $(cm^3/cm \times 10^{15})$
Mild steel on mild steel	0.62	160
60/40 brass on tool steel	0.24	25
PTFE on stainless steel	0.20	2.5
Stainless steel on tool steel	0.5	0.25
Polyethylene on tool steel	0.65	0.03
Tungsten carbide on itself	0.35	0.002

(typically 1 μm thick) provides a barrier to contact. The film may be produced by chemical reactions between additives, such as chloride or sulfide compounds, and the metals in contact; solid lubricants can be used, such as graphite, or soft metals, e.g. lead.

(c) Rolling contacts: used to separate the two surfaces, e.g. cylinders, balls, etc. Friction and wear still occur but they are often greatly reduced.

(d) Pressurized lubricant film: uses a relatively thick film of lubricant (typically 100 μm thick) to achieve complete separation of the surfaces.

(e) Elastomers: these may act to restrict sliding between components, and any motion is accommodated by the deformation of the elastomer.

(f) Other methods; e.g. use of electrostatic or magnetic fields to separate the surfaces.

Wear is manifested by a loss of surface material from one or both surfaces when they are subjected to relative motion. The wear may be clearly visible or it may require the use of elaborate measurement techniques for its detection. The rate of wear of materials may vary with time due to a combination of complex mechanisms; however, the current rate of wear cannot be changed unless the load, speed, lubrication or environmental conditions are altered. Wear may be beneficial, e.g. for 'running in' a component, and it is widely used in the workshop for grinding and other abrasive processes (see Sections 8.4.10 and 8.4.11). Wear is usually measured by the volume of material removed per unit length of a component.

Wear can be simply classified as mild or severe. Mild wear is usually associated with low loading, and the wear debris is in the form of fine particles, usually as metal oxides. Severe wear occurs with higher loads and is character-ized by a much larger particle size of the debris, and rougher worn surfaces. There is a very rapid transition from mild to severe wear as the load is increased, as shown in Fig. 6.31; the wear rate can increase by several orders of magnitude. A second transition from severe wear back to mild wear may be observed (as shown), if the increased temperatures at higher loads cause metallurgical changes which increase the material hardness, for example.

The most common mechanisms of wear are adhesive, abrasive, surface fatigue and corrosive wear. More than one mechanism may occur at the same time.

The surface of any material is rough at the microscopic level, and it is composed of a series of troughs and peaks (see Section 8.4.10). When surfaces are placed in contact they touch at the peaks. As the normal load is increased these peaks deform, first elastically and finally plastically, such that the real area of contact increases. Adhesion occurs at a certain proportion of these peak contact points. It has been found that the volume of wear is proportional to the distance of sliding and to the applied load, and it is inversely proportional to the hardness of the softer material. Adhesion does not of itself create wear particles; the adhesive region could still separate by shearing. Wear particles are probably formed where the joined regions are stronger than the underlying material so that part of the surface layer is removed.

Abrasive wear occurs by the *cutting action* of a hard surface rubbing on a softer material. This can be reduced by ensuring a high quality of surface finish,

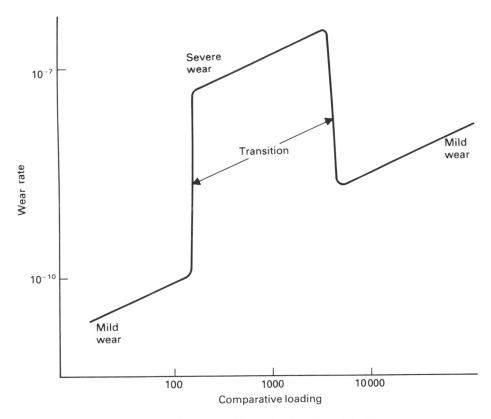

Figure 6.31 Schematic diagram illustrating the effect of loading upon type of wear.

particularly of the harder material. However, the abrasive wear caused by particles of debris is a more difficult problem, but it may be partially overcome by the use of surface grooves to collect and remove the particles.

Fatigue failure is described in Section 6.2.6; it occurs when a material is loaded and unloaded in a cyclic manner. The stress is often highest a small distance below the surface, and cracks are often produced which lead to the removal of relatively large metal particles. The surface of a specimen that has suffered fatigue failure often has a characteristic pitted appearance (see Fig. 6.58). Fatigue wear does not require surface contact between two surfaces – only that the surface material should be loaded. This is very different from the condition required for adhesive and abrasive wear. If the surfaces are separated by a lubricant film, then adhesive and abrasive wear are virtually eliminated; however, fatigue wear can still occur. The thicker the lubricant film that is applied, then the more it tends to smooth out local surface stresses.

Corrosive wear occurs when a clean metal surface reacts with the environment, these surface films are then removed by rubbing and the process is repeated. If the contaminants are harder than the pure metal, the debris can also cause abrasive wear. As already mentioned, oil additives may be used to provide protective surface layers of chlorides or sulfides.

Fretting corrosion is described in Section 5.5.7; fretting effects are also associated with small-amplitude vibrations between surfaces. This can cause adhesive wear and if the debris is trapped between the surfaces, a second stage of abrasive wear can result.

There are several factors that can effect the wear behavior of a material. The wear tends to be less the harder the material. As the normal contact pressure approaches one-third of the indentation hardness (see Section 6.2.2), the wear rate increases rapidly. It is important to determine the variation in material hardness as the temperature rises, because high local temperatures often occur in the regions of contact due to the sliding motion. In general, it has been found that wear is greatest when the mutual solubility of the materials in contact is high, e.g. identical metal surfaces. The crystal structure of a metal also affects the wear resistance. Close-packed hexagonal structures (Section 1.2.4) exhibit better wear resistance than body-centered cubic or face-centered cubic metals because they have more limited deformation characteristics, and can only deform by slip along the basal plane. However, the wear rate may increase rapidly at a particular surface temperature if the material undergoes structural changes, e.g. cobalt changes from close-packed hexagonal to face-centered cubic above 417°C (783°F).

In most engineering machinery the rate of wear is very small, e.g. microns per year. Wear tests have to be performed so that the results are available in days rather than years. For this reason the results should be treated with some caution. Some references that deal more extensively with the subjects of friction, lubrication and wear are included in the Bibliography, e.g. Halling (1975), Lansdown (1982), and Peterson and Winer (1980).

6.1.15 Introduction to statics

Statics can be regarded as a particular case of dynamics. A body will be referred to as a particle and will be represented by a point. The mass is assumed to be concentrated at a point. Statics is concerned with the relationships that exist between the forces acting on a body when the resultant force is zero. To define a force acting on a particle it is necessary to know both the magnitude and direction of the force (vector quantity).

(*Note:* For a rigid body the point of application of the force must be known.)

The magnitude of a force can be measured by its effect. In dynamics this is the change in velocity produced in a given time, i.e. the acceleration. (See Section 6.1.1 for the definition of 1 newton.) A force can also be measured by the mass it will just support. This will depend upon the force of gravity, which varies with location and altitude. Problems are usually concerned with localized situations and the relative values of the forces.

6.1.16 Parallelogram of forces

Consider Fig. 6.32, where two forces acting on a particle at point *O* are represented in magnitude and direction by the two straight lines *OA* and *OB*.

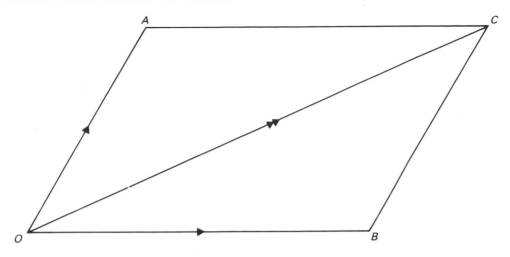

Figure 6.32 Parallelogram of forces.

They are equivalent to a single force represented in magnitude and direction by the diagonal *OC* of the parallelogram *OABC*. These are vector quantities and obey the laws of vector addition (*OC* is the resultant, *OA* and *OB* are the components).

For two forces acting in the same direction, their resultant is equal to their sum in the same direction. For two forces acting in opposite directions, their resultant is equal to their difference acting in the direction of the largest component. The resultant is zero only if the two forces are equal in magnitude and act in opposite directions. The resultant can be evaluated by scale drawing or calculation.

6.1.17 Smooth bodies

A perfectly smooth body does not exist. There is always a frictional force acting along a common surface, acting to prevent slipping of the two surfaces. For highly polished surfaces, the frictional force is very small and is often assumed to be zero. In such cases the only force between the bodies is perpendicular to the common surface. This is known as the *normal reaction*. The direction of the normal reaction is always perpendicular to the direction in which the body is capable of moving.

Examples
(1) For a rod resting against a smooth plane; the reaction is perpendicular to the plane.
(2) For a rod resting against a smooth peg; the reaction is perpendicular to the rod.
(3) For the end of a rod resting against a sphere or curved suface, the reaction is perpendicular to the curved surface and passes through the center of the sphere.

6.1.18 Tension in a string

When a string is used to suspend or move a body, the string is in a state of tension. If the mass of the string is negligible, it is called a 'light' string. For a light inextensible string vertically supporting a weight W, the tension is uniform throughout the length and is equal to W. (For a heavy string the tension varies along the length due to the weight of the string.)

If a string is passed over a small smooth pulley and supports a weight (W), then a force is required to keep the weight in position whatever the angle of the string. (This is not necessarily true for a rough pulley.)

6.1.19 Resolution of a force

A force may be resolved into two components comprising the sides of a parallelogram, in an infinite number of ways. A force is usually resolved into perpendicular components as shown in Fig. 6.33. The horizontal component (X) is equal to $F \cos \theta$ and the vertical component (Y) is equal to $F \sin \theta$. Therefore, a force F making an angle θ with the horizontal is equivalent to, or can be replaced by, forces $F \cos \theta$ and $F \sin \theta$ horizontally and vertically, respectively. Similarly the resultant of $F \cos \theta$ and $F \sin \theta$ is a force F acting in the direction which makes an angle θ with the horizontal. Resolving into angles α and β as shown in Fig. 6.34, by the sine rule:

$$\frac{R_1}{\sin \beta} = \frac{R_2}{\sin \alpha} = \frac{F}{\sin \{180° - (\alpha + \beta)\}}$$

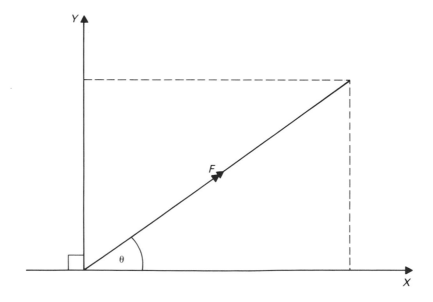

Figure 6.33 Resolution of a force into components at right angles.

Therefore:

$$R_1 = \frac{F \sin \beta}{\sin (\alpha + \beta)}$$

$$\text{and } R_2 = \frac{F \sin \alpha}{\sin (\alpha + \beta)}$$

6.1.20 Forces acting on a particle

Consider the following situations.

One force (F): the particle moves in the direction of the applied force (F) with an acceleration given by (F/m).

Two forces (P and Q): the particle moves in the direction of the resultant of the applied forces, P and Q. If P and Q are equal in magnitude and opposite in direction then, because the resultant is zero, the particle is at rest, or if it is in motion it continues with uniform velocity.

Three forces (A, B and C): if these forces are in equilibrium, then the resultant of any two forces is equal in magnitude and opposite in direction to the third force. This is shown in Fig. 6.35. This leads to the *triangle of forces*, which can be stated as follows:

If three forces acting at a point, can be represented in magnitude and direction by

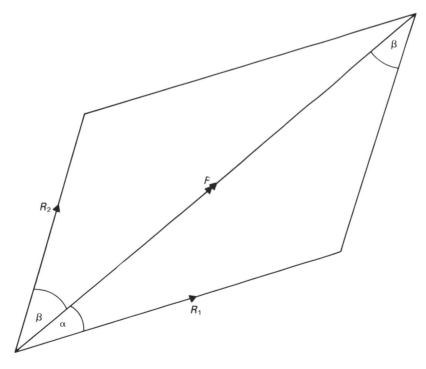

Figure 6.34   Resolution of a force into two components using the sine rule.

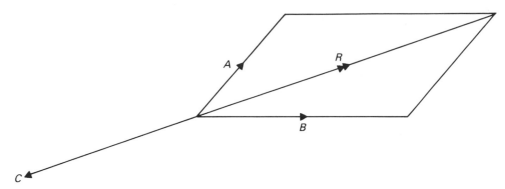

Figure 6.35 Resultant of three forces in equilibrium.

the sides of a triangle taken in order, then the forces are in equilibrium if the triangle closes as shown in Fig. 6.36.

The converse is also true: *If three forces acting at a point are in equilibrium, they can be represented in magnitude and direction by the three sides of a triangle taken in order.* This can be verified experimentally, and problems can be solved graphically or by calcualtion.

The converse is known as *Lami's theorem* and is often stated as: *If three forces acting at a point are in equilibrium, then each force is proportional to the sine of the angle between the other two.* Referring to Fig. 6.37, by the sine rule:

$$\frac{P}{\sin \alpha} = \frac{Q}{\sin \beta} = \frac{R}{\sin \gamma}$$

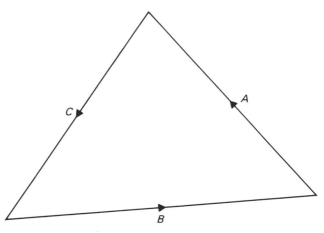

Figure 6.36 Triangle of forces.

6.1.21 Forces in a plane acting on a rigid body

If a rigid body is in equilibrium and under the action of three forces in a plane, then the lines of action of these forces must either all be parallel or all meet in a

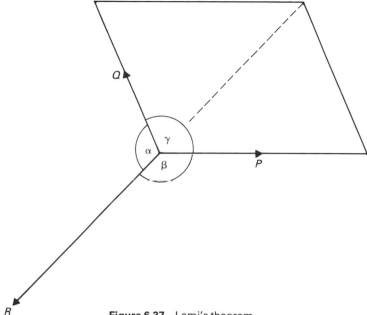

Figure 6.37 Lami's theorem.

common point. The following conditions apply:

(a) The mass of a body acts vertically downwards through the center of gravity.
(b) For a body leaning against a smooth surface, the reaction on the body is normal to the surface.
(c) For a rod resting against a smooth peg, the reaction of the peg on the rod is perpendicular to the rod.
(d) Tension in a light string is the same throughout its length; this tension is unaffected by passage over smooth pegs or pulleys. If the pulley is rough, then the tension is different on each side of the pulley.
(e) The resultant of two equal forces bisects the angle between them. For a string passing over a smooth peg, the thrust on the peg bisects the angle between the two portions of the string.
(f) When a rigid body is 'freely' suspended from a fixed point, the center of gravity of the body must lie on the vertical through that fixed point.

6.1.22 Center of gravity

Center of gravity of a number of particles
Consider particles of weights $w_1, w_2, \ldots, w_n$ in a plane at points $A_1, A_2, \ldots, A_n$, where the weights act perpendicularly to the plane (Fig. 6.38 shows the case $n = 4$).
The resultant of the weights is

$$w_1 + w_2 + \cdots + w_n = \Sigma w$$

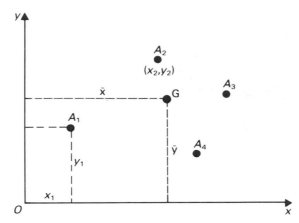

Figure 6.38 Particles in a plane.

The sum of the moments about OY is

$$w_1x_1 + w_2x_2 + \cdots + w_nx_n = \Sigma wx$$

If $\bar{x}$ is the distance of the line of action of the resultant from OY, then:

$$\bar{x}\,\Sigma w = \Sigma wx$$

$$\bar{x} = \frac{\Sigma wx}{\Sigma w}$$

Similarly:

$$\bar{y} = \frac{\Sigma wy}{\Sigma w}$$

The line of action of the resultant weight (point G) has coordinates $(\bar{x},\bar{y})$.

Since $w = mg$,

$$\bar{x} = \frac{\Sigma mx}{\Sigma m} \qquad \text{and} \qquad \bar{y} = \frac{\Sigma my}{\Sigma m}$$

The point $(\bar{x},\bar{y})$ is also called the *center of mass*.

For a rigid body there is an infinite number of particles, and it is usually considered as a set of uniform strips. If the center of gravity of the strips is known, then strips can be considered in two directions and the overall center of gravity is at the point of intersection. The center of gravity of a thin uniform rod is at the mid-point of the rod. The center of gravity of a thin uniform triangle is at the point of intersection of the medians (point G). This is one-third from each base (or two-thirds of the height from a vertex) as shown in Fig. 6.39.

For a plane uniform surface of area A, the x and y distances of the resultant weight from the axes are given by:

$$\bar{x} = \frac{\Sigma x\delta A}{A} \qquad \text{and} \qquad \bar{y} = \frac{\Sigma y\delta A}{A}$$

This determines the position of the *centroid of the surface*, sometimes called the

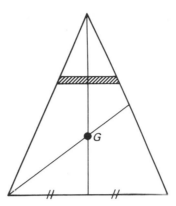

Figure 6.39   Center of gravity of a thin uniform triangle.

center of area. For a thin uniform lamina (mass proportional to area), the centroid is the same as the center of mass. (If it is not uniform then the positions are different.) The center of gravity, center of mass and centroid of a body are only the same for uniform bodies in which the directions of the weights of the particles are considered parallel.

Examples

(1) *Three rods forming a triangle*
   Weight is proportional to length; the center of gravity is at the point of intersection of the medians.
(2) *Solid cone, tetrahedron or pyramid*
   The center of gravity is at a point one-quarter of the distance between the center of gravity of the base and the vertex.

6.1.23 Problems

The numerical problems included in this chapter (Sections 6.1.23, 6.3.8 and 6.3.17) are presented in SI units only. Throughout the remainder of this book both metric and imperial units are included whenever numerical values are quoted. This is done so that the reader who works predominantly in either units can appreciate the material under discussion.

    However, for these numerical problems it was felt that two sets of units would detract from the presentation. Only a few conversion factors are required, and these are given in Table 6.3. A more extensive list is presented following the contents pages at the beginning of the book.

P.1   A car starts from rest and travels with uniform acceleration. It moves a distance of 9.5 m in the tenth second of motion. Calculate the acceleration of the car and the distance travelled in the first 5 seconds of motion.

P.2   Given that a stone takes 0.5 s to fall past a window which is 2.45 m high, calculate the height from which the stone fell.

Table 6.3 Conversion factors required for numerical problems in Chapter 6 (for a more detailed list refer to the table at the front of the book)

To convert from	To	Multiply by
cm	in.	0.3937
m	ft	3.281
m/s	ft/s	3.281
km/h	miles/h	0.6214
kg	lbm*	2.2046
metric tonne (1000 kg)	lbm	2205
kg/m^3	lbm/ft^3	0.06245
N	lbf†	0.2248
N/m^2	Pa	1.00
kN/m^2	lbf/in.^{2} (psi)	0.145
N-m	ft-lbf	0.7376
N/m	lbf/ft	0.06853
J	ft-lbf	0.7376
W	Btu/h	3.413
°C	°F	1.8, then add 32

(*Note:* Acceleration due to gravity (g) = 9.81 m/s^2 = 32.2 ft/s^2)
* lbm ~ mass of 1 lb (lb mass)
† lbf ~ weight of 1 lb (lb force)

P.3 The speed of a car on a track was measured and the following information recorded:

Time (seconds) 0 10 20 30 40 50 60
Speed (km/h) 0 34 54 66 74 78 80

Find the distance traveled and the initial acceleration. This problem requires a graphical solution, details of which are *not* given in the text.

P.4 A bullet is fired with a velocity whose horizontal and vertical components are u and v respectively. Calculate the position of the bullet at time t seconds, neglecting any air resistance.

The horizontal velocity of the bullet is 600 m/s; calculate the elevation of the gun in order to hit a mark which is 2 m above the muzzle at a distance of 500 m.

P.5 Two strings are 40 cm and 30 cm long. They are tied to a mass of 6 kg and have their other ends fastened to two nails which are 50 cm apart in a horizontal line. Calculate the tension in the longer string. The nails are replaced by smooth pulleys. The strings are passed over the pulleys and hang freely with masses of 4.5 kg and 6 kg tied to their free ends. In the new equilibrium positions, find the angles which the strings make with the vertical.

P.6 A particle is acted on by forces of 1, 2, 3 and 4 N, the angles between these forces being 60°, 30° and 60° respectively. Determine the magnitude and direction of the resultant force.

P.7 A body of mass 20 kg is placed on a rough inclined plane whose slope is $\sin^{-1} \frac{3}{5}$ and the coefficient of friction between the plane and the body is 0.2. Calculate the least force acting parallel to the plane, required to:
(a) prevent the body sliding down the plane;

(b) pull the body up the plane.

P.8   Like parallel forces of 2, 5 and 3 N act at the corners (*A*, *B* and *C*, respectively) of a triangle. The dimensions are *AB* = 4 cm, *BC* = 3 cm and *CA* = 5 cm. Find the position of the line of action of the resultant.

P.9   Masses of 3, 4 and 5 kg are placed at the corners (*A*, *B* and *C*, respectively) of an isosceles triangle, in which *AB* = *AC* = 12 cm and *BC* = 8 cm. Determine the distance of the center of gravity of the masses from *BC*, and from *AD* which is the perpendicular from *A* to *BC*.

P.10   Masses of 5, 6, 9 and 7 kg are placed at the corners (*A*, *B* *C* and *D*, respectively) of a square of side 27 cm. Calculate the distance of their center of gravity from the first mass.

P.11   A sheet of material is in the shape of a rectangle, 9 cm wide and 12 cm long. One of the shorter sides is folded over so as to lie entirely along one of the longer sides. Determine the position of the center of gravity of the shape thus formed.

P.12   A uniform rectangular board, *ABCD*, has *AB* = 10 cm and *AD* = 8 cm. Two square holes, each of side 2 cm, are cut in the board and these are filled to the original thickness with a metal of specific gravity equal to nine times that of the board. The coordinates of the centers of the holes, referred to *AB* and *AD* as the axes *x* and *y* and measured in centimeters, are (4,3) and (7,4) respectively. Calculate the coordinates of the center of gravity of the loaded board.

P.13   There are four similar planks (each 3.6 m long) stacked in a pile. The second plank projects 0.6 m beyond the first, the third 0.9 m beyond the second, and the fourth 1. 8 m beyond the third. The sides are flush with each other. Find the center of gravity of the four planks.

P.14   An engine of mass 110 tonnes is coupled to, and pulls, a carriage of mass 30 tonnes. The resistance to motion of the engine is equivalent to one hundredth of its weight. The resistance to motion of the carriage is equivalent to 1/150 of its weight. Find the tension in the coupling, if the entire tractive force exerted by the engine is equal to the weight of 3 tonnes.

P.15   A train of mass 250 tonnes is traveling up a slope of $\sin^{-1} \frac{1}{140}$ at a constant speed of 48 km/h. Frictional resistances are equivalent to $\frac{1}{160}$ of the weight of the train. Calculate the power that is being exerted.

    Calculate the maximum speed (in km/h) that a power of 450 kW could maintain on a level track, if the frictional resistances then become equivalent to $\frac{1}{150}$ of the weight of the train.

P.16   Calculate the power of an engine that is used to fill a reservoir, 500 m long and 300 m wide to a depth of 3.5 m. The water is to be pumped from a river 1.5 km away and 150 m lower in level, in a continuous operation of 15 days.

P.17   Water issues from a circular pipe of 8 cm diameter with a velocity of 5 m/s. Calculate the mass of water discharged per minute if the density of water is 1000 kg/m^3.

    Given that the water impinges directly upon a plane and its momentum is destroyed, calculate the force exerted by the jet on the plane.

P.18   A particle is tied by an elastic string of length 30 cm to a fixed point on a smooth horizontal plane. The particle is describing a horizontal circle around the fixed point at a constant speed of 20 rev/s. Given that the modulus of elasticity of the string is equal to the weight of the particle, show that the extension of the string is nearly 5 cm.

P.19   A force equal to the weight of 5 kg acts on a mass of 30 kg, originally at rest, for 10 seconds. Find the distance traveled by the mass, and the kinetic energy generated.

P.20   A hammer of mass 1 kg moving with a velocity of 6 m/s, drives a nail (mass 30 g) a distance of 2.5 cm into a fixed piece of wood. Calculate:
(a) the common velocity of the nail and hammer just after impact;
(b) the percentage loss of energy;
(c) the time of motion of the nail;
(d) the force of resistance of the wood, assuming it to be constant.

P.21   The production of a steel stamping requires a mass of 100 kg to fall on to the steel bank. The mass falls freely for 1 m, and is brought to rest after moving a further distance of 1.2 cm. Assuming that a uniform resistance is exerted by the steel, calculate the magnitude of this resistance.

P.22   A ball of mass 10 kg and moving at 5 m/s impinges directly on another ball of mass 4 kg, moving at 2 m/s in the opposite direction. The coefficient of restitution is equal to 0.5. Calculate the velocities after impact.

P.23   A sphere of mass 1 kg, moving at 10 m/s, overtakes another sphere of mass 5 kg moving in the same line at 3 m/s. Find the loss of kinetic energy during impact, and show that the direction of motion of the first sphere is reversed. The coefficient of elasticity is 0.75.

P.24   A ball moving within a velocity of 20 m/s, impinges on a smooth fixed plane in a direction making an angle of 30° with the plane. The coefficient of restitution is 0.6. Calculate the velocity of the ball after the impact.

*   *   *

KEYWORDS FOR SECTION 6.1

mechanics	tension	component
statics	resultant force	projectile
dynamics	equilibrant force	angle of projection
mass	equilibrium	angle of elevation
weight	work	trajectory
gravity	momentum	altitude
newton	speed	greatest height
force	scalar quantity	time of flight
acceleration	velocity	range
pressure	vector quantity	angular velocity
center of gravity	retardation	radian
moment of a force	law of addition of	conical pendulum
joule	displacements	retarding force
Hooke's law	parallelogram law of	friction
extension	velocities	limiting value

static friction force
coefficient of statical
   friction
dynamic (or kinetic)
   friction force
coefficient of dynamical
   friction
sliding velocity
lubrication
stick-slip
general-purpose friction
   chart
tribology
wear
surface film
rolling contact
pressurized lubricant
   film
mild wear
severe wear
wear debris
types of wear:
      adhesive
      abrasive

surface fatigue
corrosive
fretting
air resistance
mutual forces
action and reaction
impressed force
internal force
linear momentum
Newton's laws of motion
energy
power
resistance
kinetic energy
potential energy
principle of conservation
   of energy
principle of energy
conservative forces
elastic string
modulus of elasticity
Young's modulus
impulse
principle of conservation

of linear momentum
elastic bodies
direct impact
oblique impact
Newton's experimental
   law
coefficient of restitution
coefficient of elasticity
laws of friction
limiting friction
total reaction
angle of friction
rough plane
rigid body
parallelogram of forces
smooth body
normal reaction
resolution of a force
sine rule
triangle of forces
Lami's theorem
center of mass
centroid of surface

*   *   *

6.2 MECHANICAL PROPERTIES OF MATERIALS

In order to select appropriate materials and hence produce an efficient design, it is essential to measure the relevant properties of materials. Properties such as density, conductivity, resistivity are known as physical properties. The mechanical properties of a material are measured in terms of the behavior of the material when subjected to a force, and they are determined by deformation. The absolute numerical values of some mechanical properties are not easily determined, but are presented comparatively with other materials. Mechanical properties may be determined to provide either design data for the engineer or as a check on the standard of raw materials and heat treatment methods being used.

The mechanical properties of most importance to the engineer, and an explanation of each term, are listed as follows:

Strength   the ability of a material to provide an equal reaction to an applied force (tensile, compressive or shear) without rupture.

Tenacity   the ability of a material to resist rupture due to a tensile force.

Hardness   usually defined as the resistance to abrasion, deformation or indentation.

Brittleness   the tendency of a material to fracture when subjected to shock loading or a blow. There is no deformation before

Toughness

Elasticity

Plasticity

Ductility

Malleability

fracture to act as a warning of failure, e.g. a thin sheet of glass shatters immediately from a sharp blow. However, the brittleness of glass and cast iron can be demonstrated by slow load application.

the opposite of brittleness, the ability of a material to resist fracture under shock loading, usually after the elastic limit has been exceeded.

the ability of a material to return to its original shape after deformation, e.g. the extension or compression of a spring.

the opposite of elasticity, the ability of a material to retain a shape imposed by a force after that force is removed, e.g. stamping a coin.

the ability to withstand cold plastic deformation without fracture. A ductile material can be worked into a shape without loss of strength, e.g. drawing into wire form. If subjected to a shock load the material would yield and become deformed.

the ability of a material to withstand deformation in all directions without cracking. This should not be confused with ductility. The properties are similar when considering a suitable material for forming into sheet or strip, but not for wire drawing. Lead is malleable and can be beaten or rolled into sheet form, but it is not suitable as wire. Malleable materials are used for forging, stamping, pressing, etc.

Various standard tests are available for the determination of the mechanical properties of materials. The most commonly used are tensile tests and hardness tests. Depending upon the material requirements and in-service conditions to be encountered, other tests such as impact tests, creep tests, fatigue tests, malleability tests, may also be performed. A summary of the main British Standards and ASTM Standards associated with materials testing is given at the end of this section. This section is intended to provide an introduction to materials testing; practical work using a variety of machines should also be undertaken (see Complementary Activities). The figures are presented to illustrate the principles of operation of the machines.

6.2.1 The tensile test

The tensile strength of a material is the stress obtained at the highest applied force for a test piece subjected to a tensile force. A test piece of known cross sectional area and 'guage' length is held in a testing machine, and is subjected to a tensile force that can be increased by known increments. The guage length of the specimen is measured at each stage. When the test piece begins to stretch rapidly then failure is imminent.

The test piece must be accurately machined along its gauge length and have a standardized cross section – usually circular so that results obtained using different test pieces are directly comparable. The BS specification is given by the formula:

$$L_0 = 5.65 \; \sqrt{A_0}$$

where L_0 = gauge length,
and A_0 = original cross-sectional area
(both in SI units).

Typical tensile testing equipment and specimens are shown in Fig. 6.40.

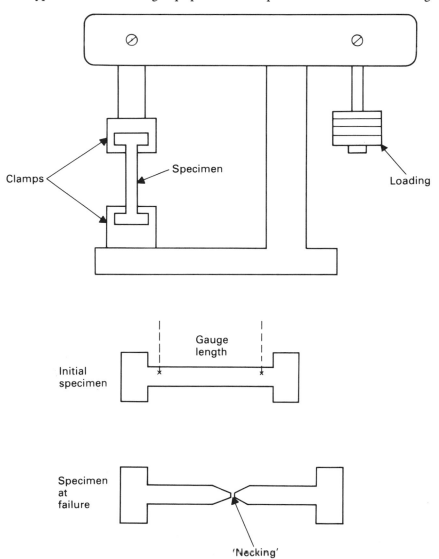

Figure 6.40   Schematic diagram of tensile testing equipment.

Tensile testing machines vary from the portable Hounsfield tensometer with a maximum force of 20 kN (2 tonf) to very large, powerful machines using forces of 1 MN (100 tonf).

The results from a tensile test can be presented as a force–extension diagram, although often only particular values are required such as maximum force, dimensions at fracture, etc. A typical tensile test curve for a material such as mild steel is shown in Fig. 6.41, which also shows the changes in shape of the test piece during testing. When tensile testing begins, the extension produced is very small and is proportional to the force (region *OP*). Within this region, if the force is removed, then the test piece will assume its original shape. The material is elastic and obeys *Hooke's law*. Point *P* on the diagram is known as the *limit of proportionality*. If any further extension is produced then the material is still elastic, but the force is no longer proportional to the extension. Point *E* is known as the *elastic limit* and is very close to point *P*. Beyond point *E* the extension is partly elastic and partly plastic, but the material will not return to its original dimensions if the force is removed. If the force is increased beyond point *E*, then

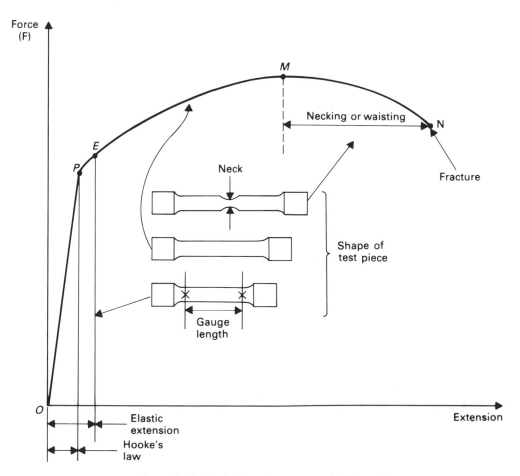

Figure 6.41   Typical tensile test curve (mild steel).

the material stretches rapidly. This extension is initially uniform along its length until the maximum force is applied at point M; the rise in force from E to M is due to work hardening. Beyond point M the material work softens causing *necking* or *waisting*: the cross-sectional area at the neck decreases rapidly until the material fractures at N. When necking occurs, the force required to fracture the material is less than the maximum force at point M.

The curve shown in Fig. 6.41 applies to mild steel, but other materials produce different-shaped force–extension diagrams as shown in Fig. 6.42. The curve for a soft carbon steel in the normalized or annealed condition possesses the particular features known as the *upper and lower yield points*; these are shown as points E and Y respectively on Fig. 6.42. Beyond the upper yield point the material suddenly extends for only a slight increase (or no increase) in the applied force; if the force is removed then a small permanent extension remains in the material. Although the material will withstand larger forces, the yield point should not be exceeded, otherwise a permanent extension will be created. Ductile materials do not generally fail on extensions beyond the yield strain unless they are strain-rate sensitive at those conditions.

The results from a tensile test are sometimes plotted as stress against strain, rather than force against extension, where:

$$\text{stress} = \text{force}/\text{cross-sectional area}$$

and

$$\text{strain} = \text{extension}/\text{original length}$$

To be strictly correct, this would mean that the cross-sectional area would have to be measured for every incremental increase in the force. In this case the equivalent stress–strain curves obtained would be the same shape as the force–extension curves in Figs. 6.41 and 6.42, except that the actual stress when necking occurs would increase rapidly as the area rapidly decreases.

Calculating the area at each point would be difficult (and measurements equally difficult); the convention usually adopted is to calculate the stress based on the original cross-sectional area. This is then referred to as the *nominal stress* or the *engineering stress*. The justification for this approximation is that materials are normally used under conditions less than the maximum force; the reduction in area is very small so that the difference between the nominal stress and actual stress is small. A plot of nominal stress against strain produces the same curve as the force–extension diagram with the solid line up to point N in Fig. 6.41. An ideal stress–strain curve for a material behaving in a ductile manner is shown in Fig. 6.43, and the actual (or true) stress region MN_1 is also shown. The tensile strength of a material, corresponding to point M on Figs. 6.41 and 6.43, is thus defined as:

$$\text{tensile strength} = (\text{maximum force}/\text{original cross-sectional area})$$

It provides an indication of the ductility of the material, as do measurements of the percentage elongation or the percentage reduction in area of the test piece at fracture. The percentage elongation can be measured more accurately and easily than the percentage reduction in area, especially as the cross section may be distorted by fracture.

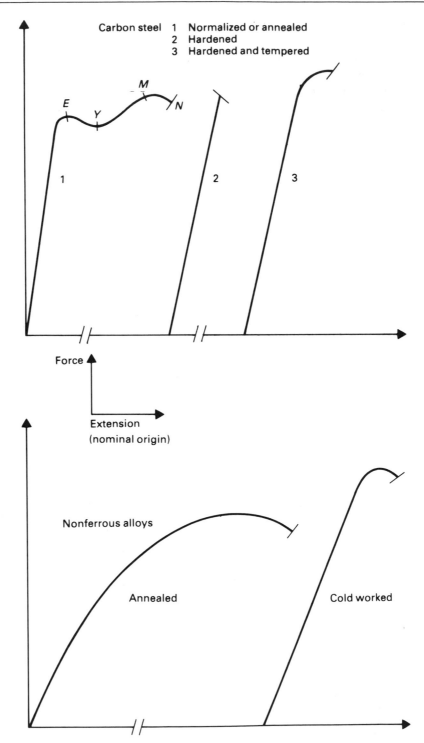

Figure 6.42 Tensile test curves for different materials.

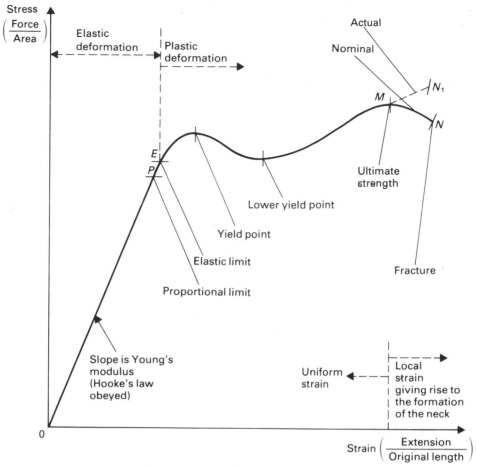

Figure 6.43   Ideal stress–strain curve.

The preceding discussion applies for small elastic deformations. However, since the actual cross-sectional area of a specimen may be only a fraction of the original area after severe plastic deformation, the *true stress* may be significantly larger than the nominal or engineering stress. The analysis of the relation between stress and plastic strain is complex and the alternative of measuring the area of the specimen at each stage of successive loadings is also not an appealing prospect. Fortunately, a useful arbitrary definition of the beginning of plastic deformation is the *0.1% offset yield strength* or *proof stress*. The method of determining this stress value is given later in this section and is shown in Fig. 6.44(b).

The following discussion applies to plastic deformation of an isotropic material. Although a metal crystal is anisotropic with respect to elastic and plastic deformation, an actual rod of metal composed of millions of small randomly oriented crystals behaves in an isotropic manner. Therefore, the discussion will be applicable to ordinary, polycrystalline metals. Just as the true stress is larger than the nominal stress, so the true strain will be smaller than the nominal value due to large changes in the length of a specimen caused by plastic deformation. The definition of strain is that it is a small increase in length (dl) divided by the length (l) that the specimen has *at that time*. That is:

$$d\varepsilon = \frac{dl}{l}$$

If the increase in length is small, say 0.1% of the original length (l_0), then the instantaneous length can be considered approximately constant (at l_0). The total strain(ε) is:

$$\varepsilon = \int d\varepsilon = \int_{l_0}^{l_0+\Delta l} \frac{dl}{l_0} = \frac{\Delta l}{l_0} = 0.001$$

If a material is only subjected to elastic deformation, this calculation is sufficiently accurate. For *uniform* plastic deformation, i.e. large Δl, the *true strain* (δ) is given by:

$$\delta = \int d\varepsilon = \int_{l_0}^{l} \frac{dl}{l} = [\ln l]_{l_0}^{l} = \ln\left(\frac{l}{l_0}\right)$$

For a large deformation, e.g. $\Delta l = l_0$, the true strain is:

$$\delta = \ln\left(\frac{l}{l_0}\right) = \ln\left(\frac{2l_0}{l_0}\right) = \ln 2 = 0.693$$

The calculated value of the nominal strain is:

$$\varepsilon = \frac{\Delta l}{l_0} = \frac{l_0}{l_0} = 1.0$$

The equation for true strain is not valid if necking occurs, i.e. nonuniform deformation, because the plastic strain then has different values at various positions in the metal. It is usually necessary to know only the maximum strain, and the strain at the necked section can be determined. It can be shown that the strain at the neck is:

$$\delta = 2\ln\left(\frac{d_0}{d}\right)$$

where d_0 and d are the initial and neck diameters, respectively.

Particular values that are calculated from tensile test measurements, or the corresponding curves (reference to Figs. 6.41 and 6.43) are:

(a) *Young's modulus of elasticity* (E) of the material, defined by:

$$E = (\text{increase in nominal stress/increase in strain})$$

The value of E is the slope of the straight line portion (O to P) of Figs. 6.41 and 6.43.

(b) *elastic limit stress* defined as F_E/A_0

(c) *proportional limit stress* defined as F_P/A_0

(d) *yield stress* defined as F_Y/A_0

(e) *ultimate stress* or *ultimate tensile strength* defined as F_M/A_0

(f) *actual stress at fracture* defined as $F_N/$(cross-sectional area at fracture)

(g) *percentage elongation* = (extension at fracture/original length) × 100%

(h) *percentage reduction in area* = (reduction in area at fracture/original area) × 100%

Most materials, especially metals and alloys, obey Hooke's law when initially subjected to a tensile load. The true stress is proportional to the true strain and the deformation of the specimen is elastic, i.e. when the load is removed the specimen returns to its original dimensions. Beyond a certain loading the deformation is no longer elastic, and the specimen exhibits plastic behavior. Consider what happens when the metal is *initially* deformed in a *plastic* manner (see Fig. 6.44(a)). This usually represents only a small portion of the entire true stress–true strain curve for the material, and deformation can continue for much larger strain values than shown on the diagram. When the material is first subjected to plastic deformation, the stress–strain relationship is no longer linear. If the load is removed, the material recovers quickly as the elastic strain is reduced. The elastic strain can be calculated by dividing true stress by Young's modulus. A residual strain remains in the material and part of this strain, the *anelastic strain*, disappears gradually. The final strain that is left in the material is the *permanent plastic strain*. These regions are shown on Fig. 6.44(a), and for initial plastic deformation the sum of the anelastic and elastic strains may be greater than the permanent plastic strain. However, for large plastic deformations (an extension of Fig. 6.44a) the plastic strain will be much larger than the associated anelastic and elastic strains.

Some materials do not obey Hooke's law, e.g. the annealed curve for nonferrous alloys in Fig. 6.42, or show only a very short region obeying Hooke's law. In such cases (i.e. nonlinear elastic behavior) the *secant* and *tangent moduli* are usually calculated instead of the modulus of elasticity (Young's modulus). The secant modulus is frequently used to describe plastic behavior, and is represented by the gradient of the straight line drawn from the origin of the stress–strain curve to a particular point on the curve, usually a given strain. The tangent modulus measures the ratio of the rates of change of stress and strain; it is represented by a tangent to the curve at a particular point, usually a given strain. For linear elastic deformation, all moduli values are identical.

The nature of fracture is dependent upon the properties of the material. Ductile materials tend to break at the waist and usually show a 'cup and cone' type of fracture, i.e. a circular crater on one side and corresponding truncated cone on the other. Fibrous materials such as wrought iron show little (if any) waist, and fracture is generally ragged and torn. Brittle materials such as cast iron rupture suddenly across a surface that is usually almost perpendicular to the direction of loading.

In service, a material should not be stressed beyond its elastic limit, otherwise it will undergo a *permanent set* or deformation. A *safety factor* or *margin of safety* is usually employed, as defined by:

$$\text{safety factor} = (\text{stress at failure/actual stress})$$

This can be based on either the yield stress or the maximum stress.

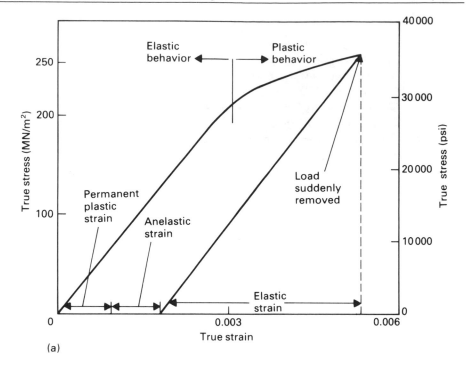

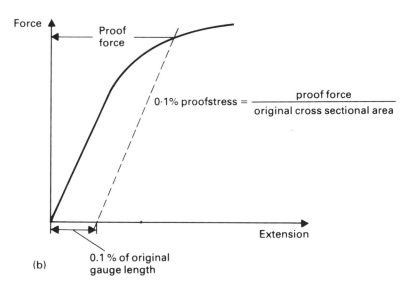

Figure 6.44 Typical tensile test curves. (a) Initial portion of a typical stress–strain curve showing the effect of suddenly removing the load; (b) determination of 0.1% proof stress.

The *proof stress* is the nominal tensile stress that is just sufficient to produce a nonproportional extension, equal to a specified percentage of the original gauge length. This value is used for materials that do not show a well-defined yield point; the proof stress is then the maximum stress which the material should experience. Usually, a value of 0.1% (or 0.2%) is used, and the proof stress is referred to as the 0.1% proof stress. This produces a nonlinear extension in the material of 0.1% of the original gauge length. The value of the 0.1% proof stress can be determined from the force–extension diagram for the material, as shown in Fig. 6.44(b). A line is drawn parallel to the proportional part of the curve and cuts the extension axis at a value equal to 0.1% of the gauge length. The intersection of this parallel line with the curve then gives the value of the proof force, and hence the 0.1% proof stress which is based on the original area.

Shear stresses act in a direction that would cause or tend to cause one section of material to slide over another. The magnitude of the shear stress depends on the value of the applied force divided by the area of the section parallel to the direction of the force. This is illustrated in Fig. 6.45. The behavior of a material subjected to shearing forces can be measured in a similar way to tensile tests; a typical arrangement of the test piece is also shown in Fig. 6.45. Shear tests are discussed in Section 6.2.7.

Compression tests are also performed on materials but it is important that the specimen is not too long compared to the thickness, otherwise buckling may occur. For a cylindrical specimen, the ratio of length to diameter is usually 2:1. The Young's modulus of a material is often the same for either tensile or compressive forces. Typical curves for different materials are shown in Fig. 6.46. The nature of failure due to compressive forces depends upon the structure of the material. Brittle materials fail diagonally, whereas ductile materials ultimately 'barrel', as illustrated in Fig. 6.46.

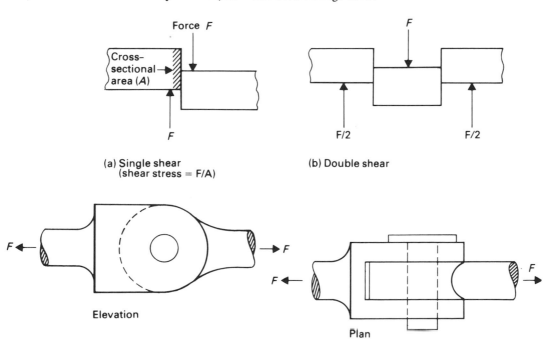

Figure 6.45   Types of shear force and typical test piece.

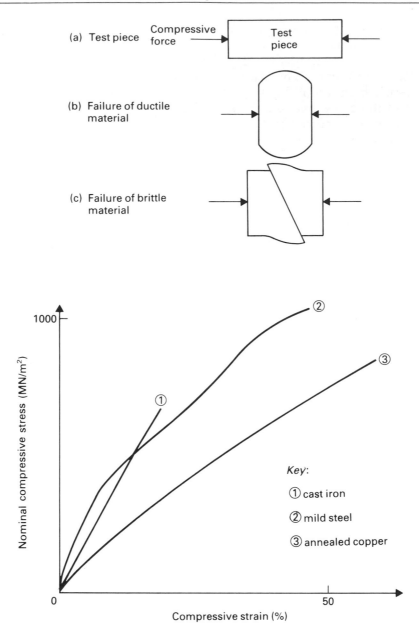

Figure 6.46 Failure due to compressive forces and examples of compression stress–strain curves.

6.2.2 Hardness testing

Hardness is usually defined as the resistance of a material to abrasion, deformation or indentation. However, the situation is more complex than this simple definition would indicate. It could be extended to include the ability to scratch another material or resist being scratched, to resist plastic or elastic deformation due to indentation, or to resist deformation due to rolling. Hardness may also be evaluated by the behavior of a material subjected to the action of cutting tools; this is dependent upon the chemical composition of the material and its mechanical properties.

Hardness depends upon the strength of bonding within a material. Polymers are relatively weak because the chains are joined together by weak van der Waals forces. Metals and ionic solids are harder, and solids possessing covalent bonding are the hardest materials known. Cold working, alloying and precipitation hardening all increase the hardness of metals.

All hardness tests are comparative and produce values compared to another material. Related to each test is a particular hardness scale, some of which are shown in Fig. 6.47. For metals the most widely used tests (and scales) are the Brinell and Vickers. The Rockwell R test is used with polymers, as is the Shore durometer test. The Shore hardness (the scale is not shown on Fig. 6.47) is measured by pushing a spring-loaded needle into the material. The Knoop scale (not shown on Fig. 6.47) is used specifically for ceramics but is becoming more popular for metals. In general, Brinell hardness cannot be converted to Rockwell, nor can the Rockwell scales be converted. For a particular material empirical conversion tables can be obtained, but they should not be used to convert hardness readings for different materials.

The hardness, shear strength and tensile strength of a material are all related. A hardness test can be used as an indication of the strength of a material, as well as its resistance to wear and its machinability. If the Brinell hardness number of carbon steels is multiplied by 500, then the value obtained is approximately equal to the tensile strength (in psi). The hardness test is a cheap and simple nondestructive test for the yield strength of a material. The yield strength is approximately one-third of the Vickers hardness value.

Scratch tests are sometimes used, but these are less useful for engineers than geologists. Tests of this type produced one of the earliest hardness scales, Moh's scale (see Fig. 6.47). Materials are ranked from the softest, talc (1), to the hardest, diamond (10). Materials are positioned in the scale such that they will scratch any substance beneath them on the scale, but not above. Machines designed to carry out these tests are not particularly successful.

Indentation tests are widely used because they are reliable, quick and easy to operate. The hardness value is usually based upon either the depth of indentation or the load divided by the *projected* surface area of indentation. The basis of the methods used in the Brinell and Vickers hardness tests are illustrated in Fig. 6.48. The results from these tests are affected not only by the original resistance of the metal to deformation, but also by the rate at which this resistance changes near the indenter during the test. The application of the indenter causes elastic deformation, followed by plastic deformation. However, the dimensions of the indentation (and hence the hardness) are measured after the load has been removed. There will be some elastic recovery, or reversed dimensional change. The different volumes of metal displaced by different-shaped indenters is one reason why hardness scales are not directly (quantitatively) comparable.

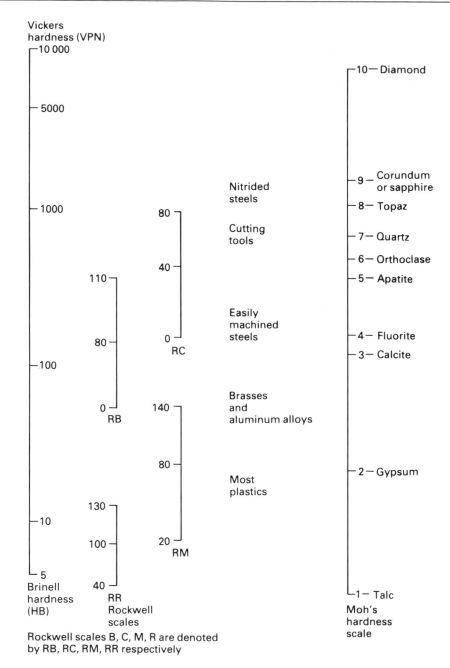

Figure 6.47 Approximate comparison of hardness scales.

The *Brinell hardness test* is carried out by pressing a hardened steel ball (usually 10 mm, 5 mm or 1 mm/0.4 in., 0.2 in. or 0.04 in. diameter) onto a flat surface of the test piece, using an appropriate force. The ball diameter and the force are related and are chosen such that the indentation is not so small as to make measurements inaccurate, or so large that the ball sinks to its full depth

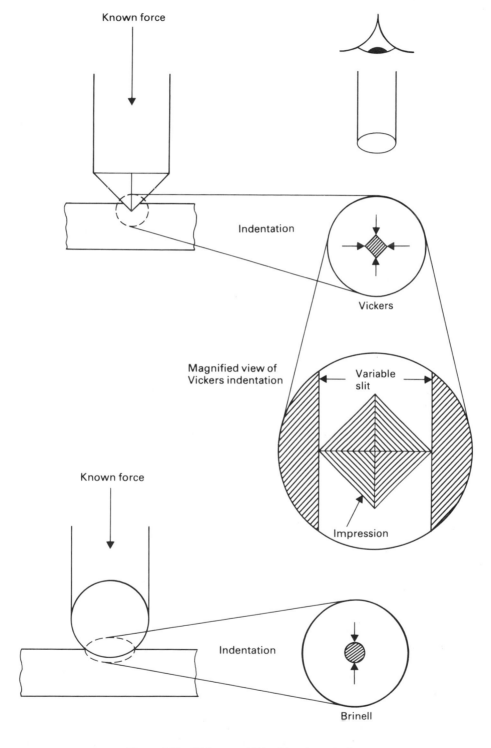

Known force

Indentation

Vickers

Magnified view of
Vickers indentation

Variable
slit

Impression

Known force

Indentation

Brinell

Figure 6.48 Vickers and Brinell hardness tests.

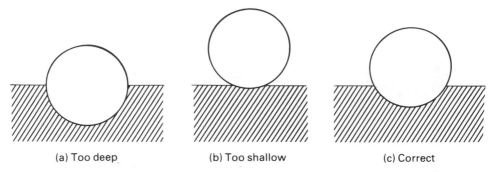

(a) Too deep   (b) Too shallow   (c) Correct

Figure 6.49   Illustration of the importance of correct ratio of force: diameter^2.

and makes results meaningless. These situations are shown in Fig. 6.49. Standard values of the ratio of force to (diameter)^2 are available so that if the ball size is chosen, the required force can be calculated. The force is applied for at least 15 s to ensure that plastic deformation occurs, with no recovery when the load is removed. To ensure that the supporting table does not absorb any of the applied force, the thickness of the test piece should be at least 7 times the depth of indentation for hard materials, and at least 15 times for soft materials. After indentation the diameter of the impression is measured using a calibrated microscope, and the area is obtained from either calculation or standard tables. This area is the value of the Brinell Hardness Number (HB) (see Fig. 6.47).

The *Vickers pyramid hardness test* is used for very hard materials (above HB of 500) where a steel ball would be deformed. The indenter is a square-based, diamond pyramid with an apex angle between opposite faces of 136°. The machine measures the diagonals of the indentation from a magnified image, using a variable slit as shown in Fig. 6.48. Measurements are usually recorded directly onto an indicator. The value of the Vickers pyramid hardness number (VPN) is then obtained from tables. The ratio of force to (diameter)^2 is not important because all impressions are geometrically similar, and the accuracy does not depend upon the depth of the impression (within certain limits). The indentation is usually small and a smoother surface finish of the test piece is required than for the Brinell test. The requirements for specimen thickness apply as already discussed for the Brinell test. The Vickers test is used for microhardness measurements but with much smaller loads.

The *Rockwell hardness test* originated in the United States, and differs from the Brinell and Vickers tests in that the hardness numbers are based upon the depth of indentation rather than the area. The indenter is brought into contact with the specimen, a small load (the minor load) is applied and equilibrium is established; this method eliminates differences due to surface conditions and zero errors. A major load is then applied and subsequently removed. The small load is maintained and the hardness number is read directly from the machine, based upon the difference between a function of depth of indentation due to the major load and a coefficient. Different hardness scales are obtained, depending upon the type of indenter used and the applied major load. Rockwell Scale C employs a diamond cone indenter and a 150 kg load; it is used mainly for very hard materials, e.g. hardened steels. Rockwell Scale B employs a steel ball and a

100 kg load for testing purposes; it is used for most other materials including nonferrous alloys and normalized steels.

Rebound tests are often used for hardness testing of large components which cannot easily be accommodated by other machines. The *Shore scleroscope* is an example of this type of dynamic test. A small diamond-tipped hammer of approximate mass 2.5 g (0.088 oz) falls freely in a cylinder (graduated from 0 to 140) from a height of approximately 25 cm (10 in.). The height of the rebound is (used as) the hardness number. The softer the material, the greater the energy absorbed on impact and the smaller the rebound height.

The *microhardness tester* is a more sophisticated machine, using a small diamond indenter and light loads (1–2000 g). The indentation produced is so small that it must be measured under a metallurgical microscope. The hardness of individual grains in the metal, particles within the grains and metal phases can be measured. Unfortunately the values obtained depend upon the load used. The microhardness tester measures absolute hardness expressed as a pressure (kg/mm²). Although this is a useful means of expressing hardness it is not usually used by engineers.

6.2.3 Fracture toughness tests and impact testing

Fracture mechanics is used to determine the plane-strain fracture toughness of a material, and the critical size of a crack or defect necessary for failure under service-loading conditions. There are many reports of catastrophic brittle failures occuring in normally ductile materials, below the maximum design loading and without extreme conditions. Part of the reason lies in our definitions (or associations) of material properties. However, there is no such thing as a brittle or a ductile material; these are types of behaviour and most materials exhibit brittle or ductile behavior, depending upon the conditions. It is often assumed that materials are hard *and* brittle, e.g. ceramics and glass, or soft *and* ductile, e.g. aluminum and structural steel. However, there are many anomalies such as bismuth which is soft and brittle, and piano wire which is hard, but ductile in small diameters. The operation of a steel file indicates that the steel is tough, yet teeth can be broken, which also makes it brittle. Many brittle failures occur in a sudden and unexpected manner. Sometimes exact causes cannot be determined, but some of the reasons are known. Brittle failure is due to a critical combination of several factors, including material structure and composition (purity), temperature, rate of loading, size and shape, type of stress, defects and dislocations.

Many materials such as glass and ceramics are considered brittle. However, as the temperature is increased, more ductile behavior is observed such as the drawing and blowing of glass. It has been shown that some brittleness is not inherent but is due to traces of impurities or dislocations. The brittleness of bulk glass was first explained by A.A. Griffith (1920) as being due to internal cracks producing excessively high local stresses at the crack tips. The strength of bulk glass is less than 1% of the theoretical strength of a silicate network. The role of a sharp crack is analogous to that of a dislocation in the shearing of a crystalline lattice. The crack is assumed to be elliptical and the applied stress is intensified at the ends of the ellipse. The *fracture stress* (σ) is given by:

$$\sigma = \sqrt{(4\gamma_s E/\pi\, a)} \qquad\qquad (6.19)$$

where γ_s = surface energy or work of fracture per unit area;
             E = modulus of elasticity;
             a = critical crack length for equilibrium.

The Griffith theory defines the failure condition in terms of the propagation of a crack, i.e. the rate of change of energy with crack length (de/da) is the condition. The *notch sensitivity* of a material is also related to the Griffith theory. This is the ability of a material to resist crack propagation due to the stress concentrations caused by a notch, and hence avoid brittle failure. However, even in the presence of a notch bulk yielding can still take place. The notch sensitivity of a material is difficult to calculate and it is usually obtained by experiment.

Equation (6.19) adequately predicts the behaviour of brittle materials, e.g. glass; however, to account for plastic flow at the crack tip the γ_s term must include the energy of plastic deformation. The term $4\gamma_s$ is then replaced by the *strain energy release rate* or the *fracture energy* (G_c), and the *stress-intensity factor (K)* is given by:

$$K \propto \sigma \sqrt{(\pi a)}$$

For an infinite plate:

$$K = \sigma\sqrt{(\pi\, a)}$$

The critical value of K, referred to as K_c, is known as the *fracture toughness*, where

$$K_c = \sqrt{(E\, G_c)}$$

or

$$K_c = A\, \sigma_{\text{crit}} \sqrt{(\pi\, a)}$$

Similarly, the critical fracture stress is given by:

$$\sigma_{\text{crit}} = \frac{K_c}{A\sqrt{(\pi\, a)}}$$

The theory of linear elastic fracture mechanics predicts that a crack will propagate and cause a specimen to fracture when the stress state near the tip of the crack (represented by the instantaneous value of K) reaches a critical value (K_c) for the material. Therefore, $K = K_c$ is a fracture criterion.

The simplest criterion for determination of the critical depth of a crack is given by:

$$h = \frac{K_c^2 Q}{1.21\,\pi\,\sigma^2}$$

where h = critical depth of a crack;
             K_c = critical stress intensity factor;
             Q = crack shape factor;
             σ = applied stress.

In metals the local stress is intensified at the tips of a fine crack and its value may exceed the yield stress, even though the average applied stress is far below this value. A small plastic region is created around the stress tip due to dislocation movements, and this acts to reduce the stress intensity by plastic flow. In ceramics, very little dislocation motion can occur and it is rarely important as a cause of failure. The local stress is then intensified and the crack propagates more easily.

Brittle fracture is particularly dangerous because there is no external warning of the approach of failure, e.g. necking or surface cracks. Also, once a crack begins to propagate through the material, progressing at a speed comparable to the speed of sound, there is no way of stopping it, even if the applied load is removed.

Impact tests are used to determine the toughness and impact strength of a material, to reveal a tendency for brittleness, and to compare the shock resistance of different materials. An impact testing machine employs a heavy pendulum with a striking knife edge, which is released from a fixed height. The striker then breaks a standard notched test piece which is firmly held. The striker continues along its arc and its highest position after fracture is measured. The energy required to fracture the mateiral is then calculated. The test piece always contains a standard notch to ensure that stress concentrations are set up, and that fracture will occur. Two types of standard test are commonly used. These are the *Izod test*, using a cantilever-type test piece, and the *Charpy test*, using a beam type test piece. Details of the test piece and the method used for each test are given in Fig. 6.50. The tougher the material, the greater the amount of energy absorbed during fracture and the smaller the subsequent height of the pendulum.

The data obtained from such tests are difficult to interpret and the results obtained by different investigators tend to be very variable. The energy required to propagate a crack depends upon the geometry of any notch machined in a specimen. The notches used are macroscopic; they do not correspond to the microscopic cracks present in materials, which give rise to brittle failure. An additional problem is the high rate of deformation that occurs during testing. For these reasons impact tests are not widely accepted and are treated with some suspicion. However, until tests that are more widely accepted are established, impact testing will continue to be used.

Typical data obtained from impact tests are shown in Fig. 6.51. Face-centered cubic (fcc) metals have high impact strengths which are practically independent of temperature. Body-centered cubic (bcc) metals, ceramics and polymers exhibit a transition temperature below which brittle behavior is observed. There is a wide variation in the transition temperature for different materials, e.g. above 500°C (932°F) for ceramics and between −100°C and +100°C (−148°F and 212°F) for metals and polymers.

There are several factors that influence whether fracture will occur. Operating at temperatures above the transition temperature is one possibility, although this temperature is not unique and depends upon the testing method used and the shape and size of the component. In alloy steels, only manganese and nickel lower the transition temperature; other alloying additions (including carbon) have a detrimental effect. The smaller the ferrite grain size the shorter the slip planes. There is less chance of building up a stress concentration to

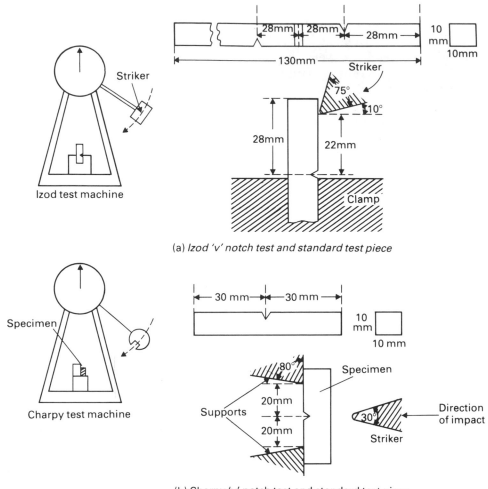

Figure 6.50   Equipment for impact testing.

nucleate a crack, and the brittleness (and transition temperature) of the steel is lowered. The dissolution of hydrogen into steels should also be avoided as this causes embrittlement, even at room temperature.

The designer can significantly reduce the possibility of impact fracture by minimizing regions where stress concentrations can occur. Machining marks and welding defects are two common causes of trouble due to poor workmanship. Local stresses can also build up in the microstructure, e.g. graphite flakes in gray cast iron act like notches and reduce the ductility.

6.2.4 Ductility tests

Ductile fracture occurs following extensive plastic deformation; it begins when necking is first observed. It is characterized by the appearance of an internal

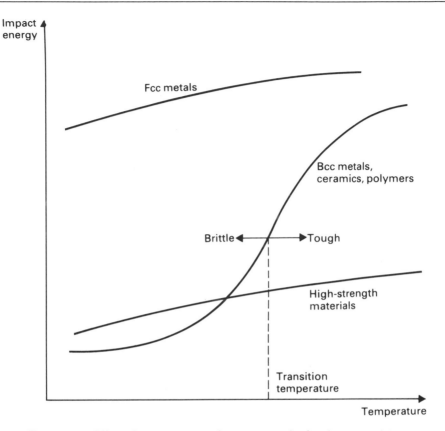

Figure 6.51 Effect of temperature on impact strength of various materials (schematic diagram).

crack at the neck. Ductile fracture can be seen when a ductile material, e.g. copper, is stretched to breaking in a tensile test machine. The fractured surfaces have a characteristic cup and cone shape as shown in Fig. 6.52. However, if the load is removed prior to fracture, then deformation stops, unlike the situation for brittle materials (see Section 6.2.3). It is thought that the cracks that give rise to ductile fracture are nucleated at impurities, e.g. slag or oxide inclusions, and that a 'pure' metal would undergo a complete reduction in area in a tensile test. When necking occurs, many small cracks are nucleated and these form a cavity which propagates outwards.

Very ductile materials are used for deep drawing operations, but an estimation of the ductility of a material from the measurements of a tensile test are often inadequate. A bend test is a simple test that does not require special apparatus or test pieces. A specimen of the material is subjected to plastic deformation, usually in one direction, although reverse bend tests are also used. Typical bend tests are shown in Fig. 6.52. After bending, a suitable material should be unbroken, free from visible surface cracks, and should not show a coarse grain ('orange peel') effect on its surface.

Details of the Erichson cupping test are given in Fig. 6.52. The test piece is clamped and a hardened steel ball is forced into it. When the test piece splits the height of the cup formed (in millimeters) is measured, and this is taken as the

Erichson value. However, results from this test are not always reproducible and it is not recommended by the British Standards Institution for specification purposes. It is useful as a comparative test.

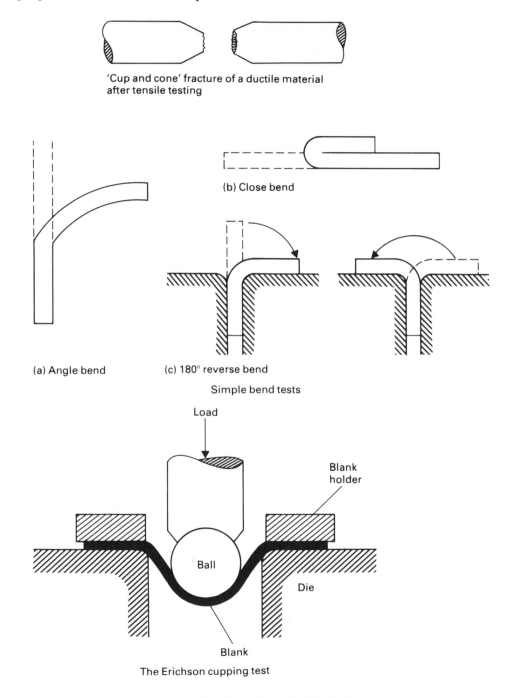

'Cup and cone' fracture of a ductile material after tensile testing

(b) Close bend

(a) Angle bend

(c) 180° reverse bend

Simple bend tests

Load

Blank holder

Ball

Die

Blank

The Erichson cupping test

Figure 6.52   Examples of ductility testing.

6.2.5 Creep testing

Creep is the extension or strain developed in a material over a long period of time, due to the application of a constant strain or a constant stress. By this definition, viscoelasticity (Section 4.14), anelasticity (Section 6.2.1), and viscous flow (Section 4.13) represent aspects of creep. This gradual extension may cause a component to fail significantly below the tensile strength of the material. When a device is loaded by the application of a fixed strain such as bolts or springs, then creep becomes apparent as a *stress relaxation*. The term *static fatigue* is sometimes used to describe the failure of plastics under sustained loading. The effect of creep is particularly important at elevated temperatures, especially near the recrystallization temperature, e.g. steel above 500°C (932°F). Creep-resistant materials should be used in the design of steam plant, chemical plant, furnaces, gas and steam turbine blades.

The creep process occurs because of two mechanisms; these are *grain boundary sliding* and *dislocation movement* by climbing past obstacles. At ordinary temperatures the plastic deformation rate is low, and most materials can withstand high stresses without deformation beyond the equilibrium elastic strain. At higher temperatures, above about half the absolute melting point (kelvin), solid state diffusion increases and lattice vacancies have considerable movement. Atomic mobility is greater at the grain boundaries than within the grains themselves, and grain boundary sliding occurs under stress by viscous flow at moderate temperatures. Vacancies can move to and from the core region of an edge dislocation, allowing the dislocation line to move up or down, out of its slip plane and to climb over obstacles. Dislocation climb increases rapidly with temperature.

Creep tests are carried out in a similar manner to tensile tests, and using a similar test piece. However, a tensile test is performed in a few minutes whereas a creep test may take several hours or weeks. The creep test is usually performed at an elevated temperature, provided by a furnace and maintained within a few degrees for the duration of the test. A constant load is applied to the test piece and maintained during the tests. The very small extensions produced in the test piece must be measured regularly and accurately. A typical diagram of stress against strain for a creep test is given in Fig. 6.53. With a constant load, as the extension increases, the cross-sectional area decreases and the true stress increases from its initial value. This is true for elastic deformation of the material; when plastic deformation occurs the decrease in cross-sectional area is reduced and the creep stress is approximately constant. For compressive loads, initially the creep stress decreases.

The typical plastic strain vs. time behavior of a specimen during a creep test is shown in Fig. 6.54(a). The initial region (*A* to *B*) is known as the *primary creep stage*. This is characterized by a steadily decreasing creep rate which is often approximated by a one-third power law:

$$\varepsilon = At^{1/3}$$

where
ε = strain;
t = time;
A = a proportionality constant.

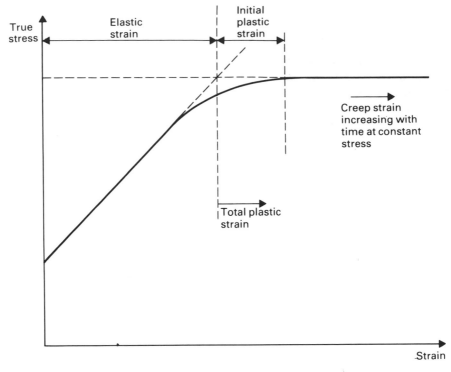

Figure 6.53 True stress against strain for creep testing.

This is followed by a period of constant strain, the *secondary creep stage* or linear creep range (*B* to *C*). The creep strain is given by:

$$\varepsilon = Bt$$

where *B* is a proportionality constant.

The secondary creep stage is of long duration and the extrapolation of results (to reduce the experiment time) should be avoided.

The strength of most materials (graphite is an exception) falls as the temperature is raised. This is because of increased dislocation mobility and a general reduction in the strength of the interatomic bonds and therefore, less rigidity of the lattice. The rate of plastic deformation is given by:

$$\frac{d\varepsilon}{dt} = B_1 \exp(-Q/kT)$$

where $d\varepsilon/dt$ = secondary creep strain rate;
  Q = activation energy for whichever dislocation or molecular mechanism is controlling the rate of deformation;
  k = rate constant;
  T = absolute temperature;
  B_1 = a constant.

At normal temperatures kT is very small, therefore deformation rates

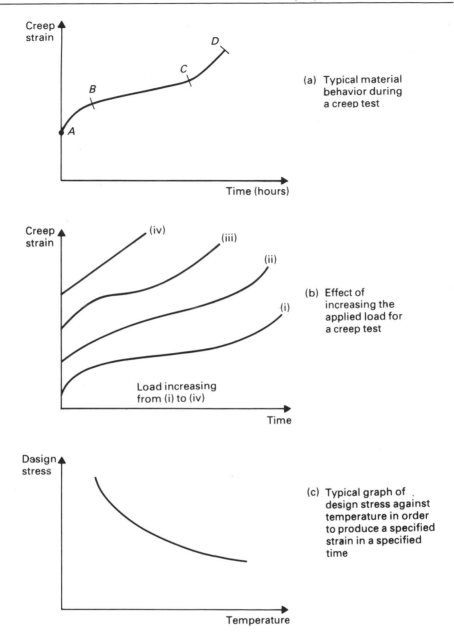

Figure 6.54   Typical creep test results.

remain very low unless the applied stress is high enough to force the dislocations through the energy barriers. Hence, most engineering materials can withstand relatively high design stresses without continuing to deform ($d\varepsilon/dt \to 0$) after they have acquired elastic strain. Elastic deformation is not thermally activated.

However, at high temperatures, the material continues to deform under sustained loads much lower than the normal yield strength or tensile strength. In engineering situations, the time–dependent creep strains are irreversible, they may be large (up to 40% even in strong alloys), and frequently lead to failure.

Accurate strain measurements are time consuming and therefore expensive. An alternative procedure known as the Larson-Miller method has been developed and is described later in this section. The creep rate of most metals and alloys (above ½ the absolute melting temperature) is highly stress sensitive and obeys an equation of the form:

$$d\varepsilon/dt = B_2\, \sigma^{m} \exp(-Q/kT)$$

where σ is the stress and B_2 and m are constants.

Finally, a *tertiary creep periodic* (C to D) is observed; the creep rate increases and ultimate rupture of the material occurs (at D). The rate obeys a cubic power law:

$$\varepsilon = Ct^3$$

where C is a proportionality constant.

The total creep curve is represented by the summation of the terms describing each region, with the addition of the instantaneous, recoverable elastic strain upon loading.

The tertiary region is always terminated by failure, this occurs along the grain boundaries and is known as intercrystalline fracture. This can be caused by the nucleation of cracks where three or more grain boundaries intersect, and where high stress concentrations are present. Alternatively if small stresses are applied for long periods of time, voids can form, grow and coalesce along grain boundaries and eventually develop into cracks. Figure 6.54(a) applies for temperatures greater than half the absolute melting point of the material. At lower temperatures rupture is unlikely to occur, the creep rate of metals then decreases with time and the total strain approaches a limiting value. In many practical situations 'failure' is deemed to have occurred when the dimensions of a component are outside the design tolerances, often long before rupture occurs.

The three creep stages (Fig. 6.54(a)) are not always apparent from experimental results and a log–log plot may be necessary to identify these regions. The constant creep region is usually considered to represent a balance between normal dislocation mechanisms of work hardening, and annealing effects such as dislocation climb, e.g. recovery.

Figure 6.54(b) illustrates the effect of increasing the constant applied load for creep tests carried out at the same temperature. Data obtained from creep tests can also be presented as the design stress required to produce an acceptable strain in a specified lifetime, as a function of temperature. This is shown in Fig. 6.54(c). Various methods have been developed to analyze experimental creep test results and provide design data. The *Larson–Miller* method assumes that the steady-state creep rate is equivalent to the reciprocal of the time for complete failure to occur. The rate equation is:

$$P = T(\log t_R + C)$$
where P = Larson–Miller parameter;
$\ T$ = absolute temperature;

t_R = time for the sample to rupture;
C = an empirically determined constant (approximate value of 20).

This procedure is less expensive and easier to perform than measuring the equilibrium creep rate at various stresses and temperatures under carefully controlled conditions. Values of P are determined over a range of stresses, they are plotted as shown in Fig. 6.55 to give the master Larson–Miller curves for different materials (they can be based upon a series of short-time tests at high temperatures). Extrapolations from these curves to estimate test results over a longer period of time should be treated with some caution as they are usually optimistic.

Creep effects can be reduced in several ways. The material chosen should have as high a melting point as possible relative to the in-service temperature, the initial loading should be small, elements should be present in solid solution and large grain size is required. If hard, dispersed second-phase particles, e.g. aluminum oxide, are present and of an appropriate size, then dislocation motion can be restricted.

Some creep-resistant alloys are dispersion-strengthened or precipitation-strengthened materials (see Sections 2.19.9 and 2.20). The precipitate must be stable in service and the average size of the precipitate must not increase rapidly

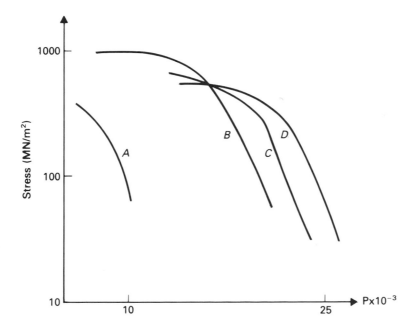

A Heat-treated aluminum alloy
B Titanium alloy
C Cobalt–molybdenum steel
D 10 0 austenitic stainless steel

Figure 6.55 Larson–Miller master creep curves.

with time. A number of nickel-based superalloys, the Nimonics (see Section 3.5.1), have been developed for such applications. They are based on an 80–20 nickel–chromium alloy that possesses high-temperature oxidation resistance but lacks strength. The strength is improved by additions of Al, Mo, Co, Ti and others. The alloys are formed by high-temperature precipitation followed by heat treatment; the final alloy consists of a strong solution-hardened fcc nickel alloy matrix. This contains a precipitated finely divided γ' phase, an ordered compound based on Ni_3Al, containing titanium and other elements to act as solution hardeners of the γ' phase. Also present are complex metal carbides which act as dispersion strengtheners and inhibit grain boundary sliding during creep. If the γ' phase is coherent with the matrix, the lattice at the interface is highly strained and dislocation mobility is reduced.

Some general rules for developing creep-resistant alloys will now be considered.

(a) Use an fcc base metal, because the dislocations are more extended and dislocation climb is more difficult than in bcc metals.

(b) Use a high melting point base metal because the rate of self-diffusion is proportional to the absolute melting temperature. However, this may not be possible with (a) above.

(c) Use a combination of strengthening mechanisms, e.g. solution hardening, precipitation hardening, addition of fine stable particles, e.g. Al_2O_3.

(d) Large grain size to minimize creep strains from grain boundary

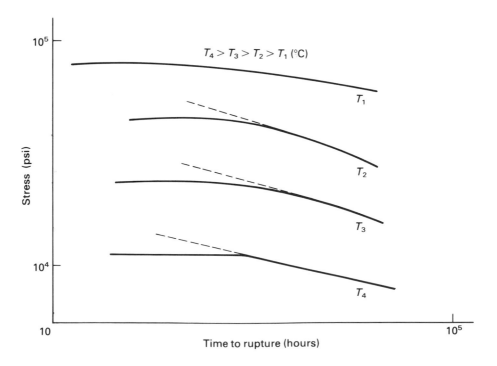

Figure 6.56 Presentation of creep data.

deformation (fine grain size makes metals stronger at low temperatures).

(e) Use a recrystallized structure rather than cold-worked material, to avoid recrystallization during creep and rapid deterioration of creep properties (this is the reason why long-time extrapolations of the Larson–Miller curves can be dangerous).

Stress rupture is the steadily applied stress that will cause fracture to occur in a material in a given time. Usually the lower the stress, the longer the time to fracture. Stress rupture may be thought of as the end point of the creep curve. In situations involving creep effects, stress rupture may be the governing factor for a design. Although creep itself may not be a problem, it is possible that stress rupture could occur. The majority of stress rupture data have been obtained from tensile and bend tests, and are usually presented as 'applied stress against time to rupture' (log or semi-log coordinates). Typical curves are shown in Fig. 6.56. The slopes of the lines change for any given condition, due to changes in the microstructure or phase changes associated with prolonged temperature or stress conditions. A design stress can be obtained by selecting an appropriate stress to avoid failure over the estimated life of the component. Alternative curves are available that show the percentage deformation as a function of the stress and the time, at a particular temperature. A prediction of stress rupture will be more difficult if the stress is not constant. However the main problem is the extrapolation of short-time data to long times, in such a case it will be safer to use a theoretical analysis such as the Larson–Miller theory.

6.2.6 Fatigue testing

Fatigue occurs when a component is 'weakened' by the initiation and growth of cracks caused by the application of continuous fluctuating or alternating stresses. The stresses that cause failure may be due to alternating tensile, compressive and shear forces, continual application and removal of a force, variations in the intensity of the load, direct stresses (e.g. piston rods), rotational bending (e.g. rotating axles), torque transmission (e.g. drive shaft of a car), or combinations of these stresses. Components often fail at stresses well below the tensile or shear strength of the material due to fatigue. The more common stress cycles are shown in Fig. 6.57.

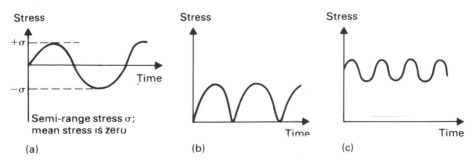

Figure 6.57 Stress cycles. (a) Reversed or alternating stress. (b) Repeated stress. (c) Fluctuating stress.

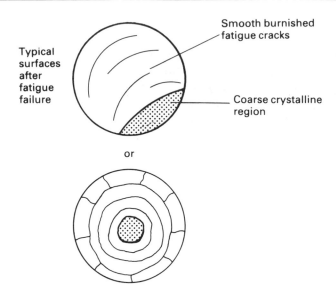

Typical surfaces after fatigue failure

Smooth burnished fatigue cracks

Coarse crystalline region

or

Figure 6.58 Typical surface after fatigue failure.

A fatigue fracture has a characteristic surface, consisting of two regions as shown in Fig. 6.58. The distinction between these regions is less clear for brittle materials than for ductile materials. The smooth region is produced by the gradual progression of the fatigue crack and rubbing of the surfaces of the crack. The coarse crystalline region indicates where final fracture occurred. A fatigue crack often starts at a point of high stress concentration; fatigue failure can sometimes be avoided by suitable design of transition shapes or elimination of tool marks, fillets, etc. For intermittent stresses, the surface of the smooth region exhibits a number of fatigue cracks as shown in Fig. 6.58. Fatigue cracks are very fine and their size can usually be detected only beyond about 90% of the component life; there is no observable distortion of the component prior to failure.

Once a crack has formed, if the material is capable of plastic deformation, then the localized stress intensities at the crack tip become work hardened. When the region is completely work hardened, the crack moves through the material until a new plastic region is reached. The crack propagates in an intermittent and brittle manner until the remaining unaffected area is unable to support the applied load and brittle fracture fails the component. Fatigue is essentially a problem of 'brittle failure' as its mechanism is crack propagation, and it is particularly sensitive to the presence of defects. However, unlike creep, fatigue does not occur by thermally activated atom movements and is therefore independent of the temperature.

If a structure is subjected to cyclic loading at high temperature, the total damage consists of fatigue damage occurring because the loading is cyclic and creep rupture damage that occurs during any hold times in the cycle. In general, these effects occur simultaneously. This situation is usually referred to as either *elevated temperature fatigue* or *creep-fatigue interaction*. Methods have been developed to analyze the problem of creep–fatigue interaction such as strain range partitioning and frequency separation. Another method has been developed by the American Society of Mechanical Engineers (ASME) in the *Boiler and Pressure Vessel Code* (see also Section 8.6.2) for the design of nuclear reactor pressure vessels. The method is based on experimental data presented in an interaction diagram (a plot of failure due to combinations of fatigue damage caused by cyclic loading and creep rupture damage caused by hold periods) for particular materials, and is presented as a Code Case by the ASME.

Fatigue can be analyzed in terms of fracture mechanics concepts. For engineering situations the crack-growth rather than the crack-initiation period is of more interest. The analysis requires knowledge of the change in the crack size during its propagation. The fatigue crack growth rate can then be measured and applied in a correlation of the form:

$$\frac{da}{dN} = C(\triangle K)^n$$

where a is the crack length, N is number of cycles, $\triangle K$ is the range of the stress intensity factor given by $(\triangle \sigma / \surd(\pi a))$ during the loading cycle, C is a constant dependent upon the mean load and rate of cycling, and n is a constant found emperically to be close to 4 for a variety of materials.

The total life of a component can be determined if the initial crack size is assumed (or known). Test specimens are usually center notched to enable a crack to grow through the material thickness, and then move laterally towards the specimen edges. It should be emphasized that material properties of fatigue are measured as semi-range stresses (zero mean stress).

Several theories have been developed to interpret fatigue results. The mathematical development of these theories will not be presented here, but the most commonly used empirical fatigue failure equations are given:

Gerber's parabolic equation

$$\left(\frac{\sigma_m}{\sigma_u}\right)^2 + \frac{\sigma_r}{\sigma_e} = 1$$

Goodman's law

$$\frac{\sigma_m}{\sigma_u} + \frac{\sigma_r}{\sigma_e} = 1$$

Soderberg's law

$$\frac{\sigma_m}{\sigma_y} + \frac{\sigma_r}{\sigma_e} = 1$$

where σ_m = mean stress;
         σ_u = ultimate tensile strength;
         σ_r = variable stress;
         σ_e = fatigue strength for complete stress reversal;
         σ_y = static yield strength.

These equations define the relationships between the mean (absolute value, not zero) and variable stresses. Equations can also be developed for the maximum and minimum stresses (σ_{max} and σ_{min}):

Goodman's law:

$$\sigma_{max} = \sigma_e + \sigma_m \left(1 - \sigma_e/\sigma_u\right)$$
$$\sigma_{min} = -\sigma_e + \sigma_m \left(1 + \sigma_e/\sigma_u\right)$$

Soderberg's law:

$$\sigma_{max} = \sigma_e + \sigma_m \left(1 - \sigma_e/\sigma_y\right)$$
$$\sigma_{min} = -\sigma_e + \sigma_m \left(1 + \sigma_e/\sigma_y\right)$$

The three laws can be compared graphically by plotting (σ_r/σ_e) against either (σ_m/σ_u) or (σ_m/σ_y) as shown in Fig. 6.59.

There are a variety of fatigue testing machines available. Bending stress machines may be of a reciprocating or rotating type. The best-known rotating type is the Wöhler reverse bend type, which operates by rotating a specimen through 180° and at the same time reducing the load to zero. This is then repeated in the reverse direction. Torsional fatigue machines subject one end of a fixed specimen to an oscillating force. Direct stress machines produce alternating or fluctuating stresses. Other more specialized machines can be used depending upon the in-service conditions. It is important to select a testing machine appropriate to the actual fatigue condition.

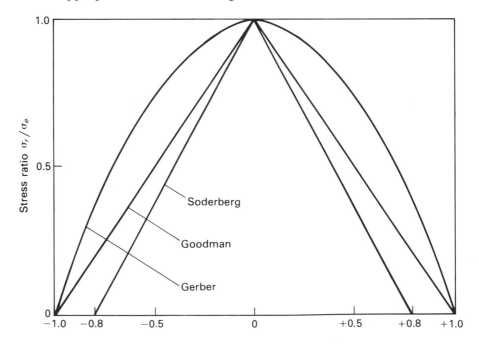

Figure 6.59 Graphical comparison of the three fatigue laws of Gerber, Goodman and Soderberg.
* For Gerber's and Goodman's laws, the stress ratio should be (σ_m/σ_u); for Soderberg's law the ratio (σ_m/σ_y) should be plotted.

The results from a reverse test machine are plotted as semi-reversed stress against the number of reversals to fracture, as shown in Fig. 6.60. This is known as an *S–N curve*; it is obtained by testing a number of identical specimens to destruction. Each specimen is subjected to different levels of semi-reversed stress and the number of reversals required to produce failure is recorded. Some materials exhibit a *fatigue limit* (or *endurance limit*); this is the stress below which the material will not fail even when subjected to an infinite number of stress cycles. Most nonferrous materials do not exhibit a fatigue limit and the curve gradually approaches the *N* axis. The *S–N* curve also indicates the expected life of a material under a given set of conditions. The criteria for failure in fatigue testing is usually taken to be either the appearance of a crack, or complete failure. The usual limiting value for tests is 10^7 cycles for structural steels and up to 10^8 cycles for other steels, nonferrous metals and alloys. Curves of probable chance of survival are plotted from the results of a large number of tests.

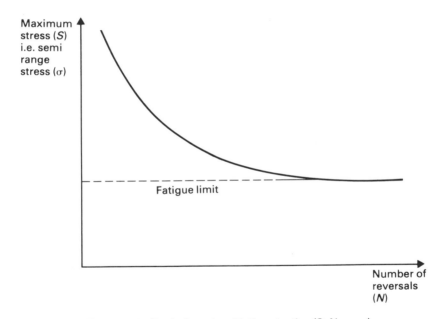

Figure 6.60 Typical results of fatigue testing (S–N curve).

Many variables and combinations of variables influence the fatigue behavior of a component, and a thorough discussion of each one is not possible here. The design engineer should be aware that they are important and be able to make estimates or predictions of their effects. The following factors need to be considered:

(a) *Surface condition*: e.g. surface finish and surface uniformity (see Section 8.4.10).
(b) *Size and shape*: the applicability of scaling-up test data to full-size components.
(c) *Reliability (reproducibility) of data*.
(d) *Temperature*: in general the endurance limit increases as the tempera-

ture decreases, phase changes may also occur at higher temperatures and increase the fatigue life.

(e) *Stress concentration*: probably the most important factor and one that is influenced by the design and materials specification. Stress concentrations may occur due to the geometry of the design, or cracks in the material caused by machining, heat treatment, impurities or weld defects.

(f) *Residual stress*: this increases the fatigue life if it is a favorable stress, e.g. a cold-working compressive stress, whereas a tensile stress decreases the life.

(g) *Environment*: a corrosive environment greatly accelerates fatigue and is known as *corrosion fatigue*. Even moist air can reduce the fatigue life of a metal compared with its life under vacuum.

(h) *Surface treatment and hardening* (see Sections 8.5 and 2.19): this should protect the surface from environmental corrosion, abrasion and wear.

(i) *Fretting*: this occurs when there is movement between two parts, producing pits and metal particles. This movement is very small.

(j) *Mean stress*: the fatigue characteristics of a material vary depending upon the mean stress value.

The fatigue life, unlike the modulus, is not a constant material property. The endurance limit depends upon many factors (listed above) and also the type of test used. The endurance limit is approximately half the ultimate tensile strength for ferrous metals, and this is probably due to the interstitial-dissolved atoms (such as carbon) present in bcc materials. These atoms strengthen the material locally near the tips of any cracks, and prevent their propagation. There is an approximate relationship between the fatigue strength (i.e. the stress corresponding to the life for a particular number of cycles) and the static tensile strength for many materials. The relationship is approximately linear for nickel and copper, but the behavior of high-strength aluminum alloys is less stable under fatigue conditions due to microstructural changes. Most steels exhibit anisotropic behavior and the fatigue properties depend upon the nature of the specimen. The major problem associated with fatigue testing is the interpretation of the data obtained. There is usually a large scatter of results, and either the data are presented between wide bands or some form of statistical analysis is required to obtain an average curve.

An alternative type of fatigue behavior known as *low-cycle fatigue* has been investigated. In this situation the material is subjected to large stresses that cause plastic deformation, and subsequent hysteresis effects that change from cycle to cycle. If a material is loaded beyond the yield point in tension, then it retains its shape only if plastic deformation occurs in compression. In general if failure occurs at less than 10 000 cycles of stress beyond the elastic limit, then this is considered to be low-cycle fatigue. The following formula has reasonable agreement with experimental data for aluminum, copper nickel, titanium, steel and stainless steel:

$$\sigma = \sigma_e + (EC/2N^{1/2})$$

where σ = stress amplitude (pseudoelastic stress);
σ_e = endurance limit;
E = modulus of elasticity;
N = number of cycles to failure;

and C is given by the formula:

$$C = \frac{1}{2}\ln\left(\frac{100}{100-A_R}\right)$$

where A_R is the percentage reduction in area in a tensile test of the material.

The fatigue analysis of pressure vessels is performed using these procedures; details are given in the ASME *Boiler and Pressure Vessel Code* (see Section 8.6.2).

6.2.7 Shear tests

For tensile testing the applied force is normal to the plane of rupture, whereas in shear tests the force is parallel to the plane. Loadings that produce shear conditions are shown in Figs. 6.45 and 6.69. The twisting action of one section of a body with respect to an adjacent section is called *torsion*. Torsional shearing stresses on circular cross-sections vary from zero at the axis of twist to a maximum at the extreme edges. If no bending occurs then it is termed *pure shear*. However, physical bodies are three-dimensional and any loading produces combinations of normal and shear stresses on the planes through any given point, i.e. certain loadings may have important three-dimensional effects.

When a shear force is applied to a material, the accompanying strain may be regarded as the effort of thin parallel sections to slide over each other. For the situation shown in Fig. 6.63(c) the shear strain is equivalent to the tangent of the angular distortion (ϕ); that is.

$$\text{Shear strain} = \tan\phi = x/L$$

If the shear strain is small then the angle may be expressed in radians.

The procedure for performing a direct shear (or transverse) test is shown in Fig. 6.72, and it provides a useful indication of the shearing resistance for items such as rivets. However, neither the elastic strength nor the modulus of rigidity can be determined because it is not possible to measure the strains produced. The results depend upon the hardness and sharpness of the equipment in contact with the test specimen, and any bending or frictional effects. An alternative direct shear test is the punching test. It is similar to the Erichson test shown in Fig. 6.52 except that it is applied to brittle materials; again the results obtained are unsatisfactory.

The theory related to torsion in shafts is presented in Section 6.3.7. It is assumed that plane sections remain plane after twisting; however only circular sections comply with this condition. For noncircular sections, correction factors need to be applied. The types of failure that can occur in a test specimen subjected to a torsional load are shown in Fig. 6.61. For a solid bar of ductile

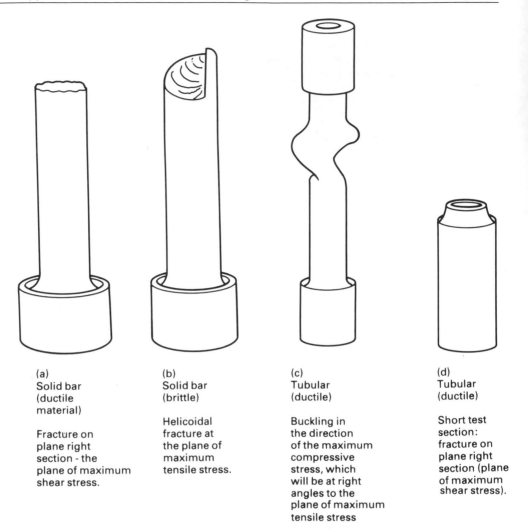

(a)
Solid bar
(ductile
material)

Fracture on
plane right
section - the
plane of maximum
shear stress.

(b)
Solid bar
(brittle)

Helicoidal
fracture at
the plane of
maximum
tensile stress.

(c)
Tubular
(ductile)

Buckling in
the direction
of the maximum
compressive
stress, which
will be at right
angles to the
plane of maximum
tensile stress

(d)
Tubular
(ductile)

Short test
section:
fracture on
plane right
section (plane
of maximum
shear stress).

Figure 6.61   Fracture specimens for torsion test.

material (Fig. 6.61(a)), there is no localized reduction in area or elongation and the break has a silky texture. If the tensile strength of a material is less than the shearing strength, then fracture occurs by separation in tension (Fig. 6.61(b)) along a helicoidal surface. This condition can be observed in cast iron and concrete testing. For tubular sections with a reduced section, buckling occurs if the length is greater than the diameter (Fig. 6.61(c)); for short reduced sections the type of failure is shown in Fig. 6.61(d).

The ratio of the torsional shear strength to the tensile strength varies from 0.8 for ductile metals to 1.3 for brittle metals. For a solid bar, the surface layer is more highly stressed and is subjected to plastic deformation first. The inner layer acts to support the outer material and the effects of shearing are not apparent. The outer material contributes toward the load-carrying capacity, but the rate of increase of torsional shear stress (τ) with shear strain (γ) has decreased. A torque (T) applied to a solid circular bar, as shown in Fig. 6.69 gives rise to a nonuniform stress distribution, as shown in Fig. 6.62(a). The torsional stress is zero at the center and has a maximum value at the outside surface. At a particular torque value, the shear stress exceeds the torsional shear strength of the material and yielding occurs. Torsional data for solid circular bars is often presented as shown in Fig. 6.62(b), with torque plotted against twist (θ) in a bar of length *l*. Point A in Fig. 6.62(b) is the torsional proportional limit, the region O to A is linear and the shear stress is calculated from:

$$\tau = \frac{Tr}{J}$$

where *r* is the radius and J is the polar moment of area.

The equation for torsional stress given in Section 6.3.7 is:

$$\frac{T}{J} = \frac{\tau}{r} = \frac{G\theta}{l}$$

where G = modulus of rigidity. (Compare this equation with the expression for simple bending of a beam derived in Section 6.3.11.)

The value of the torque at point A (Fig. 6.62(b)) can be converted to the corresponding stress; this is the torsional yield stress although the yield strength in terms of an offset is frequently used. The ultimate torsional strength is represented by point D, and fracture occurs at point F on Fig. 6.62(b). To convert values from the torque *vs.* twist graph to shear stress *vs.* shear strain, for the region O to A:

$$\tau = \frac{Tr}{J}$$

$$= \frac{2T}{\pi r^3}$$

and $$\gamma = r/G$$

Typical shear stress vs. shear strain curves for ductile and brittle materials are shown in Fig. 6.62(c). Specially designed tubular test pieces that can produce equal stresses throughout the material are available. For thin tubes, buckling can occur (Fig. 6.61(c)) and there is a critical wall thickness to ensure shear failure.

A torsion test is performed to determine the shearing properties of hollow and solid circular sections. Measurements enable calculations of the following properties to be made:

    (a) proportional limit;
    (b) shearing yield strength;
    (c) shearing resilience;
    (d) stiffness, i.e. modulus of rigidity or shear modulus of elasticity (G);
    (e) ductility, based upon the extension in the fiber length at failure;
    (f) toughness, based upon the amount of twisting.

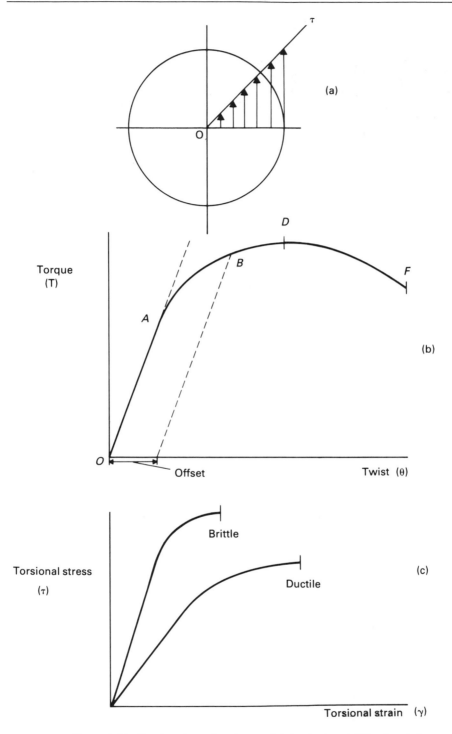

Figure 6.62 Torsional test data for a solid circular bar. (a) Torsional stress distribution in a solid circular bar. (b) Results of a torsion test for a solid circular bar. (c) Shear stress v. shear strain curve.

For brittle materials, a specimen fails diagonally (in tension) before the shearing strength is reached. The following precautions must be taken when performing torsion tests:

(a) the specimen must be of sufficient dimensions to allow accurate strain measurements to be made;
(b) the effects of gripping stresses must be absent from the region of torsion test measurements;
(c) failure must not occur due to the securing devices;
(d) bending should not occur;
(e) the ends of hollow specimens should be plugged.

The torsion test is performed using specially designed machines. A geared driving mechanism twists the grips at one end of the specimen and the applied torque (twisting action) is transmitted to the grips at the other end. This is the weighing head, and is connected to a torque indicator. Machines are available for a wide range of torque measurements in order to test bar, tube or wire. The strain or angular twist is measured by a torsion indicator (troptometer). The cross-section of the specimen should be measured to at least 1 part in 1000. The speed of the twisting head should not exceed 0.1 rev/min per 2.5 cm (1 in.) length of specimen until the yield point is reached; it may be increased after that point.

Torsion tests have not been standardized, since they are rarely included in purchasing specifications. However, they are useful for demonstrating the efficiency of components such as axles, shafts and twist drills, and also for materials lacking toughness under twisting stresses. A high-temperature modification of the torsion test is used to establish the suitability of a material for forging.

6.2.8 Summary of BS and ASTM standards (volume 03.01) covering mechanical testing of materials

BS18: Methods for tensile testing of metals:
  Part 1: Non-ferrous metals.
  Part 2: Steel (general).
  Part 3: Steel sheet and strip (less than 3 mm and not less than 0.5 mm thick).
  Part 4: Steel tubes.

Contains definitions and symbols; form and dimensions of standard test pieces; test piece preparation and testing equipment. Procedures for determining tensile properties at room temperature; proof stress, permanent stress, tensile strength, percentage elongation, with appropriate proving tests.

ASTM standard definitions of terms relating to:
  ASTM A370: Mechanical testing of steel products.
  ASTM E6: Methods of mechanical testing:

ASTM Standard methods for tension testing:

ASTM E345: Tension testing of metallic foil.
ASTM E8: Tension testing of metallic materials.
ASTM B557/B557M: Tension testing wrought and cast aluminum – and magnesium–alloy products.
ASTM E231: Static determination of Young's Modulus of metals at low and elevated temperatures.
ASTM E21: Elevated temperature tension tests of metallic materials.
ASTM E9: Compression testing of metallic materials at room temperature.

BS240: Method for Brinell hardness test:
Part 1: Testing of metals.
Test requirements and procedure, tables of hardness values and minimum recommended thicknesses.
Part 2: Verification of the testing machine.
Conditions and procedures; calibration of standard hardness blocks.

BS427: Method for Vickers hardness test:
Part 1: Testing of metals.
Tables of hardness values applicable to tests on flat surfaces; correction factors for tests on spherical or cylindrical surfaces; minimum thickness of test pieces.
Part 2: Verification of the testing machine.

BS891: Method for Rockwell hardness test:
Part 1: Testing of metals.
Test requirements and procedures using scales A to K inclusive; minimum thickness of test pieces; accuracy of hardness measurements.
Part 2: Verification of the testing machine.

ASTM standard test methods for hardness testing:
ASTM E10: Brinell hardness of metallic materials.
ASTM E140: Standard hardness conversion tables for metals (Relationship between Brinell hardness, Vickers hardness, Rockwell hardness, Rockwell superficial hardness, and Knoop hardness).
ASTM E110: Indentation hardness of metallic materials by portable hardness Testers.
ASTM E384: Microhardness of materials.
ASTM E103: Rapid indentation hardness testing of metallic materials.
ASTM E18: Rockwell hardness and Rockwell superficial hardness of metallic materials.
ASTM E92: Vickers hardness of metallic materials.
ASTM E448: Standard practice for scleroscope hardness testing of metallic materials.

BS131: Methods for notched bar tests:
Part 1: The Izod impact test on metals.
Part 2: The Charpy V-notch impact test on metals.
Part 3: The Charpy U-notch impact test on metals.

Conditions for the test; nominal dimensions and tolerances for ferrous and nonferrous test pieces. Structure; dimensions of testing machine.
Part 4: Calibration of pendulum impact testing machines for metals.
Part 5: Determination of crystallinity.

BS 5447: Methods of test for plane strain fracture toughness (K_{ic}) of metallic materials.

ASTM standard methods for impact and fracture testing:
  ASTM E23: Notched bar impact testing of metallic materials.
  ASTM E812: Crack strength of slow-bend precracked Charpy specimens of high-strength metallic materials.
  ASTM E399: Plane-strain fracture toughness of metallic materials.
  ASTM E602: Sharp-notch tension testing with cylindrical specimens.

BS 3500: Methods for creep and rupture testing of metals:
  Part 1: Tensile rupture testing.
  Determination of time to rupture of metal test pieces subjected to nominal constant tensile load, and temperature conditions in single- and multi-test piece machines, for times from 10 hours to 100 000 hours. Interrupted and uninterrupted tests; definitions, test piece dimensions, testing equipment, procedures, determination of properties and presentation of results.
  Part 3: Tensile creep testing.
  Containing information as in Part 1 (above) for determination of creep strain on metal test pieces.
  Part 5: Production acceptance tests.
  Part 6: Tensile stress relaxation testing.

ASTM E139: Standard practice for conducting creep, creep-rupture, and stress-rupture tests of metallic materials
  ASTM B95: Test method for linear expansion of metals.
  ASTM E231: Method for static determination of Young's Modulus of metals at low and elevated temperatures.

BS 4618: Recommendations for the presentation of plastics design data.
  Part 1: Mechanical properties.
  Section 1.1: Creep.
  Section 1.2: Impact behavior.
  Section 1.3: Strength.

BS 3518: Methods of fatigue testing:
  Part 1: General principles.
  Part 2: Rotating bending fatigue tests.
  Part 3: Direct stress fatigue tests.
  Part 4: Torsional stress fatigue tests.
  Part 5: Guide to the application of statistics.

ASTM standards relating to fatigue testing:
Definitions of terms:
  ASTM E513: Constant amplitude, low-cycle fatigue testing.

ASTM E912: Fatigue loading.
ASTM E206: Fatigue testing and the statistical analysis of fatigue data.
ASTM E742: Fluid aqueous and chemical environmentally affected fatigue testing.
ASTM standard practices:
ASTM E466: Constant amplitude axial fatigue tests of metallic materials.
ASTM E606: Constant amplitude low-cycle fatigue testing.
ASTM E468: Presentation of constant amplitude fatigue test results for metallic materials.
ASTM E739: Statistical analysis of linear or linearized stress-life (S–N) and strain-life (ε–N) fatigue data.

Other ASTM Standards are published (Volume 03.01) for bend and flexure testing, ductility and formability testing, machineability, shear and torsion testing, radiation effects, statistical methods and calibration of mechanical testing machines.

$*$ $*$ $*$

Exercises

1 Explain the differences between the physical and mechanical properties of materials.

2 Explain why the gauge length and cross-sectional area of a specimen must be related when performing a tensile test.

3 Is the tensile strength of a material determined by the plastic properties of the material or by its rupture strength? Explain.

4 Explain what is meant by, and distinguish between, the following terms:
   (a) true stress;
   (b) engineering stress;
   (c) nominal stress;
   (d) tensile strength;
   (e) proof stress;
   (f) yield stress;
   (g) linear strain;
   (h) true strain.

5 Define hardness.
   Is there any advantage of using either a pyramidal or spherical indenter in a hardness test?
   Why are hardness tests often used instead of tensile tests to determine the mechanical strength of a metal?

6 How is the hardness of a work-hardened metal related to its yield stress?

7 What is the relationship between the Brinell hardness number, the applied load and the indentation produced by the hardness test?

8 Which mechanical properties should be known when selecting a material? If a tensile test and a hardness test are performed on a metal, what extra information can be gained from an impact test?

9 Describe particular applications that require impact testing data to be available when selecting materials.

10 What is brittle fracture?

11 Creep and fatigue failure are both long-term effects (hopefully!). Explain the differences between tests used to determine these properties, and the limitations of the data obtained.

12 Explain what is meant by:
   (a) fatigue limit;
   (b) minimum creep rate;
   (c) creep limit.

13 Discuss briefly the material considerations when designing:
   (a) the blades of a gas turbine;
   (b) a connecting rod.

*   *   *

Complementary activities

The only way to appreciate mechanical testing is to perform the tests using different machines.

1 Prepare standard samples of a range of materials, e.g. steels, polymers, etc., for use in mechanical testing machines. Consult appropriate standards to determine the shape and dimensions of the specimens and also any heat treatment, machining, etc., that is necessary.

2 Perform mechanical tests using as many different machines as possible and determine the values of the relevant mechanical properties. Experience should be gained on:
   (a) tensile testing machines, using 'dog-bone', sheet, foil, etc., specimens;
   (b) hardness testers, e.g. indentation machines (diamond and ball indenters), rebound tests, scratch test;
   (c) impact testing machines, e.g. Izod and Charpy;
   (d) torsion testers;
   (e) creep tests;
   (f) fatigue tests;
   (g) ductility tests;
   (h) any other tests relevant to particular materials, e.g. polymers (see Complementary Activities for Chapter 4).

3 Using the measurements from the tests, calculate the mechanical properties of the materials. Identify possible sources of errors, evaluate the sensitivity of each machine and its accuracy. Determine the possible range of measurements from each machine.

4 Compare values obtained using different machines, and with published values from the literature.

5 Obtain and study manufacturers' literature for the latest mechanical testing machines.

*   *   *

KEYWORDS FOR SECTION 6.2

mechanical properties	elasticity	gauge length
physical properties	plasticity	force
strength	ductility	extension
tenacity	malleability	Hooke's law
hardness	British Standard (BS)	elastic deformation
brittleness	ASTM standard	limit of proportionality
toughness	tensile test	elastic limit

necking
waisting
fracture
yield point
stress
strain
nominal stress
engineering stress
tensile strength
Young's modulus of
   elasticity
elastic limit stress
proportional limit stress
yield stress
ultimate stress
ultimate tensile strength
stress at fracture
percentage elongation
secant modulus
tangent modulus
permanent set
safety factor
margin of safety
0.1% proof stress
compression test
hardness test
abrasion
deformation
indentation
Brinell hardness test
Brinell hardness number
   (HB)
Vickers pyramid
   hardness test

Vickers number (VPN)
Rockwell hardness test
Rockwell hardness
   scales: (RC, RB, RM,
   RR)
Shore hardness
Knoop scale
depth of indentation
projected surface area
   of indentation
scratch test
Moh's scale
rebound test
Shore scleroscope
microhardness tester
absolute hardness
impact test
shock resistance
brittle failure
Griffith theory
fracture stress
crack propagation
notch sensitivity
stress concentration
fracture toughness
critical fracture stress
critical crack depth
Izod test
Charpy test
ductility test
bend test
Erichson cupping test
creep testing
stress relaxation

static fatigue
creep resistance
grain boundary sliding
dislocation movement
   (climb)
primary creep
secondary creep
tertiary creep
intercrystalline fracture
Larson–Miller curves
stress rupture
fatigue testing
fluctuating stress
alternating stress
fatigue crack
crack growth
crack initiation
Gerber's parabolic
   equation
Goodman's law
Soderberg's law
Wöhler reverse bend test
S–N curve
fatigue limit
endurance limit
low-cycle fatigue
shear test
torsion
pure shear
angular distortion
modulus of rigidity
torque
troptometer

*   *   *

6.3 PARTICULAR SYSTEMS INVOLVING APPLIED MECHANICS

6.3.1 Stress and strain

Refer to Sections 6.2 to 6.2.7 for descriptions of mechanical tests and discussion of the mechanical behavior of materials. The load or force acting on an engineering component may be the result of various applications or physical effects, it may also take various forms as shown in Fig. 6.63. The most common situations are:

   (a) stationary or dead loads;
   (b) frictional forces;
   (c) rotation causing a centrifugal force;
   (d) bending;
   (e) change in velocity of inertia force;

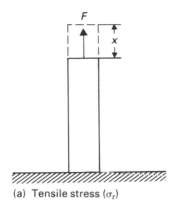

(a) Tensile stress (σ_t)

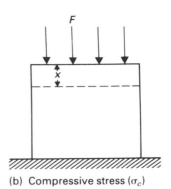

(b) Compressive stress (σ_c)

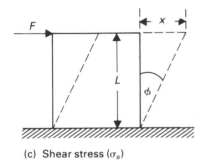

(c) Shear stress (σ_s)

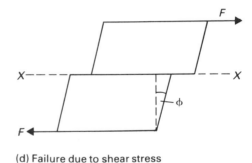

(d) Failure due to shear stress

Figure 6.63 Alternative stress situations.

(f) twisting or torsion;
(g) changes in temperature;
(h) vibration.

Rigidity
With equal and opposite forces inducing a state of shear, i.e. forces parallel and not coaxial, the resulting couple produces changes in the shape of the material. This does not apply to tensile and compressive forces which leave the shape of the body unchanged. The rigidity is a measure of the resistance of the material to this change in shape.

The material may fail at any of a number of cross-sections ($X-X$) of area A, as shown in Fig. 6.63(d). If the elastic limit is not exceeded, then the ratio of stress/strain is a constant, and is known as the *modulus of rigidity* (G). This value is a measure of the ability of a material to retain its shape under loading, and it is related to the value of the *modulus of elasticity* (E) by Poisson's ratio (Section 6.3.2).

For a Hookean material:

modulus of rigidity = shear stress/shear strain

i.e. $G = \tau/\phi$

where τ = shear stress
  and ϕ = shear strain.

Using average values of E and G, for most common metals a useful approximation is given by:

$E \simeq 2.5$ to 3 times the value of G.

Mechanical vibrations

Vibration is oscillatory motion of a *dynamic system*, such a system comprises parts that possess mass and are capable of relative motion. All bodies that possess mass and elasticity are capable of vibration. The vibration (oscillatory motion) of a system may be:

  (a) unimportant;
  (b) undesirable;
  (c) necessary.

An earthquake causing vibration may produce insignificant effects at a certain distance; machine vibration can cause fatigue failure or malfunctioning; mechanical crushers and sieve shakers use vibrations to perform their allotted tasks. The engineer should be concerned with eliminating or reducing unwanted vibrations, and designing efficient systems that utilize vibrational effects. The description and measurement of vibration is the same whether vibration is unwanted or desirable, although the design of a component or machine depends very strongly upon the types of effects that are present.

A detailed analysis of vibration will not be presented in this book. However, it will be useful and relevant to explain the meaning of some common terms, and to consider some engineering situations where vibration occurs. The elements of an ideal vibratory system include a *mass* (rigid body), a *spring* (assumed negligible mass) and a *damper* (force that opposes motion). Vibrations occur in the mass due to the application of an *excitation force*; the body gains and loses kinetic energy while the spring stores strain energy. The spring possesses elasticity and usually obeys Hooke's law, whereas the damper has neither mass nor elasticity and converts energy to heat (it is termed a nonconservative element). *Viscous damping* is often encountered in engineering; if the damping force is proportional to the velocity it is termed *linear damping*. *Forced vibrations* occur due to an excitation, e.g. the operation of unbalanced machinery, or by *periodic* excitation. If no excitation occurs after the initial disturbance, then the system experiences *free vibration*, which is the *natural mode* of vibration of a system.

The *lumped-parameter system* assumes the system contains rigid bodies, all points in a body move in phase and elastic elements have negligible mass. However, all masses possess some elasticity and all springs possess some mass. A beam (see Section 6.3.9) is a continuous combination of masses and springs inseparably distributed along its length. A body may experience transverse, compressional or torsional vibrations as shown in Fig. 6.64, depending upon the

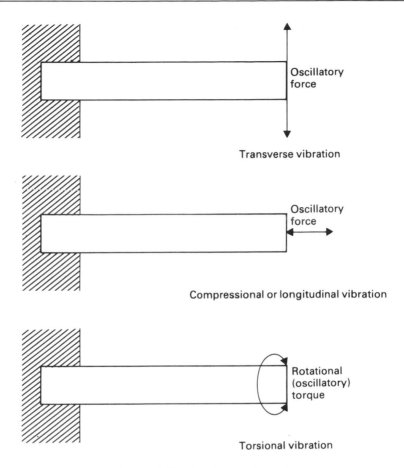

Figure 6.64   Types of vibrations in a cantilever beam.

type of excitation force. The situation may be exceedingly complex and only simple structures and loadings are capable of rigorous analysis.

When a mass connected to a spring is displaced from its static equilibrium position, then the mass oscillates periodically at its *natural frequency*. This is a case of *undamped free vibration*, as shown in Fig. 6.65. If there is no damper present then the system of forces is termed *conservative*. The periodic movement about the static equilibrium position is *sinusoidal* or *simple harmonic* and the *amplitude* of vibration does not change from cycle to cycle. If light damping occurs, then the amplitude decreases as vibration continues, and the system is said to be *underdamped*. With heavy damping, oscillation does not occur and the system is said to be overdamped. The mass slowly returns to its static equilibrium position. These effects are shown in Fig. 6.65. All systems possess some damping, although this may be very small especially for mechanical systems. Damping can be designed into a system.

Fatigue testing and the mechanism of fatigue failure are discussed in Section 6.3.6. Although fatigue is the most common cause of mechanical failure as a result of vibrations, other effects can also be significant. Failure may occur

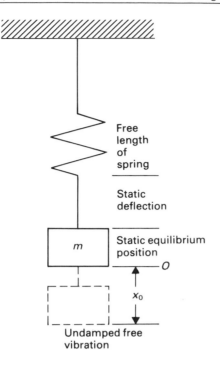

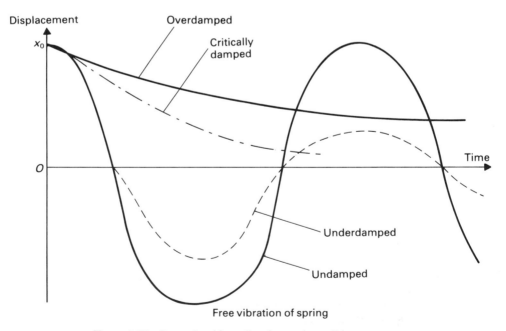

Figure 6.65 Example of free vibration and possible responses.

due to a few excessive vibrational amplitudes, e.g. brittle materials and contact failures in switches, or because a certain vibration amplitude is exceeded for too long. An *S–N* curve is obtained, showing the relationship between the average number of stress reversals and the vibration stress level. Most curves are based on pure harmonic vibrations only, a condition which is rarely achieved in practical situations. The effect on the fatigue life of the mean stress value being other than zero should also be considered. In practical situations, the mechanism of failure and the life of a component depend upon the local physical features as well as the external conditions, e.g. temperature, corrosion, etc.

For vibration studies, the velocity, acceleration (peak or root mean square values) and vibratory displacement are of interest and all three quantities are simply related. Acceleration-sensitive transducers called *accelerometers* are often used. They are smaller than velocity transducers and have wider frequency and dynamic ranges. The acceleration signal can be easily and reliably integrated to obtain the velocity and displacement. Measurements are weighted towards high-frequency situations that are of practical interest. The forces producing and resulting from mechanical vibrations also need to be measured, and piezoelectric force transducers are often used.

The most common methods of vibration testing are:

(a) sinusoidal testing;
(b) random testing;
(c) force testing.

Electronic equipment is widely used for the generation of vibration signals, and to provide essential control functions. The most common uses of vibration testing are:

(i) production control;
(ii) frequency response and dynamic performance testing;
(iii) environmental tests.

For the engineer, the control of shock and vibration or the effects produced are of prime importance. The techniques that can be employed include elimination of the source of the vibration and the use of absorbers or dampers. For natural phenomena such as earthquakes and rough seas, the only solution may be to isolate the equipment that may be damaged by the vibrations.

6.3.2 Poisson's ratio

An axial load applied to a material causes a longitudinal strain along its axis. If the volume remains constant then the dimensions normal to the direction of the load also change, due to lateral strain. A tensile load produces an increase in length along its axis and a decrease in the cross-sectional area normal to the load. A compressive load produces a decrease in length along its axis and an increase in dimensions normal to this axis due to lateral strain.

For a given material below its elastic limit, experiments have shown that the ratio of the lateral strain to the longitudinal (or axial) strain is a constant. This is known as Poisson's ratio (μ or υ).

$$\mu = \text{lateral strain/axial strain}$$

where strain (ε) is extension/original length.

Axial strain $= \sigma/E$ and therefore, lateral strain $= \mu\,\sigma/E$.

If a sign convention is used, Poisson's ratio is $-\mu$ if compressive stress is negative. This automatically shows if the strain is increased ($+$) or decreased ($-$) in bi- or tri-axial stress situations. For most metals μ has values between 0.25 and 0.33.

Engineering materials are assumed to be elastically isotropic, and the following simple relationships can be shown to exist between the elastic modulus (E), shear modulus (G) and bulk modulus (K):

$$G = \frac{E}{2\,(1+\mu)}$$

and

$$K = \frac{E}{3\,(1-2\mu)}$$

Therefore only two of the elastic constants are independent. Substituting the value $\mu = 0.33$ (for most metals), then $G \simeq 0.38E$ and $K \simeq E$. For polymers, $\mu \simeq 0.5$ and $E \simeq 3G$.

6.3.3 Bulk or volumetric strain

The effect of internal or external forces on a component is to change the dimensions. If L, A and V are the original length, cross-sectional area and volume respectively, and suffix 1 denotes the new values due to an applied load, then:

$$
\begin{aligned}
\text{Initial volume } (V) &= AL \\
\text{New length } (L_1) &= L(1+\varepsilon) \\
\text{New area } (A_1) &= A(1-\mu\varepsilon)^2 \\
\text{New volume } (V_1) &= A_1 L_1 \\
&= AL(1-\mu\varepsilon)^2\,(1+\varepsilon) \\
&= AL(1-2\mu\varepsilon + \mu^2\varepsilon^2 - 2\mu\varepsilon^2 + \varepsilon + \mu^2\varepsilon^3)
\end{aligned}
$$

Since ε and μ are both less than 1, and neglecting powers greater than the first, then:

$$
\begin{aligned}
V_1 &= AL(1 + \varepsilon - 2\mu) \\
\text{Change in volume} &= V_1 - V \\
&= AL\varepsilon(1 - 2\mu) \\
&= V\varepsilon(1 - 2\mu)
\end{aligned}
$$

Therefore, the change in volume as a fraction of the initial volume

$$
\begin{aligned}
&= (V_1 - V)/V \\
&= \varepsilon(1 - 2\mu)
\end{aligned}
$$

6.3.4 Strain energy

Strain energy is equal to the work done during the elastic straining of a component under tension, torsion, bending or a combination of these forces. It is stored by displacing atoms from original sites. It can be converted back to mechanical energy and may be released slowly, e.g. as in a clock spring, or suddenly, e.g. by vibration. The amount of strain energy, also called the *resilience*, is a measure of the ability of the material to store energy delivered by shock without permanent deformation. Three different methods of applying the load will be considered.

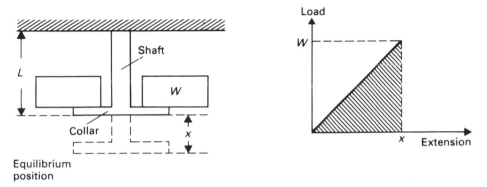

Figure 6.66   Determination of strain energy.

(a) Gradual loading
The load is initially zero and increases uniformly, e.g. liquid entering a tank. Consider the system shown in Fig. 6.66.

Let U = tensile strain energy
and σ = maximum stress.
Then:
U = (average force) × (extension)
This is the area under the graph in Fig. 6.66, e.g. the work done by the external force. Therefore:

$$U = (W/2)x$$

$$= \left(\frac{\sigma A}{2}\right)\left(\frac{\sigma L}{E}\right)$$

$$= \left(\frac{\sigma^2}{2E}\right)AL$$

$$U = (\sigma^2/2E) \text{ per unit volume}$$

The maximum possible value of U occurs at the elastic limit. If σ_{el} is the elastic limit stress, then:

$$U_{max} = (\sigma_{el}^2/2E) \text{ per unit volume.}$$

(b) Sudden loading
Assume a similar situation to case (a); the load is in contact with the collar (a distance x above the equilibrium position, i.e. the static deflection in case (a)),

but it is supported and exerts no force on the collar. The load is suddenly released and both the load and the collar travel below the equilibrium position by a distance x. The total instantaneous extension is $2x$. Therefore, the instantaneous strain and instantaneous stress are twice the values in case (a). For a static extension $2x$, the required static load is $2W$. Therefore:

$$U = \left(\frac{2W}{2}\right)2x$$

$$U = (2\sigma^2/E) \text{ per unit volume}$$

The instantaneous strain energy is increased by a factor of four.

(c) Shock loading

The load (W) is released from a height h above the collar. Therefore, the energy absorbed is the sudden load energy and the kinetic energy due to the fall:

$$U = (Wx) + \text{K E}$$

Gain in kinetic energy = loss in potential energy
$$= Wh$$

Therefore:

$$U = Wx + Wh \tag{6.20}$$

This is equal to the total loss in potential energy of the load.
The maximum force in the rod = σA
The mean force in the rod $\quad = \sigma A/2$

Therefore:

$$U = (\sigma A/2)x \tag{6.21}$$

Equating Eqs. (6.20) and (6.21):

$$\left(\frac{\sigma A}{2}\right)x = Wx + Wh$$

Since: $$x = \sigma L/E$$

Then: $$\frac{\sigma A}{2}\frac{\sigma L}{E} = \frac{W\sigma L}{E} + Wh$$

or $$\left(\frac{AL}{2E}\right)\sigma^2 - \left(\frac{WL}{E}\right)\sigma - Wh = 0$$

This problem requires the solution of the quadratic equation.

6.3.5 Thermal stresses

A change in temperature produces changes in dimensions and a thermal strain, but if this strain is prevented then stresses are created. For a given material, the linear expansion is proportional to the temperature rise and to the original length. The *coefficient of linear expansion* (α) is the change in length per degree per unit length, e.g. strain per degree.

$$\alpha = x/(l \,\triangle T)$$

Units: per °C.

As the temperature rises, x increases. If the expansion is prevented, this is equivalent to compressing a bar of length $(L + x)$ to length L.
Then:

$$\text{strain} = x/L$$
$$E = \text{stress/strain}$$

Therefore:

$$\sigma = E(x/L)$$
$$\text{where } x = \alpha L \,\triangle T.$$

6.3.6 Stresses in thin-walled cylinders

The following derivations apply for pressures inside vessels having thin walls; this assumes no stress distribution in the wall material. This is approximately true for design purposes if the thickness is less than 10% of the diameter. For vessels with thick walls, the stress is not uniformly distributed through the wall and it is then dependent on statics, geometry and properties of the material.

Consider a cylindrical vessel of length L, diameter D, wall thickness t, subjected to an internal fluid pressure P. This internal pressure acts around the circumference of the vessel, setting up hoop stress or hoop tension (σ_x) in a circumferential direction which acts to split the cylinder along its length. The internal pressure also acts on the ends of the vessel, setting up a longitudinal stress (σ_z), which acts to rupture the vessel around the circumference. These conditions are illustrated in Fig. 6.67. The hoop stress is determined with reference to Fig. 6.68.

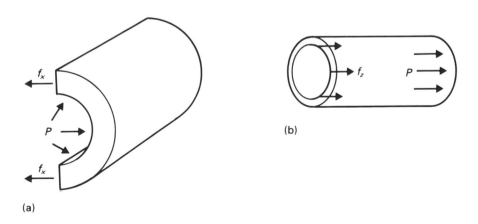

Figure 6.67 Stresses in thin-walled cylinders. (a) Hoop stress. (b) Longitudinal stress.

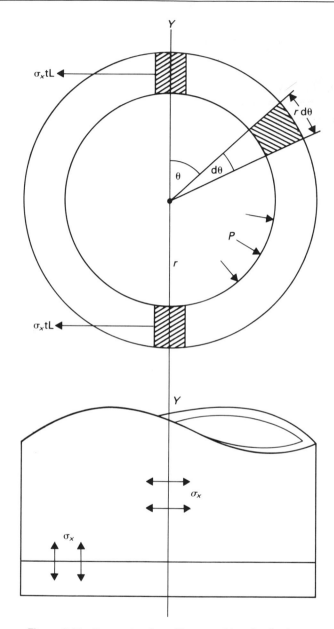

Figure 6.68 Determination of hoop and longitudinal stress.

Area of element	$= r\,d\theta\,L$
Radial force	$= r\,d\theta\,LP$
Horizontal component	$= r\,d\theta\,LP\sin\theta$

Total force on the half cylinder to the right of section *YY*:
$$F = PLr \int_0^\pi \sin\theta\,d\theta$$

Therefore:

$$F = 2PLr = PLD$$

This force is resisted by the hoop stress (σ_x) acting over the section tL on each side; therefore:

$$2(\sigma_x tL) = PLD$$
$$\sigma_x = PD/2t$$

To determine the longitudinal stress (σ_z) acting over the section πDt, this is resisted by the total force on the end:

$$\sigma_z \pi Dt = P\pi D^2/4$$

Therefore:

$$\sigma_z = PD/4t$$

The longitudinal stress is half the hoop stress and the vessel ruptures along its length. The maximum shear stress is given by ($\sigma_x - \sigma_z)/2$. The equations only give average values of the tensile stress for conditions of internal pressure. If the walls are thick or there are abrupt changes in thickness or shape, then the equations do not hold and the maximum stress is greater than that predicted. This theory does not account for the effects of external pressure or buckling, e.g. stresses due to shrink fitting a steel ring.

6.3.7 Torsion in shafts

When systems of couples that produce a turning effect are applied to a deformable material, it is necessary to calculate the torque or torsion produced.

Examples
(1) A couple consisting of two equal, opposite noncollinear, parallel forces applied to a pulley and shaft is shown in Fig. 6.69(a). The moment or torque of the couple is equal to the product of either force and the perpendicular distance between them. Therefore:

$$\text{Torque } (T) = P_1 R = P_2 R$$

Units: N-m or joule; lb-ft.

(2) Two pulleys of different radii connected by a shaft are shown in Fig. 6.69(b), with forces P_1 and P_2 acting on each pulley. A possible torque condition of equilibrium exists where:

$$T_1 = P_1 2R \text{ and } T_2 = P_2 2r$$

These torques are equal in magnitude but have opposite senses of turning, that is:

$$T_1 - T_2 = 0 \qquad \text{or} \qquad T_1 = T_2$$

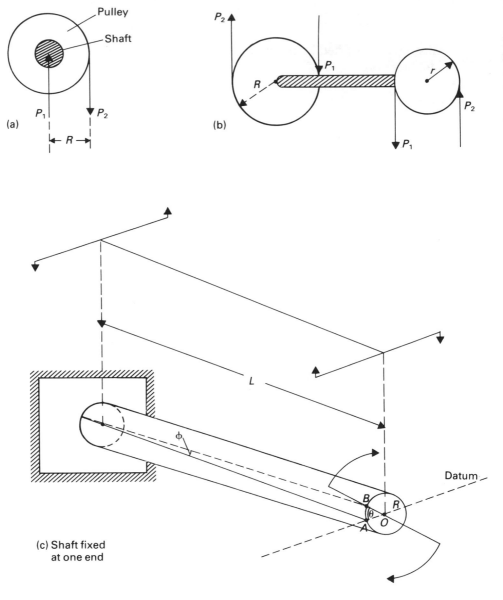

Figure 6.69 Torsion in shafts.

As shown in Fig. 6.63(c), for a shearing force (F), the shear strain is equal to ϕ radians.

Within the elastic limit,

the modulus of rigidity (G) = shear stress/shear strain
$$= \tau/\phi$$

Consider the solid shaft shown in Fig. 6.69(c), rigidly fixed at one end and subjected to a turning force at the other end. This is a system of equal and opposite couples producing a state of pure shear.

$$\text{Arc } AB = R\theta = L\phi$$

Therefore:

$$\text{shear strain } (\phi) = R\theta/L$$

Therefore:

$$\frac{\tau}{G} = \phi = \frac{R\theta}{L}$$

$$\tau = \frac{G\theta R}{L}$$

or $\tau = kR$ for any given value of θ.

For a given twist on a given shaft, τ is proportional to R, i.e. zero stress at point O. Then for $r = 0$ to $r = \text{R}$:

$$\frac{\tau'}{r} = \frac{G\theta}{L}$$

if $\tau' = $ stress at any radius r,
and $\tau = $ maximum stress at radius R
then:

$$\frac{\tau'}{r} = \frac{\tau}{R}$$

or $$\tau' = \frac{\tau r}{R}$$

The shear force on an elemental ring

$$= \tau' \, 2\pi r \, dr$$

$$= \frac{\tau r}{R} \, 2\pi r \, dr$$

$$= \frac{\tau \, 2\pi r^2}{R} \, dr$$

The moment of this force about the polar axis

$$= \frac{\tau}{R} \, 2\pi r^2 \, dr$$

Total moment of resistance to the external torque (T) by the shear stress set up in the material is given by:

$$T = \frac{\tau}{R} \int_0^R 2\pi r^3 \, dr$$

$$T = \frac{\tau}{R}\left(\frac{\pi R^4}{2}\right)$$

$$\text{or } T = \frac{\tau}{R}\left(\frac{\pi D^4}{32}\right)$$

Let $J = \pi R^4/2$, which is known as the *polar moment of area* of the cross-section of the shaft. The torque that can be transmitted by a given diameter shaft for a given maximum shear stress is calculated from:

$$T = J\left(\frac{\tau}{R}\right)$$

Therefore:

$$\frac{\tau}{R} = \frac{T}{J}$$

Since:

$$\frac{\tau}{R} = \frac{G\theta}{L}$$

Then:

$$\frac{\tau}{R} = \frac{T}{J} = \frac{G\theta}{L}$$

This is known as the *torsion equation.*

Relationship between speed and shaft diameter

It is necessary to determine the relationship between speed and shaft diameter for a given maximum stress and power. Consider a pulley, acted on by two forces connected to a shaft as shown in Fig. 6.69(b). In one revolution, the total work (W_T) performed by each force is given by:

$$W_T = P_1 2\pi R + P_1 2\pi R$$
$$= 2\pi (2 P_1 R)$$
$$W_T = 2\pi T$$

At N rev/s:

$$\text{Power} = 2\pi T N$$

$$\text{and } T = K_1 \left(\frac{1}{N}\right) \text{ if the power is constant.}$$

For a particular case of a *solid shaft*:

$$T = \frac{\pi d^3 \tau}{16}$$

where τ = maximum shear stress
and d = shaft diameter

Therefore:

$$d^3 = \frac{16T}{\pi \tau}$$

Alternatively:

$$d^3 = k_2 T$$

if τ is constant;

or

$$d^3 = k_1 k_2 \left(\frac{1}{N} \right)$$

or

$$d^3 = k \left(\frac{1}{N} \right)$$

Therefore $N\, d^3$ = a constant (k) for a given power and maximum stress. For higher speeds, a smaller shaft diameter is required.

Hollow shafts
Consider a hollow shaft with outside diameter d_1 and inside diameter d_2. From the torsion equation:

$$\frac{\tau}{R} = \frac{T}{J} \quad \text{and} \quad R = d_1/2$$

Then:

$$J = \frac{\pi d_1^4}{32} - \frac{\pi d_2^4}{32}$$

$$= \frac{\pi}{32} (d_1^4 - d_2^4)$$

$$= \frac{\pi}{2} (r_1^4 - r_2^4)$$

Therefore, the safe torque is given by:

$$T = \frac{\tau}{(d_1/2)} \left(\frac{\pi}{32} \right) (d_1^4 - d_2^4)$$

$$T = \frac{\pi \tau}{16} \left(\frac{d_1^4 - d_2^4}{d_1} \right)$$

Torsional strain energy

Considering Fig. 6.69(c) and using the torsion equation:

$$T = \frac{GJ\theta}{L}$$

The torsional strain energy (U_t) is the area under a graph of torsion against θ. This represents the work done in twisting one end of a shaft through an angle θ relative to the other end. Within the elastic limit, the work is stored in the material as strain energy. The torsional strain energy is the product of the average torque and the angle of twist.

$$U_t = \frac{T\theta}{2} \text{ and } T = \frac{GJ\theta}{L}$$

Therefore:

$$U_t = \frac{GJ\theta^2}{2L}$$

Alternatively:

$$U_t = \frac{T\theta}{2} \text{ and } \theta = \frac{TL}{GJ}$$

Therefore:

$$U_t = \frac{T^2 L}{2GJ}$$

Since $T = \dfrac{\tau J}{r}$ and $\theta = \dfrac{\tau L}{rG}$

Then:

$$U_t = \frac{T\theta}{2}$$

$$= \frac{1}{2}\left(\frac{\tau J}{r}\right)\left(\frac{\tau L}{rG}\right)$$

$$= \left(\frac{\tau^2}{G}\right)\left(\frac{LJ}{2r^2}\right)$$

However:

$$J = \frac{\pi d^4}{32}$$

Therefore:

$$U_t = \frac{\tau^2}{G}\frac{L}{2r^2}\frac{\pi d^4}{32}$$

$$U_t = \frac{\tau^2}{4G}\frac{\pi d^2 L}{4}$$

$$U_t = \frac{\tau^2}{4G} \text{ per unit volume of the shaft}$$

Coil springs

Coil springs have many engineering applications; these can be broadly classified according to the nature of the loading:

(a) Compression loading – the spring acts for prevention of shock by absorption of kinetic energy, e.g. car springs.

(b) Extension forces – the spring causes restoration of a deflection to a null position, e.g. indicating instruments.

(c) Axial twisting – the spring acts to release energy under controlled conditions, e.g. clock springs.

Shaft couplings

Consider the assembly of the shaft coupling shown in Fig. 6.70, which is to be subjected to axial twisting. The pulley has a couple applied to it, with a torque value of $2P_1R$. The assembly is in a state of equilibrium with the applied torque balanced by a torque of P_2r (one of the P_2 forces is acting on the shaft surface, the other is acting at the center of the shaft). Therefore:

$$2P_1R = P_2r$$

Hence, P_2 can be calculated if P_1 is known. The shear stress on the key is given by:

$$\tau = \frac{P_s}{A} = \frac{P_2}{A}$$

where A is the shaded shaft area shown in the diagram; usually the arc of the surface is neglected. Hence, τ can be calculated. The torque is transmitted by the flanged coupling and there is a shear stress in each bolt of diameter d.

For one bolt: $P_s = A_s\tau = \dfrac{\pi d_1^2}{4}\,\tau$

For n bolts: $\Sigma P_s = \dfrac{\pi d_1^2}{4}\,\tau n$

Resisting the moment of each bolt is the moment of the shearing strength about the center of the shaft, i.e. $P_s r_1$. The resisting moment of the shearing strength of all n bolts is given by

$$nP_s r_1 = n\frac{\pi d_1^2}{4}\,\tau\, r_1$$

The external torque (T) transmitted by the bolts in the coupling is equal to the resisting moment, this is:

$$T = n\frac{\pi d_1^2}{4}\,\tau\, r_1$$

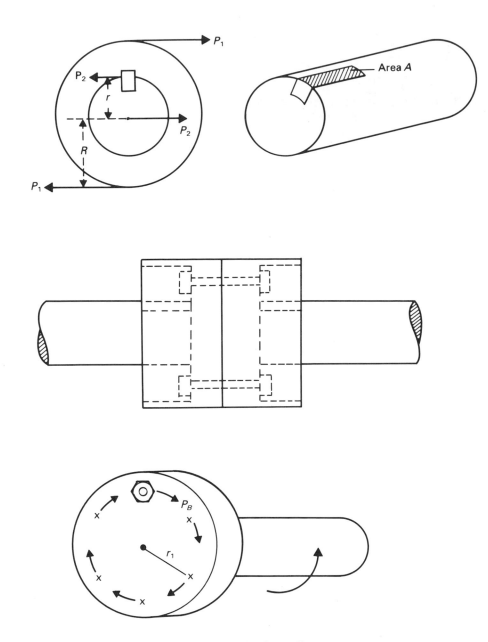

Figure 6.70   Shaft coupling.

6.3.8 Problems

(See the note regarding units in Section 6.1.23.)

P.25   A metal bar of diameter 25 mm and length 4 m is subjected to an axial pull of 50 kN. The Young's modulus for the material is 207 GN/m². Determine the tensile stress produced and the total elongation of the bar.

P.26 A straight tube of 16 mm uniform bore and 2 mm radial thickness is subjected to an axial tensile stress of 78.5 MN/m². Determine the tensile load carried by the tube and the elongation over a 5 m length. Young's modulus for the material is 100 GN/m².

P.27 A steel rod, 40 mm diameter and 7.5 m long, is tightened by means of a nut at one end and stretches by 2.5 mm. Assuming that the material is not stressed beyond its elastic limit, determine:
(a) the tensile stress produced;
(b) the tensile load imposed on the rod.
Young's modulus for the material is 207 GN/m².

P.28 Estimate the total reduction in length of the strut shown in Fig. 6.71 when carrying an axial compressive load of 1·1 MN. Assume the Young's modulus of the material to be 175 GN/m².

P.29 The following data were obtained from a tensile test on a hardened brass specimen:

Stress (MN/m²)	0	70	140	170	200	250	300	330	350	fracture
Strain (%)	0	0.05	0.10	0.20	0.40	0.80	1.60	3.00	5.00	–

Draw a stress–strain curve and determine:
(a) the elastic limit stress;
(b) the ultimate strength;
(c) Young's modulus;
(d) appropriate safety factors for an operating stress of 85 MN/m².

P.30 A mild steel test piece, of gauge length 75 mm and original diameter 13.35 mm, was subjected to a standard tensile test and the following results obtained:

Load (kN)	20	40	50	53	54	56	78	90	95.5	87	78	fracture
Extension (Divisions)	1.82	3.65	4.56	4.83	5.92	34.3	110	210	470	900	960	–

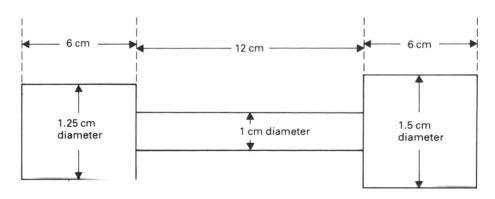

Figure 6.71 Refer to Problem P.28.

1 extension division = 0.25 mm
Final diameter at the point of fracture = 9.68 mm
Determine:
(a) elastic limit stress;
(b) Young's modulus;
(c) ultimate stress;
(d) stress at fracture
(e) 0.1% proof stress;
(f) proof resilience;
(g) percentage elongation;
(h) percentage reduction in area.

P.31 A knuckle joint, as shown in Fig. 6.72, carries an axial tensile load of 90 kN. Determine the stress in the pin which has a diameter of 18 mm.

P.32 A 25 mm diameter bolt has an effective length of 300 mm. The nut is tightened until the strain energy is 0.5 J. Calculate:
(a) the stress induced;
(b) the extension produced;
(c) the force in the bolt.
Young's modulus of the bolt material is 200 GN/m².

P.33 A steel bar, 6 m long, is suspended vertically from a rigid support. A mass of 100 kg is threaded on the bar and a collar is fitted to the lower end of the bar. The mass is allowed to fall freely from rest through a height of 150 mm before coming into contact with the collar. Given that the stress in the bar is not to exceed 80 MN/m², calculate the minimum diameter of the bar.
Young's modulus for steel = 207 GN/m².

P.34 A steel pipe, nominal bore 100 mm and thickness 10 mm, is 20 m long at 20°C. Calculate the elongation due to a rise in temperature of 70°C.

The pipe is fixed between a rigid support at one end and a support which yields under load at the other end. Given that the compressive load on the pipe is not to exceed 100 kN, determine the amount by which the support must yield per kN if the temperature rises to 120°C.
Young's modulus = 207 GN/m²
$\alpha = 12 \times 10^{-6}$ per °C.

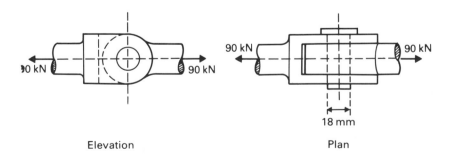

Elevation   Plan

Figure 6.72 Knuckle joint, refer to Problem P.31.

P.35 Two alloy plates are riveted together at $10°C$ using copper rivets. Neglect the compressive strain in the plates and estimate the stress increase in the rivets resulting from a rise in temperature to $21°C$.
Young's modulus for copper $= 120 \text{ MN/m}^2$.
For copper, $\alpha = 5 \times 10^{-6}$ per $°C$.
For alloy, $\alpha = 7 \times 10^{-6}$ per $°C$.

P.36 A hollow bullet is 5 cm diameter and 2 mm thick. Estimate the hoop stress induced by an internal pressure of 14 MN/m^2, assuming the stress to be uniform.

P.37 A steam boiler of 2.3 m diameter is to operate at a pressure of 2.1 MN/m^2 above atmospheric pressure. Estimate the plate thickness required to limit the hoop stress to 60 MN/m^2.

P.38 A test piece 5 cm wide and 1.5 cm thick has a gauge length of 10 cm as shown in Fig. 6.73. Assume Young's modulus is 200 MN/m^2 and Poisson's ratio is 0.28. Calculate for an axial tensile load of 14 kN:
(a) longitudinal strain;
(b) lateral strain;
(c) volumetric strain;
(d) change in volume.

P.39 A solid steel shaft having a diameter 75 mm is subjected to twisting. Given that the angle of twist is not to exceed $1°$ in 2 m length, determine:
(a) the maximum torque which the shaft may carry;
(b) the maximum shear stress produced by this torque.
The modulus of rigidity of steel $= 80 \text{ GN/m}^2$.

P.40 Determine the required diameter of a solid shaft to transmit 112 kW at 270 rev/min, if the maximum shear stress due to torsion must not exceed 70 MN/m^2. Take the maximum torque as 1.3 times the mean torque.

P.41 A hollow steel shaft, 6.5 cm outside diameter, is to be connected via a clutch to a solid alloy shaft of the same diameter. Given that the torsional rigidity of the steel shaft is to be 0.8 of that of the alloy shaft, calculate the

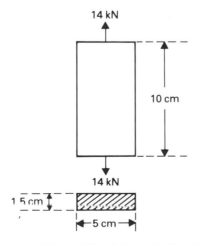

14 kN

10 cm

14 kN

1.5 cm

←5 cm→

Figure 6.73 Refer to Problem P.38.

required inside diameter. The modulus of rigidity of the alloy material is 0.4 of that of the steel

P.42   Calculate the strain energy stored in the material of a hollow shaft, 120 cm long, 5 cm inside diameter and 10 cm outside diameter when subjected to a torque of 1200 J. The modulus of rigidity of the material is 70 MN/m^2.

6.3.9 Bending of beams

A *beam* is a length of material subjected to forces acting perpendicular to its long axis. There are three main types of beam in general use, as shown in Fig. 6.74. These are:

(a) Simple beam – supported at each end.
(b) Overhanging beam – supported at two points.
(c) Cantilever beam – supported at one end.

(The arrows in Fig. 6.74 show possible point or concentrated forces acting on the beam. At each support there will be a force or reaction on the beam; in the

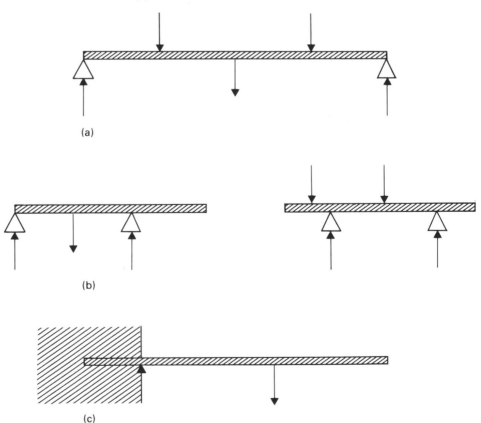

Figure 6.74   Main types of beams. (a) Simple beam. (b) Overhanging beam. (c) Cantilever beam.

last case (c) a moment is required to maintain equilibrium.) Uniform loads are assumed to be evenly distributed along the length of the beam.

Bending moment (*M*) is the product of a force (causing bending) and its perpendicular distance from the line of action. Units: N-m or joules; lb-ft.

Convention: To the left of any section, an anticlockwise moment is considered to have a negative sign.

Examples

(1) See Fig. 6.75 showing a cantilever beam of negligible mass, with a single concentrated load at the free end. The bending moment at any point (M_x) is equal to $-Wx$, as shown in the bending-moment diagram of Fig. 6.75. The maximum bending moment (M_{max}) is equal to $-WL$, and is equal to an opposite moment that has to be provided by the support at the fixed end. The bending moment causes tensile stress, and therefore an increase in length in the upper layers of the beam. The lower layers are subjected to a compressive stress, and at an intermediate point there will be a plane of zero stress.

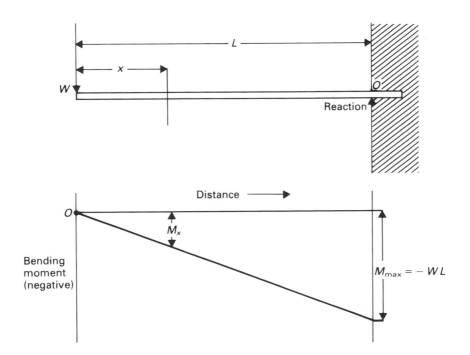

Figure 6.75   Refer to Example 1.

(2) See Fig. 6.76 showing a cantilever beam of negligible mass, with three concentrated loads along its length. The bending moment diagram is also shown in Fig. 6.76.

(3) See Fig. 6.77 showing a simple beam supported at each end and with a central load; also shown is the bending moment diagram.

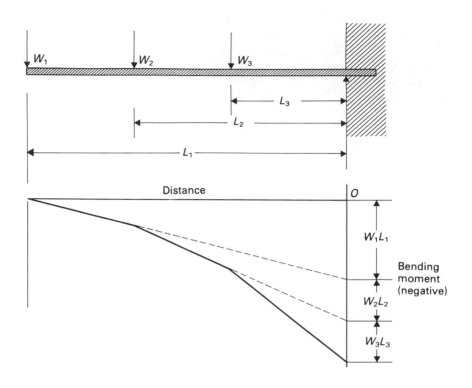

Figure 6.76 Refer to Example 2.

At any section (X–X):

$$M_x = \frac{W}{2} x$$

At the center:

$$M_{max} = \frac{W}{2} \frac{L}{2}$$

$$= \frac{WL}{4}$$

At any section (Y –Y):

$$M_y = \frac{W}{2} Y - W\left(Y - \frac{L}{2}\right)$$

$$= \frac{WL}{2} - \frac{WY}{2}$$

$$= \frac{W}{2}(L - Y)$$

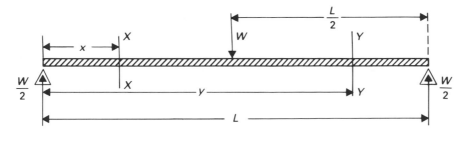

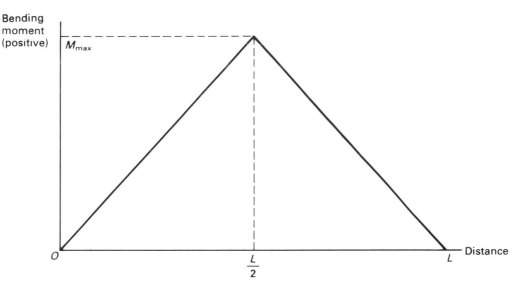

Figure 6.77   Refer to Example 3.

(4) See Fig. 6.78 showing a light beam of negligible mass, supported at each end and subjected to three forces; also shown is the bending moment diagram. Taking moments about point O:

$$16R_2 = (6 \times 2) + (8 \times 4) + (10 \times 12)$$
$$R_2 = 10.25 \text{ N}$$

Therefore:

$$R_1 + 10.25 = 6 + 8 + 10$$
$$R_1 = 13.75 \text{ N}$$
$$M_A = 13.75 \times 2 = 27.5 \text{ N-m}$$
$$M_B = (13.75 \times 4) - (6 \times 2) = 43 \text{ N-m}$$
$$M_C = (13.75 \times 12) - (6 \times 10) - (8 \times 8) = 41 \text{ N-m}$$

Overhanging supports
An example is shown in Fig. 6.79, with the bending moment diagram. The *general case* is as shown in Fig. 6.80, where:

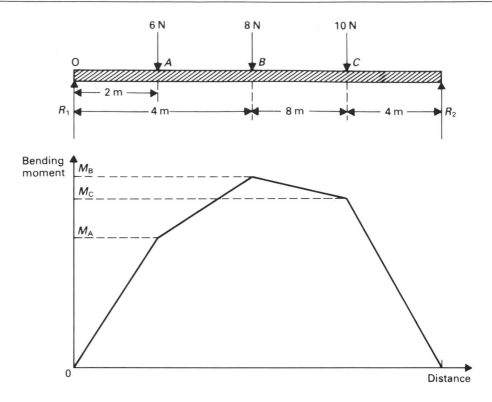

Figure 6.78 Refer to Example 4.

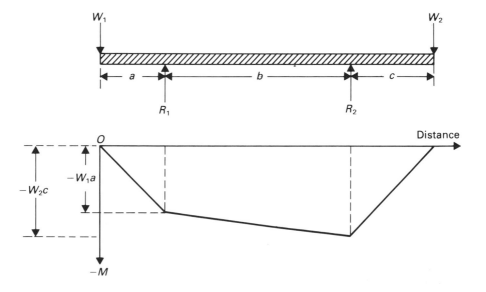

Figure 6.79 Bending moment diagram for an overhanging beam.

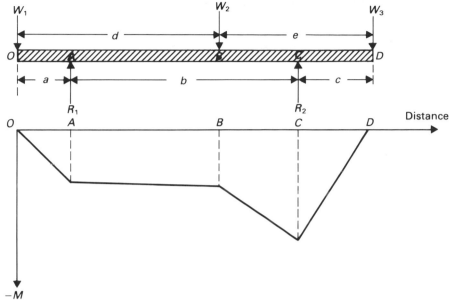

Figure 6.80   Bending moment diagram - general case.

$$M_A = -W_1 a$$
$$M_B = -W_1 d + R_1(d - a)$$
$$M_C = -W_1(a + b) + R_1 b - W_2(e - c)$$
$$= -W_3 c$$

The force W_2 reduces the (negative) bending moment between the supports. For a large load W_2, then the bending moment can become positive as shown in Fig. 6.81.

If there is no load change between sections, then the bending moment is assumed to change uniformly.

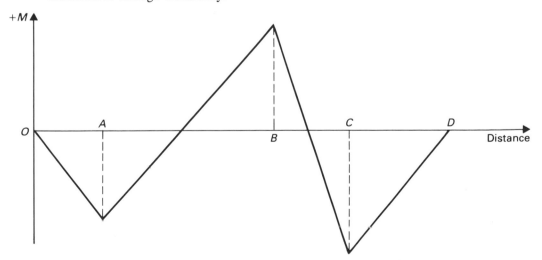

Figure 6.81   Bending moment diagram - large central load.

6.3.10 Bending moment with uniform loads

The weight of a beam is usually assumed to be evenly distributed throughout its length. It is often considered to be concentrated (i.e. to act) at the center of gravity.

Consider a beam of total weight W, which is distributed as w per unit length as shown in Fig. 6.82. To the left of section X–X, the weight (wx) acts at a distance $x/2$. Therefore:

$$M_x = -wx(x/2)$$
$$M_x = -wx^2/2$$

The graph of M against x is a parabola, as shown in Fig. 6.82.

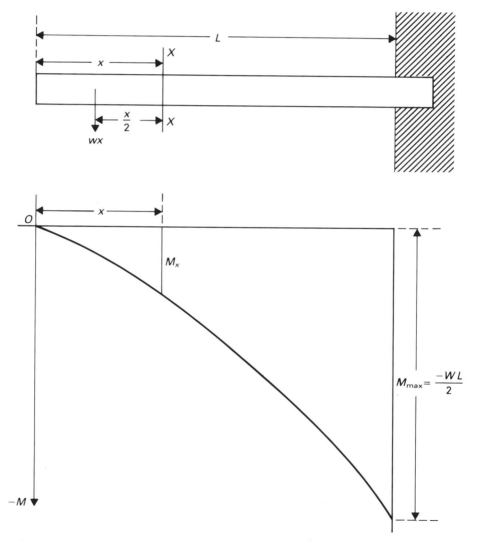

Figure 6.82 Cantilever beam with uniform loading.

At $x = L$:
$$M_{max} = -wL^2/2$$
Because $wL = W$:
$$M_{max} = -WL/2$$

Examples

(5) A simply supported beam with a uniform load, as shown in Fig. 6.83.

$$\text{Total load } (W) = wL$$
$$\text{Load per support} = wL/2$$

At section X–X:

$$M_x = \left(\frac{wL}{2}x\right) - \left(wx\,\frac{x}{2}\right)$$

$$= \frac{wLx}{2} - \frac{wx^2}{2}$$

This is a parabola as shown in Fig. 6.83. To find the value of x that makes the bending moment a maximum:

$$\frac{dM_x}{dx} = \frac{wL}{2} - wx = 0$$

Therefore, at $x = L/2$:

$$M_{max} = \frac{wL}{2}\frac{L}{2} - \frac{w}{2}\left(\frac{L}{2}\right)^2$$

$$= \frac{wL^2}{8}$$

or
$$M_{max} = \frac{WL}{8}$$

(6) Simple supports not at the ends, as shown in Fig. 6.84. Point B is the center of the beam; $a > 0.5b$; weight is w per unit length.

$$M_A = M_C = -wa(a/2) = -wa^2/2$$

The reactions at A and C are equal, and of magnitude:

$$\frac{w}{2}(a + b + c) = \frac{w}{2}(2a + b)$$
$$= w\,(a + b/2).$$

Then:

$$M_B = -w(a + b/2)\left(\frac{a + b/2}{2}\right) + w(a + b/2)(b/2)$$

$$= \frac{-wa^2}{2} + \frac{wb^2}{8}$$

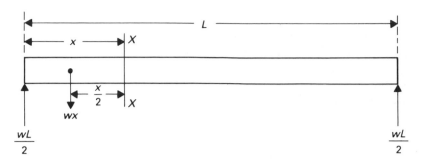

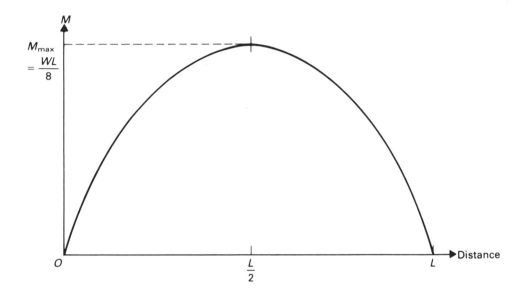

Figure 6.83   Refer to Example 5.

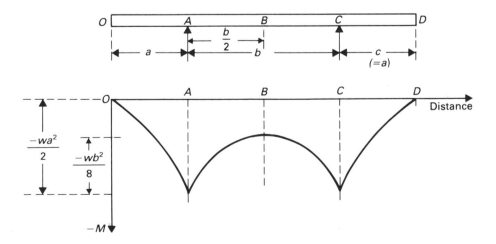

Figure 6.84   Refer to Example 6.

The maximum value of M between the supports is less than that at the supports, by an amount equal to $wb^2/8$.

For M_B to be zero:

$$0 = \frac{-wa^2}{2} + \frac{wb^2}{8}$$

$$a^2 = b^2/4$$
$$a = b/2$$
$$= L/4$$

The bending moment diagram in this situation is shown in Fig. 6.85. If $a < b/2$, then M_B becomes positive and the best use is made of the beam when $M_B = -M_A$, as shown in Fig. 6.86. Then:

$$\frac{-wa^2}{2} + \frac{wb^2}{8} = \frac{wa^2}{2}$$

Therefore:

$$a^2 = b^2/8$$
$$a = b/2\sqrt{2}$$
$$\simeq 0.356b$$

If the supports are not symmetrically spaced, i.e. $c > a$, then the bending moment diagram is shown in Fig. 6.87. The deflected shape of the beam (greatly exaggerated) is also shown in Fig. 6.87. The bending moment is zero at points P and Q and here the beam is straight. The sign of the bending moment changes at these points and they are known as *points of contraflexure.*

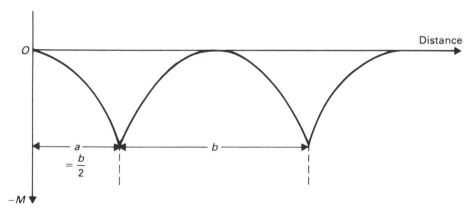

Figure 6.85   Refer to Example 6.

6.3.11 Simple bending theory

A bending moment applied to a beam produces tensile and compressive strains separated by an unstrained or neutral plane, as shown in Fig. 6.88. For a

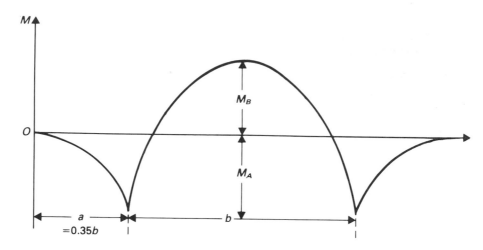

Figure 6.86 Refer to Example 6.

symmetrical cross-section, the neutral axis bisects the cross-sectional area. In order to calculate the stress at any point in an elastic beam, assume:

(a) the material is homogeneous and isotropic;

(b) the effects of shear force can be neglected, i.e. a plane section perpendicular to the axis remains plane under bending conditions;

(c) in neither tension or compression is the limit of proportionality exceeded, and the Young's modulus is the same in both regions.

Consider the beam with relevant dimensions as shown in Fig. 6.89. The sections AB and CD are a distance dx apart, considered straight not arc. Let R be the radius of curvature of the neutral axis of the beam due to bending. Elements above the neutral axis, e.g. EF, are reduced in length; those elements below are extended. For the element EF:

Original length $dx = R\,d\theta$ (the same as that at the neutral axis).

Length after bending $= (R - y)\,d\theta = R\,d\theta - y\,d\theta$
Reduction in length $= y\,d\theta$
Therefore, compressive strain $= y\,d\theta / R\,d\theta$
$= y/R$
and compressive stress (σ_c) $= E\,(y/R)$

For elements below the neutral axis, the tensile stress (σ_t) is:

$$\sigma_t = -E(y/R)$$

Since R is constant over a small length (dx) of the beam, σ_t is proportional to y. The maximum values of the stress are obtained at the upper and lower surfaces (these maximum values are not necessarily equal). At the neutral plane i.e. $y = 0$, the stress is zero.

For section AB to be in horizontal equilibrium, the resultant compressive force above the neutral plane must be equal and opposite to the resultant tensile force below. Therefore:

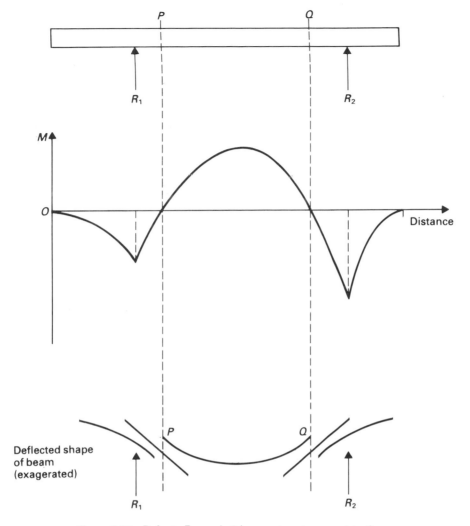

Figure 6.87 Refer to Example 6 (supports not symmetrical).

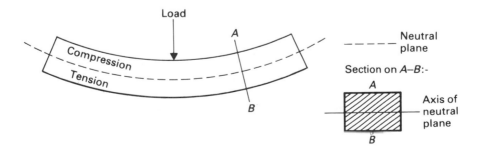

Figure 6.88 Bending moment of a beam.

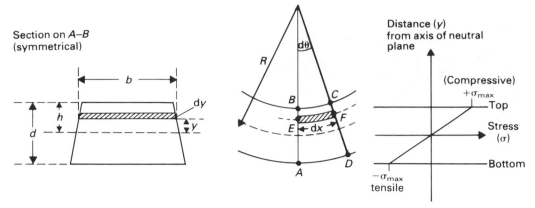

Figure 6.89 Stresses in a beam due to bending.

$$\Sigma\sigma\,\triangle a = 0$$

Since:

$$\sigma = E(y/R)$$

$$\Sigma\frac{Ey}{R}\,\triangle a = 0$$

$$(E/R)\Sigma y\,\triangle a = 0$$
$$\text{Since } E/R \neq 0$$
$$\Sigma\triangle a\,y = 0$$

By definition, $\Sigma\triangle ay$ is the *first moment of area* of the section about the neutral axis. This can only be zero if this axis passes through the centroid of the section.

Let the transverse area of the element *EF* be $\triangle a$, where $\triangle a = b\,dy$.

Compressive force on this element = stress × section
$$= (Ey/R)\,\triangle a$$

The moment of this force about the neutral axis:

$$= (Ey\,\triangle a/R)y$$
$$= (E/R)\,\triangle a\,y^2$$

The summation of all moments for tensile and compressive forces is the *moment of resistance* of the beam:

$$= \Sigma\frac{E}{R}\,\triangle a\,y^2$$

$$= (E/R)\Sigma\triangle ay^2$$

By definition, the *second moment* of area of the section about the neutral axis is $\Sigma\triangle a\,y^2$. This is denoted by the symbol *I*. Therefore:

$$I = \Sigma bdy\,y^2$$
$$= \Sigma by^2\,dy \text{ (or } \int_{y_1}^{y_2} by^2\,dy)$$

(*b* must be constant, or a known function of *y*).

Therefore, the moment of resistance to bending is $(E/R)I$.

At any section, within the limit of proportionality of the material, the internal moment of resistance is equal and opposite to the applied bending moment (M). Therefore:

$$M = \frac{E}{R} I$$

$$\frac{M}{I} = \frac{E}{R}$$

Since

$$\sigma = \frac{E}{R} y$$

Therefore

$$\frac{\sigma}{y} = \frac{E}{R} = \frac{M}{I}$$

This is the fundamental equation for simple bending. Since $\sigma = (My)/I$, for a given beam section (i.e. for a given value of I) the stress varies along the beam in proportion to M, and across the section in proportion to y. The bending-moment diagram for a cantilever beam is shown in Fig. 6.90.

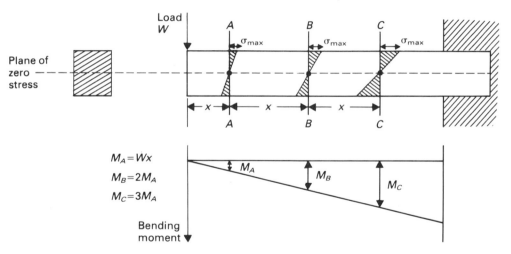

Figure 6.90   Bending moment for a cantilever beam.

Beam of circular section

The equation for simple bending is still applicable but it is necessary to calculate the second moment about the neutral axis, i.e. about a diameter. The value of y is the greatest radius. Considering the beam shown in Fig. 6.91, since:

$$I_z = I_x + I_y \qquad \text{and} \qquad I_y = I_x$$

then

$$I_x = I_z / 2$$
$$= I_{NA}$$

For the circular element,

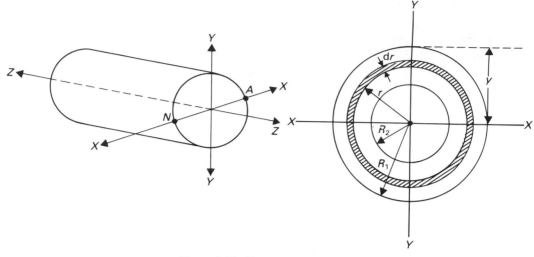

Figure 6.91 Beam of circular section.

$$\text{area} = \text{length} \times \text{width}$$
$$= 2\pi r \, dr$$

All points on this element are a distance r from the axis ZZ, then second moment about ZZ

$$= 2\pi r \, dr \, r^2$$
$$= 2\pi r^3 \, dr$$

Second moment of whole section:

$$I_z = \int_{R_2}^{R_1} 2\pi r^3 \, dr$$
$$= 2\pi \left[\frac{r^4}{4} \right]_{R_2}^{R_1}$$
$$= \frac{\pi}{2} (R_1^4 - R_2^4)$$
$$I_z = \frac{\pi}{32} (D_1^4 - D_2^4)$$
$$I_x = \frac{\pi}{64} (D_1^4 - D_2^4)$$

For a solid shaft, $R_2 = 0$ and $R_1 = R$. Hence:

$$I_x = I_z/2$$
$$= \pi D^4/64$$

6.3.12 Shear force in beams

Convention: If the resultant force to the left of a section is upwards, then the shear force is taken as positive as shown in Fig. 6.92. Let F be the symbol for shear force.

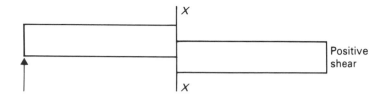

Figure 6.92 Convention for shear force.

Examples of shear force diagrams

I *Cantilever beam of negligible mass*
    (a) *Concentrated load at the free end* (Fig. 6.93)

        The concentrated load (W) causes a bending moment ($-Wx$) at section XX, and also produces a shear force (F, net traverse).

    (b) *Several concentrated loads* (Fig. 6.94)

        Shear force increases in a series of steps:

$$O{\rightarrow}A: F = -W_1$$
$$A{\rightarrow}B: F = -(W_1 + W_2)$$
$$B{\rightarrow}C: F = -(W_1 + W_2 + W_3)$$

II *Simply supported beam*
    (a) *Single concentrated load*
        (i) At the center as shown in Fig. 6.95.

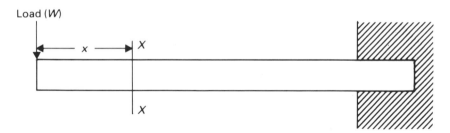

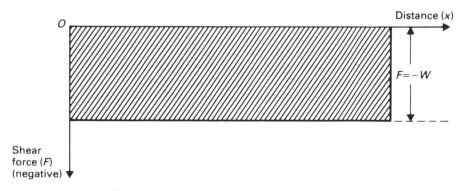

Figure 6.93 Cantilever beam with load at free end.

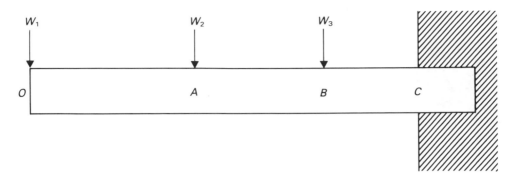

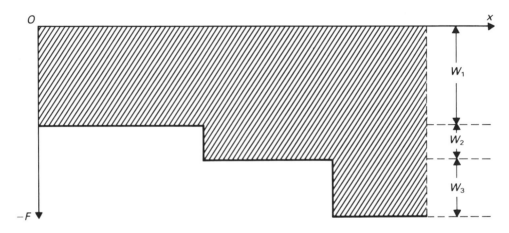

Figure 6.94 Cantilever beam with several concentrated loads.

(ii) Not at the center, see Fig. 6.96.
Taking moments about O:

$$Wa = R_2 L$$

$$R_2 = Wa/L$$

Similarly,

$$R_1 = Wb/L$$

(b) *Several concentrated loads* as shown in Fig. 6.97. R_1 and R_2 can be calculated by taking moments about each end.
For each section:

$$O{\rightarrow}A: F = +R_1$$

At A, the shear force is reduced by W_1, therefore:

$$F = R_1 - W_1$$

Similarly, between C and D:

$$F = R_1 - (W_1 + W_2 + W_3) = -R_2$$

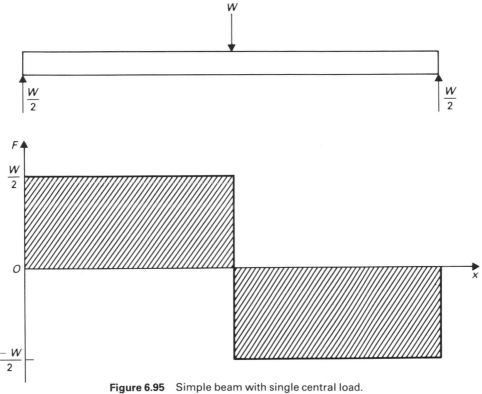

Figure 6.95 Simple beam with single central load.

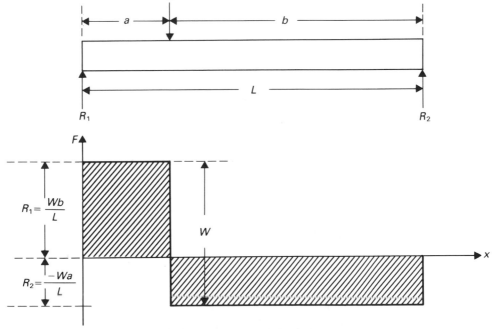

Figure 6.96 Simple beam with load not at center.

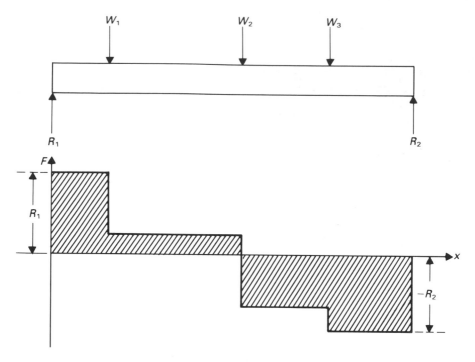

Figure 6.97 Simple beam with several loads.

(c) *Beam with loaded ends, overhanging the supports*
  (i) *Symmetrically placed supports* as shown in Fig. 6.98. Between O and A the net force is downwards and the shear force is negative $(-W)$. At A the shear force is reduced by W; therefore, between A and B the shear force is zero. Similarly, between B and C shear force is $+W$.
  (ii) *Supports not symmetrical* as shown in Fig. 6.99. Calculate R_1 and R_2 by taking moments about each end. The shear force between the supports is zero only if the bending moments at A and B are equal; i.e. $W_1 a = Wc$. Then $R_1 = W_1$ and $R_2 = W_2$.

III *Shear force with uniform load*
The shear force may be caused by the weight of the beam (W) being assumed to be uniformly distributed along its length (i.e. w per unit length).

Examples of shear force diagrams
  (a) *Cantilever beam* shown in Fig. 6.100. At the section XX, the weight (wx) causes a bending moment equal to $-wx(w/2)$ and a shear force $(-wx)$ acting vertically downwards. Since w is constant, the shear force (F) is proportional to the distance (x).
  (b) *Simply supported beam* shown in Fig. 6.101.

  Load per support $= wL/2$
  $ = W/2$
  At any section XX: $F_x = (wL/2) - wx$

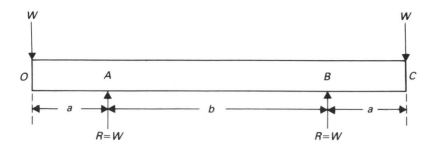

Figure 6.98 Beam with loaded ends overhanging the supports.

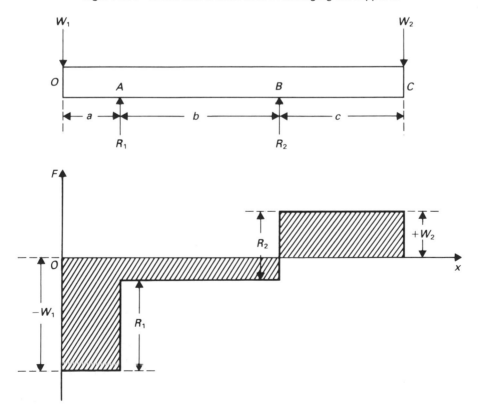

Figure 6.99 Beam with loaded ends, unsymmetrical supports.

At the center, $F = O$. This is the point subjected to the maximum bending moment.

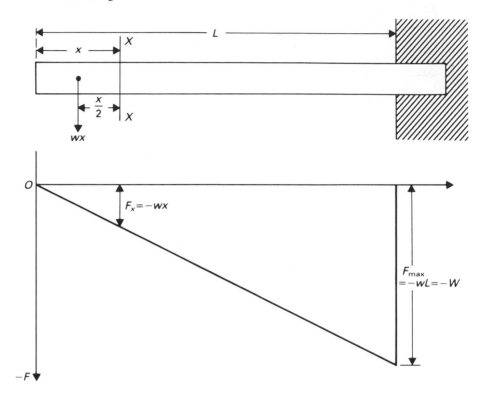

Figure 6.100 Cantilever beam with uniform load.

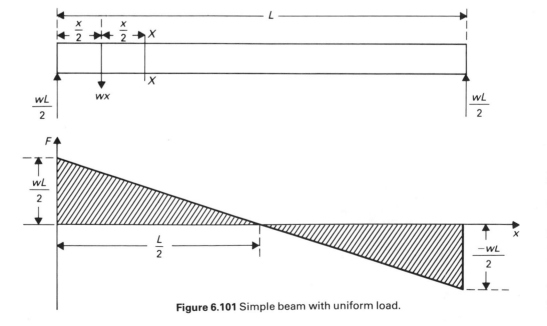

Figure 6.101 Simple beam with uniform load.

6.3.13 Shear stress distribution

The distribution of shear stress across a section is frequently calculated by dividing the shear force at a point by the cross-sectional area of the member at that point. However, this method provides only a value of the average shear stress for the section. The true shear stress has different values at different locations in the section.

The beam shown in Fig. 6.102(a) has the constant cross-section illustrated in Fig. 6.102(b). Consider the planes CE and DF that are a distance δx apart, the neutral axis is at a depth y_1 from the top surface of the beam.

The bending moment and shear force at section CE are denoted by M and F, respectively. Similarly, at section DF the values are $(M+\delta M)$ and $(F+\delta F)$. Complementary to the vertical shear stress, there is a horizontal shear stress (σ) acting across the layer AB of width b and at a distance y_2 from the neutral axis. Let F' be the resultant force on plane AC due to the bending moment M, and $(F' + \delta F')$ the resultant force on DB due to $(M+\delta M)$.

Consider a thin strip above AB of thickness δy, distance y_3 from the neutral axis and of width b_3. Let σ' be the stress due to bending on this strip at section CE.

$$\text{Area of cross-section at } AC = \int_{y_1}^{y_2} b_3 \, \delta y$$

From the simple bending equation:

$$\sigma' = \frac{My}{I}$$

Therefore,
$$\frac{F'}{\int_{y_1}^{y_2} b_3 \, \delta y} = \frac{My_3}{I}$$

where I = second moment of area of the whole section about the neutral axis.

$$F' = \frac{M}{I} \int_{y_1}^{y_2} b_3 \, \delta y \, xy_3$$

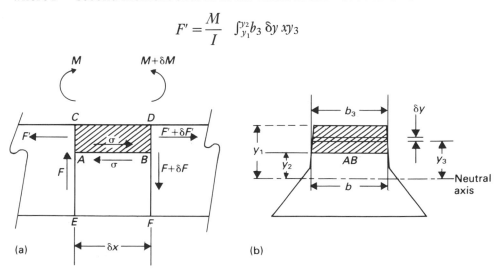

Figure 6.102 Distribution of shear stress in a beam. (a) Short length of a beam. (b) Cross section of the beam.

The term $\int_{y_1}^{y_2} b_3 \, \delta y \, xy_3$ is the sum of the first moment of area of all the strips above AB about the neutral axis. Therefore:

$$F' = \frac{M}{I} \, a\bar{y}$$

where a = area of cross-section above AB;

$\qquad \bar{y}$ = distance of centroid of area above AB from the neutral axis.

Similarly, considering section DF (assuming constant cross-section):

$$(F' + \delta F') = \frac{(M + \delta M)}{I} \, a\bar{y}$$

Subtracting these equations for each section:

$$\delta F' = \frac{\delta M}{I} \, a\bar{y}$$

The element ABDC must be in equilibrium:

$$F' + \sigma b S = F' + \delta F'$$
$$\delta F' = \sigma b \, \delta x$$

Therefore,
$$\sigma b \, \delta x = \frac{\delta M}{I} \, a\bar{y}$$

$$\sigma = \frac{\delta M}{\delta x} \frac{a\bar{y}}{Ib}$$

Since
$$\delta M / \delta x = F:$$

$$\sigma = \frac{Fa\bar{y}}{Ib}$$

6.3.14 Relationship between uniform load, shear force and bending moment

Let w = weight per unit length,

$\quad F$ = shear force,

$\quad M$ = bending moment.

Consider the cantilever beam shown in Fig. 6.103. Consider the element in equilibrium under the action of the forces exerted by the remainder of the beam. From a force balance, upward forces equal downward forces:

$$F + \mathrm{d}F = F + w \, \mathrm{d}x$$
$$\mathrm{d}F = w \, \mathrm{d}x$$

$$\frac{\mathrm{d}F}{\mathrm{d}x} = w$$

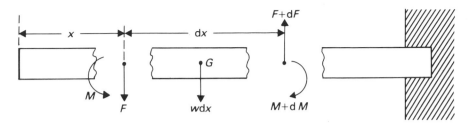

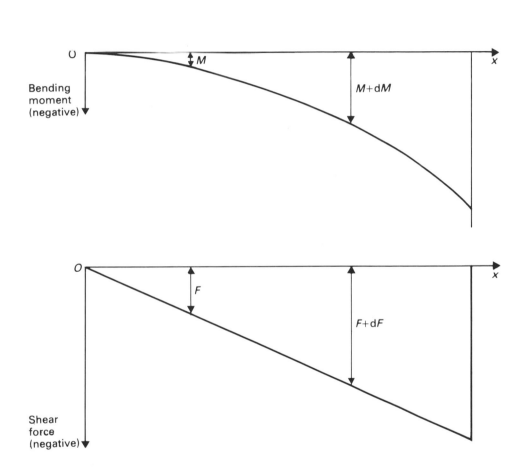

Figure 6.103   Bending moment and shear force in a uniformly loaded cantilever beam.

The slope of the shear force *v.* distance graph is constant for a constant applied load. Since this is negative, a downward acting load must be negative:

$$\frac{\mathrm{d}F}{\mathrm{d}x} = -w$$

Taking moments about point *G*:

anticlockwise moments = clockwise moments

$$M + F\,(\mathrm{d}x/2) + (F + \mathrm{d}F)(\mathrm{d}x/2) = M + \mathrm{d}M$$

Neglecting a product of two small quantities:

$$F\,\mathrm{d}x = \mathrm{d}M$$

$$F = \frac{\mathrm{d}M}{\mathrm{d}x}$$

This is the slope of the bending moment against distance graph, i.e. equal to F (negative sign). Then:

$$F = w\,\mathrm{d}x = \text{area of load-intensity diagram}$$
$$\text{and } M = F\,\mathrm{d}x = \text{area of shear-force diagram.}$$

(Note: If w is not constant, then the relationship between w and x must be known.)

6.3.15 Composite beams

A relatively weak beam such as timber can be stiffened as shown in Fig. 6.104, by clamping with steel plates (shown shaded). The beam is then said to be 'flitched'. When a load is applied both components bend to the same radius (R), where:

$$R = \frac{E_t I_t}{M_t} = \frac{E_s I_s}{M_s}$$

where t denotes timber and s denotes steel. Also:

$$M_t = \frac{\sigma_t I_t}{y_t} \text{ and } M_s = \frac{\sigma_s I_s}{y_s}$$

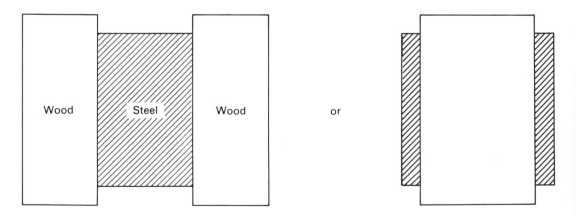

Figure 6.104 Composite beams.

The total allowable bending moment (M_T) for the composite beam is given by:

$$M_T = M_t + M_s$$

The structure should be symmetrical about the neutral axis and a vertical axis.

6.3.16 Reinforced concrete beams

Concrete is a material that is relatively weak when subjected to tensile loads. Steel reinforcing rods are embedded in the concrete where it is subjected to tensile strain, such as the upper layer of a cantilever beam or the lower layer of a simply supported beam. When designing reinforced beams the following simplifying assumptions are often made:

(a) there is perfect adhesion between the reinforcement and the surround-
     ing concrete;
(b) the tensile load is only carried by the reinforcement;
(c) the tensile stress in the reinforcement is uniformly distributed over its
     section;
(d) in the concrete, strain is proportional to stress;
(e) sections that are plane before bending are plane after bending.

Such beams are composed of two materials that possess different values of Young's modulus. Therefore, simple formulae (such as $\sigma = My/I$) cannot be used because the neutral axis does not pass through the centroid of the section. Consider a composite beam as shown in Fig. 6.105. Assuming that the strain at any point is proportional to the distance from the neutral axis, then:

$$\frac{\text{compressive strain in concrete at } A}{\text{tensile strain in steel}} = \frac{h}{d-h}$$

But $$\varepsilon_c = \frac{\sigma_c}{E_c} \text{ and } \varepsilon_s = \frac{\sigma_s}{E_s}$$

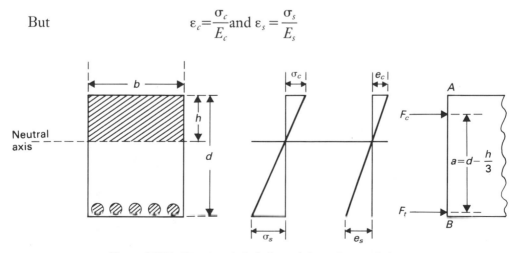

Figure 6.105 Stress and strain in a reinforced concrete beam.

Therefore

$$\frac{\varepsilon_c}{\varepsilon_s} = \frac{E_s \sigma_c}{E_c \sigma_s}$$

or

$$\frac{h}{d-h} = \frac{E_s \sigma_c}{E_c \sigma_s}$$

where (E_s/E_c) is known as the *steel–concrete modular ratio*. For steel, tensile force $F_t = \sigma_s A_s$
where A_s = cross-sectional area.

The average stress in the concrete above the neutral axis is $(\sigma_c/2)$, where σ_c is the maximum stress distributed over the area A_c $(=bh)$. Therefore in concrete, thrust $F_c = (\sigma_c/2)bh$. Since there is no resultant force normal to the section, the total compressive force in the concrete is equal and opposite to the total tensile force in the steel:

$$F_t = F_c \quad \text{or} \quad \sigma_s A_s = (\sigma_c/2)bh$$

For concrete, the stress diagram is triangular and the resultant force (F_c) acts at the centroid, i.e. $(2h/3)$ from the neutral axis. Similarly, the tensile force in steel acts at a distance $(d-h)$ from the neutral axis. These forces are equal, forming a couple of arm length a, where:

$$a = (2h/3) + (d - h)$$
$$= d - (h/3)$$

Therefore, the magnitude of the couple, i.e. the moment of resistance of the beam (M), is given by:

$$M = F_t\, a$$
$$= \sigma_s A_s\, [d - (h/3)]$$
$$\text{or } M = F_c\, a$$
$$= (\sigma_c/2)bh\, [d - (h/3)]$$

While the beam remains in equilibrium under the forces acting on it, the moment of resistance of the beam is equal and opposite to the bending moment acting on the beam.

6.3.17 Problems

(See the note regarding units in Section 6.1.23.)

P.43 A light beam of negligible mass is simply supported and carries concentrated loads of 4 kN, 20 kN and 6 kN at distances of 3 m, 5 m and 8 m respectively from one end. The beam is 8 m long. Calculate the bending moments at positions along the beam and draw the bending-moment diagram.

P.44 *ABCDE* is a beam which is supported at A and D. There are concentrated loads of 10 kN and 7 kN at B and E, respectively. The load due to the mass of the beam is 8 kN and may be assumed to act at the center of the beam (C).

The dimensions of the beam are $AB=4$ m, $BC=6$ m, $CD=3$ m and $DE=7$ m. Calculate the bending moments along the beam and draw the bending-moment diagram.

P.45 A cantilever beam of unsupported length 18 m carries concentrated loads of 3 kN and 5 kN at distances of 15 m and 8 m respectively from its supported end. The mass of the beam may be neglected. Draw to scale the bending-moment diagram.

P.46 A joist (AB) with a span of 25 m is supported at points 10 m from A and 8 m from B. The mass of the joist is equivalent to a load of 16 kN and acts at the mid-point of the beam. There are concentrated loads of 6 kN at A and 30 kN at 18 m from A. Determine the load at each support and draw to scale the bending-moment graph for this system.

P.47 The mass of a cantilever beam that is equivalent to a load of 80 kN is uniformly distributed over a span of 20 m. Draw to scale the bending-moment diagram.

P.48 The mass of a simply supported beam that is equivalent to a load of 15 kN/m is uniformly distributed over a span of 12 m. Draw to scale the bending-moment diagram. Draw the bending-moment diagrams for the following situations:

(a) the supports are situated at distances of 2 m from the ends of the beam;

(b) the supports are situated at a distance of 4 m from each end of the beam;

(c) one support is at the end of the beam and the other support at a distance of 4 m from the other end of the beam;

(d) the supports are situated at distances of 2 m and 4 m from the ends of the beam.

P.49 *ABCDE* is a beam supported at points B and D, 10 m apart. At points A, C and E there are concentrated loads of 1.5, 2.0 and 1.0 kN, respectively. Between B and D there is a uniform load of 0.5 kN/m. The remaining dimensions are $AB=2$ m, $BC=5$ m, $DE=3$ m. Draw to scale the bending-moment diagram.

P.50 The beam of cross-section 12 cm by 6 cm shown in Fig. 6.106 carries a load of 2 kN concentrated at the center of a 12 m span. Find the greatest bending stress in the beam when the longer side is (a) upright; (b) horizontal.

P.51 A horizontal beam is simply supported over a span of 20 m. It is 18 cm deep and the value of the second moment about the neutral axis is 1150 cm^4. Determine the value of the maximum stress given that the load per meter (including the load due to the mass of the beam) is 1.5 kN.

P.52 A beam has the cross-section shown in Fig. 6.107. Calculate the value of the second moment about the neutral axis, and the bending moment required to induce a maximum stress of 100 MN/m^2.

P.53 Find the depth of the neutral axis for the T-section shown in Fig. 6.108. Determine the maximum tensile and compressive stresses when a bending moment of 4.75 kN-m is applied, assuming the beam is supported as a cantilever.

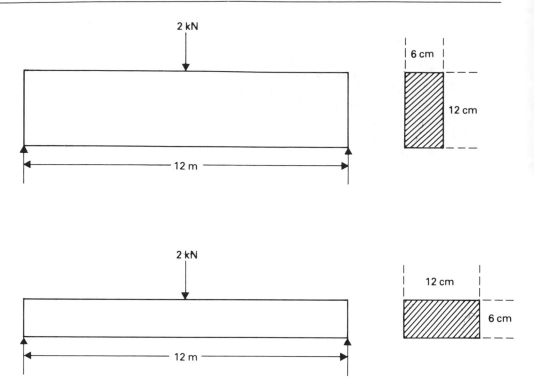

Figure 6.106 Refer to Problem P.50.

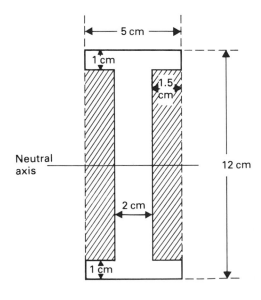

Figure 6.107 Refer to Problem P.52.

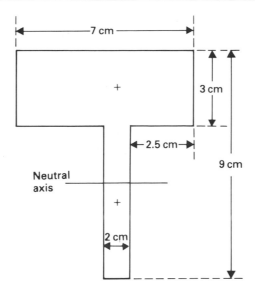

Figure 6.108   Refer to Problem P.53.

P.54 Find a suitable spacing for the supports of a 14 m long beam carrying a load of 15 N/m^2 as shown in Fig. 6.109, given that the maximum bending stress must not exceed 7 MN/m^2. The supports have a rectangular cross-section 12 cm high and 4.5 cm wide.

P.55 Find the maximum bending stress induced by its own weight in a 9 cm diameter shaft, 30 m long, when it is simply supported at the ends. The density of the material is 5500 kg/m^3.

P.56 Draw the shear force graph for a cantilever beam of length 10 m that has concentrated loads of 1.5, 4 and 6.5 kN at distances of 2, 5 and 10 m respectively from its support.

P.57 Calculate the values of the support loads and draw the shear force diagram for a 16 m long, simply supported beam, carrying loads of 6, 8 and 10 kN at distances of 2, 4 and 12 m, respectively from one end.

P.58 A beam of length 25 m having negligible mass is supported at one end, and at a distance of 5 m from the other end. The beam carries a load of 4 kN at

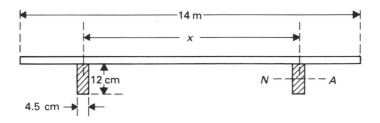

Figure 6.109   Refer to Problem P.54.

the unsupported end and loads of 2, 5 and 3 kN at distances of 8, 17 and 23 m respectively, also from the unsupported end. Draw to scale the shear force diagram for this system.

P.59 For a beam of rectangular cross-section, show that the maximum shearing stress is 1.5 times the mean shearing stress.

A beam of rectangular cross-section is simply supported over a span of 5 m, and is required to carry a uniformly distributed load of 9 kN/m on the entire span. Calculate suitable cross-sectional dimensions for the beam if the maximum allowable values for the stresses in the beam are 8 MN/m^2 in tension and 0.8 MN/m^2 in shear.

P.60 For the beam shown in Fig. 6.110, calculate the maximum horizontal shear stress when subjected to a vertical shear force of 150 kN/m. Draw the shear stress distribution diagram for the section.

P.61 The mass of a beam is equivalent to a load of 0.5 kN/m and is uniformly distributed over a span of 40 m. The beam is supported at distances of 6 m and 10 m from each end. Draw to scale the shear-force diagram and the bending-moment diagram.

P.62 *ABCDE* is a beam, length 40 m, which is supported at *A* and *D*. *AB* is 25 m and carries a uniformly distributed load of 1 kN/m. *BC* is 5 m and at *B*

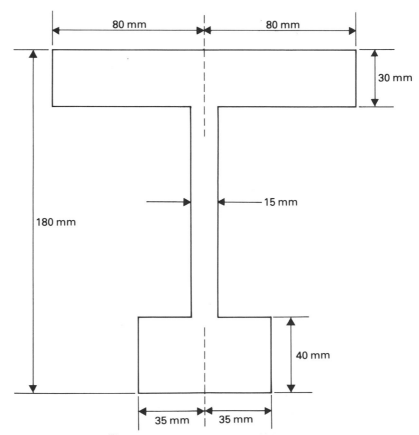

Figure 6.110   Refer to Problem P.60.

there is a concentrated load of 3 kN. Draw the shear-force diagram and the bending-moment diagram for this system.

P.63 A steel plate, 1.5 cm thick, is rigidly held between two timber beams, each 8 cm wide and 24 cm high, as shown in Fig. 6.111. The overall span is 20 m. The ratio of the Young's moduli for steel:timber is 20:1 and the

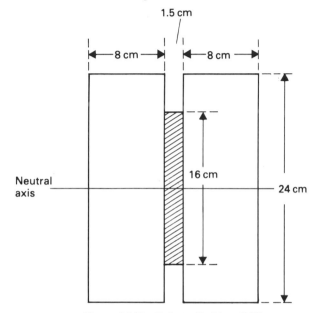

Figure 6.111   Refer to Problem P.63.

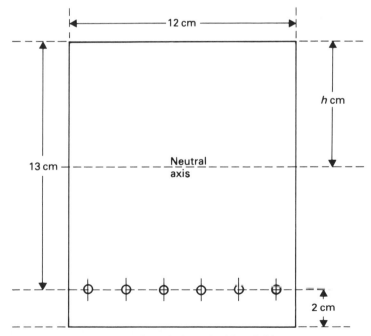

Figure 6.112   Refer to Problem P.64.

maximum stress in the timber is not to exceed 7 MN/m². Calculate:
(a) the corresponding stress in the steel;
(b) the permissible uniform load (per meter).

P.64 A concrete beam, 12 cm wide and 15 cm deep, is simply supported over a span of 180 cm. The beam is reinforced by six equally spaced steel rods of 0.75 cm diameter. The centers of the steel rods are 2 cm from the bottom of the beam. The maximum stress in the concrete is 600 kN/m² and the ratio of the Young's moduli for steel:concrete is 15:1. Determine
(a) the depth of the neutral axis;
(b) the moment of resistance of the beam;
(c) the stress in the steel;
(d) the total permissible uniform loading.
The system is shown in Fig. 6.112.

* * *

KEYWORDS FOR SECTION 6.3

stress	Poisson's ratio	torsional strain energy
strain	longitudinal strain	angle of twist
rigidity	lateral strain	coil spring
modulus of rigidity	axial strain	shaft coupling
modulus of elasticity	bulk (volumetric) strain	simple beam
mechanical vibration	strain energy	overhanging beam
oscillatory motion	resilience	cantilever beam
dynamic system	gradual loading	uniform load
spring	sudden loading	bending moment
damper	shock loading	negligible mass
excitation force	thermal stress	concentrated load
viscous damping	coefficient of linear	supports
linear damping	expansion	points of contraflexure
forced vibration	thin-walled cylinder	first moment of area
free vibration	hoop stress	moment of resistance
lumped-parameter	hoop tension	neutral axis
system	longitudinal stress	second moment of a
natural frequency	torsion	section
undamped free vibration	torque	shear force in a beam
simple harmonic	couple	composite beam
amplitude	shaft	clamping
underdamped	pulley	reinforcing rods
overdamped	moment	steel–concrete modular
accelerometer	polar moment of inertia	ratio
force transducer	torsion equation	

* * *

BIBLIOGRAPHY

Bassin, M.G., Brodsky, S.M. and Wolkoff, H., *Statics and Strength of Materials*, 2nd Ed., McGraw-Hill Book Co., Inc., New York (1969).

Baxter-Brown, J. McD., *Introduction to Solid Mechanics*, John Wiley and Sons, Inc., New York (1973).

Bruch, C.D., *Strength of Materials for Technology*, John Wiley and Sons, Inc., New York (1978).

Buckley, D.H., *Surface Effects in Adhesion, Friction, Wear and Lubrication*, Elsevier Scientific Publishing Co., Amsterdam, The Netherlands (1981).

Cordon, W.A., *Properties, Evaluation and Control of Engineering Materials*, McGraw-Hill Book Co., Inc., New York (1979).

Davis, H.E., Troxell, G.E. and Hauck, G.F.W., *The Testing of Engineering Materials*, 4th Ed., McGraw-Hill Book Co., Inc., New York (1982).

Den Hartog, J.P., *Strength of Materials*, Dover Publications, Inc., New York (1961).

Dieter, G.E., *Mechanical Metallurgy*, 2nd Ed., McGraw-Hill Book Co., Inc., New York (1976).

Flinn, R.A. and Trojan, P.K., *Engineering Materials and their Applications*, 2nd Ed., Houghton Mifflin Co., Boston, Massachussets (1981).

Griffith, A.A., *Phil. Trans. Royal Soc. (London)*, **221**(A), 163–198 (Oct.1920).

Halling, J.(Ed.), *Principles of Tribology*, The Macmillan Press Ltd, London (1975).

Harris, W.J., *The Significance of Fatigue*, Engineering Design Guide No. 14, Oxford University Press, Oxford (1976).

Harris, W.J. and Syers, G., *Fatigue Alleviation*, Engineering Design Guide No. 32, Oxford University Press, Oxford (1979).

Higdon, A., Ohlsen, E.H., Stiles, W.B., Weese, J.A. and Riley, W.F., *Mechanics of Materials*, 3rd Ed., John Wiley and Sons, Inc., New York (1976).

Higdon, A., Stiles, W.B., Davis, A.W. and Evces, C.R., *Engineering Mechanics: Statics and Dynamics*, Prentice-Hall, Inc., Englewood Cliffs, New Jersey (1976).

Kempster, M.H.A., *Materials for Engineers*, Hodder and Stoughton Ltd, Sevenoaks, Kent (1975).

Kragelsky, I.V., Dobychin, M.N. and Kombalov, V.S., *Friction and Wear: Calculation Methods*, Pergamon Press Ltd, Oxford (1982).

Lansdown, A.R., *Lubrication: A Practical Guide to Lubricant Selection*, Pergamon Press Ltd, Oxford (1982).

Lewis, G., *Properties of Engineering Materials. Theory, Worked Examples and Problems*, Macmillan Publishers Ltd, London (1981).

Meriam, J.L., *Engineering Mechanics. Statics and Dynamics*, John Wiley and Sons, Inc., New York (1978).

Mott, R.L., *Applied Strength of Materials*, Prentice-Hall, Inc., Englewood Cliffs, New Jersey (1978).

Neely, J.E., *Practical Metallurgy and Materials of Industry*, John Wiley and Sons, Inc., New York (1979).

Peterson, M.B. and Winer, W.O., *Wear Control Handbook*, American Society of Mechanical Engineers (ASME), New York (1980).

Pollack, H.W., *Materials Science and Metallurgy*, 3rd Ed., Reston Publishing Co., Inc., Reston, Virginia (1981).

Pugh, B., *Friction and Wear*, Butterworth and Co. (Publishers) Ltd, London (1973).

Redford, G.D., *Mechanical Engineering Design*, 2nd Ed., Macmillan Publishers Ltd., London (1973).

Sarkar, A.D., *Wear of Metals*, Pergamon Press Ltd, Oxford (1976).

7

Joining Processes

7.1 INTRODUCTION

The three main processes used to join engineering materials can be classified as:

Metallurgical processes
Mechanical methods
Adhesives

However, the choice of a particular joining method depends upon consideration of several factors. These should include:

(a) composition and mechanical, chemical and physical properties of the parent materials;
(b) whether permanent or temporary joints;
(c) whether heat can be applied;
(d) required strength;
(e) size and shape of components;
(f) costs involved;
(g) corrosion resistance;
(h) sealing efficiency of the joint;
(i) surface finish.

7.2 METALLURGICAL PROCESSES

These processes require heat in order to join the components. It is important that the engineer has an understanding of the metallurgical changes that can occur, and the effects upon the properties and structure of the materials being joined. A manufacturing operation involves the design, production and inspection of the component. The design stage requires detailed knowledge of the alternative methods that are available, and their advantages and disadvantages. Also, some knowledge of the suitable joints that can be employed and the relevant codes of practice. The production of a component requires comparisons to be made of the costs involved in different processes and any necessary safety measures. Finally, the component should be inspected in order to identify any defects that are present and to deduce their cause.

Before considering the different processes that are available, it will be useful to identify the various welds and joints in common use.

7.2.1 Welds and joints

The terms weld and joint are often confused. The weld (or welded joint) is taken to mean the weld inclusive of adjacent areas of the workpiece. There are two basic (or common) types of weld as shown in Fig. 7.1.

It is the configuration of the connected parts that determines the type of joint (or unwelded joint). The joint is the space between the surfaces to be

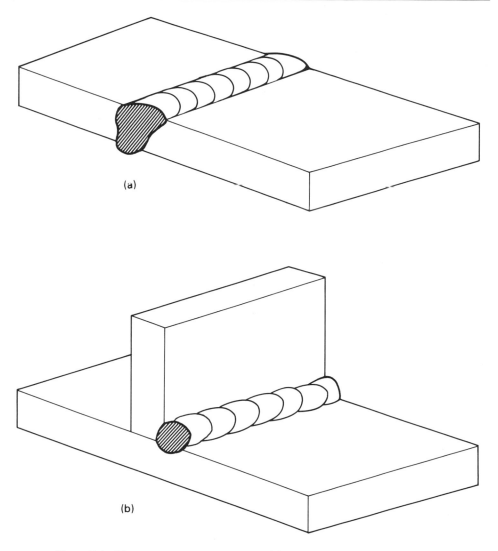

Figure 7.1 The two most common welds. (a) The butt weld. (b) The fillet weld.

joined by welding. Common types of joint are shown in Fig. 7.2.

It is possible to make one particular type of joint using different types of weld, e.g. the corner joint shown in Fig. 7.3.

* * *

Self-assessment exercises

It be will necessary to refer to other material in order to answer these exercises, books by Cary (1979), Giachino *et al.* (1973), Gray and Spence (1982), Hicks (1979), and Stewart (1981) would provide a useful starting point, although there are many other excellent welding handbooks available.

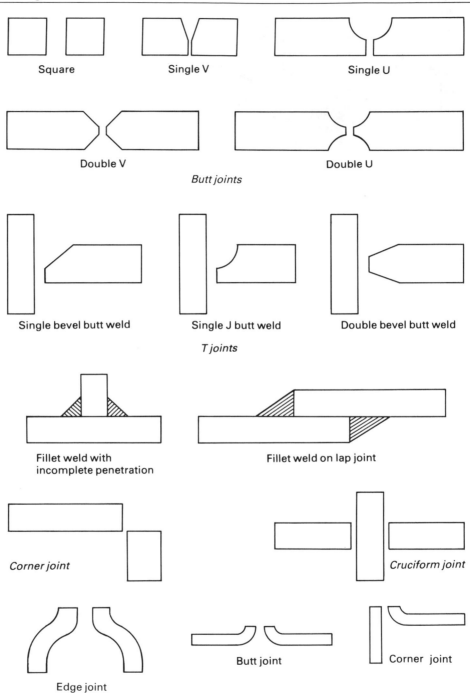

Figure 7.2 Common types of joint.

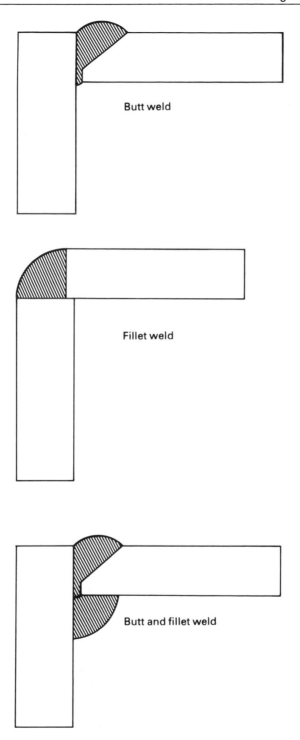

Figure 7.3   Corner joint formed by different welds.

1 Explain the meaning of, and distinguish between, a weld and a joint.

2 In which welding situations is edge preparation required, and why? Describe on diagrams various types of edge preparation.

3 Describe, using diagrams, the weld and joint in the following situations:
   (a) two strips of metal joined to form a flange;
   (b) the edges of a metal sheet joined to form a seamed tube;
   (c) two joined overlapping plates;
   (d) three metal sheets joined along their length to form a hollow closed channel;
   (e) tubes secured into the holes of a perforated plate;
   (f) a solid rod fixed to the center of a circular metal plate.

* * *

7.2.2 Definition of terms

Weld	A union between two pieces of metal at faces made plastic or liquid, by heat or pressure or both. This is more correctly the welded joint, of which there are various types. The weld itself is the joining metal and there are only two types of weld, the butt weld and the fillet weld.
Joint	This would be correctly termed the unwelded joint and is the space between the surfaces to be joined by welding. There are various types, e.g. butt joint, T joint, lap joint, etc.
Parent metal	Metal to be joined, surfaced, hardened or cut.
Filler metal	Metal that is added to make the union.
Soldering	Joining of components using a molten filler metal that has a melting point below 500 °C (932 °F). It is performed at temperatures below the melting point of the parent metal.
Brazing	Joining of components using a molten filler metal that has a melting point above 500 °C (932 °F), but performed at temperatures below the melting point of the parent metal.
Braze welding	Brazing used for single and double-V butt joints or fillet joints.
Welding	Joining metals that are molten or plastic by using heat or pressure, with or without a filler metal.
Surfacing	Deposition of a surface layer of metal.
Solder or brazing alloy (spelter)	Metallic bonding agent used in soldering or brazing.
Flux	Powder, paste or liquid applied to the surfaces to be joined to remove surface oxides by chemical reaction and to act as a wetting agent; it also prevents further oxidation during heating.
Slag	Product of flux and surface oxides.
Arc	Flow of electrons between two electrodes, through either an ionized air gap or a conducting flux (submerged arc welding).
Penetration	Depth of the fusion weld pool below the surface of the parent metal.
MMA	Manual-metal-arc welding.

MIG	Metal-inert gas-arc welding.
MAG	Metal-active gas-arc welding.
TIG	Tungsten (electrode) – inert gas-arc welding.
Arc time factor	Relationship between length of time the arc burns and the total time for the welding operation.
Deposition rate	Weight of metal deposited per unit time at 100% arc time factor.
Effective value	Proportion of filler metal transformed into weld metal.
Change value	Number of electrodes used per unit weight of weld metal (for MMA).
Metal recovery	Weight relationship between weld metal and core wire used (for MMA).

7.2.3 British Standards defining welding terms, and for approval testing

BS499: Welding terms and symbols.
Part 1: Welding, brazing and thermal cutting glossary.

Seven sections (terms common to more than one section) relating to pressure welding, fusion welding, brazing, testing, weld imperfections, thermal cutting.

Part 2: Specification for symbols for welding.

Part 2C: Chart of BS welding symbols (based on Part 2).

Part 3: Terminology of and abbreviations for fusion weld imperfections as revealed by radiography.

BS 4870: Approval testing of welding procedures.
Part 1: Fusion welding of steels.

Items to be recorded in welding procedure tests for various processes, changes affecting approval; types of test weld, test pieces; nondestructive and destructive testing; statement of results, recommended procedure, test record form.

BS 4871: Approval testing of welders working to approved welding procedures.
Part 1: Fusion welding of steel.

Information to the welder, other items similar to BS 4870: Part 1.

BS 4872: Approval testing of welders when welding procedure approval is not required.
Part 1: Fusion welding of steel.

Approval testing of welders for manual and semi-automatic fusion welding of ferritic or austenitic stainless steels. Choice of 9 test welds covering sheet, plate and pipe.

7.2.4 The principles of soldering and brazing

The joining processes that require heat are soldering, brazing and welding. The main differences between these processes is the temperature at which the operation takes place and the joining mechanism. Welding is the joining of metal that has become plastic or liquid due to heating or applied pressure. For fusion welding the temperature employed is the melting point of the metal, or slightly above. For pressure welding much lower temperatures are used. Fusion welding employs a filler metal with a melting point that is approximately the same as that of the parent metal, except for thin sheet in the case of gas welding.

Soldering and brazing both occur at temperatures below the melting point of the parent metal. Joining is achieved by allowing molten filler metal (lower melting point than the parent metal) to be drawn in (by capillary action), and retained in the space between the closely adjacent surfaces of the parts to be joined. The difference between soldering and brazing is usually said to be that for soldering a filler metal (or solder) with a melting temperature below 500°C (932°F) is used, whereas for brazing the filler metal melts above 500°C (932°F). To be more accurate, soft soldering occurs in the melting range of 180°C to 250°C (356°F to 482°F), hard (or silver) soldering between 600°C to 800°C (1112°F to 1472°F), and brazing usually requires temperatures of between 800°C and 900°C (1472°F and 1652°F). Soldering and brazing are often used for joining nonferrous metals, especially copper and copper alloys.

With soldering and brazing the joint is heated to the melting temperature of the filler metal. The parent metal remains as a solid phase throughout the joining process. The molten filler metal then wets the surfaces to be joined, and becomes alloyed with the parent metals within a narrow region adjacent to the respective surfaces (the fusion zone). The surfaces are joined by a metallic bond and as diffusion occurs between the filler metal and the parent metal, the fusion zone has a higher melting point than the filler metal. The strength of the brazed or soldered joint depends upon the composition and bonding within the fusion zone. Maximum bonding strength is achieved if thin layers of solidified filler metal occur within the joint. The gap between components should not exceed 0.5 mm (0.02 in.) and usually gaps of 0.05–0.25 mm (0.002–0.01 in.) are used; a small gap permits the filler metal to penetrate by capillary action. In gaps of identical width, filler metals with a lower melting point usually penetrate further than those with a higher melting point. Using a solder of 56% tin–44% lead, the maximum joint strength will be obtained for the conditions shown in Table 7.1.

7.2.5 Soldering

Sources of heat

  (a) Electrically heated soldering iron – mainly for soft solders.
  (b) Torch flame.
  (c) Dip soldering – parts to be joined are submerged in molten filler metal (they may first be dipped in molten flux).
  (d) Furnace heating – for parent metal and pre-placed filler metal.
  (e) Induction coil – using high-frequency current around the workpiece.
  (f) Resistance soldering – electric current passed through the workpiece; heating due to electrical resistance.

Fluxes

Before heating, the surfaces to be joined must first be mechanically cleaned, e.g. using a wire brush. Although they appear clean, the surfaces must then be chemically cleaned with a flux to remove any remaining thin film of grease and oxides. The flux is a paste or liquid which by means of chemical reaction reduces the metal oxides present on the surfaces, and also prevents further oxidation during the heating cycle. The flux will float oxides and other impurities to the surface where they can be easily removed, and it reduces the surface tension of the solder. Fluxes in common use are non-organic, weak acid solutions and sometimes resins dissolved in organic solvents. The acid and salt fluxes are corrosive, and all residues should be removed after soldering. However, resin fluxes do not corrode the parent metal. Highly active fluxes, such as zinc chloride, are used to obtain clean surfaces with metals such as stainless steel and aluminum that have a stable oxide film. Borax is a flux used for hard soldering and brazing.

Table 7.1 Conditions required to produce maximum joint strength using 56% tin – 44% lead solder.

Joint thickness mm (in.)	Soldering temperature °C(°F)
0.025 (0.001)	400 (752)
0.10 (0.004)	270 (518)
0.175 (0.007)	230 (446)
0.25 (0.010)	220 (428)

Filler metal (solder)

Copper and its alloys are the most common metals to be joined by soldering. Lead–tin alloys are the most common filler metals used with these parent metals. For aluminum joining, filler metals consisting of alloys of either zinc and cadmium or zinc and aluminum are used. The most common alloys used as filler metals, and their typical applications, are detailed in Table 7.2. Reference to the thermal equilibrium diagram for tin and lead may show differences in the characteristics of these alloys. Up to 3% antimony can be added to some solders, mainly to reduce the cost. This is not recommended for zinc or galvanized work.

Hard solders

For high-strength joints, higher temperatures must be used. For brazing, brass is often used as the filler metal. This requires a high melting temperature and can be disadvantageous, particularly with copper and its alloys. When silver is added to the brass, a lower-temperature silver brazing alloy is produced. This alloy is known as silver solder, or hard solder, and is an alloy of silver, copper and zinc. Hard soldering is an intermediate stage between low-temperature soft soldering and high-temperature brazing.

Hard solders may be divided into two groups. The first group consists of low-temperature silver-brazing alloys composed of silver, copper, zinc and

Table 7.2 Soft solders.

Tin (%)	Lead (%)	Melting characteristics	Typical applications
60 (60–65)	40 (40–35)	Lowest melting point of the series	Components liable to damage by heat or requiring a free-running solder, e.g. electrical, radio and instrument assemblies.
50 (44–50)	50 (56–50)	Moderately low melting point; short pasty range	Bit soldering and general machine soldering.
40 (30–40)	60 (60–70)	Moderate pasty range	Torch soldering of high-speed, body-forming machines.
30 (29–35)	70 (65–71)	Long pasty range	Plumber's solder, for wiping of cable and lead pipe joints. Dipping baths.

Table 7.3 Hard solders.

Grade	Silver (%)	Copper (%)	Zinc (%)	Approximate melting range °C (°F) (solidus–liquidus)
Hard	80	20	—	778–825 (1432–1517)
Medium	75	20	5	750–775 (1382–1427)
Easy	50	30	20	690–740 (1274–1364)
BS206: Grade A	61	29	10	690–735 (1274–1355)
Grade B	43	37	20	700–775 (1292–1427)

cadmium (up to 50% silver), that have a melting point range of 600–800°C (1112–1472°F). These alloys are used for most general engineering applications where strength combined with a low melting point are required. The second group consists of silver solders that are predominantly silver and are mainly used for silversmithing work. Hard solders are available for a wide range of applications and in a variety of forms, e.g. rod, strip, wire, washer, disk, powder, powder and flux, and paint. The compositions and melting points of some common hard solders are given in Table 7.3.

The joint gaps for hard soldering should be between 0.05 mm and 0.15 mm (0.002 in. and 0.006 in.); if possible the joints should be self-locating. For build-up work requiring more than one join, then high melting point joints are usually made first in order to avoid melting the previous joints. All hard solders have a similar appearance and it is usual to mark them along the length for distinguishing purposes. Both copper and silver have high coefficients of expansion and this is important when joining or subsequently reheating the workpiece. 'Easy' silver solder (see Table 7.3) should not be used on a seamed joint because, when annealed, the zinc volatilizes, leaving a porous residue. This is known as *fretting* and can be partially overcome by thinly painting with borax and water.

7.2.6 Brazing

This process produces a stronger joint than is obtained by soldering. The brazing alloy (or spelter) has a higher melting point, usually 800–900°C (1472–1652°F), but it is below the melting point of the parent metal. Typical brazing alloys and their melting points are listed in Table 7.4.

Table 7.4   Brass alloys for brazing.

Metal to be brazed	Copper (%)	Zinc (%)	Approximate melting points °C (°F)
Brass	40	60	840 (1544)
Copper	50	50	880 (1616)
Iron and steel	60	40	890 (1634)

Small amounts of tin and silicon are often added to brass alloys (see Table 7.4); these are then used for joining mild steels. Alloys of silver–copper–zinc have a low melting point and low viscosity and can be used with most parent metals, except aluminum and the light metals. Alloys of copper and phosphorus (and sometimes silver) are used for brazing copper or copper alloys; the phosphorus reduces any copper oxides and a separate flux is not required. These alloys are not suitable for use with steel or nickel because brittle phosphides are formed. Alloys of aluminum–silicon and aluminum–silicon–copper are used for brazing aluminum alloys.

Because of the higher temperature required for brazing, electrically heated irons are not normally used, but otherwise the same sources of heat are used as for soldering. Fluxes are usually compounds of various metal salts and are applied as viscous liquids, pastes or powders; borax is in common use.

Braze welding, or bronze welding, is a similar process, relying on the adhesive strength of the fusion zone. The difference is in the technique and application, which is illustrated in Fig. 7.4. For brazing the filler metal is applied by capillary action, whereas for braze welding a fillet of deposited metal is built up to facilitate the joint.

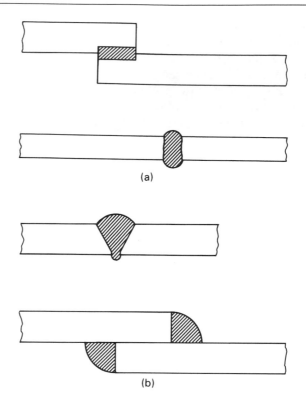

Figure 7.4   Joining by (a) brazing and (b) braze welding.

*   *   *

Self-assessment exercises

1 Describe the main differences between soldering and brazing.

2 Describe practical applications of soldering and brazing.

3 What is a flux, and why is it used?

4 What fluxes and filler rods are used for soldering and brazing?

5 Explain why the joint clearance is important for brazing.

6 What are the relative strengths of the joints obtained by soldering, brazing and welding?

7 What are the relative heating costs of soldering, brazing and welding?

8 List applications where soldering or brazing are not acceptable.

*   *   *

7.2.7 The metallurgy of welding

Welding involves melting areas of the parent metal and (usually) a filler metal, bringing the molten metal into intimate contact and then allowing joining to occur by solidification. Soldering and brazing are lower-temperature joining processes that do not require the parent metal to be molten. The majority of welding is made in mild steel or low-carbon steels, although increasing amounts of alloy steels and nonferrous metals are being used in welded constructions. The properties of the weld metal and the adjacent parent metal depend upon the thermal conditions and the alloying elements present. Successful joining requires knowledge and understanding of the metallurgical changes that occur. Most of the discussion presented in this section will be concerned with low-carbon steel.

Before studying this material it is essential that the work covered in Chapter 2 has been followed. In particular, the regions present on the iron–carbon phase diagram (Section 2.4), time–temperature transformation diagrams (Section 2.5), the effects of alloying additions (Section 2.8) and heat treatment processes and effects (Sections 2.18 and 2.19).

During welding there is a steep temperature gradient from the molten weld pool (above 1500° C; 2732° F) to the cold section of the parent metal. The region adjacent to the weld pool is known as the *heat-affected zone* (HAZ). If cross-sections are examined from the weld pool to the parent metal, then several regions can be observed. For a low-carbon steel containing approximately 0.3% carbon, typical regions and their grain size are shown in Fig. 7.5. Each of these regions will now be discussed.

The *weld pool* consists of molten parent metal mixed with molten filler metal. As this region is cooled, or at a short distance away from the pool, the metal solidifies and forms large columnar austenitic crystals. This is the *weld deposit*. No structural change occurs until the temperature falls to about 800° C (1472° F), i.e. the upper critical point. Ferrite then precipitates out of the austenite, at the grain boundaries, as long continuous plates. A coarse structure is obtained with low impact strength. Grain refiners may be added to the filler

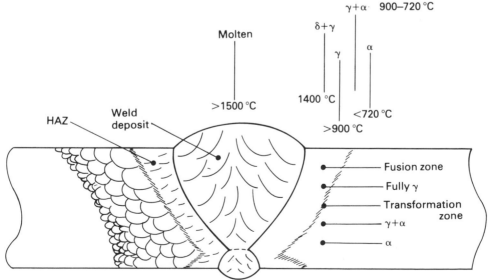

Figure 7.5   Regions within a weld and the heat affected zone (HAZ).

metal to act as nuclei, and also to promote a finer distribution of impurities throughout the grain boundaries. This increases the ductility and reduces the possibility of weld cracking.

The region where the weld metal has just solidified is known as the *fusion zone*. If the HAZ is very short, it may consist of columnar grains on the weld pool side and normalized pearlite and ferrite grains on the parent metal side (or at the outer surfaces). When the structure is fully austenitic, the carbon is present in an fcc lattice and an intermediate grain size is obtained. Recrystallization occurs above the upper critical temperature, i.e. before the transformation of austenite to ferrite occurs.

Below the lower critical temperature (the transformation zone), a mixture of ferrite and pearlite exists. Some spheroidizing and softening may take place. If the outer regions of the metal are heated to temperatures below the critical temperature, then this will have little effect upon the microstructure unless the metal was previously cold worked. In that case the cold-worked metal may undergo recrystallization in the HAZ. After welding, the properties are similar to those possessed by a hot-rolled steel. It is possible for two refined regions to exist in the HAZ, due to phase transformation and as a result of recrystallization.

During welding, temperature gradients exist in all three dimensions. Along the length of the weld, in the section that has just been welded, it is cooling down. The weld being formed is molten, and the section to be welded is comparatively cold (or receiving some preheating). There is also a temperature gradient through the weld itself, the top surface being hotter than the underside if welding is performed from one side only. A temperature profile also exists along the parent metal normal to the weld. The structures obtained depend upon the welding method used and the rate of cooling (see Section 2.5.1). Since the microstructure determines the mechanical properties of the finished component, e.g. hardness, impact strength, ductility, etc., the desired properties should be decided and then the appropriate operating conditions can be specified.

The changes that occur in the microstructure during welding depend upon:

(a) the composition of the parent metal;
(b) the type and composition of the filler metal, or the electrode coating for arc welding;
(c) the process environment;
(d) the welding process used.

For arc welding, the temperature of the molten metal is high and the metal at the edges of the weld pool cools quickly, forming small chill crystals. Long columnar crystals are formed as the molten metal solidifies towards the (hotter) center of the weld pool. The high temperatures mean that the crystals have sufficient time to grow. In the HAZ above the recrystallization temperature, grain growth occurs.

If welding is performed as a single-pass operation, then the weld deposit contains coarse grains in the columnar structure. If subsequent welding runs are required, then the structure obtained depends upon the temperature of the previous weld when the next pass is performed. For mild steel, the maximum grain refinement caused by subsequent welding is obtained if the metal is cooled below the critical temperature, but still well above room temperature (typically 550°C; 1022°F). If the first weld is cooled to room temperature before the next pass, then less grain refinement occurs. If the weld is deposited onto metal above the critical temperature, then no grain refinement can take place. Subsequent weld runs form columnar crystals as they cool. Grain growth occurs in the HAZ where the temperature is above the recrystallization temperature.

Compared to gas welding, the HAZ for arc welding is narrower. Although the arc temperatures are higher, the heating is more localized and the temperature rises more quickly. Grain growth is less pronounced for arc welding.

7.2.8 Welding alloy steels

Alloy steels have many applications because of the range of properties that can be achieved, e.g. increased strength, hardenability, corrosion resistance, etc. However, these steels present particular problems if welding is required. For example, austenitic alloy steels undergo significant changes in their structure when subjected to the heat required for welding. It is the austenitic nature of the steel that is responsible for the characteristic properties; therefore some form of post-welding heat treatment is usually required.

The range of alloying elements present in an alloy steel means that it is difficult to obtain a weld with the same composition (and properties) as the parent metal. Not only can the quantities and distribution of alloying elements vary within the weld, but the structures produced by heating these alloys may also differ considerably from those present in the parent metal. Therefore the welding of alloy steels must be considered on an individual basis (or for a particular type), taking into account the properties and structure required and the in-service conditions. The selection of a suitable electrode or filler rod is particularly important.

Alloy steels are sometimes easier to weld than plain carbon steels, because their carbon content can be lower to produce comparable properties. A low carbon content reduces the tendency for brittleness in the weld and for underbead cracking (see Section 7.21). However, the tendency for cracking increases as the quantity of alloying elements is increased, and low-alloy steels are easier to weld (successfully) than the high-alloy steels possessing special properties.

Several formulas have been proposed for calculating the *carbon equivalent* (CE%) of various alloying additions, i.e. the carbon steel equivalent to an alloy steel. One such formula is given by:

$$CE\% = C\% + \frac{Mn\%}{6} + \frac{Cr\% + Mo\% + V\%}{5} + \frac{Ni\% + Cu\%}{15}$$

For low-alloy steels the CE is less than 1.0%, and often less than 0.7%. The CE of a high-alloy special steel, such as 18–8 stainless steel, is obviously much higher. An alternative formula sometimes used is:

$$CE\% = C\% + \frac{Mn\%}{4} + \frac{Ni\%}{20} + \frac{Cr\%}{10} \pm \frac{Mo\%}{50} \pm \frac{V\%}{10} + \frac{Cu\%}{40}$$

The terms for molybdenum and vanadium can be added or subtracted, depending upon the form in which these elements are present. If present in solid solution, they aid hardenability and should be added. When present as complex carbides, the hardenability is unaffected and they are taken as negative values. CE values are useful for selecting appropriate welding conditions; detailed precautions corresponding to particular values of the CE are available in welding handbooks, e.g. if the CE exceeds 0.55% then preheating and approved low hydrogen electrodes are required. However, the CE value gives no indication of, and does not take into account, prior heat treatment or the grain size.

7.2.9 Weld absorption

The structure of a weld depends upon any elements absorbed during welding. The main elements of concern are oxygen, nitrogen, hydrogen and carbon; these will be discussed briefly.

Oxygen may be absorbed from the air, from excess oxygen in gas welding (Section 7.4) or from oxides present on the parent metal or the filler rod. If oxygen is absorbed, then oxides of iron, silicon, etc., form in the weld; these cause grain growth and hence lower the tensile strength, ductility and corrosion resistance. If oxides are present in the grain boundaries, the impact strength and fatigue resistance are reduced. Iron oxide can also react with carbon in the weld to form carbon monoxide, giving rise to blow holes (Section 7.21). The problem may be alleviated by introducing deoxidizing agents that either prevent iron oxide formation (by reacting to form other oxides) or by removing the iron oxide as a slag.

Nitrogen may be present in the weld mainly from arc welding (only a small amount due to gas welding). The nitrogen may occur as fine needle-shaped crystals of iron nitride (Fe_4N), the effect is to make the weld brittle and reduce the ductility. Nitrogen may also be present in solution in the iron, and forms Fe_4N only if heated above 850°C (1562°F). Although nitrogen does not form blowholes it can be trapped inside them.

Hydrogen is present in many flux coatings and can also be absorbed from any associated moisture. It is mainly a problem with arc welding of mild (ferritic) steel, occurring in the HAZ as transgranular (cold) cracks. These cracks appear at low temperatures (less than 150°C; 302°F), several hours after welding. The hydrogen diffuses from the weld pool into martensitic structures, causing embrittlement. Local strains initiate the cracks, thus reducing the tensile strength of the weld. This type of defect occurs only in martensitic structures and can be controlled by reducing either the hydrogen available or the stress levels.

Carbon can be absorbed by the weld either from the filler rod, or by diffusion from the parent metal in medium- and high-carbon steels. The carbon absorbed directly can form a porous deposit, if it is not oxidized to carbon monoxide. The carbon that diffuses forms carbides where the weld metal and the parent metal are joined, rapid cooling produces hardness and brittleness.

7.2.10 Effect of heat on mechanical properties

The mechanical properties are defined in Section 6.2, and the more common methods of mechanical testing are described in Sections 6.2.1–6.2.7. The general effects on the mechanical properties of metals caused by heating will be mentioned briefly.

The tensile strength usually decreases as the temperature rises; there is an increase in elongation and the proportionality limit is reduced. Internal stresses can be relieved by heating, so that at the lower limit of proportionality the stresses cause plastic deformation and are relieved. However distortion may also occur.

The hardness of a metal decreases as the temperature rises, increasing the carbon content of a steel, or cold working can be used to raise the hardness. Most metals, e.g. carbon steel, become less brittle when heated, although there are exceptions such as copper. Metals usually become less ductile when shaped, e.g. by drawing (Section 8.2.9). The brittleness created can be reduced by heat treatment such as annealing (Section 2.19.1). Plasticity generally increases as heat is applied, although there are exceptions, e.g. wrought iron.

Metals such as iron and steel become more malleable as the temperature is increased. Other metals such as copper are not malleable near the melting point, and zinc is malleable only between $140°C$ and $160°C$ ($284°F$ and $320°F$). The effects of temperature on the creep and fatigue properties are discussed in Sections 6.2.5 and 6.2.6 respectively.

7.2.11 Welding

The welding process has previously been defined as the union of metallic faces that have been made plastic or liquid due to heat or pressure. From this definition there arise two basic categories of welding methods: fusion welding and pressure welding. Within each category different methods can be used. These are listed below:

Fusion welding (Sections 7.3–7.9)
   7.3 Arc welding – including metal-arc, gas-arc, gas-metal-arc welding, carbon-arc welding and stud welding.
   7.4 Gas welding.
   7.5 Electro-slag welding.
   7.6 Consumable nozzle process.
   7.7 Electron-beam welding.
   7.8 Laser welding.
   7.9 Thermit welding.
Pressure welding (Sections 7.12–7.20)
   7.12 Forge welding.

Descriptions of the most common welding processes are included here, and a summary of the relevant features of the less common methods.

FUSION WELDING PROCESSES (Sections 7.3–7.9)

7.3 ARC WELDING

A large number of fusion welding methods can be classified as arc welding, in most cases the heat is provided by creating an electric arc between an electrode and the workpiece.

7.3.1 Introduction

An electric current is a flow of electrons, which is prevented if an air gap appears in an electric circuit. If the air gap is ionized, then electrons can travel across the air gap, forming an electric arc, and the current is re-established. For safety reasons, the maximum open-circuit voltages for single-operator arc welding are 120 V for d.c. and 80 V for a.c. These voltages are too small to initiate an electric arc, which requires approximately 5000 V/mm (125 kV/in.) gap. The arc is ignited by short-circuiting the electrode on the workpiece. This occurs at small points of contact, the current density increases and the energy developed causes melting at the points of contact. Following this stage, short-circuiting ceases and the voltage increases. Electrons flow from the cathode (electrode) to the anode (workpiece) and the air gap is ionized. Short-circuiting can be avoided by using a high-frequency voltage unit that produces sparks which ionize the air gap (between 4 mm and 7 mm; 0.16in. and 0.28in.). This process extends the life of the electrode. It is sometimes necessary to avoid the magnetic effects caused by having two conductors parallel to each other, because the magnetic field around the workpiece can push the arc away from the earth (arc blow). This problem occurs only when using a d.c. power supply and particularly so when welding metals that are good conductors. The problem can sometimes be overcome by inclining the electrode at an angle to the workpiece, or by using carefully controlled magnetic fields called arc oscillators to influence the direction of the arc.

7.3.2 Manual-metal-arc welding (MMA)

The most widely used fusion welding method is probably the manual application of the metal-arc process using coated electrodes. This will be abbreviated to MMA. A schematic diagram for the process is shown in Fig. 7.6. An arc is

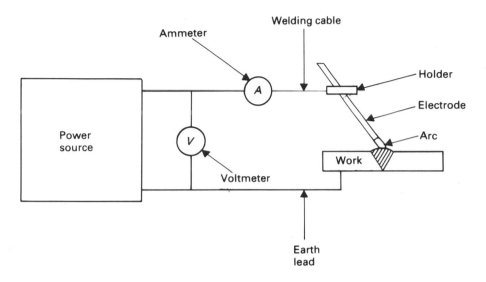

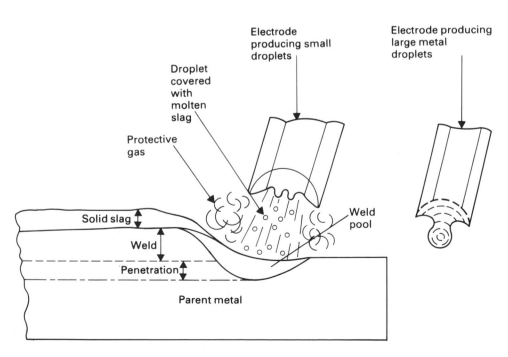

Figure 7.6 Principle of metal transfer in MMA welding.

formed in an a.c. or d.c. circuit, between the coated electrode and the workpiece. The arc melts the edges of the surfaces to be joined, forming a weldpool. The tip of the electrode also melts, and metal droplets (filler metal) are transferred through the arc to the workpiece. The electrode is consumed during the welding operation; the molten metal is protected from the surrounding air by a shielding gas that is released by the melting of the electrode coating. The weldpool is protected by a molten slag that covers the electrode metal droplets. This slag floats on the molten metal, solidifies at the same rate and forms a protective covering. The principle of metal transfer in MMA welding is shown in Fig. 7.6.

The welding electrode (the cathode) is composed of a metal wire core that provides the weld filler metal, and a coating of extruded flux on the outside. The metal core usually has approximately the same composition as the parent metal. One end of the electrode is left uncoated (approximately 20 mm/0.8 in.) to ensure good electrical contact with the electrode holder. For manual welding the electrode is usually 300–450 mm (12–18 in.) long, depending upon the electrode diameter, i.e. wire core diameter. The coating consists of various finely pulverized chemicals and minerals such as ferromanganese, rutile, fluorspar, carbonates or quartz, mixed with a suitable bonding agent, e.g. sodium and potassium silicates. The coating must be capable of promoting ionization in the arc, it must readily emit electrons and ions when heated, and it must produce sufficient gases to protect the metal droplets and the weld from the surrounding air. The flux must be capable of deoxidation of the weld metal, and be able to form a protective slag which solidifies at the same speed as the weldpool, thus influencing the shape of the surface of the solidified weld.

Different categories of electrode are available, depending upon the chemical properties of the slag being formed; the main differences are outlined below. Basic electrodes are generally used with a d.c. (+) supply and are suitable for welding in all positions. The solidified weld produced has low levels of impurities and good mechanical properties. Large droplets are produced, approximately 60–80% of the wire core diameter, the slag is thick and viscous and this makes these electrodes difficult to use. Basic electrodes must be protected from moisture otherwise porous welds, or cracks due to hydrogen embrittlement, can result. Rutile electrodes can be used with both a d.c. (+ or −) and an a.c. supply but they are normally limited to horizontal welding. They are easily reignited (good for intermittent use) and produce small droplets (approximately 30–40% of the wire core diameter). The slag has a low viscosity, producing a smooth weld surface. The purity and mechanical properties are lower than those obtained with basic electrodes. In particular, the weld metal often contains a high hydrogen impurity which may lead to hydrogen embrittlement. Alloyed rutile electrodes are often used for welding various types of stainless steels.

Acid electrodes are used only with d.c. (−) supply and very small metal droplets are produced (approximately 10–40% of wire core diameter). The weldpool is overheated and there is a relatively high level of impurity that can lead to hot cracking and porosity. The electrodes are easy to use, usually confined to horizontal welding, and they produce a smooth weld. However, the mechanical properties are lower than for basic electrodes, and they are often confined to less qualified mild-steel welding. Synthetic electrodes usually contain all the alloying elements in the coating, e.g. welding 18–8 austenitic stainless steel using a mild-steel electrode core and a coating containing chromium (18%) and nickel (8%). If the coating is damaged, then the composition of the weldpool may vary. Certain electrodes are produced with intermediate properties and are known as rutile-basic or rutile-acid electrodes. Some metallic material is transferred to the weldpool from the coating and a *metal recovery factor* is often quoted. This is the weight of weld metal produced as a percentage of the weight of wire core used. For high-recovery electrodes this is usually greater than 130%, and can be as high as 240%; the coatings used are relatively thick.

MMA welding can be performed using either d.c. or a.c. supply; in each case the power source must supply current and voltage suitable for the type of electrode used. The arc voltages for the most common types of electrodes are: normal type electrodes ∼ 20–30 V, high recovery electrodes ∼ 30–50 V, and deep penetration electrodes ∼ 60–70 V. The power source for welding can be used by either a single operator or a multi-operator unit (usually for 8–12 welders, depending upon the electrode diameter). For single operator d.c. power sources, either motor generators or welding rectifiers are used. A motor generator converts electrical power (from mains) using mechanical energy to a suitable d.c. voltage. The electrical efficiency is approximately 50–55%. A welding rectifier consists of a transformer and a rectifier bridge and converts a.c. mains to d.c. directly, with an electrical efficiency of 70–75%. For welding with a.c., a welding transformer is used that transforms mains voltage to a suitable open-circuit voltage (60–70 V). The electrical efficiency is 80–85%. Power sources are available that can provide either a.c. or d.c. supply.

The choice of d.c. or a.c. for welding depends upon the type of electrode to be used. Although d.c. is the more commonly used, there are many electrodes produced for a.c. welding. The energy cost is lower with a.c. welding, but this is only a small part of the total welding costs. All types of electrodes can be used with d.c. but this is not so for a.c.; also welding of thin sheet is more difficult with a.c. Magnetic effects are negligible with a.c., but with modern electrodes the effects with d.c. are also small. The transformer is less expensive than the corresponding rectifier, and its efficiency is higher than the motor generator or the rectifier. Maintenance costs are lowest for the transformer.

MMA welding is used in a wide range of applications, e.g. ship building, general machine construction, frameworks, tanks, pressure vessels, bridges, etc. MMA is particularly versatile, and is used with a large variety of different materials in most sizes above 1.5 mm (0.06 in.) thickness. The process can be used for workshop or site welding, either inside or outside. The availability of suitable electrodes for a particular application normally decides whether MMA welding can be used. The choice of electrodes determines the quality of the weld joint and the welding cost incurred. Manufacturers' information concerning available electrodes is often presented in terms of application areas, i.e. the types of materials to be joined. The usual classification of electrodes for welding particular materials is as follows:

Mild steel   – used for structural applications and pressure vessels.

Low-alloy steel, stainless and heat-resistant steels, tool steels.	– especially for high-temperature use.
Nonferrous metals	– the most important are nickel and nickel alloys.
Cast iron	– generally unsuitable for welding, but it may become necessary to perform repair welding.

Gravity welding is a particular type of semi-automatic, metal-arc welding. The electrode moves by gravity (or spring force), and the arc is extinguished when the electrode is consumed. The equipment is simple with low investment costs, the method is particularly suitable for long, straight weld runs and repetitive work. It is used in shipyards for fillet welding in the horizontal and vertical positions.

7.3.3 Submerged-arc welding

An electric arc is struck between a continuously fed electrode and the workpiece. Several electrodes and arcs can be used simultaneously. The electrode is fed by an automatic feeding device, powdered flux is supplied to the weld joint and totally covers the arc and the weldpool, as shown in Fig. 7.7. During welding, part of the powder is converted into slag which floats on top of the weld. Excess powder is recovered for further use. Relatively high current densities are used, providing a high rate of metal (weld) deposition and deep penetration. Therefore for certain applications the weldpool must be supported from the underside; this can take various forms, as shown in Fig. 7.8. The high rate of metal deposition normally means that this process is economical compared with other welding methods. If welding is to take place other than in

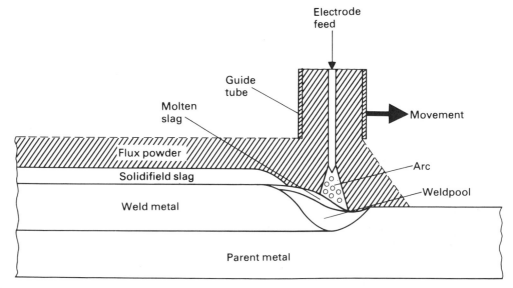

Figure 7.7   Submerged-arc welding process.

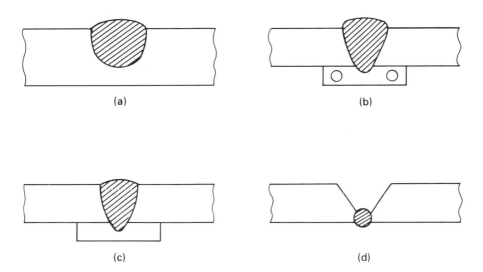

Figure 7.8 Different types of underside support (backing). (a) Thickness of parent metal. (b) Water-cooled copper shoe. (c) Backing plate (temporary or permanent). (d) Preliminary backing weld run (by another process).

the horizontal position, then a special device is required to support the flux in the required position.

The filler metal is fed continuously from a spool and can be in the form of bare wire between 2.4 and 6.0 mm diameter (0.096 and 0.24 in.), or as strip electrodes 0.5 mm by 60 mm (0.02 in. by 2.4 in.) wide, although strip up to 200 mm (8 in.) wide can be used. The filler metal usually has approximately the same chemical composition as the parent metal; the choice of flux decides the mechanical properties and composition of the weld metal. For welding stainless steels and nonferrous metals, e.g. nickel, wires of similar composition to the parent metal are used.

The flux is required to protect the molten metal from the effects of the atmosphere and to aid ionization of the arc. The molten flux should provide a slag of suitable viscosity and surface tension to ensure that solidified welds of suitable shape and surface finish are obtained. The slag produced by a flux may be either basic or neutral. The flux may be in the form of granules or agglomerations. The granulated fluxes obtain their grain size by crushing; they are homogeneous when molten and have the appearance of glass. Only silicon and manganese can be transferred to the weldpool by these fluxes. Agglomerated fluxes are heterogenous in that each grain is composed of smaller grains consisting of the chemical components; a bonding agent is used. These fluxes are dried at relatively low temperatures, and elements such as ferro-chromium, ferro-nickel, etc., can be added without oxidation occurring. Thus chromium and nickel can be introduced into the weldpool. Both types of flux consist mainly of quartz, limestone and manganese silicates.

Both d.c. and a.c. may be used for submerged-arc welding. Therefore, either welding rectifiers, motor-driven generators or welding transformers can be used as the power source. D.C. is used with the electrode connected to either the positive or negative pole, depending upon the application. Connection to the positive pole provides better penetration, and the negative pole connection enables the highest deposition rate. The process is generally carried out mechanically or by automatic control.

Submerged-arc welding is widely used with mild-steel structures, particularly for butt welding. The process is also used for welding stainless steels, and nonferrous metals such as nickel. Many applications can be found within the ship building industry. Submerged-arc welding is used for surfacing on mild and low alloy steels, e.g. pressure vessels, to provide the parent metal with a layer of corrosion resistant stainless steel.

7.3.4 Metal-gas-arc welding

This type of arc welding includes various categories depending upon the type of shielding gas used. The two main types are:

Metal-inert gas-arc welding (MIG)
Metal-active gas-arc welding (MAG)

The abbreviations shown in brackets will be used in the text. For MIG welding the inert gas is argon or helium. For MAG welding the active gas is usually carbon dioxide or mixed gases.

In MIG/MAG welding an electric arc is struck between the workpiece and a continuous-feed, consumable electrode. A contact tube within the welding gun is connected to the d.c. supply (+) and provides the current for the electrode. The electrode is fed continuously by an automatic unit, it melts in the arc and is transferred to the weldpool in the form of droplets. If a large current and voltage are used, then metal transfer occurs as large droplets which momentarily short circuit the arc; this is known as *dip transfer* or *short-arc welding*. For a lower current and voltage the metal transfer is in the form of finely dispersed, non-short-circuiting droplets; this is known as *spray transfer* or *spray-arc welding*. An intermediate current and voltage provide a mixed mode of metal transfer. The gas that shields the arc, and the molten metal, from the effects of the atmosphere is supplied through the welding gun. The welding process is illustrated in Fig. 7.9.

The filler metal is wire of between 0.6 mm and 2.4 mm diameter (0.024 in. and 0.096 in.) fed from spools containing 15 kg (33 lb) of wire. For MAG welding, the wire may be copper-coated to prevent corrosion during storage, and to provide better electrical contact between the electrode and the contact tube. Solid wire electrode is most commonly used, although tubular flux-cored wires are available. The flux has the same effect as in MMA welding. For MIG welding the shielding gas is argon or helium, or mixtures of these or other gases. Up to 5% oxygen may be added to reduce the metal droplet size in the arc, by reducing surface tension and viscosity, which is beneficial for spray transfer. In MIG welding there is no reaction between the gas and the molten metal, the filler rod has a chemical composition similar to that of the parent metal. For MAG welding the shielding gas is either carbon dioxide or argon containing

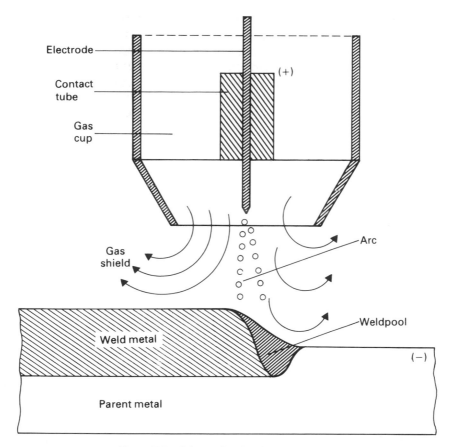

Figure 7.9   Schematic of MIG/MAG welding.

20% carbon dioxide. Dissociation of carbon dioxide occurs in the arc, and ferric oxide can be formed in the weld pool. This could lead to brittleness in the completed weld. Any carbon in the parent metal tends to reduce the ferric oxide and produce carbon monoxide, resulting in a porous weld. This can be prevented by using a filler metal containing silicon or manganese, which have a greater affinity for oxygen. The oxides formed then tend to float on top of the weld as light slags which can be removed.

For MIG/MAG welding, a d.c. supply is used with the electrode connected to the positive pole and the workpiece to the negative pole. The use of motor-driven generators has now mainly been replaced by rectifiers. A pulsating-current d.c. supply has been used, mainly for MIG welding. The welding gun has various functions including the supply of electrode wire to the weld from either a push or pull supply, providing current to the electrode through the contact tube, and providing the shielding gas supply. The gun may be gas cooled, for currents up to 250 A, or water cooled for higher currents. The electrode, power, gas and cooling fluid are all fed to the gun through a flexible sheath.

The MIG/MAG welding process has a low heat input which makes it particularly suitable for welding thin sheets, e.g. less than 3 mm (0.12 in.) thickness, due to the lack of distortion. It is commonly used for welding car bodies with the dip transfer process. For thicker plates, between 3 mm and 8 mm (0.12 in. and 0.32 in.), spray transfer is used with higher current densities. This provides a higher deposition rate and is a particularly economic process. MAG welding is mainly used with mild steels, and MIG welding with stainless and heat-resistant steels and nonferrous metals, e.g. aluminum, nickel and copper. The process is not suitable for outdoor applications due to the problem of maintaining the gas shield. It is used for welding in the flat position and is a semi-automatic process, although it can easily be adapted for full automation.

7.3.5 Tungsten-inert gas-arc welding (TIG)

In TIG welding, an electric arc is struck between a tungsten electrode and the workpiece, which is shielded by an inert gas. The electrode only acts as the focal point for the arc and does not provide molten metal to the weldpool. Therefore, either pure tungsten (melting point $3370°C/6098°F$) or tungsten alloyed with 2% thorium is used. The filler metal is provided separately, as shown in Fig. 7.10. The arc is initiated by a spark, generated by a high-frequency voltage across the air gap. The spark ionizes the air gap and enables the main arc to be struck, without short-circuiting the arc column. A d.c. power supply is normally used with the electrode connected to the negative pole, thus keeping the tip cooler and permitting the use of smaller diameter electrodes. For aluminum welding, the use of a positive electrode helps break down the surface oxide film

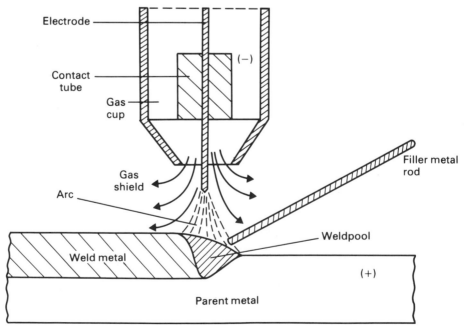

Figure 7.10 Schematic of TIG welding process.

and a.c. supply is often used. Gas cooling is used for currents up to 130 A; water cooling for higher currents. Rectifiers, motor-driven generators and transformers are all used as the power supply, often supplying either a.c. or d.c.

The shielding gas is usually a pure inert gas, e.g. argon containing less than 0.05% impurity. Even higher-purity gas (e.g. 0.01% maximum impurity) can be required when welding titanium or zirconium. Helium is sometimes used as the inert gas or argon with 5% hydrogen, both of which increase the intensity of the arc. Inert gas shielding should also be used on the underside of the weld to prevent oxidation. The inert gas prevents reactions occurring with the molten metal, and the chemical composition of the filler wire should be approximately the same as the parent metal. The wire may be heated by passing an electric current through it.

TIG welding is mainly used as a manual process although it can be easily automated. The process produces high-quality welds and is used in the chemical and food industry, and with metals susceptible to oxidation, such as titanium. TIG welding is mainly used with stainless and heat-resisting steels and with aluminum, nickel and nickel alloys.

7.3.6 Plasma-arc welding

Plasma is the name given to a gas that has undergone a process of dissociation, and has a very high energy content. The process is similar to TIG welding. A tungsten electrode, normally alloyed with thorium, is recessed within a water-cooled nozzle and an arc is struck with the workpiece. Gas surrounds the electrode and acquires a high velocity by passing through one or more small holes in the tip of the nozzle. When heated the gas expands, and then contracts as it passes through the holes; it has high energy and a temperature between 20 000°C and 40 000°C (36 000°F and 72 000°F). The gas dissociates to plasma as it accelerates through the arc. On reaching the workpiece the plasma recombines to form gas, and the energy creates the weldpool. The plasma arc and the weldpool are protected by shielding gas, this is often the same composition as the plasma gas, e.g. pure argon or argon with 5% hydrogen (to increase the heat content). The use of filler metals is the same as with TIG welding.

A high-frequency spark is used to initiate the plasma arc. This ionizes the gas in the gap between the electrode and the inner nozzle, forming a pilot arc. The pilot arc then initiates the main arc. This is known as *transferred-arc welding* and is used for joint welding and surfacing applications. For *plasma spraying* and sometimes surfacing, a non-transferred arc is used where the main arc is struck between the electrode and the nozzle (the workpiece does not form part of the electrical circuit). The principle of plasma welding is shown in Fig. 7.11. The plasma arc produced is cylindrical, rather than the TIG conical shape. A d.c. supply is used with a negative electrode to reduce its rate of erosion. Plasma-arc welding is used for stainless steels and nonferrous metals. The range of applications is similar to TIG welding but it can be used with a wider range of material thicknesses, especially light gauge material, e.g. 0.02–1.0 mm (0.008–0.04 in.).

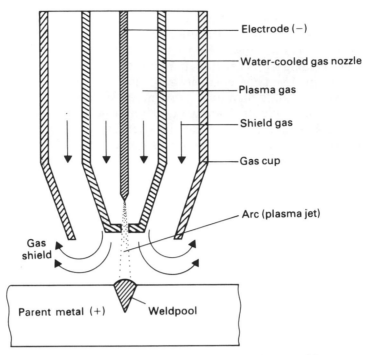

Figure 7.11 Schematic of transferred-arc plasma welding.

7.3.7 Nonshielded metal-arc welding

In this process, a separate flux or shielding gas are not used to protect the arc and the molten metal from the atmosphere. Instead a tubular, flux-cored, wire electrode is used, which is melted in the electric arc. The wire provides filler metal for the weld, and the flux core evolves a shielding gas. The equipment and the processes that are used are very similar to those described under submerged-arc welding and MIG/MAG welding. Power sources can be either a.c. or d.c., but they must have a minimum capacity of 600 A to obtain the high deposition rates that are a feature of this process. It is mainly used for welding mild steels above 8 mm (0.32 in.) thickness, although electrodes are available for welding alloy steels.

7.3.8 Carbon-arc welding

This process is mainly used for copper welding, and uses a thick carbon electrode connected to the negative pole of a d.c. supply. The electric arc acts as the heat source and large currents are required.

7.3.9 Stud welding

This process is used for welding studs, bolts, pins, etc., to thick plates that form large structures. The stud is held within a welding gun connected to a power supply. One end of the stud is fluxed; this end is then made to short circuit with the plate to which it is to be welded. It is then withdrawn from the plate, causing a rise in voltage and thus initiating an arc. When the plate has formed a weldpool and the end of the stud has melted, then a spring forces the stud into the weldpool and it is allowed to cool in position. The flux evolves a shielding gas, and the stud position can be aligned using pre-fixed ceramic rings, which are broken away upon completion. Figure 7.12 shows the principle of stud welding by the method described above, for studs up to 25 mm (1 in.) diameter. Capacitor discharge stud welding is used with studs up to 8 mm (0.32 in.) diameter, that are welded onto sheets between 1.0 mm and 1.5 mm (0.04 in. and 0.06 in.) thick without damage or discoloration to the reverse side of the sheet. This method uses a localized heat input.

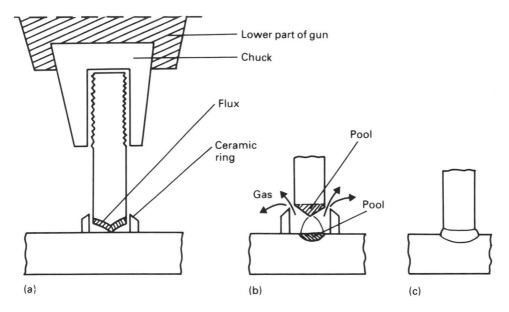

Figure 7.12   Stages in stud welding. (a) Setting up. (b) Welding (c) Completed weld.

*   *   *

Self-assessment exercises

1 Prepare a table to compare and contrast the main features of alternative arc-welding methods.

2 What is the advantage of using a titanium electrode?

3 What is plasma? How is it produced? Is plasma arc welding an expensive process compared with MMA, submerged arc, MAG or MIG?

4 Is a nonshielded metal-arc process susceptible to atmospheric oxidation?

5 List the metals that should not be joined using carbon-arc welding. Explain briefly the reasons.

6 In what applications is stud welding used? Is this a fusion or pressure process?

7 For MMA welding, what shielding gas is used and how is it supplied?

8 For the submerged arc process, by what is the arc submerged and why?

9 What active gases can be used for MAG welding? Why are they selected, and do they react with the heated or molten metal?

*   *   *

7.4 GAS WELDING

Heat is provided by burning a combustible gas with oxygen, which is used to melt the filler wire and the edges of the metal to be joined. Acetylene (C_2H_2) is the most commonly used combustion gas, although other types of hydrocarbon compounds are now used. The gas can be burnt with either air or pure oxygen depending upon the temperature required. For air–acetylene mixtures the flame temperature is between 2300°C and 2500°C (4172°F and 4532°F), compared to temperatures of 3100–3300°C (5612–5972°F) for oxy-acetylene flames. A comparison of gas-welding flame temperatures is given in Table 7.5; by comparison the metal-arc temperature is over 6000°C (10832°F) depending upon the type of arc used. The flame is also used to heat the workpiece. Gases are supplied from high-pressure portable steel cylinders or from liquid gas storage containers. Connections to the welding torch are via a pressure-reducing valve and flexible hoses. Both gases supplied to the torch are above atmospheric pressure.

Table 7.5   Gas-welding flame temperatures

Gas mixture	Flame temperature °C (°F)
Oxy-acetylene	3100 (5612)
Air-acetylene	2325 (4217)
Oxy-butane (Calor gas)	2820 (5108)
Oxy-propane (liquified petroleum gas)	2815 (5099)
Oxy-methane (natural gas)	2770 (5018)
Air-methane	1850 (3362)
Oxy-hydrogen	2825 (5117)
Metal-arc (for comparison)	> 6000 (> 10832)

A *neutral flame* is used for welding, with a gas mixture (by volume) of 48% acetylene and 52% oxygen. The regions present in the flame are shown in Fig. 7.13. There is an inner cone (A) which is a dazzling luminous white and provides an unburnt mixture of oxygen and acetylene, although some acetylene is broken down into carbon and hydrogen. There is a second region (B) where the primary reaction occurs:

$$C_2H_2 + O_2 \rightarrow 2CO + H_2$$

Secondary combustion occurs in the outer cone (C); the reaction is:

$$2CO + H_2 + \frac{3}{2} O_2 \rightarrow 2CO_2 + H_2O_{(g)}$$

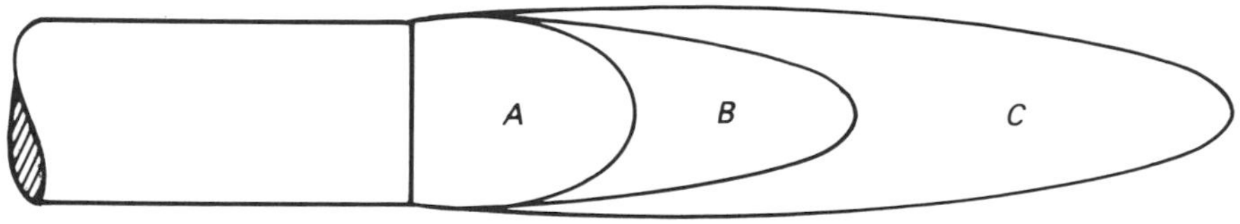

Figure 7.13　Regions in a neutral oxy-acetylene flame.

If there is an excess of acetylene in the gas mixture then the flame becomes *carburizing* (or *carbonizing* or *reducing*). In Fig. 7.13 there will be an extra region between B and C – a blue reducing zone containing CO and H_2. A large excess of acetylene produces a smoky flame. An excess of oxygen produces an oxidizing flame and up to 60% of the required oxygen is provided by the surrounding air.

A flux is only required for welding cast iron and stainless steels. Filler rods should contain enough silicon to remove excess oxygen, and sufficient manganese to produce a self-fluxing weldpool, e.g. 0.3% Si, 1% Mn, C < 0.15%. The slag should rise to the surface for a good quality gas weld. The slow heating and cooling cycles are useful if the material tends to harden or become brittle, but can also lead to distortion and undesirable metallurgical changes.

The flame can be used for metal cutting and brazing. Gas welding is usually a manual process and has partly been replaced by other methods that can be automated, which produce an improved quality weld and create less distortion by using a less intense heat supply. Gas welding is now mainly used for assembly and installation work, especially unalloyed pipe systems, and for cast iron repair work.

* 　 * 　 *

Self-assessment exercises

1　For gas welding, compare the thermal energy that can be provided by:
　(a)　oxy-acetylene;
　(b)　air-acetylene;
　(c)　oxy-propane;
　(d)　oxy-natural gas;
　(e)　any other gases.

2　What 'types' of flames can be obtained, and what effects do they have upon the weld that is formed?

3　Comment upon the statement: 'Gas welding has become a redundant process because of its dependence upon a human operator and problems of quality control of the welds'.

4　How would you ignite a gas torch?

5　What is the purpose of a flux?

* 　 * 　 *

7.4.1 British Standards for fusion welding

BS 638: Arc welding power sources, equipment and accessories (4 parts).

BS 639: Covered electrodes for the manual-metal-arc welding of carbon and carbon manganese steels.

BS 1453: Filler materials for gas welding.
Sizes, condition, packing, marking. Ferritic steels, cast iron, austenitic stainless steels, copper and copper alloys, aluminum and aluminum alloys, magnesium alloys.

BS 1723: Brazing.

BS 1724: Bronze welding by gas.
Copper, mild steel, galvanized mild steel, cast iron, malleable iron and their combinations. Joints for each type of parent metal, and other information.

BS 1821: Specification for class I oxyacetylene welding of ferritic steel pipework for carrying fluids.

BS 2633: Class I arc welding of ferritic steel pipework for carrying fluids.

BS 2901: Filler rods and wires for gas-shielded arc welding (5 parts covering different materials).

BS 2971: Specification for class II arc welding of carbon steel pipework for carrying fluids.

BS 3019: General recommendations for manual inert gas-tungsten-arc welding.
Part 1: Wrought aluminum, aluminum alloys and magnesium alloys.
Part 2: Austenitic stainless and heat-resisting steels.

BS 3571: General recommendations for manual inert gas-metal-arc welding.
Part 1: Aluminum and aluminum alloys.

BS 4165: Electrode wires and fluxes for the submerged-arc welding of carbon steel and medium-tensile steel.

BS 4206: Methods of testing fusion welds in copper and copper alloys.

BS 4570: Fusion welding of steel castings.
Part 1: Production, rectification and repair.
Part 2: Fabrication welding.

BS 5135: Metal-arc welding of carbon and carbon manganese steels.

BS 5465: Specification for electrode wires and fluxes for the submerged-arc welding of austenitic stainless steel based on weld metal composition.

Reference should also be made to BS 4871 (Part 2) and BS 4872 (Part 2) for approval tests of TIG and MIG welding of aluminum and its alloys (see Section 7.2.3).

7.5 ELECTRO-SLAG WELDING

This process is a form of mechanized welding, it is usually carried out in the vertical position and was originally developed for welding thick plates (greater than 40 mm/1.6 in.). It can also be used for thinner plates (18 mm/0.72 in. or more) the main advantages being high productivity and considerable cost reductions.

The joint is set up with a wide gap, between 25mm and 36mm (1 in. and 1.5 in.), depending upon the plate thickness. A mould is formed between the plates and a pair of tight fitting, water-cooled copper shoes (one each side of the joint). A backing plate (starting block) is fitted at the base of the plates to be welded, in order to contain the weldpool when welding commences. The consumable wire electrode (filler metal) and a flux are fed to the weldpool from above. Welding is commenced by striking an arc between the electrode and the workpiece. As the flux melts, a slag of increasing depth is formed. The composition of the flux is such that it obstructs ionization of the arc atmosphere. As the temperature and therefore the conductivity of the slag increases, then the arc is quenched and conduction of the current occurs through the slag. The heat is generated by electrical resistance. As part of the weld is completed, then the welding head and the copper shoes rise up the plates at an even speed. The speed is directly related to the rate of solidification of the weld, in order to prevent spillage of the weldpool or the slag. The principle of electro-slag welding is shown in Fig. 7.14.

The entire cross-section of the workpiece is welded in a single pass. The number of wires to be used depends upon the plate thickness; one head with three wires can weld a plate 450 mm (18in.) thick. With thick plates the wires may be made to oscillate in order to provide uniform heat generation. The electrode is normally copper-coated 3 mm (0.12 in.) diameter wire, fed at a constant speed. The flux is fed separately, or flux-cored wire can be used. The electrode compositions are similar to those used for submerged-arc welding, but are usually available in extra long coils so that the weld can be completed in a single run. A high proportion of parent metal is used to make the weld, an approximate ratio of filler metal to parent metal of 1:2. The parent metal should therefore contain less than 0.25% carbon, to obtain a low carbon content in the weld and thus avoid hot cracking.

The fluxes produce complex silicate slags containing SiO_2, MnO, CaO, MgO, Al_2O_3. The addition of calcium fluoride increases the electrical conductivity, and lowers the viscosity of the slag. Slags based on calcium fluoride and calcium oxide have a strong desulfurizing effect. This is useful with steels containing more than 0.25% carbon, for reducing the possibility of hot cracking in the weld. However, the flux must be kept dry.

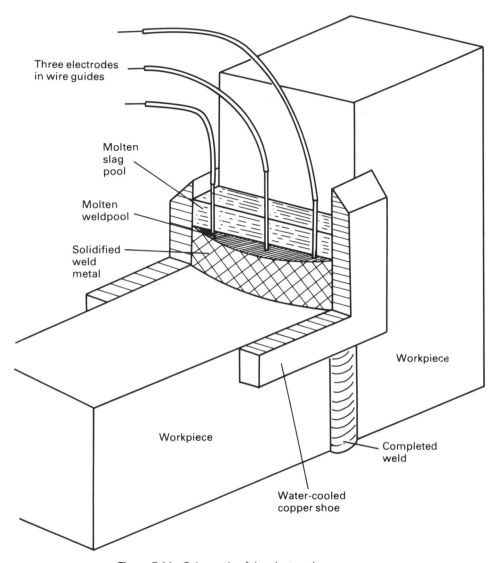

Figure 7.14 Schematic of the electro-slag process.

A d.c. positive supply is used with one or two wires; for three wires a.c. supply from a three-phase transformer is used. The process produces good homogeneous joints, but the weld and the heat-affected zone may have a larger grain size than is produced by other welding methods. It is a high-speed process, but takes time in setting up and is therefore good for repetitive work.

7.6 CONSUMABLE NOZZLE (OR GUIDE) PROCESS

This process is a variant of the electro-slag process. One or more wires are fed to the weldpool through a consumable guide, which itself melts at the same rate as the welding proceeds. The electrode and the tube both act as filler metal for the weld. Welding is carried out with fixed, water-cooled, copper shoes because it is possible to maintain a greater distance between the contact tube and the slag bath. The equipment is cheaper and simpler than when movable shoes are used, and the setting-up time is reduced. The process is not used for long weld runs, and is easier to manage with metal thicknesses of less than 150 mm (6 in.). Four wires (each 3 mm/0.12 in. diameter) can be used to weld plate which is 200 mm (8 in.) thick. Submerged-arc equipment can be adapted for this process; it can be used to weld large components having a complex profile if a portable nozzle can be mounted on the joint. The properties of the weld depend upon the parent metal composition. Diagrams of the consumable tube and plate welding processes are shown in Fig. 7.15.

7.7 ELECTRON-BEAM WELDING

For electron-beam welding, a beam of electrons is generated under a high vacuum and accelerated to approximately 60% of the speed of light. The beam is focused by a magnetic lens to a focal spot (diameter less than 0.1 mm/0.004 in.) at the center of the joint to be welded. Approximately 99% of the kinetic energy of the electrons is released at the workpiece and is transformed into heat. The surface of the parent metal vaporizes, thus exposing a new focal spot where vaporization again occurs. This allows the beam to penetrate deep into the workpiece, and welds with a width:depth ratio of 1:25 can be made. The vaporized metal condenses on the walls of the crater, forming a thin-walled cylinder of molten metal through which the beam passes. As the beam moves along the joint, the pool solidifies behind the beam. The gap across the joint should be smaller than the diameter of the beam.

The beam has a very high power density, approximately 2 to 3 times that of an electric arc, but the heat input to the workpiece is low. The best results are obtained if the joint is welded under vacuum conditions; for welding in gases account must be taken of the electron-beam deflection. The process is used for welds not requiring a filler metal, and the workpiece is usually moved in relation to the electron beam.

The process is readily automated and produces high-quality welds, in terms of composition and dimensional tolerances, e.g. for nuclear reactors and aircraft components. It is particularly useful for long weld runs (high productivity) and for materials and situations where other fusion-welding methods cannot be used.

7.8 LASER WELDING

A laser beam is used to supply energy to the joint. This can be from a solid (crystal) laser with a low overall efficiency (only 1% of the energy is transferred to the weld), or from a gas laser with an efficiency of between 10% and 30%. Gas lasers can be used for welding stainless steel between 3 mm and 5 mm (0.12 in.

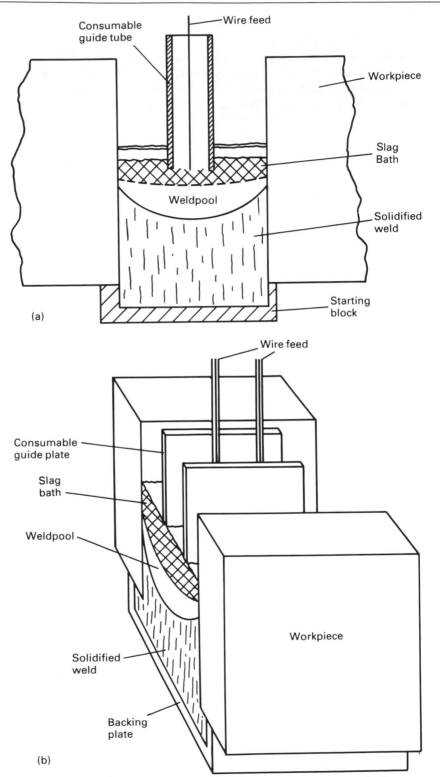

Figure 7.15 Consumable guide welding. (a) Tube guide. (b) Plate guide.

and 0.2 in.) thick. The process is similar to electron-beam welding with a similar high-energy density (approximately 10^7 W/mm^2; 6.1×10^6 Btu/h in^2.) and a focal spot focused to diameters less than 0.1 mm (0.004 in.) The heat released melts the metal to a depth depending upon the energy and the type of metal; filler metals are not normally used. The process is simpler and cheaper than electron-beam welding because a vacuum chamber is not required. It is used for precision work, for cutting, and also for joining dissimilar precious metals.

7.9 THERMIT WELDING

The components to be welded are held in position in a graphite mold, as shown in Fig. 7.16. A welding powder (thermit) consisting of metal oxide and metal, is held in a reaction crucible (as shown) and is ignited by a spark. The oxide is reduced to its metal, large amounts of heat are evolved (by chemical reaction) and the powder melts. The melt then flows down to surround, and alloy with, the workpiece. No extra energy is supplied. The process is mainly used for welding copper alloys or aluminum alloys.

7.10 THERMAL CUTTING

Thermal cutting techniques are widely used in industry, particularly for edge preparation before welding, for mild steel in ship building and for heavy plate thicknesses.

With gas cutting, a hydrocarbon fuel gas (e.g. acetylene or propane) is burnt in oxygen to produce a heating flame similar to that used in gas welding. After pre-heating the surface, extra oxygen is introduced through a separate tube and the cutting process commences. The pressure of this extra cutting oxygen removes the combustion products, and forms a hole in the workpiece. A narrow cut is formed by moving the blowpipe. The cutting oxygen also acts to remove any metal oxides, which should be liquid and of low viscosity. This method can be used to cut mild steel between 1 mm and 1.5 mm (0.04 in. and 0.06 in.) thick, also steel castings and low-alloy steels. Copper and aluminum and their alloys cannot be gas cut. Cast iron, steels with carbon contents greater than 2.5%, stainless and heat-resisting steels can only be cut by this method if chemicals or iron powder are added to the oxygen stream. These provide additional combustion energy and the process is then known as *powder cutting.* The process can be mechanized, which then requires less gas and cuts at higher speeds. It is then possible to produce finished faces comparable to those obtained by machining.

An electric arc, struck between a consumable electrode and the workpiece, can also be used as the heat source. A high current and high arc voltage are used and the melted material is removed, sometimes using compressed air. This process is usually manual and produces a relatively poor-quality finish.

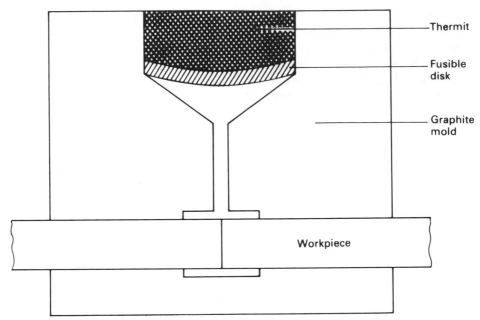

Thermit

Fusible disk

Graphite mold

Workpiece

Figure 7.16   Mold prepared for thermit welding.

Plasma cutting is another fusion process using a plasma arc (transferred arc) as the heat source. Argon is used as the plasma gas, hydrogen or nitrogen are added when cutting begins so that the arc fuses the material. The metal then partially vaporizes and is removed. Very high temperatures are generated by the arc and this method can be used for cutting mild steel, low-alloy steel, stainless and heat-resistant steels, aluminum and copper and their alloys. The cut is usually slightly larger on the upper side of the workpiece. The process is usually mechanized, which permits higher cutting speeds, although the associated costs are higher.

7.11 METAL SPRAYING

A metal spraying technique is often used to apply corrosion- and wear-resistant coatings, especially to mild steel and for worn components. The coatings may be metallic, or ceramic for electrical insulation, and can vary from 10 μm to several millimeters.

For metallic coatings, either flame spraying or arc spraying is used. With flame spraying the heat source is an oxy-acetylene or oxy-propane gas flame and the source metal is powder or wire. For arc spraying the heat source is an electric arc, struck between two consumable wire electrodes which also provide the coating material. Arc spraying provides better adhesion of the metal coating and is generally a more economic process, except for zinc-alloyed coatings. However, in this process certain elements gasify and it is particularly unsuitable for coatings with high carbon contents.

The principle of the process is that the wire or powder is heated to its melting point, and is then projected at the surface by a stream of compressed gas or air. The surface is usually prepared by grit blasting, and may also be grooved. The metal is transferred as spherical particles having diameters between 50 and 200 μm, forming a spray cone between the heat source and the workpiece. At the edge of the cone, the particles cool and solidify and are lost as metallic dust; this amounts to approximately 20–30% of the supplied metal. The particle surface becomes coated with an oxide film during transfer; this breaks up on impact and the surface layer is formed by adhesion and diffusion. The coatings are porous for approximately 8–15% of their volume and contain between 10% and 15% metal oxide. The spraying is carried out with a spray gun that is often fixed while the workpiece is moved and rotated. The mechanical properties of the coating depend upon the surface preparation, the spray technique employed and the composition of the coating.

A plasma spraying technique is used for ceramic or metal oxide coatings that have a high melting point. The costs involved are higher, although the porosity and oxide content of the coating are reduced. The heat source is a nontransferred arc with a gas in the plasma phase (see Section 7.3.6, Plasma–arc welding), resulting in very high arc temperatures. The coating materials are usually in powder form, although wire and rod can be used. The material is transferred by pressure from the plasma arc.

* * *

Self-assessment exercises

1 Describe, using diagrams only, the electro-slag and consumable guide welding processes.

2 List the advantages of each process and the metals that can be joined.

3 State the normal limits of metal thickness, and the gap, for these methods.

4 Describe the main feature in *one* sentence for each of the following processes:
(a) electron-beam welding;
(b) laser-beam welding;
(c) thermit welding;
(d) thermal cutting.

5 Explain why metal spraying is more economical and provides better adhesion, if an arc rather than a flame is used. Why is this method not suitable for producing coatings containing a high carbon content?

* * *

PRESSURE WELDING PROCESSES (Sections 7.12–7.20)——————

7.12 FORGE WELDING

There are several different methods of providing the heat and pressure required for forge welding, and these constitute a number of alternative processes. These are:

(a) blacksmith welding;
(b) hammer welding;
(c) roll welding;
(d) pressure-gas welding.

The principle of forge welding is that the metal is heated to a temperature below its solidus temperature, and then joining is achieved by the application of pressure. The heat source is typically an electric furnace or a combustion flame, and the pressure is due to hammering or clamping. The process involves deformation of the metallic crystals within the joint, followed by recrystallization and grain growth. Forge welding has now largely been replaced by other pressure welding techniques for industrial applications.

7.13 RESISTANCE WELDING

Resistance welding is an electric-pressure welding process that is employed to form lap joints, butt joints or T-joints, usually without using a filler metal. The surfaces to be joined must be cleaned, either mechanically or chemically, to remove any surface films that may have a high electrical resistance. The presence of a thin protective oil film is acceptable, and the material properties should be consistent along the regions to be joined. The joint surfaces must be properly aligned and brought into intimate contact. Heat is provided by an electric current that is passed through the workpieces via electrodes. Local temperatures are close to the melting point of the metal, and a force is applied which causes plastic deformation of the metal to be joined. The surfaces are then joined by atomic forces. An a.c. supply is used and a transformer is required. High currents (up to 100 kA) are used, due to the small electrical resistance at the weld (less than 500 µohms) and the short welding time cycles. There are various processes which employ the principle of resistance welding and these will be mentioned briefly.

7.13.1 Spot welding

The temperature of the weld region is close to the melting point of the metal; when this temperature is reached the current is cut off. The clamping force due to the electrodes is maintained during the cooling cycle to avoid damage to the weld. The variables which affect the spot-welding process are:

(a) shape and size of electrodes;
(b) electrode force;
(c) welding current;
(d) clamping time;
(e) welding time;
(f) cooling time.

The electrodes are rods with a high electrical conductivity. They are usually made from copper alloyed with cadmium, chromium or beryllium, for increased strength. The electrodes are water-cooled and the size of the tip area (contact area) is related to the parent metal thickness, i.e. the weld area should be approximately the same as the electrode tip. The force exerted by the electrodes is between 1 kN and 5kN (225 lbf and 1125 lbf) depending upon the sheet thickness. A small force is preferable in order to produce a high contact resistance, but this results in material loss from the weld and poor joining of the workpieces. Spot welding is commonly used for lap joints, particularly for mild steel sheet up to 3 mm (0.12 in.) thick. The process is used for sheet welding from about 0.1 mm (0.004 in.) thickness, and up to 15 mm (0.6 in.) for steel and 4 mm (0.16 in.) for aluminum. The maximum thickness depends on the mechanical properties of the metal. The process is particularly suited to mass production and is a high-speed operation. Typical applications are car bodies, domestic appliances and office machinery. The principle of spot welding is shown in Fig. 7.17.

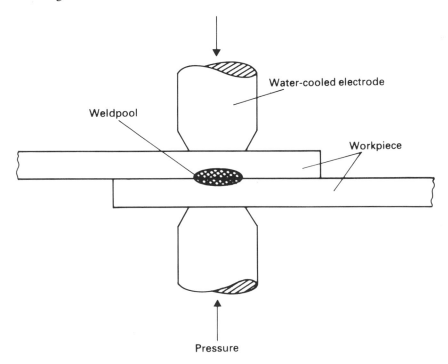

Figure 7.17   Principle of spot welding.

7.13.2 Projection welding

A series of projections or ridges is produced along the surfaces to be joined, by machining or pressing. The welding current is then passed through these projections, thus localizing the applied pressure and maintaining equal current distribution along the joint. The technique is similar to spot welding except that the electrode covers a larger area of the work. The important variables affecting the weld are electrode force, weld time and the current used. Projection welding can be used for long weld runs on uniform components, where the component can easily be fed to the machine. However, after welding the component will have permanent marks as a result of the projections. The principle of projection welding is shown in Fig. 7.18.

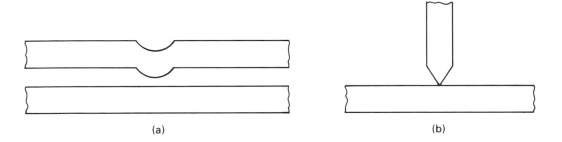

(a)   (b)

Figure 7.18   Principle of projection welding. (a) Lap joint with projection.
(b) Projection T-joint.

7.13.3 Seam welding

The electrodes are in the form of rollers and provide both the current and the force to the workpiece. The force remains constant during welding with an electrode wheel situated on each side of the weld, one of which may be power driven to feed the workpiece. The current can be intermittent or continuous. A continuous current produces a continuous weld, providing the weld speed is not too high. A series of weld spots can be produced which may overlap and form a gas-tight joint; with roller spot welding the spots do not overlap. Seam welding is limited to sheet welding and is mainly for lap joints. The process has been applied to butt-seam welding for tubes, but the metal fin formed on both sides may be difficult to remove from the inside of a tube. For continuous welded joints the sheet thickness is usually less than 3 mm (0.12 in.) for steels. The main applications of seam welding are illustrated in Fig. 7.19.

7.13.4 Resistance butt welding

This process is used for butt welding of wire and rod, mainly for welding cross-sectional areas of 100 mm^2 (0.155 in.^{2}), although areas up to 300 mm^2 (0.47 in.^{2}) can be joined. The surfaces must be squarely cut, and cleaned to provide a uniform electrical contact. The parent metal is held in clamps, one fixed and one movable, and is aligned and forced together. An electric current is passed through the clamps and the workpiece, and an axial force is applied by the movable clamp. The temperature rises at the weld region and in the length of metal between the clamps; plastic deformation occurs, resulting in a relatively

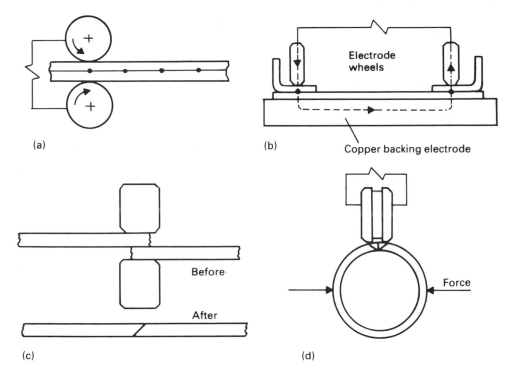

(a)   (b)

(c)   (d)

Figure 7.19   Principle of seam welding. (a) Roller spot welding. (b) Series seam welding. (c) Mash seam joint. (d) Butt seam welding.

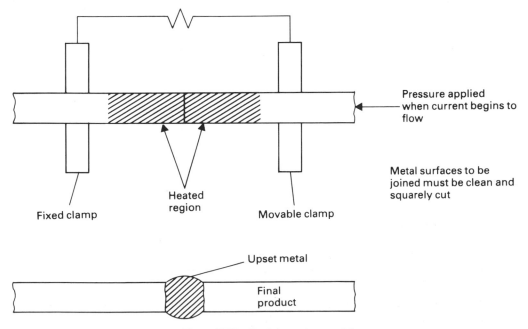

Figure 7.20   Resistance butt welding.

large amount of upset metal around the joint. The principle of this process is illustrated in Fig. 7.20.

7.13.5 Flash welding

Flash welding is used to join bar, sections and plate with cross-sectional areas between 150 cm^2 and 400 cm^2 (23 in.^{2} and 62 in.^{2}). The process is similar to resistance butt welding using a fixed and movable clamp for the workpiece. Initially a gap is left between the workpieces, and an electric potential acts across the gap. The surfaces are brought into contact and metal melts at the irregular points of contact. The molten metal is thrown out and new points of contact are formed. This is known as the *flashing operation* and continues until both surfaces are at a suitable temperature. An axial force is applied and the current is turned off. The surfaces fuse and cool and a fin or flash of metal surrounds the joint; this is less than the upset metal in resistance butt welding. The process is shown in Fig. 7.21.

*   *   *

Self-assessment exercises

1 Describe the process of resistance welding. State the temperatures, filler metals and fluxes that are used.

2 State the main distinctive feature of each of the following joining methods:
   (a) spot welding;
   (b) projection welding;
   (c) seam welding;
   (d) resistance butt welding;
   (e) flash welding.
   List the metals and particular applications for which these processes are used.

3 Describe the differences in crystal structure and phases formed when fusion welding or resistance welding are used for comparable joining situations.

*   *   *

7.14 HIGH-FREQUENCY INDUCTION PRESSURE WELDING

The current from a high-frequency generator flows through an induction coil, close to or surrounding the joint. The metal must be electrically conductive and is heated, due to its electrical resistance, by the induced current. When the temperature is near the melting point of the metal, the surfaces are forced together. Filler metal is not normally used and magnetic and nonmagnetic materials can be welded. The process is used for continuous-tube welding from mild-steel strip (at high speeds up to 1.5 m/s; 4.9 ft/s), copper or aluminum. High-frequency heating is also used for metal hardening, soldering and brazing.

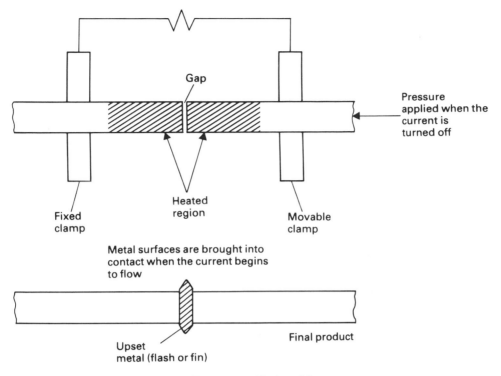

Figure 7.21   Flash welding.

7.15 COLD-PRESSURE WELDING

Welded joints can be formed without any heating by the application of a sufficiently high pressure. The principle of this method is shown in Fig. 7.22. The surfaces to be joined must be absolutely free from grease and any oxide film. The surfaces are forced together, using fixed and movable clamps, and atomic forces (metallic bonds) join the metals. The weld strength is usually higher than that of the parent metal, due to the cold working that has occurred. The process is limited to the formation of butt and lap joints mainly of copper and aluminum, although in some cases of mild steel. The pressure to be applied depends upon the mechanical properties of the parent metal and the cross-sectional area to be joined, which may be up to 10 cm^2 (1.5 in.^2).

7.16 FRICTION WELDING

The workpieces are brought together under loading and are made to move relative to each other; generally one is stationary and the other rotates. Heat is generated due to friction, and when the temperature approaches the melting point of the metal the pressure is increased to form the joint. Any oxides or

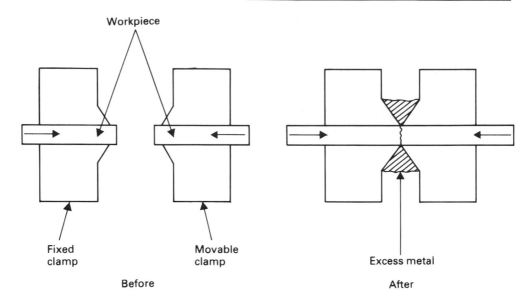

Figure 7.22 Principle of cold-pressure welding.

impurities are expelled from the joint to form a fin of upset metal. A filler metal is not normally used, although an intermediate piece of metal between the workpieces can be included. The process is used for joining bar or tube of a wide range of metals, and for joining dissimilar metals.

7.16.1 British Standards for pressure welding

BS 1140: Specification for resistance spot welding of uncoated and coated low carbon steel.
For the fabrication of assemblies comprising two thicknesses of metal.

BS 2630: Specification for resistance projection welding of uncoated low-carbon steel sheet and strip using embossed projections.
Requirements for fabrication of assemblies of single sheet thickness 0.4–3 mm. Recommendations for equipment and guidance on welding conditions.

BS 2996: Projection welding of low carbon wrought steel studs, bosses, bolts, nuts and annular rings.

BS 3065: Rating of resistance welding equipment.

BS 4204: Specification for flash welding of steel tubes for pressure applications.

BS 4215: Spot welding electrodes and electrode holders.

BS 4577: Materials for resistance welding electrodes and ancillary equipment.

Specifies requirements for wrought, sintered and cast metal products.

BS 6223: Specification for friction welding of butt joints in metals for high duty applications.

BS 6265: Specification for resistance seam welding of uncoated and coated low carbon steel.

7.17 EXPLOSIVE WELDING

The energy from an explosive charge is used to bond the workpieces together; the process is shown in Fig. 7.23. A flyer plate is situated a few millimeters above the parent metal (target plate), and a plastic buffer is usually placed between the flyer plate and the explosive charge. The charge is made to detonate in a certain direction and the two plates impact at an angle of between 5° and 20°. High pressures are generated at the bonded surfaces and any oxides or impurities are continuously ejected, although the surfaces should have been thoroughly cleaned. The process is similar to cold-pressure welding, plastic deformation of the surfaces occurs and the bonded region has a wavy form. However, the fusion zone is very much smaller than for most other welding methods. The point of impact of the surfaces moves forward at the same speed as the detonation front, i.e. 2000 m/s to 2500 m/s (6500 ft/s to 8200 ft/s). Appropriate safety measures must be taken and most explosive welding is undertaken outdoors in restricted areas. The process is used for cladding metals and welding tubes to tube sheets in heat exchangers.

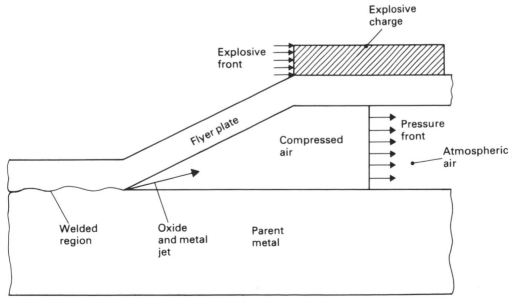

Figure 7.23 Principle of explosive welding.

7.18 ULTRASONIC WELDING

With ultrasonic welding no electric current or heat is used to form the weld. The workpieces are tightly clamped together and are made to vibrate at high frequencies. The outer layers of the metal surfaces are broken down, and the surfaces are joined by atomic forces in the form of metallic bonding. Frequencies between 1 kHz and 100 kHz are used and loads between 1 and 5000 N (0.2 lbf and 1125 lbf), depending upon the type of metal and the size of the workpieces. The temperature at the joint does not rise significantly. The process is used for spot or continuous welding, but only for overlap joints. It can be used with a wide range of metals and for joining dissimilar metals, and also for joining thin metals and thick and thin sections together.

7.19 PERCUSSION WELDING

Percussion welding uses the energy supplied by capacitors. These are discharged by an arc directly across the gap between the workpieces, as they are forced together at high speed. A transformer is not used. An extremely concentrated heated region is created, which allows dissimilar metals to be joined and also thick and thin sections.

7.20 DIFFUSION BONDING

This is not a common process and the other methods already described can often be used to produce the weld required. The process is confined to lap joints, the surfaces are joined by diffusion due to heating and pressure. The pressure used is less than for most other welding processes in order to prevent deformation of the workpieces; therefore relatively long welding times are required. The surfaces must be thoroughly cleaned and the atmosphere free from pollution; it is normally carried out in vacuum.

*   *   *

Self-assessment exercises

1 Assess the importance of the newer or alternative welding methods.
2 What new methods are likely to be developed? What are the disadvantages of these methods that have restricted their present use or development?

*   *   *

7.21 DEFECTS IN WELDS

There are many defects that can occur in welded structures, either on the surface, within the weld metal or in the heat-affected zone (HAZ) of the parent metal. Defects may be due to either poor welding technique, e.g. slag inclusions or lack of penetration, incorrect joint design or poor weldability, e.g. using incompatible welding materials. Some of the main types of welding defects are shown in Fig. 7.24.

Defects such as slag inclusions and cavities are due to poor workmanship. This may be due to the nature of the welding process or the result of human errors. Other defects are caused by technological factors. These can often be corrected or avoided by an adequate understanding of either the principles of different welding processes or the metallurgical changes that occur during welding (see Sections 7.2.5–7.2.8).

Cracking can occur for many reasons, but all depend upon the microstructure and the stress level. Hot cracking (or solidification cracking) occurs at the grain boundaries just before complete solidification is achieved. It is caused by the metal having low ductility over a particular temperature range, and being unable to accommodate the localized strains created by welding. The possibility of hot cracking can be reduced by:

(a) appropriate metal composition control;
(b) minimizing stresses;
(c) selecting a weld metal with the minimum freezing range, i.e. minimum low ductility period;
(d) control of harmful elements, e.g. low levels of sulfur and phosphorus;
(e) addition of beneficial elements, e.g. manganese which reacts with sulfur;
(f) promote the formation of ferrite rather than austenite;
(g) low current density;
(h) avoid pre-heating.

Cold cracking occurs in both the weld metal and the HAZ and may be due to hydrogen embrittlement (Section 7.2.8), excessive joint restraint, insufficient cross-sectional area of the weld, high welding speeds and low current density.

Reheat cracking almost exclusively occurs in high- and low-alloy creep-resistant steels. This problem is particularly serious and difficult to correct. One form of reheat cracking occurs within the creep temperature range due to insufficient creep ductility. This is caused by carbide precipitation and impurity segregation which strengthen the microstructure; creep strain is then confined to the grain boundaries. The selection of appropriate heat treatment and control of the metal composition can help prevent reheat cracking. Carbide-forming elements, e.g. chromium, and the formation of coarse grains in the HAZ should be avoided. A low temperature (approximately 300°C; 572°F) type of reheat cracking is caused by excessive thermal stresses; these occur during post-welding heat treatment and initiate cracks at existing defects. Heating rates and temperature distributions should be carefully controlled, and stress concentrations should be avoided.

Lamellar tearing is a defect occurring in plate metal. The defect is exposed by welding and it is not strictly a weld defect. Welding causes excessive strain in the metal and because the bonding between inclusions and the base metal is weak, tearing occurs. Brittle inclusions should be avoided and the level of

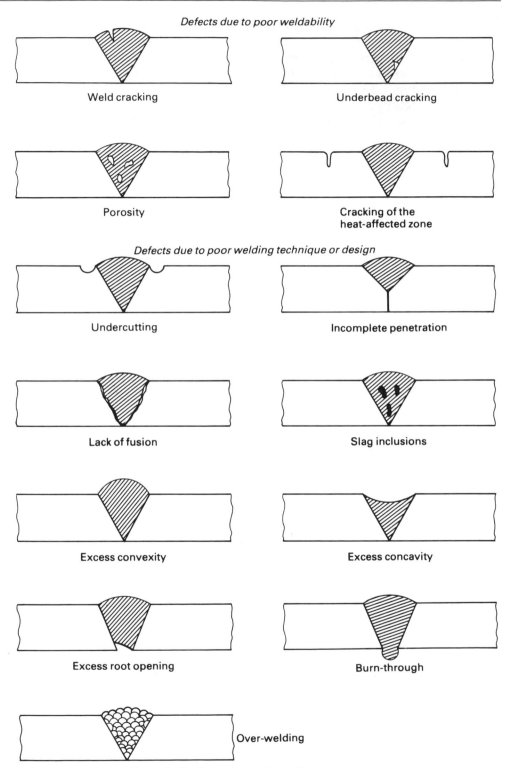

Figure 7.24 Common welding defects.

inclusions should be kept low.

The following weld defects occur due to poor workmanship or human error. Solid inclusions in the weld are the most common and most serious type of defect. These include slag inclusions, often due to poor slag detachability, oxide inclusions, particularly for aluminum joining and often as a result of inadequate precleaning, and tungsten inclusions due to poor TIG welding operation (Section 7.3.5), e.g. excessive current.

Porosity is due to supersaturation of solid weld metal with a particular gas, e.g. oxygen, nitrogen, hydrogen (see also Section 7.2.8). The pores are formed by nucleation of gas bubbles on discontinuities within the metal, e.g. at grain boundaries. The gases may be present due to air entrainment, moisture or grease on the filler rod, or chemical reaction, e.g. the formation of carbon monoxide.

Defects such as lack of fusion or insufficient penetration are usually due to either electrode manipulation, poor joint preparation or poor joint design. Undercutting is caused by use of a high welding current or slow welding speed, and excessive penetration is due to excessive heat. The causes of some defects are obvious, whereas in other cases incorrect operating procedures, e.g. incorrect current or voltage, may be to blame.

Other types of welding defect can occur. If a steel is heated to too high a temperature, or is maintained above the upper critical point for too long, then a very coarse (Widmanstätten) structure is obtained (see Section 2.4.1). Upon cooling, a coarse structure of ferrite and pearlite is obtained, possessing low impact strength and fatigue resistance. This type of microstructural defect (or undesirable condition) can be corrected by appropriate heat treatment. However, if oxygen is absorbed at the high temperature, then the boundaries of the crystals become oxidized. The steel is weakened and is said to be 'burnt'; this condition cannot be corrected by heat treatment.

If an austenitic steel is heated to between 600°C and 850°C (1112°F and 1562°F), then chromium carbide is precipitated along the grain boundaries. Adjacent areas are depleted in chromium and their resistance to corrosion is reduced. A region of the parent metal parallel to the weld is heated within this temperature range during welding, and may be subject to corrosive attack. Although corrosion does not occur in the weld itself, this effect is known as *weld decay*. The problem can be alleviated if the carbon is redistributed in solid solution by heating to 1100°C (2012°F), followed by water quenching. If the component is too large to be heat treated, carbide-forming elements such as titanium and molybdenum can be added to the steel. These elements restrict the amount of carbon available to react with the chromium; the steels are known as *stabilized steels* and contain low levels of carbon (0.03–0.1%). The carbide-forming elements should not be present in excess or ferrite may form in the weld; this reduces the corrosion resistance and may produce a brittle sigma phase (see Section 2.11.1).

The metal comprising the welded joint rarely has the same composition as the parent metal. If an electrolyte is present, corrosion can occur because of the contact between dissimilar metal regions. It is preferable if the weld metal has a (more) positive electrode potential (Table 5.1) with respect to the parent metal.

Pitting corrosion then occurs in the larger area of parent metal, rather than in the weld. The effects of corrosion within the weld itself are reduced if the composition of the weld metal is homogeneous. Detailed descriptions of corrosion mechanisms and effects are included in Chapter 5.

7.22 STRESSES AND DISTORTION

During welding a molten weldpool is formed from the melted (or fused) parent metal at the edges of the unwelded joint, and from molten filler rod or electrode. The welded joint is formed when the heat source and other welding equipment travels along the unwelded joint, and previous molten metal is allowed to cool. During welding a phase change occurs and phase transformations take place. The weld and the adjacent parent metal are subjected to conditions of expansion and contraction, and the forces that do not disappear upon cooling are called *residual stresses*. The process of removing these residual stresses is termed *stress relieving* (see also Section 1.4.9). If the joint assumes its original shape after stress relief, then it was subject to elastic deformation. If permanent distortion remains then it has been subjected to plastic deformation.

The creation of residual stresses depends upon the temperature used for welding, the cooling rate and the expansion and contraction of the metal. Any stresses present in the metal before welding, possibly due to fabrication, may be partially relieved by heat during welding. Any subsequent stresses may then be smaller. The quality of the parent metal and the filler metal, size and shape of the weld and mass of weld metal all influence the stresses set up during and after welding. Residual stresses also depend upon the type of unwelded joint that is used, the welding procedure, e.g. positioning by tack welds, multiple pass welding, and the presence of any neighboring joints. Residual stresses may be avoided or reduced if welding is performed from the fixed end of the joint to the free end, rather than in the reverse direction. This enables the parent metal to retain some movement during welding.

The decision to perform stress relieving depends upon the type of welded structure, its intended use and associated in-service conditions. The most common method is by heat treatment, where the component is heated to between 600°C and 650°C (1112°F and 1202°F), maintained at that temperature (soaked) and then slowly cooled (Section 1.4.9). *Peening* is an alternative method which consists of lightly hammering the weld and surrounding area to relieve stresses. Peening is a subject of some controversy and, if performed, it should be carried out by an experienced operator while the weld is still hot, or has just cooled.

When metal plates are welded together, expansion occurs. If the plates are rigidly held and the expansion is resisted, then deformation occurs. Upon cooling some contraction takes place; if it is prevented then stresses are created. If these stresses cause movement then distortion results; alternatively they can remain as residual stresses within the material. Distortion can be reduced by clamping the plates, or by setting the plates at a slight angle to each other, so that when cooling occurs they are in the correct alignment. Decreasing the welding speed, using a small flame, using a large-diameter electrode and low current are all advantageous for reducing distortion. However, the correct penetration and fusion of the welded joint must also be achieved. If distortion is prevented by restricting movement of the plates, then care should be taken that large residual stresses are not created.

Sheet metal and plates may be joined using the various joints shown in Fig. 7.2. The most common joints are the butt and lap joints. Butt welds (or groove welds) are generally described by the way in which the edges are prepared (see Fig. 7.2). The objective of joint design is to achieve complete penetration using the minimum amount of weld metal. These conditions need to be achieved so that the residual stresses and distortion are minimized. The welding process becomes more economical the less preparation that is required. However, an open-square butt joint has the minimum preparation, but requires excessive weld metal compared to a vee joint, as shown in Fig. 7.25. Square-edge butt joints often require the use of a backing plate, and care should be taken that its use does not cause residual stresses. The strip should be removed when welding is completed. Butt welds should be designed so that the width-to-depth ratio of the weld exceeds the critical value for crack formation. This is shown in Fig.

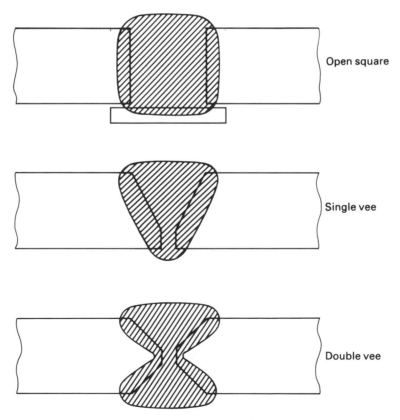

Figure 7.25 Comparison of weld metal required for square or vee butt joints.

7.26. Solidification begins along the weld surface adjacent to the cold base metal, and it is completed at the centerline of the weld. If the weld is deeper than it is wide, the surface may solidify before the center. Shrinkage forces then act on the center core and cause internal cracking.

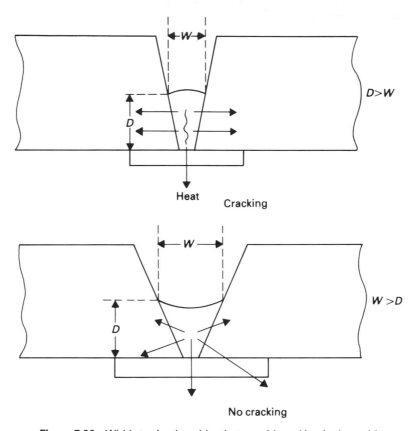

Figure 7.26   Width-to-depth weld ratio to avoid cracking in the weld.

Fillet welds are illustrated in Figs. 7.1 and 7.2. The minimum size of a fillet weld depends upon the amount of heat, and the size of the weld required to ensure fusion. Most of the heat supplied is absorbed by the plates being joined; this heat will be removed from the weld more quickly with thick plates. The surface of a fillet weld may be convex, concave or flat. The stress concentrations that occur in a fillet weld depend mainly upon whether the applied load acts parallel or transverse to the weld. In either situation shear failure occurs on the plane of maximum shear stress. The ends of manual fillet welds tend to be tapered and can cause excessive stress concentrations, especially in short weld runs. The weld should be continued 'around the corner' to avoid this situation.

7.23 INSPECTION OF WELDS AND APPROVAL

Welds should be inspected in order to detect any faults that will reduce the service life of the component. The effects of defects on the static strength are usually negligible, unless the defects are of the size caused by incorrect welding conditions. It is important to detect faults that will increase the possibility of low-stress brittle fracture, and hence reduce the fatigue life of a component. The designer must decide which faults can be allowed to remain in a structure and which must be rectified. Porosity and slag inclusions are not considered too serious, although limits are often placed upon their length. However, these types of defect can obscure more serious faults from detection. Planar defects, e.g. hydrogen cracking, lack of penetration, etc., are more serious as they can reduce the fatigue life or initiate brittle fracture. In many components planar defects are not allowed because they can cause very high local stress concentrations, and may cause premature fatigue cracking. It is usually the service conditions of the component that determine not only the defects that are allowable, but also the degree of inspection required. Safety factors, e.g. for explosive and toxic materials, are usually decisive, and so are any relevant legal requirements. When the design requires the operation of a component close to its maximum limit, then inspection becomes more important.

The terms *quality control* (QC) and *quality assurance* (QA) are often considered to be synonymous; however they represent different aspects of welding inspection and quality. The term quality control has been used since the 1940s and is often taken to include all methods used to control the quality of a product, e.g. inspection and nondestructive testing. To be strictly correct, QC includes the practical techniques and procedures, and their use, which sustain the quality of a product. The term quality assurance has only been in common use since the 1960s and includes all planning necessary to provide adequate confidence that a product will be satisfactory. QC is *used* by the supplier or manufacturer, whereas QA is *provided* for the purchaser.

If it were required to improve the quality of welds produced in a workshop (i.e. reduce the number of defects), then QC would include:

(a) deciding how to reduce the occurrence of defects; and then
(b) implementing the necessary techniques.

This might be achieved by introducing computer-controlled automatic welding machines.

QA includes these aspects, (a) and (b), as they provide some assurance of quality. However, QA also includes inspection of the product, appropriate testing and provision of adequate records and documentation relating to the inspection and testing performed. A QA (or QC) approach to a problem or a design has inherent asssociated costs. Some of these are beneficial, such as reduced scrap components or maintenance requirements, but other costs necessitate actual expenditure, mainly for quality control equipment and for appraisal.

The QA aspects of a welding process are concerned with estimating the probability that the welded structure will perform satisfactorily for its working life, and the consequences of failure. The QC procedures that are implemented relate to all aspects of the materials used, the welding processes and procedures, stresses and environmental conditions. In addition to these activities, QA includes the inspection and testing of the welded component. These are separate aspects of controlling welding quality and should be seen as such.

Inspection should be carried out by a qualified and experienced person; what is not always clear is who has the responsibility to ensure that inspection is performed. It should be the manufacturer's responsibility to ensure that a weld conforms to the specification, and to provide evidence of the standard achieved. The customer should take some responsibility for checking that a weld is satisfactory.

A welding inspector needs to be qualified and experienced in many aspects of the work, including:

(a) knowledge of codes and standards;
(b) identifying and verifying the parent metal and consumables;
(c) approval of welders;
(d) witnessing of welder and procedure approval tests;
(e) pre-heating and post-weld heat treatment;
(f) pre-weld and post-weld inspections;
(g) welding observation;
(h) nondestructive testing.

The inspection process should be performed in several stages, starting at the design stage. The inspector should check the specifications against the welding procedures to be used. Periodic inspections of welds should be performed, and the final inspection of the completed component. The acceptance of a product depends upon the level of particular defects that are present; hence the use of nondestructive testing techniques. If the level of defects is unacceptable, then the necessary repairs have to be carried out.

The acceptance of a welded structure depends upon:

(a) the expected frequency of defects;
(b) the significance of particular defects;
(c) the samples required for inspection.

The occurrence of defects due to poor workmanship (Fig. 7.24) is random, arising from local variations in the welding conditions, e.g. incomplete slag removal. Major defects occur due to incorrect welding technique such as the use of incorrect electrodes, and these are not random events. The reliability of methods for the detection of defects depends upon the size distribution of the defects present. It is difficult to describe size distribution quantitatively, but it is usually found that the majority of defects are small and large defects are less frequent, as would be expected.

The ideal situation is the production of welds containing no defects; however the costs incurred would be prohibitive and it may not be technically possible. Therefore tolerance levels of particular defects are defined for different types of constructions, e.g. pressure vessels. From a technical viewpoint this is an unsatisfactory approach and an assessment of the significance of particular defects is preferable. This includes an assessment of the possibility of a defect causing fatigue failure or brittle failure (see Sections 6.2.3 and 6.2.6); the methods used show some variation in their predictions because of the range of parameters to be considered.

When a structure contains a large number of welds, usually only a sample is inspected and tested. An exception is the extensive testing required in certain nuclear installations. The sample chosen should be representative of the welding work performed, and should include regions where failure would have the most serious consequences. Requirements for a particular sample size, e.g. 10%, should be defined in an unambiguous manner, so that it is clear whether this is 10% of the total weld runs, 10% of butt welds and/or 10% of fillet welds or both, 10% on a daily basis, etc.

After welds have been inspected they should be clearly marked. Certification should identify the regions inspected, defects discovered, repairs performed, subsequent examination, who accepted the welds, etc.

7.24 DETECTION OF WELDING DEFECTS

Various methods are used to detect faults in welds and to determine the internal soundness of components in general. Some of the techniques are expensive, and others are more involved than they may appear from the descriptions presented here (at least as far as obtaining reliable information is concerned!). Some methods require the use of a properly trained operator, and the examination of welds is often best left to experienced personnel. The designer would be wise to seek advice before commencing the design of a structure that will require detailed examination and testing.

Several methods of nondestructive testing are described in Sections 7.24.1–7.24.8. The techniques described in Sections 7.24.1–7.24.4 are used to detect surface defects such as cracks, and they require increased visual contrast between the defect and its background. The disadvantage of these methods is that the depth of the defect cannot be determined. It is usually necessary to detect internal flaws in a welded structure, and some techniques that are available are described in Sections 7.24.5–7.24.8. The two established techniques are radiography (since the 1920s) and ultrasonic testing (since the 1940s); other methods have been developed and are now widely available.

7.24.1 Visual examination

Visual examination of a weld, while being made and when completed, is usually the first stage of the inspection process. After some experience has been gained it can be used to provide a good indication of the probable strength of the weld. The naked eye with magnifying lenses and suitable illumination can establish whether the weld is the correct shape, and is also used to detect surface flaws. The dimensions of the weld can be tested using gauges, as shown in Fig. 7.27. Visual examination is usually used in conjunction with other methods that are both quicker and more reliable.

7.24.2 Dye penetration

The joint is cleaned with a solvent and allowed to dry. It is then sprayed with a

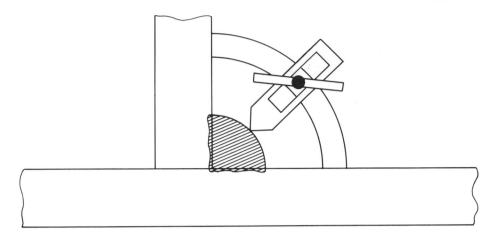

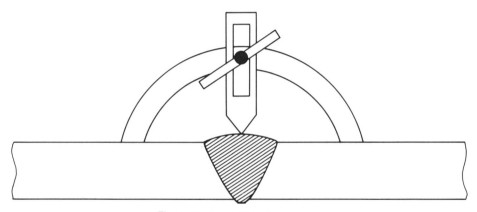

Figure 7.27 Weld testing gauges.

low viscosity, strongly colored oil or liquid. The dye penetrates any cracks or flaws, and excess dye is removed with a surface cleaner. A white chalk developer is sprayed on the surface and dye that has seeped into the defects is drawn to the surface. A variant of this technique uses a fluorescent dye, and ultraviolet light is shone onto the weld instead of a developer.

This method is less sensitive than the use of magnetic particles (Section 7.24.3), but it can be applied to any material and it is quick and easy to use. It cannot be used with porous materials, and any grease or machining particles that 'close' the defects will affect the results.

7.24.3 Magnetic particles

This method is similar to the dye penetration technique, but it is more sensitive and can detect flaws up to 0.5 mm (0.002 in.) below the surface. It is more complicated and is suitable only for ferro magnetic materials. The area for inspection must be magnetized to an appreciable depth and fine (magnetic) iron particles are applied, either dry or in a liquid-dye suspension. The particles form irregular patterns that are visible to the naked eye, or when illuminated by ultraviolet light. These patterns indicate the positions of surface cracks and sub-surface flaws, although defects at a significant depth will not be detected. It is a quick cheap method, but the component must be cleaned after testing.

7.24.4 Eddy-current testing

Eddy-current testing can be used to detect surface or subsurface defects. However, it is not widely used for welded joints because of its sensitivity to metallurgical differences and the surface conditions. The method consists of subjecting a conducting material (e.g. a metal) to a magnetic field, and observing the induced eddy currents. The equipment required includes a probe, a test sample and a calibrating instrument. Instruments are available to make either absolute or differential measurements; they can be used to detect metallurgical properties such as case hardness or changes in alloy composition. The change in eddy-current distribution around a defect is shown in Fig. 7.28. The size of the defect is indicated by the magnitude of this change, and is seen as a deflection in the impedence plane projected onto the screen of a cathode ray tube. The method is fast and relatively inexpensive; it can be used to locate defects or provide information regarding material properties. The extreme sensitivity of the method is a disadvantage, both for the operation of equipment and the interpretation of results.

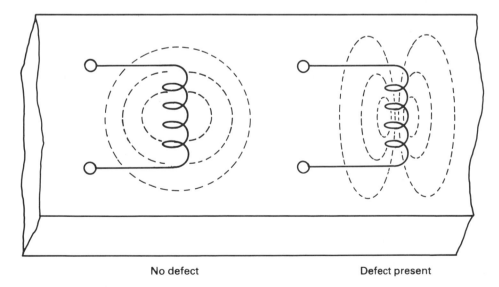

No defect   Defect present

Test probes using alternating current

Figure 7.28   Eddy current testing.

7.24.5 Radiography

Radiographic methods involve passing a penetrating beam of radiation through the object to be examined. Different sections or defects absorb varying amounts of radiation, the intensity of the emergent beam varies and is recorded on a radiographic film. Large slag inclusions and pores in the weld metal absorb little radiation and show up as darker regions on the film. X-ray radiography is used for detecting voids, porosity, cracks and inclusions in ferrous and nonferrous metals. Hard X-rays (using high voltages) are more penetrating and are used for thick plate; soft X-rays are used for thinner materials. Radiography is an old established technique, it is still widely used because it is accurate and provides a permanent record of the test. Radiographic equipment and techniques are being continually developed, mainly to provide smaller, portable equipment using shorter exposure times.

The principle of using X-rays to detect defects in welds is illustrated in Fig. 7.29. To ensure that a correctly exposed negative is obtained, and containing details of the smallest defects, a penetrometer (image quality indicator) is used (Fig. 7.29). This is a small strip of the same material as the component being tested, and it is placed on the upper surface of the component (closest to the X-ray tube). It is approximately 1% of the component thickness, and will be shown as a shadow on the negative. Therefore any defects of at least that thickness will also be recorded.

The time required to obtain results from an X-ray film can be significantly reduced by using a display system involving fluoroscopy. An image intensifier is used: the X-rays fall on a phosphor coating which fluoresces and excites the photocathode. Electrons emitted by this excitation are accelerated and focused on an output phosphor coating. The intensified image can be viewed continuously using a television monitor. This technique is used for examination of longitudinal welds in submerged–arc welded pipelines. The fluoroscopic image loses some sensitivity, i.e. contrast and resolution, compared with a good radiograph but it is very much quicker.

For testing thicker material (over about 10 cm; 4 in.), the greater penetrating power of gamma rays is required. They have a shorter single wavelength which is characteristic of the isotope used, and they are continuously emitted. The source of X-rays can be turned on and off, and their wavelength depends upon the applied voltage. Cobalt-60 is a common source of gamma rays. Gamma radiography is portable and of low initial cost compared to X-ray equipment.

Neutron rays (N-rays) have also been used as a source of penetrating radiation. They are absorbed only by certain materials and the absorption is not related to the density of the material. The N-rays that are not absorbed after passage through the test material are neutral and will not expose an X-ray film. They are projected onto a conversion screen, which produces ionized radiation which exposes the film. Hydrogen has the highest absorption coefficient for N-rays and it is virtually impossible for them to pass through. Neutron radiography is a particularly useful testing method for hydrogen embrittlement, or checking for voids that have been filled with water. Therefore, X-rays and N-rays are

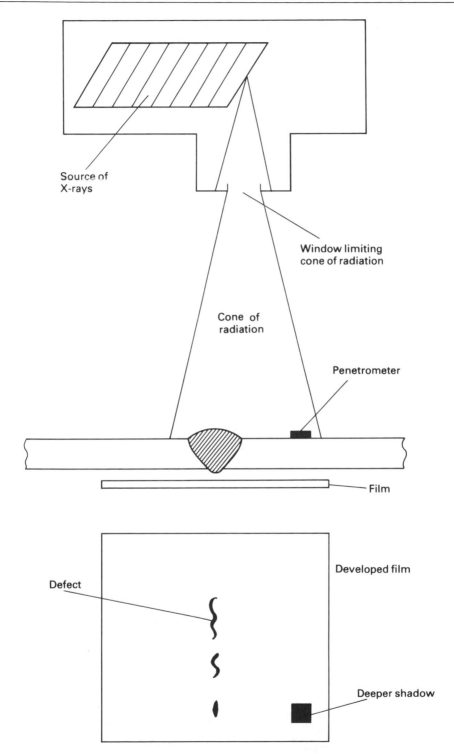

Figure 7.29 X-ray examination of welds.

complementary testing methods. Unfortunately there is no low-cost small source of N-rays available; the best alternative is the use of a radioisotope, although it will be continually decaying.

The X-ray method is quicker than using gamma rays (minutes instead of hours), it is also more accurate for the detection of small defects in sections less than 5 cm (2 in.) thick. Both methods produce good results for plate thicknesses between 5 cm and 10 cm (2 in. and 4 in.). Gamma rays have less scatter and are preferable for components of varying thickness. Gamma rays can be used for testing pipe and tube by placing the film around the outside and the radiation source inside (at the center), in order to expose all surfaces together.

The main disadvantage of using the radiographic technique is the possibility of health hazards and the associated safety precautions required. However, this method provides a permanent test record, it is accurate (despite some differences of expert opinion!) and it will identify all significant defects, their size and location.

7.24.6 Ultrasonic testing

Ultrasonic testing involves the use of a transducer producing frequencies between 1 and 25 MHz (usual operating range of 2–6 MHz), to generate mechanical vibrations. These vibrations are well-defined small-diameter beams which obey the laws of reflection and refraction. The presence of all types of defects can be detected due to the reflection or refraction of the beam at the discontinuity. Two types of wave motion are used for testing; these are compression (or longitudinal) waves and shear (or transverse) waves. The basis of both types is shown in Fig. 7.30; a combined transmitter and receiver is shown although they can be separate devices. The transducer is usually a hand-held probe coupled to the test metal by a layer of water or oil. A short-pulse technique is most frequently used, although resonant instruments are available. The reflected beam is displayed on a cathode ray oscilloscope.

All types of flaws can be detected, but for planar defects the beam must be normal to the plane of the defect. Although the location of defects can be determined, the size and type of a defect is more difficult to determine accurately. It may be necessary to take readings in two directions, and because of the small diameter of the beam the entire surface must be scanned. This problem is partially alleviated by using the delta method of testing. The transmitter directs the beam as a shear wave adjacent to the weld, and a separate receiver is used to detect any feedback that has a frequency different from the natural frequency of the material. Movement of the transmitter towards and away from the weld allows vertical scanning in the section through the weld and provides a complete and accurate coverage.

Ultrasonic testing is unsatisfactory for coarse-grained austenitic steels, or in very thin materials (less than 6 mm; 0.25 in.) where the depth of metal inspected is less than the beam diameter. The equipment needs to be checked against standard reference blocks containing different sized holes. However, ultrasonic testing is a relatively inexpensive technique, it is quick and can usually be carried out with the component *in situ*. Materials up to 30 cm (12 in.) thick can be examined with negligible loss of accuracy.

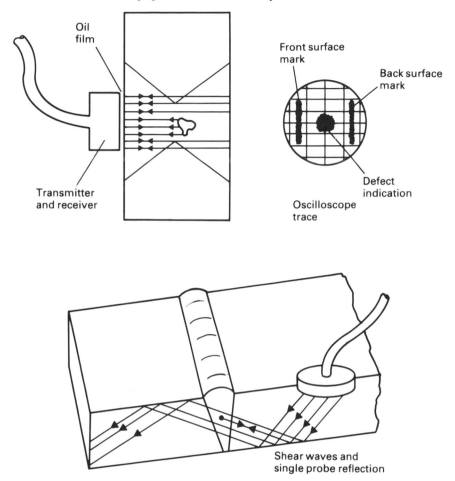

Figure 7.30 Principles of ultrasonic testing.

7.24.7 Holography

Holographic testing combines laser photography and laser interferometry. A holograph is a three-dimensional image, and an interference pattern is composed of bands of equal intensity or temperature. The technique that is used is known as holographic interferometry. The detection method involves comparing a holographic image of the object in its 'standard' state and superimposing it on the image when an energy disturbance, e.g. ultrasound, is applied. The differences between the images are used to identify defects that are present. Holography is a very rapid technique and does not require special surface preparation, attachments or coatings. It provides a visual presentation and can be used to detect stress corrosion and thermal fatigue cracking. However, the environment must be free of vibration and it is difficult to identify the types of defects that are detected.

7.24.8 Other methods

Microwave testing uses electromagnetic radiation to detect flaws. Internal and external defects can be identified, and changes in thickness. The method can be used with polymers and ceramics but is unsuitable for some metallic conductors, e.g. steel. A permanent film record can be obtained, but initial equipment costs are high and health and safety precautions must be observed.

Sonic testing uses variations in the resonance of the workpiece to indicate the presence of defects, rather than the interpretation of returning sound waves. It is portable and less expensive than ultrasonic testing, but reference standards are required and the surface geometry influences the test results.

Acoustic emission testing measures the stress waves present in a structure soon after it is formed, to indicate the presence of defects. There is a time limit for the detection of flaws, e.g. only up to 20 minutes in welds, and the method is not suitable for very ductile materials that have low-amplitude emissions.

Thermal testing requires the application of heat over the specimen; this heat diffuses uniformly unless defects are present. A defect can be identified as a hot spot at the surface of the specimen. These changes in the surface temperature can be detected by applying heat-sensitive coatings. *Infrared methods* consist of focusing a visible radiation beam across the sample, and measuring the surface temperature remotely using an infrared radiometer.

7.24.9 Destructive tests

The tests described in Sections 7.24.1–7.24.8 are all nondestructive tests. The specimen should not be damaged by the testing method and the defects that are detected should be capable of repair or removal. Tests should also be performed where the detection of defects leads to failure of the component.

Laboratory tests are conducted on samples of the material that was used to produce the component, or on actual components if they are small mass-produced items. Material samples should be carefully chosen so that they are representative of the actual material conditions, e.g. the effect of rolling on the properties in a particular direction or hardness variations due to different cooling rates. Samples should be prepared for testing so that the test results are not affected by the way in which the sample was obtained, e.g. by sawing or thermal cutting.

Mechanical tests are performed on samples; these tests have been described in Section 6.2. The tensile test (Section 6.2.1) is the most common but tests appropriate to the in-service conditions, and the properties required of the material, should be performed. These tests usually result in destruction of the specimen, except for hardness tests (Section 6.2.2). It will be necessary to describe how many samples need to be tested, and reference should be made to the appropriate standards.

Chemical tests should be performed on material specimens. These may be analytical tests to determine the chemical composition of the weld metal, or corrosive tests to determine the effects of particular environments. Other laboratory tests include examination of carefully prepared specimens using a metallurgical microscope, in order to determine the microstructure of the weld and the parent metal. Specimens can also be examined using a magnifying glass to identify cracks, inclusions, blowholes, etc. This macrographic examination will not identify smaller defects, but the sample preparation time is shorter.

Workshop tests should be performed, such as 180° bend tests on welded sections, and examination for cracks. Drop tests using large loads (45 kg; 100 lb) from a height of 2–4m (6–12 ft) can be used to simulate impact loading. A stop block should be positioned to limit the amount of induced strain. Sample sections of welds can be broken open in a vice and examined for the presence of defects.

Tests should be carried out on the completed (or semi-finished) component under conditions more severe than are normally (to be) encountered. An example is the hydraulic test on boilers, where water is pumped into the welded boiler to a pressure of 1.5–2 times the working pressure. If a fault develops the hydraulic pressure rapidly falls, without the danger to personnel that would be caused by a compressed gas. Similarly, compressive or tensile loads can be applied to a structure to observe the resultant behavior.

7.24.10 BS and ASTM Standards for testing of welds

BS 709: Methods of destructive testing fusion welded joints and weld metal in steel.
Object of tests; preparation and dimensions of test specimens; macro-examination and tests for bend, nick break, fillet weld fracture, hardness across the welded joint, Charpy V notch impact, intercrystalline corrosion and fracture toughness.

BS 2600: Methods for radiographic examination of fusion welded butt joints in steel.
Part 1: 5 mm up to and including 50 mm thick.
Part 2: Over 50 mm up to and including 200 mm thick.
General requirements for techniques, films, gamma ray sources, protection, surface preparation and processing. Specific requirements for particular techniques.

BS 3451: Methods of testing fusion welds in aluminum and aluminum alloys.
Reporting of results for visual examination aided by penetrant methods, radiography, transverse tensile test, macro-examination, bend tests, etc.

BS 3923: Methods of ultrasonic examination of welds.
Part 1: Manual examination of fusion welds in ferritic steels.
Part 2: Automatic examination of fusion welded butt joints in ferritic steels.
Part 3: Manual examination of nozzle welds.

BS 4360: Specification for weldable structural steels.
Requirements for general and weather-resistant weldable structural steels,
including chemical composition and mechanical properties.

BS 4416: Method for penetrant testing of welded or brazed joints in metals.

*PD 6493: Guidance on some methods for the derivation of acceptance levels for defects in
fusion welded joints.*
Simplified treatment of the use of fracture mechanics methods to establish
acceptance levels based on fitness for purpose. Applies to fusion welded joints in
ferritic steels, austenitic steels and aluminum alloys at least 10 mm thick.
ASTM Standards: volumes 03.01, 03.02 and 03.03.

E190: Standard method for guided bend test for ductility of welds.
For use with ferrous and nonferrous metals, the test is developed for plates and
is not intended to be substituted for other methods of bend testing. When a
specimen is subjected to progressive localized overstressing, defects not shown
by X-rays may appear in the surface.

*E328: Standard recommended practices for stress-relaxation tests for materials and
structures.*
A broad range of testing activities is covered. A general section is applicable to
all stress-relaxation tests for materials and structures, e.g. temperature control,
humid ty control, etc. This is followed by sections that apply to tests or material
characteristics when subject to specific, simple stresses, e.g. uniform tension,
uniform compression, bending or torsion.

*G58: Standard practice for the preparation of stress corrosion test specimens for
weldments.*
This practice describes procedures for making and using test specimens for the
evaluation of weldments in stress corrosion cracking environments. Test
specimens are described where stresses are developed either by the welding
process only, by an externally applied load in addition to welding stresses, or by
an externally applied load only with residual welding stresses removed by
annealing.

E164: Standard practice for ultrasonic contact examination for weldments.
This practice covers techniques for the ultrasonic A-scan examination of
specific weld configurations, joining wrought ferrous or aluminum alloy
materials, to detect weld discontinuities. The reflection method using pulse
waves is specified. Manual techniques are described employing contact of the
search unit through a couplant film or water column.

*E273: Standard practice for ultrasonic examination of longitudinal welded pipe and
tubing.*
Describes general ultrasonic test procedures for the weld and adjacent HAZ of
pipe and tubing > 50 mm (2 in.) diameter, and wall thicknesses of 3–27 mm
(0.125–1.0625 in.).

E390: Standard reference radiographs for steel fusion welds.

E749: Standard practice for acoustic emission monitoring during continuous welding.

E750: Standard practice for measuring the operating characteristics of acoustic emission instrumentation.

E751: Standard practice for acoustic emission monitoring during resistance spot-welding.

7.25 COMPARATIVE WELDING COSTS

It is often possible to obtain the required quality of a welded joint by a variety of welding methods. In this case it will be necessary to perform calculations to determine the comparative costs between different processes. The design of the component and the size of the weld run will determine whether a manual, semi- or fully automatic process needs to be employed; this has a significant effect on the production costs. Equations for the calculation of welding costs are not presented here, but these are published in various manufacturers' literature and in specialized welding handbooks. However, a list of the main factors that affect the final production cost for most fusion welding techniques will be given. These are usually evaluated per meter (or foot) of a weld run, or sometimes per hour.

7.25.1 Machine costs

This requires consideration of the capital cost, depreciation period and interest charges involved. It also includes the maintenance costs, the expected number of hours each year for which the machine will operate, and the estimated machine life.

7.25.2 Material costs

This includes the cost of electrodes, flux and gas, and any other ancilliary materials which may be required, e.g. cooling water. Both the long- and short-term availability of these materials should be considered.

7.25.3 Wages costs

These costs depend upon the quantity of welding to be performed and whether the welding operator is a permanent employee or is employed on an hourly contract basis. A permanent employee is subject to certain fixed costs, such as National Insurance payments. Wages costs obviously depend upon whether the process is manual or automatic, although even automatic processes incur labor costs for setting up the equipment prior to welding.

7.25.4 Energy costs

These are the costs most likely to change during the life of the machine or even the period of production, often without warning due to world economic factors. These costs include not only the cost of energy required for the welding operation and for the surrounding metal, but also any personal heating required by the welding operatives.

7.25.5 Other costs

Often costs incurred due to edge preparation, tack welding, transport, etc., are independent of the welding process used. It may be that transport and erection costs are significantly different for a welded structure produced in a workshop using a fully automatic process than for manual welding on site.

* * *

Self-assessment exercises

1 Prepare a list of welding defects in order of:
   (a) seriousness;
   (b) ease of detection.

2 State how each welding defect could be avoided.

3 Prepare a list of detection or testing methods for welding defects, according to:
   (a) costs incurred;
   (b) ease of operation;
   (c) reliability.

4 Obtain equations that can be used to calculate welding costs. Assess the usefulness of these equations, and try to use them to compare the economics of different welding methods in the same situation.

5 Which welding cost is the most significant for different processes?

* * *

7.26 MECHANICAL METHODS

These methods are sometimes classified as joining with residual stresses. The two main methods used are riveting and fastening. When using mechanical fasteners the parts can usually be easily separated for maintenance or repair; this is not the case with riveted structures.

7.26.1 Riveting

The shapes of the most common rivets are shown in Fig. 7.31. These are obtainable in a wide range of sizes and in the more common metals such as aluminum, brass, copper and steel. Before riveting, the rivets are headed at one end (see Fig. 7.31) and the shank is longer than the combined thickness of the plates to be joined. The plates are drilled to accept the rivet. The joint is made by forming a second head on the end of the shank using special equipment,

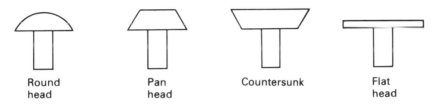

Round head   Pan head   Countersunk   Flat head

Figure 7.31 Rivet shapes.

either by machine or by hand, and the shank is made to fill the hole as shown in Fig. 7.32. Steel rivets are used hot in a plastic condition and are stressed upon contraction to draw the plates together. Aluminum alloy rivets are solution treated to become soft, and then undergo precipitation hardening after joining. The joints used are generally either lap joints or butt joints (using one or two cover plates). The rivet size and spacing is usually calculated such that the rivets will fail before the plates, these being easier and cheaper to replace.

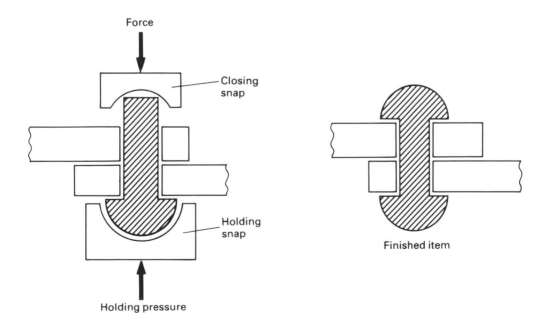

Figure 7.32 Riveting.

7.26.2 Fastening methods

(a) Screws
Cap screws are used to hold two or more pieces of metal together, they are shown in Fig. 7.33. One part is drilled to a suitable clearance size and the other part is screwed to accept the screw shaft. The screw may be tightened by either a hexagon socket and wrench, by a hexagon and spanner, by a Pozidriv (cross recess) or a slot and screwdriver.

Figure 7.33 Hexagon socket head cap screw.

Set and grub screws are used to prevent motion between the two parts, the methods of tightening them are the same as for cap screws. A typical set screw is shown in Fig. 7.34, and types of screwdriver slot heads in Fig. 7.35. Screws are useful when there is no room for a nut, or two spanners cannot be used. Easy removal is possible by using a thumb screw as shown in Fig. 7.36.

(b) Nuts and bolts
Both parts to be joined are drilled to a suitable clearance size. There must be easy access to both sides of the parts to be joined, so that the bolt can be assembled in the holes and also for tightening and releasing the parts. Bolts are

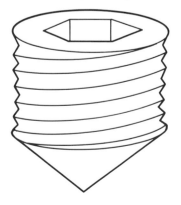

Figure 7.34 Hexagon socket set screw.

usually either square or hexagonal headed. They are specified by a diameter or number, type of thread, length to underside of head, and in special cases the threaded length is also given. Nuts are made to suit all threads and are usually either plain (square or hexagonal) nuts or special locking nuts. A typical assembly is shown in Fig. 7.37.

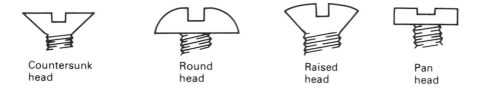

| Countersunk head | Round head | Raised head | Pan head |

Figure 7.35   Screw heads with screwdriver slot.

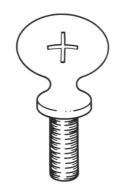

Figure 7.36   Thumb screw.

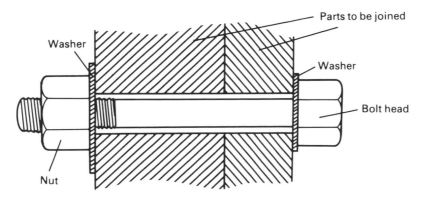

Figure 7.37   Typical nut and bolt assembly.

(c) Studs

A tap bolt or a stud and nut are used when joining two parts together, if it is preferable not to have a nut or bolt head projecting from one side of the structure. The stud is either screwed into the larger component, or is attached to it during casting or molding. The smaller component is then joined by a nut at the end of the stud. Only one spanner is required to tighten and release the parts. This arrangement is used when the two pieces are to be frequently separated. A tap bolt is usually square or hexagonal headed, and is threaded at the opposite end. The bolt is screwed directly into the larger component, with the head securing the smaller component. Figure 7.38 shows the typical assembly of a tap bolt and stud and nut.

When a nut must be removed easily and without the use of tools, then a wing nut is used as shown in Fig. 7.39.

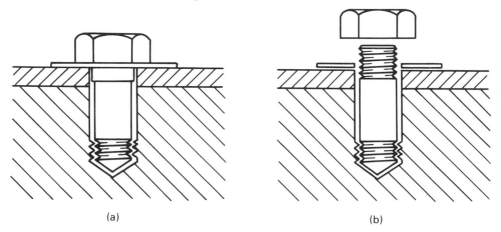

(a)   (b)

Figure 7.38   (a) Tap bolt and (b) Tap stud and nut.

Figure 7.39   Die-cast wing nut.

With these fastening methods, use is made of washers and locking devices. These may be coil spring washers, tabwashers, tooth lock washers, locking plates or locking nuts. Two thin lock nuts may act together, or a slotted nut can be used in conjunction with a cotter pin (split pin). These devices are shown in Fig. 7.40.

Fastening methods are used when it is necessary to separate the parts for maintenance or repair, without damage to the parts or the fastener. It is usual to replace tabwashers and locking plates before re-assembly.

7.27 ADHESIVES

Recent advances in the field of adhesive bonding have led to applications in areas that have traditionally employed soft soldering, riveting and welding

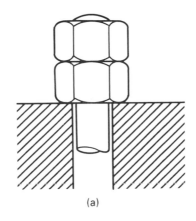

(a)

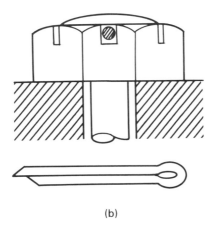

(b)

Figure 7.40   Lock nuts and slotted nut. (a) Lock nuts. (b) Slotted nut and split pin.

techniques. Advantages may be in terms of reduced costs, reduced weight, quicker bonding and in some cases greater strength. Some adhesives now possess strengths greater than certain metals. Before selecting an adhesive a number of factors must be considered. These include the service conditions for the joint, particularly the type and magnitude of the applied stress. The bond may be subjected to vibratory or cyclical loads, in which case a flexible bond would be more appropriate. The in-service temperature and the duration of exposure of the bond must also be considered. The presence of any chemicals in the environment which may affect the performance of the adhesive should be known. Many other factors need to be considered, particularly the economics of the process compared to the alternatives available, including the cost of labor, equipment and materials. The method of applying the adhesive and the conditions required for application must be considered.

    An adhesive must be selected that is suitable for use with both of the materials to be joined. Table 7.6 lists some of the common types of adhesives that may be used to join certain materials. If no suitable adhesive is available, it

Table 7.6 Principal properties of adhesive types.

To select the most suitable adhesive type(s) for bonding two materials, search these lists for a bonding agent compatible with both adherends (e.g. for rubber/polystyrene the only suitable adhesive would be an epoxy; for metal/polyamide there is a choice between epoxies, phenolics or rubber-based adhesives).

Adhesives for metals in structural applications
Cyanoacrylate
Epoxy
Epoxy-phenolic
Epoxy-polyamide
Epoxy-polysulfide
Epoxy-silicone
Phenolic-neoprene
Phenolic-nitrile
Phenolic-polyamide
Phenolic-vinyl
Polyurethanes

Adhesives for metals in non-structural uses
Acrylics
Polyvinyl acetate
Rubber-based – natural
Rubber-based – synthetic
Polyamides

Adhesives for plastics
Acrylics such as Perspex:
  Acrylic in solvent solution
  Epoxy
  Resorcinol formaldehyde
  Rubber-based – synthetic

Cellulose acetate:
  Cellulose acetate in solvent
  Cellulose nitrate
  Epoxy
  Polyvinyl acetate
  Resorcinol formaldehyde
Cellulose nitrate:
  Epoxy
  Resorcinol formaldehyde
Polyamides such as nylon:
  Epoxy
  Phenolics
  Resorcinol formaldehyde
  Rubber-based – synthetic
Polystyrene:
  Epoxy
  Polystyrene in solvent solution
Polyethylene and PTFE:
  Phenolics
  Silicone
PVC:
  Acrylics
  Neoprene rubber
  Nitrile rubber
  Phenolic-neoprene
  Phenolic-nitrile
  Phenolic-vinyl

Thermosetting plastics including phenolics, urea, melamine and polyester laminates:
  Epoxy
  Furane
  Phenolic-neoprene
  Phenolic-nitrile
  Phenolic-vinyl
  Phenol formaldehyde
  Polyester
  Resorcinol formaldehyde
  Rubber-based – synthetic
  Urea formaldehyde

Adhesives for rubbers
Chlorinated rubber
Epoxy
Epoxy-polyamide
Natural rubber solution and latex
Neoprene rubber
Nitrile rubber
Phenolic-neoprene
Phenolic-nitrile
Phenolic-vinyl
Resin-rubber formulation
Silicones

may be possible to include an intermediate layer of material between the materials to be joined, and then use two adhesives suitable for each joint. However, the overall bond strength and other in-service conditions must be considered.

The following classifications are often used for adhesive materials. First, natural resins including rubber-based materials and proteins, e.g. animal glues, these have been used for many years and are relatively cheap. Second, inorganic materials based on silicates. Third, synthetic resins based on thermoplastic or thermosetting polymer resins. These include vinyls, cellulose esters, alkyds, epoxies, phenolics, etc. Many rubber-based adhesives are cured by solvent evaporation. The surfaces to be joined must be clean and free from dirt, grit and grease. Surface oxides can also be removed using abrasives, thus leaving a roughened surface with improved bonding. The application of the adhesive is usually by brushing, rolling, spraying, coating or dipping. Both surfaces should be thoroughly wetted and the recommended setting procedures observed; these may require heat and/or pressure to be applied.

The mechanism of adhesion is probably a combination of mechanical locking between the adhesive and the surface, and molecular (or van der Waals) forces between the molecules in the adhesive and the atoms in the adherent surface. Four basic types of stress are encountered in bonded joints, as shown in Fig. 7.41. For optimum performance, structural joints should be designed so

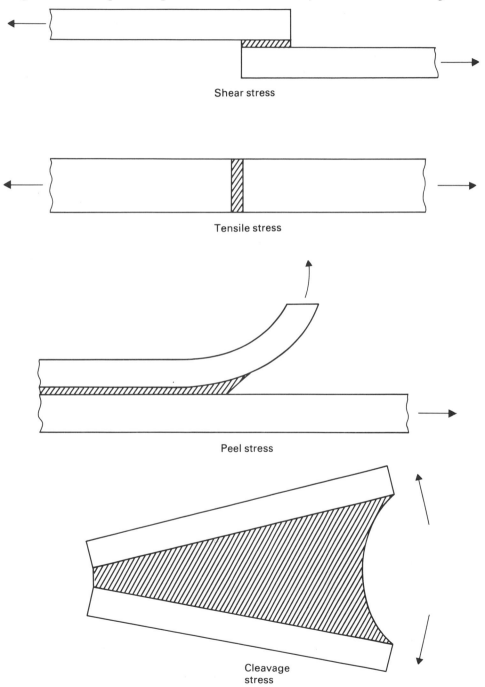

Figure 7.41 Basic stress situations with adhesive bonding.

that the applied load is evenly distributed over the bonded area. This will occur only if the bond is subjected to either tensile or shear stress, where all the adhesive works together and contributes to the strength of the joint. A joint stressed by tearing or cleavage would be less suitable because the stress is concentrated in only a small portion of the adhesive. With peel stress only a very thin line of adhesive at the edge of the bond is subjected to the stress, and this type of joint should be avoided.

The strength of an adhesive joint can be greatly increased by using appropriate design features. The lap joint is probably the most commonly used joint for shear stress. However, if the shear forces do not act in the same plane, then the joint will distort to try and achieve this situation. This may cause cleavage stress or peel stress at the edges, depending upon the magnitude of the load. In Fig. 7.42, it can be seen that the use of a joggle lap joint brings the bonded area and the shear stress into the same plane and provides a stronger joint. The tapered edges of the parent metal allow bending of the joint edge rather than bending of the adhesive film. The use of the double butt or double scarf lap joints greatly increases the resistance to bending, but these joints need machining and are therefore more costly. The rigidity of corner joints can be increased by the designs shown in Fig. 7.43, and the angle joints as shown in Fig. 7.44. Butt joints can be strengthened by using recesses as shown in Fig. 7.45. The common types of cylindrical joint are shown in Fig. 7.46. Stiffeners are often bonded to thin metal sheet in order to increase their rigidity; the main shapes used are shown in Fig. 7.47.

It should be remembered that technical advice is usually freely available

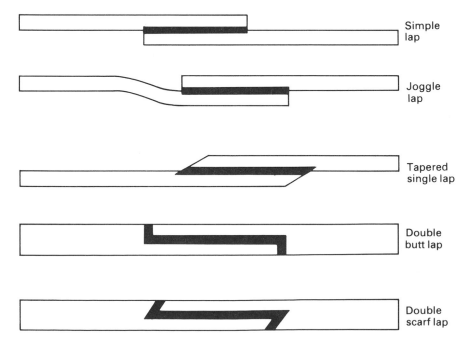

Figure 7.42 Adhesive bond lap joint design.

from adhesive manufacturers, and consultation in the early stages of a design problem can often prevent expensive mistakes.

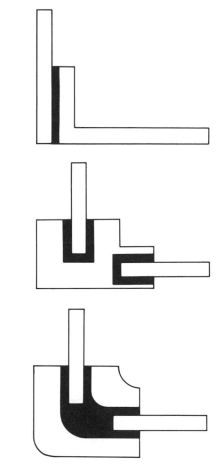

Figure 7.43   Corner joints.

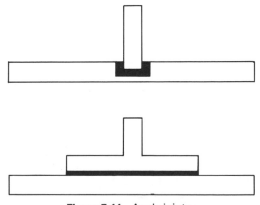

Figure 7.44   Angle joints.

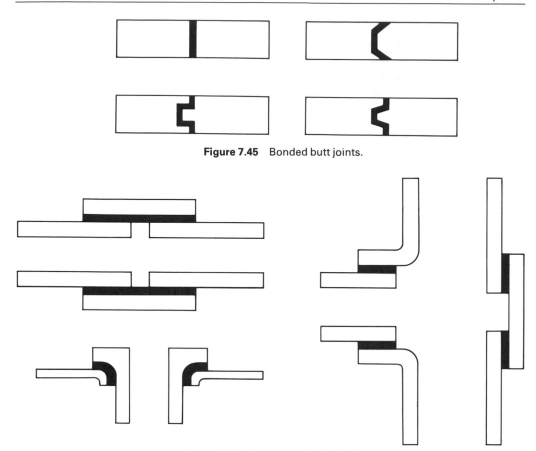

Figure 7.45 Bonded butt joints.

Figure 7.46 Bonded cylindrical joints.

7.27.1 Weldbond

The process known as *weldbonding*, or *glue welding*, combines the techniques of spot resistance welding and adhesive bonding. The process is mainly used for lap joints, and two alternative techniques are applied, as illustrated in Fig. 7.48. With the *flow in* technique, the parts are welded together and then a low-viscosity adhesive penetrates the overlap joint by capillary action. This is followed by curing. For the *weld-through* technique the sequence is adhesive bonding, spot welding and curing. The process is mainly used in the production of trucks and lorries, although it has also been evaluated for aircraft and space-vehicle manufacture. Compared to spot welding, the weldbond process has the following advantages:

> higher static strength;
> improved fatigue strength;
> avoids sealing operations;
> improved corrosion resistance;
> applicable to aircraft design.

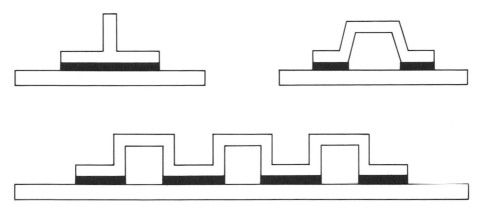

Figure 7.47   Stiffener joint design.

Compared to mechanical joining methods, the weldbond process has the following advantages:

> higher static strength;
> improved fatigue strength;
> eliminates holes;
> no alignment or fitting problems;
> improved corrosion resistance;
> avoids sealing operations;
> eliminates noise during manufacture;
> enables design with thinner materials;
> reduced manufacturing costs;
> can be adapted to mechanized and automatic processes.

Compared to adhesive bonding, the weldbond process has the following advantages:

> reduced labor;
> simplified tooling;
> eliminates expensive equipment;
> can be used for large structures.

* * *

Self-assessment exercises

1 What are the main advantages of mechanical joining?

2 If rivets of the same material are not available, what material should be chosen when joining parts made of:
   (a) aluminum;
   (b) mild steel;
   (c) galvanized iron?

3 Compare the cost of riveting with a metallurgical joining process.

4 What are the most common adhesives used for adhesive bonding?

5 What happens if a structure joined by adhesive bonding, is used at an in-

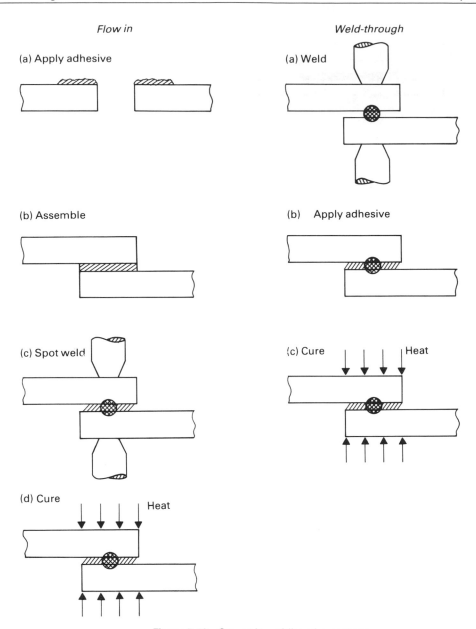

Figure 7.48 Stages in weldbond processes.

service temperature that is higher than the recommended temperature?

6 Describe the main applications and disadvantages of epoxy adhesives.

7 For weldbonding, describe the problems that can occur if spot welding is used *after* the adhesive is applied.

8 What effects may occur in an adhesive bond if it is subjected to low temperatures?

* * *

* * *

Exercises

1 Prepare a table of arc-welding methods, designating them as:
  (a) a.c. or d.c.;
  (b) automatic or manual;
  (c) constant current or constant potential;
  (d) flux or gas shielded.

2 Explain the significance of *fusion* in welding .

3 What are the basic metallurgical characteristics of the deposited weld metal in a fusion weld?

4 Select a shielding gas for the following MIG processes:
  (a) joining mild steel;
  (b) joining two pieces of thin aluminum sheet;
  (c) welding thick aluminum plate.

5 Why are carbon electrodes rarely used in arc welding?

6 What is the weakest part of a fusion weld? Why?

7 Why is a crack in a welded structure more serious than in a riveted structure?

8 Explain why a flux is not required for resistance welding processes.

9 What are the differences between projection, seam and spot welding?

10 Explain why shallow penetration is required when cladding mild steel plate with Monel.

11 Does any melting occur during explosive welding? Why is this process used for joining dissimilar metals?

12 Explain why touch starting is not allowable when welding titanium using a TIG process.

13 Explain why the corrosion resistance of welded 304 stainless steel sheet is reduced if 304 stainless steel wire is used as the filler.

14 Describe the difference in appearance of the welded joint when using percussion welding or flash butt welding.

15 What are the advantages of brazing compared to welding? What is the main disadvantage of brazing?

16 Describe the characteristics of copper that make it suitable for use as spot welding electrodes.

17 Compare the maximum in-service temperatures of adhesives with those of solders.

18 How are epoxy resins cured?

19 What is a structural adhesive?

20 For microjoining, explain why the plasma needle-arc method is unsuitable for welding aluminum or magnesium.

21 Describe the use of an electron beam as a fast, microjoining riveting machine.

22 A gas exhaust duct is prefabricated from aluminum sheet joined with aluminum rivets. Suggest why corrosion occurs around each rivet after a period of prolonged operation.

* * *

*   *   *

Complementary activities

1 Observe soldering, brazing and welding operations performed by trained operators. Observe fusion-welding and pressure-welding operations. Examine equipment used for different welding operations.

2 Prepare metal surfaces so that they are suitable for joining. Perform joining operations (under supervision), including soldering, brazing, gas welding, forge welding.

3 Test the strength of the welded joints obtained.

4 Examine the welds visually with the naked eye, and prepared specimens using the metallurgical microscope.

5 Identify the nature and causes of any defects that are present.

6 Break open some of the joints and examine their internal condition.

7 Perform metal cutting operations.

8 Examine some common welded structures, e.g. tanks, support frames, stair rails, car bodies, etc., and evaluate the quality of the welds.

9 Perform some riveting operations.

10 Study the range of mechanical fasteners that are available, visit the local hardware store or engineering supplier.

11 Prepare some simple structures joined together using different mechanical fasteners. Dismantle the structures and assess the performance of the different fasteners.

12 Prepare some specimens and join them using different adhesives. Test the strength of these joints using mechanical testing machines (Section 6.2).

13 Compare the strengths of joints made by welding, mechanical fasteners and adhesives.

*   *   *

KEYWORDS

joining processes	solder	soldering iron
metallurgical	brazing alloy (spelter)	torch flame
mechanical	flux	dip soldering
adhesive	slag	furnace heating
weld (welded joint)	arc	induction coil
joint (unwelded joint)	electrode	resistance soldering
butt weld	penetration	acid flux
fillet weld	weldpool	salt flux
butt joint	MMA	resin flux
T joint	MIG	borax
corner joint	TIG	fretting
flanged joint	arc time factor	HAZ
workpiece	deposition rate	weld deposit
parent metal	effective value	fusion zone
filler metal	change value	columnar grains
soldering	metal recovery	recrystallization
brazing	soft solder	transformation zone
braze welding	hard (silver) solder	carbon equivalent (CE)
welding	metallic bond	weld absorption
surfacing	capillary action	fusion welding

arc welding
gas welding
electro-slag welding
consumable nozzle
  process
electron beam welding
laser welding
thermit welding
pressure welding
forge welding
resistance welding
high-frequency pressure
  welding
cold-pressure welding
friction welding
explosive welding
ultrasonic welding
percussion welding
diffusion bonding
air gap
ionization
electrode coating
shielding gas
basic electrode
rutile electrode
gravity welding
submerged-arc welding
active gas (CO_2)
inert gas (Ar, He)
plasma-arc welding
carbon-arc welding
stud welding
air-acetylene flame
oxy-acetylene flame
thermal cutting
plasma cutting

metal spraying
  flame spraying
  arc spraying
spot welding
projection welding
seam welding
resistance butt welding
flash welding
weld defect
welding technique
weldability
workmanship
hot cracking
cold cracking
reheat cracking
lamellar tearing
porosity
weld decay
stabilized steels
residual stress
stress relief
peening
distortion
inspection of welds
quality control (QC)
quality assurance (QA)
weld acceptance
certification
nondestructive testing
visual examination
dye penetration
magnetic particles
eddy current
radiography
X-ray method
penetrometer

gamma rays
neutron rays
ultrasonic testing
holographic
  interferometry
microwave testing
sonic testing
acoustic emission
thermal testing
infrared
destructive testing
laboratory test
mechanical test
chemical test
workshop test
pressure test
welding cost
machine cost
materials cost
wages cost
energy cost
mechanical fastener
residual stress
riveting
screw
nut and bolt
stud
adhesive bonding
shear stress
tensile stress
peel stress
cleavage stress
lap joint
weldbond

*   *   *

BIBLIOGRAPHY

Cary, H.B., *Modern Welding Technology*, Prentice-Hall, Inc., Englewood Cliffs, New Jersey (1979).

Easterling, K., *Introduction to the Physical Metallurgy of Welding*, Butterworth and Co. (Publishers) Ltd, Sevenoaks, Kent (1983).

Giachino, J.W., Weeks, W. and Johnson, G.S., *Welding Technology*, American Technical Society, Chicago, Illinois (1973).

Gourd, L.M., *Principles of Welding Technology*, Edward Arnold (Publishers) Ltd, London (1980).

Gray, T.G.F. and Spence, J., *Rational Welding Design*, 2nd Ed., Butterworth and Co. (Publishers) Ltd, London (1982).

Griffin, I.H., Roden, E.M. and Briggs, C.W., *Basic Welding Techniques: Three Books in One. Arc, Oxyacetylene, TIG and MIG*, Van Nostrand Reinhold Co., New York (1979).

Gurney, T.R., *Fatigue of Welded Structures*, 2nd Ed., Cambridge University Press, Cambridge (1979).

Hicks, J.G., *Welded Joint Design*, Granada Publishing Ltd, London (1979).

Houldcroft, P.T., *Welding Processes*, Engineering Design Guide No. 6, Oxford University Press, Oxford (1975).

Kenyon, W., *Basic Welding and Fabrication (1979), Welding and Fabrication Techniques (1982)*, Pitman Books Ltd, London.

Lancaster, J.F., *Metallurgy of Welding*, George Allen and Unwin, London (1980).

Lucas, W. (Ed.), *Exploiting MIG Welding Developments*, The Welding Institute, Abington, Cambridge (1983).

Parkin, N. and Flood, C.R., *Welding Craft Practice*, Volumes 1 and 2, 2nd Eds., Pergamon Press Ltd, Oxford (1979).

Pratt, J.L., *Introduction to Welding of Structural Steelwork*, Constructional Steel Research and Development Organisation (Constrado), Croydon, Surrey (1979).

Schell, F.R. and Matlock, W., *Industrial Welding Procedures*, Van Nostrand Reinhold Co., New York (1979).

Schmidt, R.P., *Welding Skills and Techniques*, Reston Publishing Co., Inc., Reston, Virginia (1982).

Schwartz, M.M., *Metals Joining Manual*, McGraw-Hill Book Co., Inc., New York (1979).

Shields, J., *Adhesive Handbook*, 2nd Ed., Newnes-Butterworths, Sevenoaks, Kent (1976).

Smith, D., *Welding: Skills and Technology*, McGraw-Hill Book Co., Inc., New York (1984).

Stewart, J.P., *The Welder's Handbook*, Reston Publishing Co., Inc., Reston, Virginia (1981).

Tylecote, R.F., *The Solid Phase Welding of Metals*, Edward Arnold (Publishers) Ltd., London (1968).

8
Fabrication Processes

─── CHAPTER OBJECTIVES ───

To obtain an understanding of:
1 how castings are made;
2 how components are produced by mechanical shaping methods, including hot and cold working techniques;
3 the application of powder metallurgy as a fabrication technique;
4 the machining operations that may be performed on a component to produce the finished product;
5 the important design codes relating to pressure vessels.

─── IMPORTANT TERMS ───

casting   powder metallurgy   shaping
hot working processes   machining processes   planing
    pressing and rolling   sawing   grinding
    extrusion   turning   grain and bond
cold working processes   drilling   surface preparation
    pressing and deep   boring   surface treatment,
    drawing   flame cutting   hardening and
    impact extrusion   milling   coating

577

ASME Boiler and	BS5500	numerical control (NC),
Pressure	bursting disk	CNC and DNC
Vessel Code		CAM

SEMI-FINISHED PRODUCTS——————————————————

8.1 CASTING PROCESSES

Most metals are produced in molten form and are cast either as ingots or directly into the final required shape. The ingots can then be either mechanically worked or remelted and cast. Ingots can be produced by a continuous casting process where the molten metal is solidified in a short water-cooled mold, and the solidified material is withdrawn from the base at the same rate as molten metal is added. Many nonmetals can only be shaped by casting because they are unsuitable for mechanical working. Casting consists of heating the metal to make it molten, pouring into a mold and removal after solidification. The material should melt at a low temperature and be fluid when molten; during cooling the material should pass through a narrow temperature solidification range thus avoiding a cored structure. Considerable improvements have been made in casting technology, particularly since the use of permanent molds became more common in the early twentieth century. High rates of production and lower labor costs are now possible. Processes are available that provide a good surface finish, good dimensional accuracy and improved material properties for large shapes and small intricate castings.

8.1.1 Sand casting

A wood or metal pattern of the required object must first be obtained. Patterns made of soft wood, e.g. white pine, can be easily worked. For small shapes, hardwoods such as mahogany are used. Metal patterns have less wear and are used for mass-produced items. The stages in a simple casting process will now be described; this is illustrated in Fig. 8.1.

   (a) The bottom half of the molding box (the drag) is placed upside down on a firm flat surface. This is packed with foundry sand. If a split pattern is to be used, then the section with the sockets is placed face down and dusted with parting powder (a dry, clay-free sand). This is because better detail is obtained in the bottom half of the box, since gases and impurities (dross) rise.

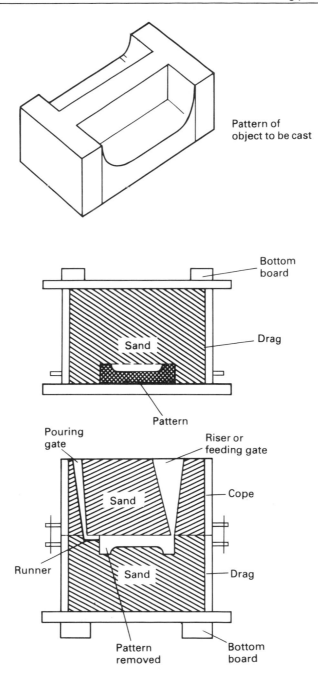

Pattern of
object to be cast

Bottom
board

Drag

Sand

Pattern

Pouring
gate

Riser or
feeding gate

Cope

Sand

Runner

Sand

Drag

Pattern
removed

Bottom
board

Figure 8.1   Principle of sand casting (other stages may be included).

(b) The drag is turned over and the upper molding box (the cope) is fitted
onto the drag. A parting powder is used between the boxes, and the
cope is packed with sand.

(c) The cope is carefully lifted off and the pattern is removed. The runner and riser channels are cut so that molten metal can be fed in and impurities can escape, respectively. The runner also supplies molten metal to the casting as it solidifies and shrinks. (Alternatively runner and riser molds can be positioned before the cope is packed with sand.)

(d) Molten material is then poured into the mold via the runner channel.

The drag and the cope are aligned using dowels and sockets. The packing sand is either traditional 'green sand', i.e. a mixture of sand and clay (Mansfield or Erith sands), or more commonly clean foundry sand mixed with oils or synthetic resins to act as binders. The properties of the sand should be a compromise between permeability and smoothness, also incorporating cohesion. The moisture content should be such that a sample remains unbroken after a firm hand grip, but will break when thrown down. If the sand is too wet, then bubbling may occur during casting and result in a porous finish.

The pattern may be a complex shape and it may be necessary to split the mold into several sections using a multi-part box. As the hot solidified casting cools, contraction occurs and the pattern must be oversized to account for this. The following values are typical contraction allowances:

Metal	*Contraction allowance* (cm/m or %)
Lead, zinc	2.6
Tin	2.1
Steel (carbon)	1.6–2.1
Brass	1.3–1.6
Aluminum alloys	1.0–1.6
Gray cast iron	0.8–1.3

Patterns also provide a taper (or draft) to allow for easy removal, e.g. a taper of 10 mm/m (approximately 1/8 in./ft) on each vertical side. The metal cools at a different rate from the sand and all corners should be rounded for maximum strength and easy removal, as shown in Fig. 8.2. Abrupt thickness changes weaken the structure, and an extra 3 mm (0.12 in.) of metal is often allowed for finishing, and sometimes a metal lug for holding during machining.

For more complicated shapes, especially hollow structures, a core is used. This is the internal shape of the object to be made; it is used to restrict the space into which the molten metal can flow. The core is made in a core box which splits into two halves. The use of a core is illustrated in Fig. 8.3. The core is inserted into the mold cavity when the pattern has been removed. The core is made from silica sand, specially bonded with oil and cereal binders to give cohesion. It is then baked in an oven at approximately 200°C (392°F). It can be reinforced with wire and usually has a central vent. The core should be strong, yet brittle enough to collapse when the casting contracts; it should also be permeable to allow gases to escape. It must be possible to remove the core material from the finished object.

Sand casting is used for a wide variety of metals and alloys, e.g. steel, cast iron, brass, bronze, aluminum. It can be used to produce very intricate shapes.

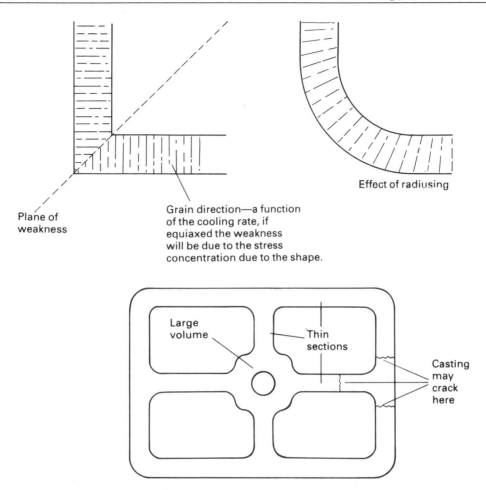

Effect of radiusing

Plane of
weakness

Grain direction—a function
of the cooling rate, if
equiaxed the weakness
will be due to the stress
concentration due to the shape.

Large
volume

Thin
sections

Casting
may
crack
here

Figure 8.2   Casting defects.

The equipment is cheap and the process is economical for small numbers of castings. However, the castings produced are not very accurate and the process is unsuitable for thin-walled sections. Sand casting is a slow process, but it can be speeded up by using machine molding.

8.1.2 Die casting

Die casting employs a permanent metal mold known as the die. The die is split so that the casting can be removed, but this also results in some loss of accuracy. The equipment, especially the die, is more expensive than that used for sand casting but it becomes more economical when large numbers are produced.

Two methods of die casting are commonly used. *Gravity die casting* (or *permanent mold casting*) is similar to sand casting in that the molten metal enters the die or mold due to gravity. For hollow shapes either sand cores or split metal cores are used. Large castings can be produced by this method because no

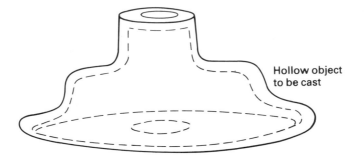

Hollow object
to be cast

Make a core of the internal shape (cavity)
in a split core box

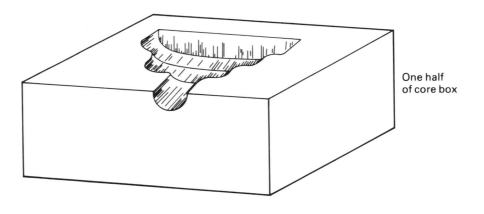

One half
of core box

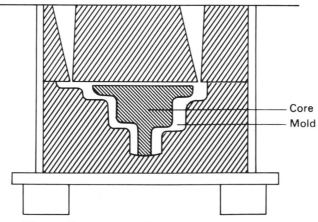

Core

Mold

When pattern removed, insert core;
mold is then ready for molten metal

Figure 8.3 Use of a core in sand casting.

special equipment is required to feed in the metal.

For *pressure die casting* the molten metal is forced into the die by a plunger; when the casting is solid the mold opens automatically and the casting is ejected. Sand cores cannot be used and metal cores tend to wear, causing inaccuracies. Often the use of cores can be avoided by careful design. The most common process employs a cold chamber (die) and this is shown in Fig. 8.4. It is a rapid process but can be used only for small castings. It cannot be used for all metals, although it is especially suited to zinc-based alloys. Because the metal cools much more quickly in a metal mold some metals would crack due to rapid contraction. However, rapid cooling produces a more uniform structure and a finer grain size. Most pressure die castings require little or no machining due to their improved accuracy.

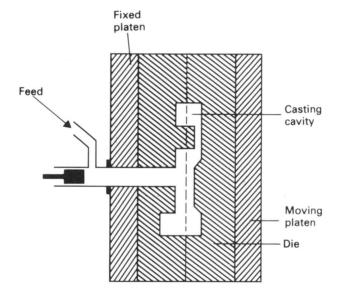

Figure 8.4   Cold chamber-pressure die casting.

8.1.3 Shell molding

This method produces very accurate castings and can be used with most metals. The cost of the pattern is high but only semi-skilled labor is needed to produce the mold. It is basically a sand casting process, where clay-free sand is mixed with a thermosetting resin binder to produce the mold. The method is illustrated in Fig. 8.5. The half-pattern plate is first heated to about 250°C (482°F) and placed in a dump box containing the sand–resin mixture. The box is inverted so that the pattern is covered by this mixture. The resin melts and a hard shell of bonded sand quickly forms on the pattern. The box is then inverted so that any surplus mixture falls back to the bottom of the box. The pattern plate and shell are removed from the box and the shell is hardened in an oven (2 minutes at 315°C/600°F), the shell is then removed from the pattern. Both halves of the mold are produced in this way, and they are assembled as a complete mold with

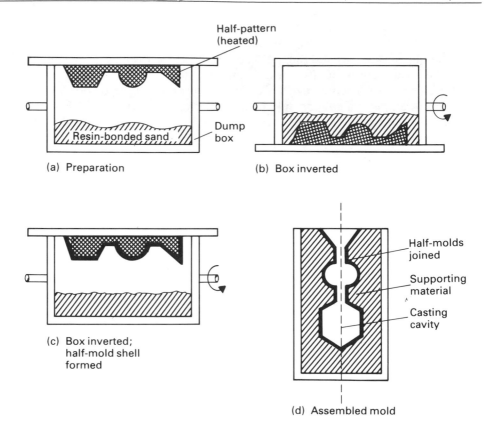

Figure 8.5   Shell molding.

a surrounding support material. The castings produced have a higher accuracy and a superior finish than sand castings or pressure die castings. The process can be carried out at a lower temperature due to the lower heat capacity of the shell mold. The molds can be easily stored.

8.1.4 Centrifugal casting

This process is used to produce hollow castings such as cast iron pipes, and castings that are symmetrical about one axis. A metal mold is rotated at high speed and molten metal is poured in. A hollow cylindrical-shaped casting is produced that has a uniform thickness and a uniformly fine-grained surface. It is also accurate in shape and structurally sound. The product is superior to that produced by sand casting. A core situated at the axis of rotation can be used and this method is known as *semi-centrifugal casting*. The equipment used is shown in Fig. 8.6. *Centrifuging* is the name given to the process where molten metal is forced into the mold through runner channels, rather than being thrown directly to the wall.

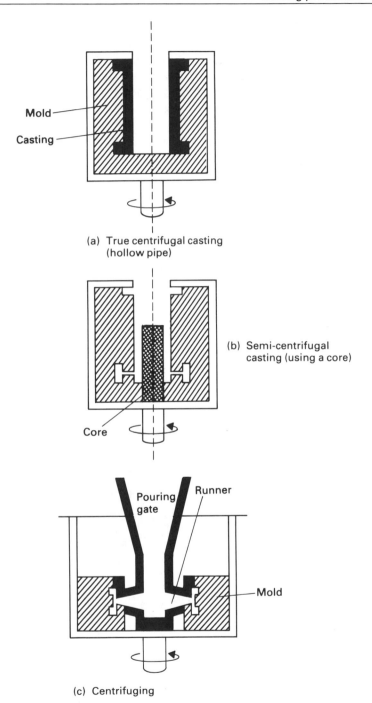

(a) True centrifugal casting (hollow pipe)

(b) Semi-centrifugal casting (using a core)

(c) Centrifuging

Figure 8.6   Centrifugal casting processes.

8.1.5 Investment casting

In this process a separate pattern and mold are made for each casting. This method is sometimes known as the *'lost wax'* process because the pattern is made from a material with a low melting point, e.g. wax. A master mold is therefore required in order to produce the wax pattern. The pattern (or several patterns) are covered (or invested) with a suitable investment material. This may be fine silica sand and plaster of Paris for low temperatures, or a suitable refractory-type material. The investment is heated to harden the mold and melt out the wax. The mold is then left to dry. Molten metal is fed into the mold, either by gravity or under pressure (for thin sections). The solidified casting is removed by breaking open the mold.

High accuracy is achieved because a rigid, one-piece mold is used. This means that there is no line along the casting as appears with two-piece molds. This method is particularly advantageous for materials or shapes that cannot be obtained by other fabrication methods or by machining, and where the design requires easy removal of the pattern or the casting. The disadvantages of investment casting are that it is an expensive process and the size of the component is normally limited to about 5 kg (11 lb). Investment casting is used to produce a wide range of parts including valves, nozzles, cams, levers, machine parts and machine tools. Vacuum casting of high-grade alloy steels, etc., for turbine blades uses this system, giving good dimensional stability and high purity.

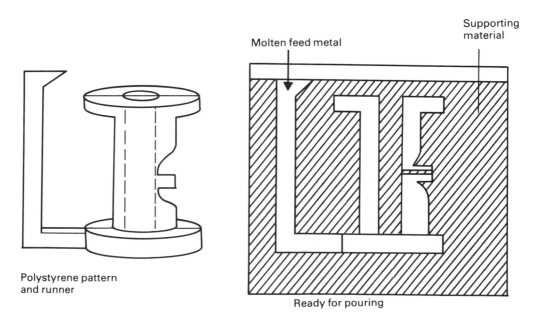

Molten feed metal

Supporting material

Polystyrene pattern and runner

Ready for pouring

Figure 8.7 Full mold process.

8.1.6 Full mold process

This process is similar to investment casting and a single part pattern is used. However, it is a one-off process as the pattern is made of consumable expanded polystyrene (see Section 8.1.5). The pattern and the runners are cut from the polystyrene and surrounded by packing material, as shown in Fig. 8.7. When molten metal is introduced the polystyrene burns very quickly, evolving carbon dioxide and water vapor. This mixture diffuses into the sand and a sticky skin holds the mold in shape. Only 2% of the pattern is actual polystyrene and no residue remains. The process is well suited to the manufacture of large, one-off components.

8.1.7 Summary

The choice of a particular casting process depends upon various factors. Initially these are probably economic considerations, number of components to be produced, materials of construction, equipment required and the complexity of the components. Table 8.1 makes a comparison of the main features of different casting processes.

8.1.8 Casting characteristics of particular metals and alloys

The compositions and properties of common metals and alloys are described in Chapters 2 and 3.

Cast iron. Cast irons are shaped by casting processes; their high carbon content (1.5% to 5.0%) lowers the melting point. This means that casting can be performed at liquid temperatures much higher than the melting point, with subsequent increased fluidity. Thin sections and complicated shapes can be cast. Shrinkage during solidification depends upon the particular alloy and the amount of graphite present. Malleable irons require prolonged heat treatment, which can take place as the casting cools, rather than reheating the cooled product. The heat treatment also relieves internal stresses.

Cast steel. Cast steel can be easily obtained and provides an alternative to the wrought form. The properties of cast steel do not depend upon the direction in which they are measured. Quenched and tempered castings are used when high strength, abrasion resistance and sliding friction are required.

Alloy steel. Complicated castings made from alloy steels can be air hardened, and high tensile strengths obtained without quenching. Nickel, molybdenum and manganese increase the ability for air hardening; chromium and vanadium are also added to improve the strength and wear resistance.

Copper and copper-based alloys. Copper is difficult to cast, does not possess good flow characteristics and is not suitable for intricate shapes. The design should allow for solidification shrinkage to avoid cracking. Alternatively, copper-based alloys are easily cast.

Table 8.1 A comparison of casting methods

	Sand casting	Gravity die casting	Pressure die casting	Centrifugal casting	Investment casting
Alloys that can be cast	unlimited	copper-based, aluminum-based & zinc-based alloys	copper-based, aluminum-based & zinc-based alloys	unlimited	unlimited
Approximate maximum possible size of casting	unlimited	50 kg (110 lb)	15 kg (33 lb)	several tonnes	5 kg (11 lb)
Thinnest section normally possible in mm (in.)	3 (0.12)	3 (0.12)	1 (0.04)	10 (0.4)	1 (0.04)
Relative mechanical properties	fair	good	very good	best	good
Surface finish	fair	good	very good	fair	very good
Possibility of casting a complex design	good	good	very good	poor	very good
Relative cost for production of a small number of castings	lowest	high	highest	medium	low
Relative cost for large-scale production	medium	low	lowest	high	high
Relative ease of changing the design during production	best	poor	poorest	good	good

Aluminum and aluminum alloys. Aluminum can be easily cast by sand casting, permanent mold and die-casting methods. The low density and low melting point are particularly advantageous and eliminate many common problems. However, care must be taken to avoid gas absorption and to design for shrinkage, e.g. uniform sections. Silicon, magnesium and copper are added to aluminum to make casting and heat treatment easier. Silicon increases the fluidity, magnesium (above 8%) makes the alloy heat treatable but more difficult to cast, and copper increases the hardness and strength.

* * *

Self-assessment exercises

1 Describe (using diagrams) the stages in sand casting processes to produce:
   (a) a solid object that is thinner at the centre than at the ends;
   (b) a hollow object with two openings of diameters less than the component diameter.

2 Explain the terms: (a) drag; (b) cope; (c) parting powder; (d) green sand; (e) dry sand; (f) riser; (g) pattern; (h) core; (i) pouring gate; (j) binder.

3 Explain what is meant by:
   (a) shrinkage allowance;
   (b) machining allowance;
   (c) taper.

4 Explain why castings should be designed with the minimum wall thickness, and (if possible) without long thin projections.

5 What is plaster mold casting?

6 Explain why die casting is the fastest casting production process, and describe the use of fixed and movable cores. Why should finishing operations be kept to a minimum for die-cast components?

7 Describe the differences between the patterns used for shell molding and sand casting.

8 What are the main advantages of shell molding?

9 Why is the porosity of shell-molded products less than for sand-cast items?

10 What materials are used for the molds and the cores in investment casting?

11 Why does an investment casting not possess a parting line?

12 Describe the process of permanent mold casting.

13 Explain why the mechanical properties of most metals are significantly improved by using centrifugal casting.

14 How can laminates be produced by centrifugal casting?

15 Explain whether centrifugal or semi-centrifugal casting is best suited to the production of intricate shapes.

16 Describe the process and outline the advantages of die casting. List the materials that can be die cast.

* * *

8.2   MECHANICAL SHAPING METHODS

Introduction————————————————————————

The mechanical shaping of a material when in the solid state is known generally as *working*. Metals to be worked are in the wrought condition (see Section 2.2.1) and they must be malleable or ductile, although these properties can be acquired by suitable heat treatment. Malleable metals can be considerably deformed by compression before cracking occurs. Ductile metals can be considerably deformed by tension before fracture occurs. All ductile metals are malleable, although the opposite is not always true (see Section 6.2).

Working processes are usually classified as either *hot working* or *cold working processes*. The former are carried out above the recrystallization temperature and the latter at a lower temperature. With hot-working processes, large deformations can be produced without causing residual stresses. This is not the case with cold working, and frequent intermittent heat treatment must be carried out (see Section 2.18). However, the stresses induced by cold working generally increase the strength and hardness of the material. The surface finish may be of a higher quality for cold-worked materials, and they possess a fibrous structure. The structure of a casting changes from fine grain due to rapid cooling at the surface, through laminar crystals to coarse crystals at the center. Hot working of a metal in the plastic state refines the grain structure, whereas cold working distorts the grain and has little effect on its size. Some processes and equipment can be used for both hot and cold working, but the forces required and the methods of removing the heat are usually quite different. Hot-working processes are usually preferable because less power is required in the shaping process and it is usually quicker, although factors such as the properties of the finished product are also influential.

Hot-working processes————————————————————

8.2.1   Hand forging and machine forging

Hand forging is one of the oldest metal-working methods and can be used to produce complex shapes using simple tools. For large shapes a machine hammer or press may be used and the process is then known as machine forging. With either process the tools used and the operations performed are the same, the only difference being the scale of operation. Only a small region of the metal is heated and worked at any time.

When shaping hot metal by hand, the accuracy of the finished component will depend upon the skill of the operator who is known as *the smith*. The equipment required includes a forge (or furnace), an anvil (work surface), a selection of hand tools to be used by the smith and tools to be used by an assistant, e.g. a sledgehammer. The basic operations that are carried out and the tools used will now be described.

Upsetting is an operation in which the thickness of a section is increased at the expense of the length. This is achieved by hammering, usually one end, when the material is placed on an anvil or held in a vice. The process is shown in Fig. 8.8.

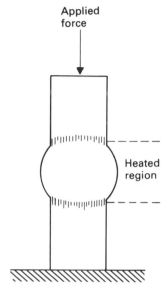

Figure 8.8   Upsetting.

Drawing down is an operation where the length is increased at the expense of the section thickness. This can be achieved using special hammers called *fullers* that are used for *necking down* and reducing a piece of material, or *swages* which reduce the material and produce a final round or hexagonal cross-sectional area. Details of these operations are shown in Figs. 8.9 and 8.10. The finishing operations performed on reduced surfaces are carried out using *flatters*, which are perfectly flat-faced hammers approximately 8 cm (3 in.) square or 8 cm (3 in.) diameter, as shown in Fig. 8.11. Flatters are also used for *setting down* operations where localized thinning down of the material is required. If material is to be shaped by a bending operation, then the metal should first be prepared as shown in Fig. 8.12, so that local thinning down does not occur at the bend.

With machine forging, the metal is suspended by a heavy chain ('burden' chain) which is also used to maneuver the workpiece. The metal is counterbalanced by a 'porter bar'.

8.2.2 Drop forging

If mass production of a large number of identical items is required, then drop forging is used, being quicker, easier and more economical. A two-part die made

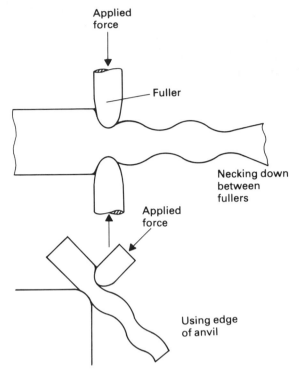

Figure 8.9 Drawing down.

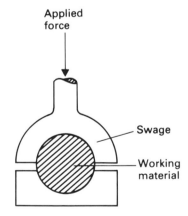

Figure 8.10 Reducing and finishing with swages (may be hexagonal cross section).

of hard steel is required: the heated metal sits in the lower die and then the top die is dropped (or forced) onto it. This is continued until the two dies make contact. The component must be of a simpler shape than can be produced by casting, although more complicated shapes can be made using a number of dies and more blows. Usually an excess of metal is used and this is forced into a gutter, as shown in Fig. 8.13. This leaves a fin of metal attached to the

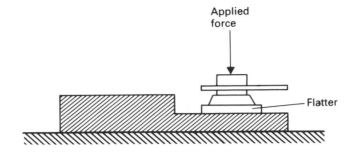

Figure 8.11 Setting down (localized thinning) or surface finishing using a flatter.

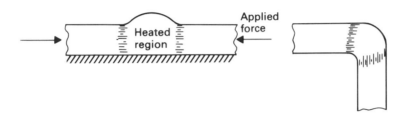

Figure 8.12 Metal preparation prior to bending.

component; when this is trimmed off it leaves a flash line around the object. It is difficult to produce accurate square corners on components by this method. Therefore, a draft of metal (Fig. 8.13) is often incorporated in the die and is then machined from the final forged component. The positions of the flash and the draft must be carefully considered in the design of the forging, as shown by

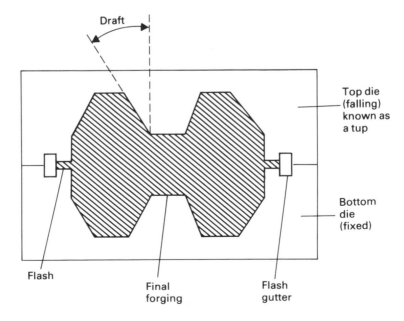

Figure 8.13 Drop forging process.

the two alternative forgings in Fig. 8.14. The strength of the component depends upon its in-service position and the direction of its fibrous structure, which will be determined by the direction of the metal flow during formation. Drop forgings can be produced from nearly all steels, some aluminum alloys and certain brasses and bronzes.

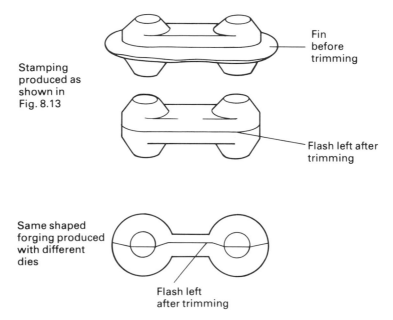

Stamping produced as shown in Fig. 8.13

Fin before trimming

Flash left after trimming

Same shaped forging produced with different dies

Flash left after trimming

Figure 8.14   Alternative forgings of the same component.

8.2.3 Hot pressing

Hot pressing is similar to drop forging except that the component is produced by a single compressive thrust of a hydraulically driven ram rather than by a series of rapid blows. The advantage of this process is that a more uniform internal structure is obtained because of the gradual deformation which occurs throughout the material, rather than in the surface layers. Hot pressing is normally used only for simple shapes.

8.2.4 Hot-rolling

A hot-rolling process reduces the cross-section of white-hot feed metal, which is usually in ingot form. The metal is rolled into long lengths by passing through a powerful 'two-high' reversing mill, as shown in Fig. 8.15. These lengths are passed through a series of rolls of gradually reducing section to obtain the final product dimensions. The number of reduction stages is known as the number of passes. The product may be strip, sheet, rod, channel or other sections. Hot rolling can be applied to most nonferrous alloys for the initial 'breaking down'

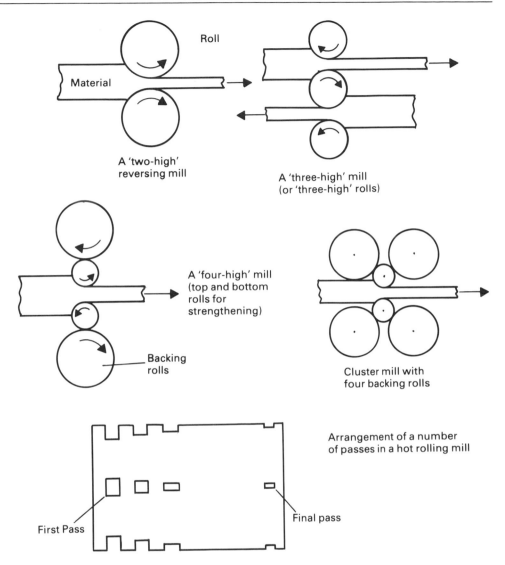

Figure 8.15 Equipment for hot rolling.

stages, but the final reduction stages are more likely to be carried out by cold rolling (see Section 8.2.8). The full process is most often used with steels. Bar, strip or sheet steel possessing a nonpolished, reddish-blue surface has been hot rolled. The ingots are typically 2 m (6 ft) long, slightly tapered, with approximately square cross section of between 30 cm and 50 cm (12 in. and 20 in.). The size of the mill is expressed as the distance between the centers of the rolls. When a mill is used for plate rolling, the length of the roll is also specified. Different arrangements of the rolls are shown in Fig. 8.15. The effect of hot rolling on the grain structure of the metal is shown in Fig. 8.16. The typical number of passes required to reduce a square billet to round bar stock is illustrated in Fig. 8.17.

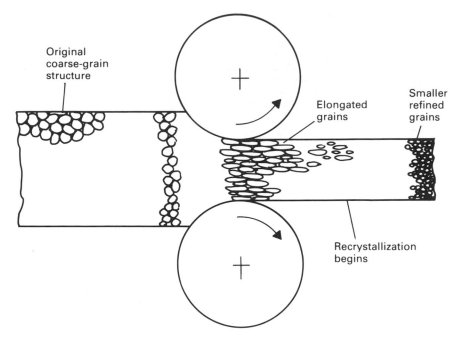

Figure 8.16 Effect of hot rolling on grain structure.

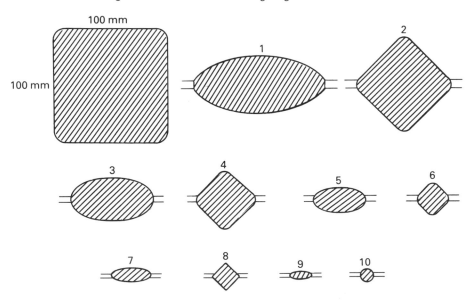

Figure 8.17 Reduction of square billet to round bar stock by hot rolling.

8.2.5 Extrusion

Extrusion is usually a hot-working process and is performed at the temperature at which the material becomes plastic, although lower pressures are needed than with other processes. The process can be used with most ferrous and nonferrous metals; the temperatures required are 350–500°C (662–932°F) for aluminum alloys, 700–800°C (1292–1472°F) for copper or brasses, and 1100–1250°C (2012–2282°F) for steels.

Steel is more difficult and expensive to extrude because of the higher temperatures and pressures required. The advantage of extrusion is that a heated metal billet can be worked to produce complex sections of reasonable accuracy in a *single* operation. This is achieved by forcing the metal through a hard-alloy steel die of the required size and cross-section. Extrusion can be either a direct process where the metal is forced through the die, or an indirect process where the die is forced through the metal: both processes are illustrated in Fig. 8.18. Indirect extrusion requires less power and produces a more uniform structure in the metal, but a more complex machine is required.

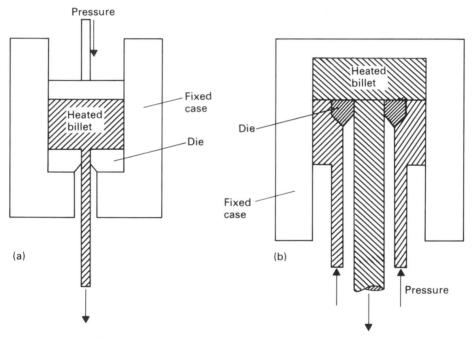

Figure 8.18   Principles of extrusion. (a) Direct extrusion. (b) Indirect extrusion.

8.2.6 Tube manufacture

Large tubes are usually made by shaping metal sheet either by hot or cold working, and then welding it along the edges. This produces 'seamed' pipe. Smaller diameter tubes are made either by hot rolling or extrusion methods. The method of tube manufacture by hot rolling is shown in Fig. 8.19(a); this is known as the *Mannesmann process*. Hot metal is spun between two rolls, which rotate in the same direction and draw in the rod. This action draws the metal away from the center of the rod, thus forming a cavity. Cylindrical tube is formed by a nose piece on a mandrel. The tube thickness and bore can be varied by adjustment of the rolls and the position and size of the nose piece. The strength and appearance of the tube can be improved by cold drawing through a die, which also acts to reduce the diameter.

Tube making by a direct extrusion process is shown in Fig. 8.19(b). The tube bore is determined by a fixed mandrel, which is positioned inside the billet and the ram, and which protrudes through the die opening.

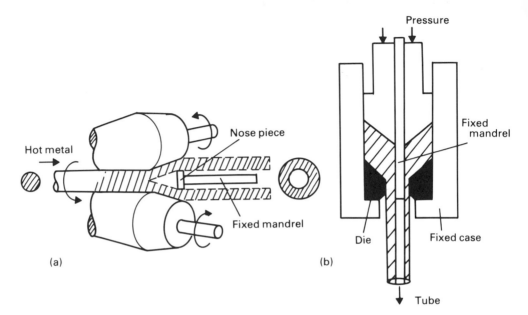

Figure 8.19   Principles of seamless tube manufacture. (a) Mannesmann tube process. (b) Extrusion process.

8.2.7 The effects of hot working

Hot working is usually performed on metals and alloys that are not very ductile at normal temperatures, or which harden rapidly when cold worked. Typical hot-worked materials are aluminum alloys containing copper, magnesium and zinc, copper alloys with a high zinc or aluminum content, pure zinc and magnesium and their alloys, irons and steels. Some cold working may be performed on these metals, especially for production of wire products.

Hot working is performed at temperatures above the recrystallization temperature (see Section 1.4.9). As the metal is deformed, recrystallization and softening are continually taking place which offset the hardening effects of shaping. The properties of the structure depend upon the applied pressure and the temperature. The working pressure can prevent the growth of crystals at the higher temperature, or the crystals may fracture. The ductility of the metal depends upon the amount of deformation.

If hot working is terminated when the temperature is just above the recrystallization temperature, then fine, uniform and unstressed grains will be obtained. The properties of the metal will be similar to those possessed by cold-worked metal that has been annealed at the recrystallization temperature. If working stops well above the recrystallization temperature, then grain growth occurs and the material properties are inferior. The hardness and strength of a metal can be improved (at the expense of ductility) by continued working after cooling below the recrystallization temperature. The grains are then fine but partially distorted. The properties of a hot-worked material depend upon the direction of elongation of insoluble constituents.

Hot working is used for the initial size reduction of large metal ingots, because of the lower power required compared to cold working, for a particular degree of reduction. The disadvantages of hot working include:

(a) formation of oxide scale which may be forced into the surface;
(b) difficult to control final dimensions and tolerances;
(c) impurities with low melting points can prevent effective hot working, e.g. excess sulfur in steels.

*   *   *

Self-assessment exercises

1 Explain what is meant by the plastic range of a material.

2 What mechanical properties should a forged component possess?

3 Explain the difference between closed-die and open-die forging.

4 Explain why, and in what ways, forgings generally perform 'better' in-service than castings. (First decide what is meant by 'better' in this case: e.g. wear, creep, fatigue, strength, etc.?)

5 Explain why a forged component is often produced in progressive stages.

6 What are the advantages of hot rolling a metal, in terms of the structure and mechanical properties produced?

7 List the materials that can be shaped by extrusion.

8 Can the sizes of metal tubes produced by extruding hot metal be reduced by hot or cold rolling?

*   *   *

Cold-working processes———————————————

8.2.8 Introduction

There are more cold-working processes for shaping metals than hot-working processes. This is partly due to the greater dimensional accuracy that can be obtained by cold working, and also because the tensile strength and toughness of a metal are increased at lower temperatures. Although the malleability of metals increases at higher temperatures, the loss of toughness means that drawing (or pulling) operations cause the metal to tear more easily. Cold-working influences the mechanical properties of the metal, and is used as a final shaping operation to produce the required strength and hardness. The surface condition of a metal is also improved if it is subjected to cold working. Metals that have been hot worked are usually heavily oxidized or scaled, this must be removed either by sand blasting or by 'pickling' in acidic solutions to obtain a satisfactory surface finish.

8.2.9 Cold rolling

Cold rolling is used to produce strip, section and foil. The feed metal is hot-rolled strip that has been 'pickled' to remove surface scale, which would otherwise be rolled into the metal. Cold-rolled strip can be produced to a high degree of uniformity and accuracy for a thickness within 40 μm (1.6 mils). The rolling mills are the same as for hot rolling (Fig. 8.15); a four-high mill is often used to produce higher pressures due to the increased metal resistance. The backing rolls act as support for the smaller main rolls if the material is very wide. A cluster mill with many backing rolls is sometimes used to produce high pressures. For thicker materials a 'two-high' mill is used. Cold rolling makes the metal become work-hardened and springy. The grains become distorted and elongated (as shown in Fig. 8.16 for hot rolling) and as they are deformed, the number of dislocations increases. These dislocations concentrate at the grain boundaries, reduce slip and increase the strength of the metal; unfortunately it becomes more difficult to form. If the material is unsuitable in this condition then it can be stress relieved by annealing (see Section 2.19.1), either after the final pass or after two or three passes. During annealing oxygen must be excluded; otherwise metal oxide forms on the surface.

8.2.10 Drawing

All wire is made by cold drawing through dies; rod and tube is also produced by this method. This method can only be used as a cold-working process because it is dependent upon the ductility of the material. The metal is pulled through a tapered hole (between 12° and 15°) in the die; the pulling and squeezing effect causes an increase in length and reduced cross-sectional area. The feed material is usually black rolled bar from a hot-rolling mill. The contact area between the metal and the die needs to be lubricated because of the large pressures that are created. A solid lubricant such as grease or soap can be used, but this is not usually a convenient method. The feed metal can be 'pickled' with dilute acid and then washed with lime water. This leaves a layer of lime that acts as a lubricant when dried; this is a dry lubrication process. Alternatively a wet process can be employed, the metal being passed through copper sulfate solution and then soap solution. The deposited copper and the soap act as the lubricant. Cold-drawn steel has a smooth bright surface. Drawing hardens the metal and annealing may be necessary. The principle of drawing is shown in Fig. 8.20. The die must be very hard and is usually made of tool steel, tungsten carbide or diamond.

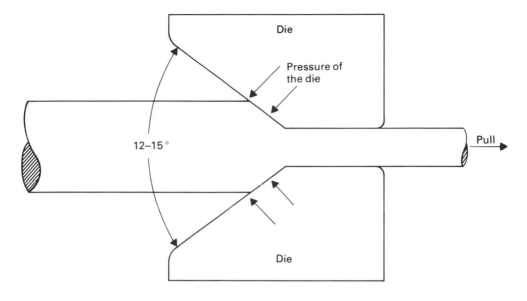

Figure 8.20   Principle of drawing.

8.2.11 Pressing

Simple cold-pressing is used for the shaping of alloys that lack the ductility required by other processes. The metal is pressed into shape between formers as shown in Fig. 8.21.

8.2.12 Deep drawing

A deep drawing process can be used to shape very ductile metals, e.g. 70–30 brass, copper, aluminum, cupro-nickel, and also steels. The principle of the process is shown in Fig. 8.22. Sheet metal is first stretched into a cup-shape using a punch and die, and is then subjected to deep drawing with the subsequent wall-thinning.

8.2.13 Stretch forming

This process is used for the shaping of sheet metal and sections, particularly in the aircraft industry. It is used mainly for light alloys, although stainless steel and titanium are shaped by this method. The process is illustrated in Fig. 8.23. A tensile load is applied to the workpiece by a 'rising block', while the ends of the metal are held in fixed clamps. The metal must be stressed beyond the elastic limit so that permanent deformation occurs. An alternative procedure can be used where the block remains stationary and the clamps are retracted. It is necessary for the block to be lubricated.

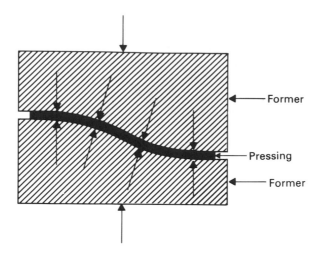

Figure 8.21   Pressing.

8.2.14 Coining and embossing

Both processes use a punch and die set and the principle of each operation is shown in Fig. 8.24 and Fig. 8.25. The coining process is really cold forging, and is performed in a closed die. Compressive forces cause metal to flow and fill the die cavities by local indentation and extrusion. Very large pressures (up to 1500 MN/m^2; 218×10^3 psi) are applied; the size of the metal blanks must be accurately controlled so that the die is not damaged by any excess metal that cannot be removed.

8.2.15 Spinning, flow forming and flow turning

Spinning is a process used to shape soft and ductile sheet metal into hollow shapes. A former, corresponding to the internal dimensions of the finished component, is rotated with the metal blank or disk in the chuck of the lathe. During rotation a suitable tool forces the sheet metal against the former, thus producing a hollow solid of revolution. Lubrication is required between the former and the metal. The process is shown diagrammatically in Fig. 8.26.

Flow forming is similar to spinning; the metal blank has approximately the same diameter as the finished component but a greater thickness. A heavier machine is required and the principle of the operation is shown in Fig. 8.27.

Flow turning is also similar to spinning and flow forming; the final component thickness is much less than that of the original blank. The metal is subjected to pressure rolling by a rotating tool. Both the metal (in a plastic condition) and the tool move in the same direction. The operation shown in Fig. 8.28 would produce a component with a base thicker than the walls.

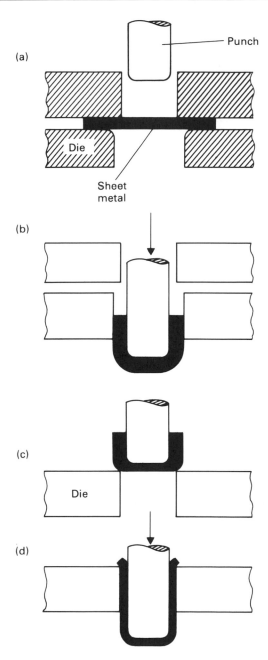

Figure 8.22   Stages in deep drawing. (a) Preparation of sheet metal blank.
(b) Cupping operation. (c) Preparation for deep drawing. (d) Deep drawing.

8.2.16 Impact extrusion

The principle of extrusion has been described previously as a hot-working
process (see Section 8.2.5). However, small quantities of metal can also be

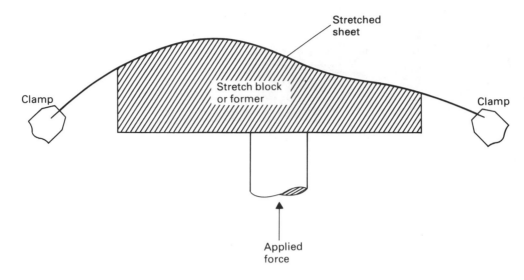

Figure 8.23   Principle of stretch forming.

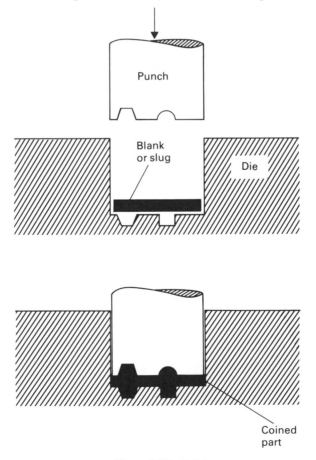

Figure 8.24   Coining.

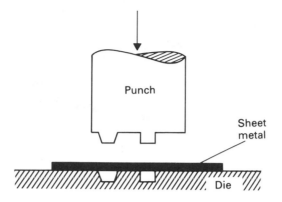

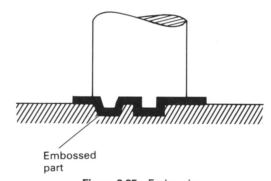

Figure 8.25   Embossing.

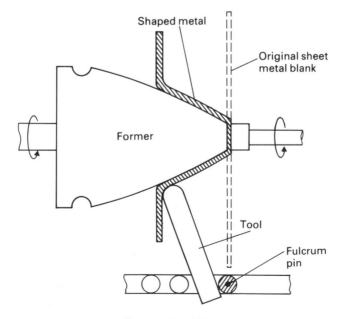

Figure 8.26   Spinning.

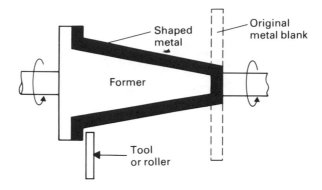

Figure 8.27 Flow forming.

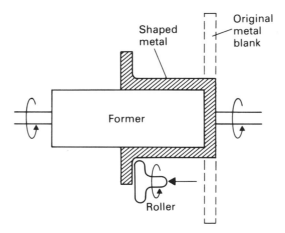

Figure 8.28 Flow turning.

shaped by a method involving drop forging and extrusion, known as impact extrusion. A flat or cupped blank of metal is placed in the die cavity and is subjected to very high pressure by a punch. If the cavity is closed as shown in Fig. 8.29, then the metal is forced upwards through the gap between the punch and the sides of the die. A tube-shaped shell is thus formed around the punch.

If the die cavity has an orifice below the metal blank, then the metal will be extruded downwards through the orifice; this is also shown in Fig. 8.29.

8.2.17 The effects of cold working

Plastic deformation occurs in all shaping processes and the metal remains deformed even when the applied stress is removed. Cold working has several effects upon the structure and properties of metals. The crystal grains become progressively elongated in the direction of working; with sufficient working the structure may break up and become fibrous. If a second constituent is present, then this will also assume a fibrous character. As deformation proceeds, the

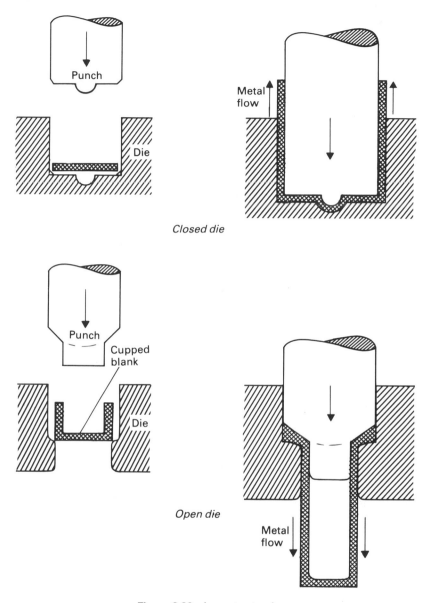

Figure 8.29 Impact extrusion.

strength and hardness increase at the expense of the ductility. The resistance to deformation also increases. Internal stresses may be created within the material if deformation is not uniform, and this can cause stress corrosion cracking (see Section 5.5.3). If deformation is continued long enough, then brittle fracture occurs.

After a certain amount of cold working has been performed, metals are usually heat treated to restore some of the ductility (at the expense of the hardness and strength), to relieve internal stresses and to prevent brittle failure.

The processes of stress relief, recrystallization and softening have been described in Section 1.4.9. When a cold-worked metal is heated to the recrystallization temperature, then small equiaxed grains are formed which allow the distorted structure to rearrange itself into an unstrained structure. If the metal was severely distorted, then the recrystallized metal may exhibit different properties depending upon the direction in which they are measured. Fabrication usually involves alternate deformation and heat treatment (softening) to obtain the desired shape and properties.

The recrystallization temperature of a cold-worked metal depends upon the melting point, composition and impurities. The shorter the reheating time, the higher the recrystallization temperature. However, as the amount of the cold working increases so the recrystallization temperature is lowered. If the temperature is raised above the minimum required for recrystallization, then grain growth occurs and the strength decreases. In general a fine grain size is advantageous, and this can be achieved by using grain-refining additions or by heating to the minimum temperature. The recrystallized grain size decreases as the amount of prior cold working increases.

8.2.18 Fabrication characteristics of particular metals and alloys

Reference should be made to the appropriate sections of Chapters 2 and 3, relating to the composition and properties of particular metals and alloys.

Cast iron. As their name implies, cast irons are usually shaped by casting. Their properties and behavior are largely governed by the high carbon content (approximately 1.5–5%), and the form in which the carbon is present. The main types of cast iron are gray, white, malleable and ductile. Cast irons are not usually hot or cold worked, but the ductility of malleable irons can be advantageous for some sizing and straightening operations.

Plain carbon steel. Low-carbon steel contains less carbon and alloying elements to prevent distortion; it can be easily formed by hot or cold working. Medium carbon steels that have been cold rolled lack the ductility for subsequent shaping, although hot-rolled steels are more ductile.

Alloy steel. These steels are used for gears, bearings, crankshafts, etc., but they are only usually shaped by forging. They possess good strength and toughness, and are often induction-hardened.

Stainless steel. The chromium–nickel stainless steels (200 and 300 series) are easier to shape than the straight chromium stainless steels (400 series), although they undergo more work hardening. However, chrome–nickel stainless steels can be cold worked by drawing or deep drawing to a far greater extent than chromium stainless steels or plain carbon steels.

Copper and copper-based alloys. Pure copper is very ductile and work hardens less rapidly than brasses or bronzes. Deoxidized and oxygen-free coppers can withstand more deformation than tough-pitch coppers. The copper alloys that are easiest to shape are the 70–30 cartridge brasses; if less than 63% copper is present then the brittle β phase occurs. In general the ductility of copper alloys increases as the grain size increases. The tensile strength, elongation, hardness and other properties are all related to the grain structure.

Aluminum alloys. Most aluminum alloys can be cold worked; however, there is considerable variation in the degree of working that is possible between different alloys. The condition of the material, i.e. previous treatment, also influences the suitability for cold working. The non-heat-treatable alloys will work harden, but they can be easily shaped when annealed. Heat-treatable alloys can be shaped in the annealed condition. Shaping must be performed before aging or precipitation occurs in solution-treated alloys.

* * *

Self-assessment exercises

1 List the advantages and disadvantages of hot- and cold-working processes.

2 How can desirable mechanical properties be obtained in cold-rolled products?

3 What is the common feature of all drawing processes?

4 Explain why wire is not produced by hot drawing, which would reduce the pressures required and extend the life of the die.

5 What is the main difference between coining and embossing?

6 Would there be any benefit to performing stretch forming between two dies?

7 Explain the difference in operation between spinning, flow forming and flow turning. Sketch a typical part produced by each method. What are the differences between these final products?

8 What is the difference between open-die and closed-die impact extrusion?

9 Explain what is meant by residual stress and how it can be removed.

* * *

8.3 POWDER METALLURGY

Powder metallurgy is a process of compacting and sintering powdered metals, ceramics and other compounds in order to produce a particular component. This method is used if the component cannot be easily or economically fabricated by an alternative process, either because the metal has a high melting point or the melt does not form a homogeneous solid solution. Many simple shapes are produced by this method. Sometimes the part is porous (with a volume porosity less than 65%) and may lack strength and toughness. Most metals are not sufficiently brittle to be powdered by crushing or grinding methods, and special techniques are used. These involve the reduction of metal oxides and salts, chemical reaction, electrolytic deposition or atomization. Atomization is now the major method, and involves forcing molten metal through an orifice where a jet of air or water disperses the metal into very small particles. Solidification occurs rapidly, with the minimum of oxidation if the metal is not particularly reactive.

Metal powders may be classified as dense or porous. Dense powders have minimum porosity and are used for the production of cutting tools. Porous powders are mainly used to produce bearing materials. Different materials, or different particle sizes, can be mixed together to achieve the desired properties. Organic binders are sometimes added during mixing; these volatilize during subsequent heating, which increases the porosity. The operations to be performed when producing a component by powder metallurgy are shown in Fig. 8.30.

The shaping process begins by compacting the powder in a steel die, either using a mechanical or hydraulic press, or by compressing between two rotating rolls. This stage is carried out at a suitable (low) temperature and a lubricant may be added to assist the pressing. The high pressures (at least 150 MN/m^2; 21 800 psi) used in the compacting stage cause deformation of the metallic particles. They become joined together at their points of contact by 'cold welding', i.e. an exchange of atoms between the particles. However, metal powders have problems of internal friction and when compressed the pressure is not distributed evenly throughout the material. The hardness and density decrease with distance from the punch. This problem can be avoided by the addition of a lubricant and using punches at both ends of the die. Lubricants should be removed by slow heating before the sintering stage, to avoid reactions occurring. The variation in density caused by compacting in two directions can be avoided by using an isostatic pressing method, where pressure is applied from the sides as well as the ends. Some variation in density still exists from the surfaces of the part to its center.

The compacted shape is then heated, or *sintered*, above the recrystallization temperature, at a temperature near, but below, the melting point. The atmosphere is controlled to avoid oxidation or carburization (neutral or reducing). Sintering produces a change in particle shape, accompanied by an increase in density. Grain boundaries are formed and recrystallization occurs, particularly in the regions where the particles are joined. Grain growth takes place across these boundaries and plasticity is increased. The final product can be drawn down to wire or rolled, which reduces the porosity. The microstructure, porosity, strength, etc., of the product all depend upon the sintering temperature and the heating time; the degree of compacting has less effect. The presence of a liquid can produce a near pore-free, high-density product by forming a solid solution, e.g. copper melts with iron powder at the sintering temperature. After sintering, the component may be quenched in oil which is absorbed into the pores. This is particularly useful for bearing materials which are then self-lubricating in service.

If an accurately shaped component is required, then a coining operation may be used as the compacting process. Hot pressing techniques are used for the refractory materials, e.g. tungsten, molybdenum, tantalum and niobium. Ceramics are also shaped by this method, using inert graphite or ceramic molds to avoid contamination. Beryllium is a very reactive powder and it is compressed and sintered at 1100°C (2012°F) in an argon or nitrogen atmosphere. This process is particularly useful for beryllium, which becomes brittle when cast or hot worked and is difficult to fabricate. Powder metallurgy processes have been

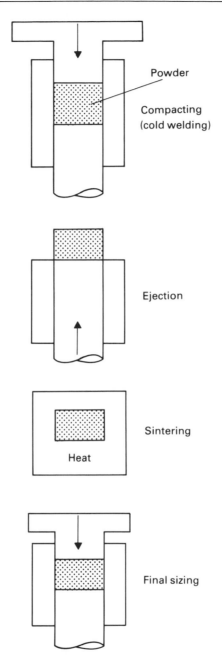

Figure 8.30   Stages in powder metallurgy.

successful both technically and economically for the production of cutting tools and refractories from cemented carbides, despite high manufacturing costs. The process does have great versatility for producing structures, components and properties in a controlled way.

*   *   *

Self-assessment exercises

1 What materials are shaped by powder metallurgy? Why are these materials not shaped by more conventional methods?

2 Describe the stages in the production of a part by powder metallurgy.

3 Powder metallurgy is an expensive process. Explain why and when it is used.

4 What factors determine the size and mass of parts produced by powder metallurgy?

5 What shapes cannot be produced by powder metallurgy?

*   *   *

FINISHED PRODUCTS

8.4 MACHINING PROCESSES

8.4.1 Introduction

After the basic component shape has been obtained by either casting or hot- or cold-working methods, it is often necessary to machine the part in a final finishing operation. Such operations as drilling, turning and grinding are used to produce the required dimensional accuracy, and to obtain the necessary surface quality of the finished part. During a machining operation, metal is removed from the component in the form of small chips. The range of materials that are available and the sizes and shapes of components which need to be machined have meant that a wide diversity of basic machine tools are available. Common machine tools will be described (briefly), followed by descriptions of the more common machining processes.

8.4.2 Basic machine tool elements

Most machine tools are built up from a number of elements or components. These will have different functions depending upon the machining operation to be performed, but they will also possess certain common features. The basic elements from which the machine tool is built are listed below.

(a) The machine frame
This will be specified in terms of the material of construction and the particular structure required. Factors to be considered at the design stage include the shape and rigidity, accessibility, safety and ease of metal chip removal.

(b) The drive unit

An electric motor is used to operate the gears or pulleys (contained in the headstock) that are used to drive the workpiece.

(c) The spindle

This is used to hold and rotate the cutting tool or the workpiece.

(d) Positioning table

Used to hold either the tool or the workpiece.

(e) Methods of machine control

Machine tools may be mass-produced machines such as lathes, drills, etc., or they may be expensive, specially assembled machines. These tools differ not only in the number of cutting edges which they employ, but also in the way in which the tool and the workpiece move in relation to each other. When using a lathe the work is made to rotate and the tool is stationary, whereas with a boring machine the tool rotates and the work is stationary. This description is not absolutely correct, as when using a lathe the tool is moved along the carriage during shaping of the workpiece. The basic structural elements of a machine lathe are shown in Fig. 8.31. The shaper and the planer are both single point cutting tools; however, the shaper is a more efficient machine than the planer for small workpieces. The milling cutter is an extremely versatile machine that can be used for many machining operations, e.g. sawing, milling, gear cutting. Prior to the introduction of the milling cutter, the boring machine was the only machine employing a rotating tool. Details of the work and tool movements of the more conventional machines are given in Table 8.2.

8.4.3 Principles of metal cutting

Machining operations that produce a finished product require the removal of metal as small chips. So that these operations can be carried out economically, both in terms of the work performed and the life of the cutter, it is important that the principles of metal cutting are understood. Factors such as tool selection, operating speed and feed rate all affect the efficiency of a machine in terms of machine utilization and operator time. However, an increase in machine throughput greatly reduces the tool life. Only the operation of a single-point cutter, where the cutting edge acts perpendicular to the cut, will be considered in this chapter. A typical single-point cutter is shown in Fig. 8.32, including the associated terminology. The cutting action and the cutting angle should be such that the minimum power is required. The cutting edge must be able to withstand the forces acting on it and also conduct the heat created. The choice of cutting angle and cutting edge depend upon the hardness of the material to be cut; softer materials require a smaller cutting angle.

Metal cutting is more in the nature of a tearing process. Schematic diagrams showing the way in which metal chips are removed and the terminology

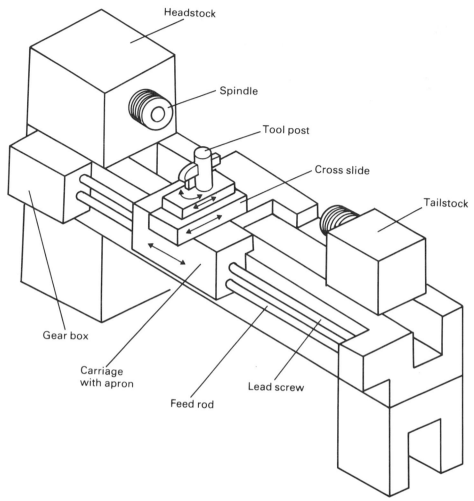

Headstock

Spindle

Tool post

Cross slide

Tailstock

Gear box

Carriage
with apron

Lead screw

Feed rod

Figure 8.31 Basic structural elements in a lathe machine.

used are given in Fig. 8.33. With reference to Fig. 8.33, case (a) assumes that the chip is severed by a shearing action and the deformed chip is in compression against the tool face, thus creating a large frictional force. The larger the frictional force, the greater the chip thickness (t), but the smaller the shear angle (α).

The *machinability* or ease of metal removal can be measured by the shear angle α in Fig. 8.33. The closer the shear angle is to 45° the better the machinability. A value can also be determined from measurements of the chip thickness (t) and the original depth of the cut (d). The ratio of t to d is known as the chip thickness ratio (r), and the shear angle can be calculated from the expression:

$$\tan \alpha = \frac{r \cos \theta}{1 - r \sin \theta}$$

Table 8.2   Principles of operation of conventional cutting machines.

Machine	Cutting movement	Feed movement	Operations performed
Saw	Tool	Tool	Cut off
Broaching	Tool	Tool	External and internal surfaces
Drill	Tool rotates	Tool	Drilling, boring, facing and threading
Cylindrical grinder	Tool rotates	Table and/or tool	Grinding cylindrical surfaces
Shaper	Tool traverses	Table	Shaping flat surfaces
Planer	Table traverses	Tool	Planing flat surfaces
Lathe	Work rotates	Tool and carriage	Cylindrical surfaces, drilling, boring, reaming and facing
Boring machine	Tool rotates	Table	Drilling, boring, reaming and facing
Horizontal boring	Tool rotates	Tool traverses	Flat surfaces
Horizontal milling machine	Tool rotates	Table	Flat surfaces, gears, cams, drilling, boring, reaming and facing

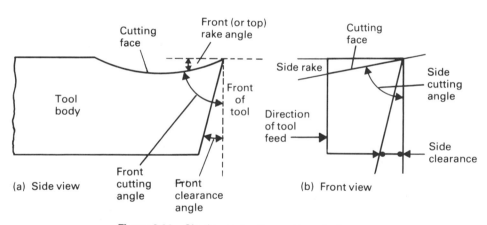

Figure 8.32   Single point cutter and terminology.

Materials that build up in front of the tool are more difficult to machine. To reduce wear on the tool there must be clearance between the tool and the machined surface, as shown in Fig. 8.33, case (c). The rake on the tool gives the chip a wedge-like form, and this rake is usually larger for soft and ductile metals than for hard and brittle metals.

Three types of metal chip can be formed during cutting operations. When cutting ductile materials possessing a low coefficient of friction, a simple deformed continuous chip that slides up the tool face is obtained (Fig. 8.33, case a). This type of chip is formed at higher cutting speeds; small cracks may appear in the chip but they are not sufficiently developed to cause fracture. The chip does not cling to the tool and a good surface finish is obtained.

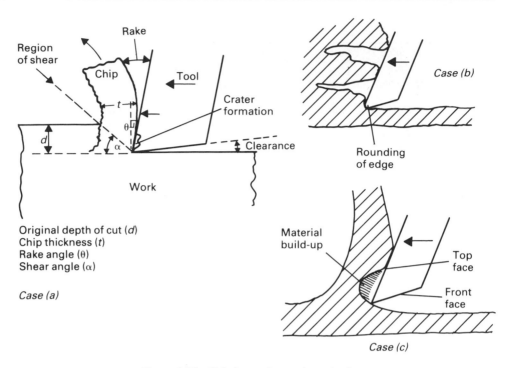

Figure 8.33 Chip formation and terminology.

For brittle materials such as cast iron and bronze, discontinuous (segmental) chips are produced. As metal is removed severe distortion occurs adjacent to the tool face, forming a crack that runs ahead of the tool. The chip fractures and separates when the shear stress across the chip becomes equal to the shear strength of the material. This situation is illustrated by case (b) in Fig. 8.33. When machining ductile materials that possess a high coefficient of friction, some material builds up ahead of the cutting edge as shown by case (c) in Fig. 8.33. The chip moves up the tool face and periodically some of the built-up material is removed with the chip, or becomes embedded in the turned surface. The formation of built-up material depends upon the cutting speed and the associated heat and friction. At low cutting speeds, insufficient heat and friction are created for the build-up to occur. At very high speeds, the heat causes annealing of the material in the shear zone and strain hardening is prevented. This type of chip formation is observed with mild steel and produces a poor surface finish.

For a production operation it is important that long curling chips are not formed. A 'chip breaker' can be used to curl and stress chips so that they break into small pieces for easy removal. The chip breaker may be a plate fixed onto the tool so that the chip is forced against it, or it may be achieved by special preparation of the tool cutting edge.

It is possible that as much as 97% of the work required to carry out a cutting operation is dissipated as heat. This heat is generated in the three regions shown in Fig. 8.34, and the approximate heat distribution is also shown. Approximately twice as much heat is generated in the shear plane than in the

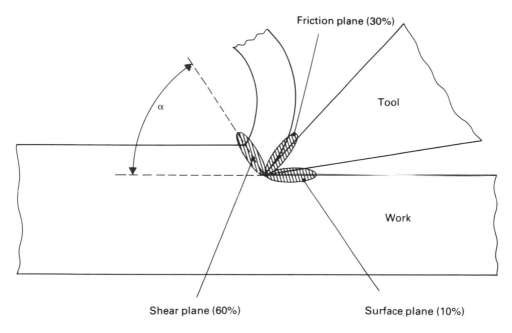

Figure 8.34   Regions of heat generation during cutting, and approximate distributions (%).

frictional plane, and together they account for 90% of the total heat. The amount of heat in the shear plane can be reduced by increasing the shear angle (α), but this requires the use of a coolant and correct grinding of the tool. To increase the rate of metal removal it is preferable to increase the feed rather than the cutting speed, as the latter has a far greater effect on the temperature produced.

During cutting there are forces acting on the tool in the longitudinal, radial and tangential directions; these are shown in Fig. 8.35. The approximate

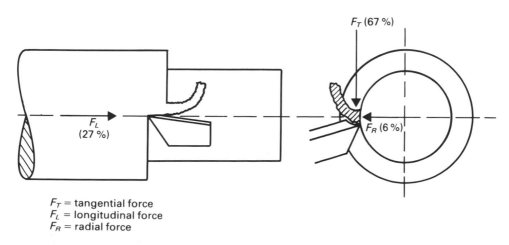

F_T = tangential force
F_L = longitudinal force
F_R = radial force

Figure 8.35   Location and distribution of forces during cutting.

percentage distribution of these forces is also given. The tangential force is the largest in most operations. These forces are reduced if either the tool feed or the depth of cut is reduced. The use of a coolant slightly reduces the forces but greatly increases the tool life. Changes in the cutting speed do not alter these forces. The tangential force is increased as the chip size increases, but is reduced by increases in the back rake angle (approximately 1% per degree). The longitudinal force can be decreased by increasing either the nose radius or the side cutting edge angle.

Cutting tools are manufactured from many different materials because of the wide range of operating conditions which occur. In the selection of a tool material it is important to consider the ability to resist abrasive forces, softening at high temperatures and fracture, as well as the need to possess a low coefficient of friction. Materials that are in common use include high-carbon steels (0.8–1.2% carbon), high-speed steels, cast nonferrous alloys (mainly containing chromium, cobalt and tungsten), carbides (mainly tungsten), diamond (mainly for high speeds and light cuts) and various ceramic materials.

Coolants may be used during a machining operation for a variety of reasons. These include reductions in temperature, power consumption, friction and corrosion. The coolant may also assist the removal of metal chips, increase the tool life, prevent welding of the tool to the workpiece and improve the surface finish. A coolant should possess the following properties: good heat-transfer, high flash temperature, nonvolatile, nonfoaming, stable, lubricating, harmless to the machine and the operator. Coolants may be solid such as soap or graphite (which is present in gray cast iron), or gases and vapors, e.g. compressed air, although water-and oil-based solutions are most commonly used. The reason is that liquids are easier to direct and recirculate. The coolant may also act as a lubricant and certain chemical agents are added to improve their effectiveness, e.g. amines for rust prevention, germicides to control bacterial growth. A coolant has maximum effectiveness when directed to the region of metal–tool contact: the main advantages are a reduction in friction and cooling of the tool. Some typical metal–coolant combinations are cast iron and compressed air (for dust removal) or soluble oil, aluminum and kerosene, brass and paraffin oil, steel and mineral oil, wrought iron and lard oil or a water-soluble oil.

The machinability of a material is a relative term, evaluated almost entirely on the basis of tool life. Although other considerations such as power consumption and surface finish may be taken into account, these are of far less importance. The machinability of a metal is affected by its hardness and ductility. Hard metals are difficult for the cutting tool to penetrate and their machinability is low. Discontinuous chips are produced when hard metals are cut and the tool wear occurs by a rounding of the sharp edge and wearing back of the tool tip, as shown in case (b) of Fig. 8.33. As previously discussed, ductile metals cause a build-up of metal ahead of the cutting edge and machinability is reduced. A material possessing low values of ductility and hardness has superior machinability, although in practice as one of these values decreases the other increases. A compromise is usually established between the requirements of machinability and the in-service conditions of the finished component.

Good machinability does not mean good surface finish. It is a rating for the metal associated with low tool forces and long tool life, thus ensuring metal removal at the lowest cost and a satisfactory surface finish. The surface finish of a product is generally improved by using high cutting speeds, light cuts, small feeds, round-nosed tools, increased rake angle, cutting fluids and correctly ground tools.

A cutting tool must produce a satisfactory cut, e.g. good surface finish, and a large amount of material removed. Whenever excessive wear has occurred the tool must be reground and this is both expensive and inconvenient. In general, the tool life decreases as the cutting speed is increased. The tool life is reduced if the tool is reground incorrectly or if excessive heat causes loss of tool hardness. If too heavy a cut or too small a lip angle is used, then the tool edge may break, and the tool may fracture if too heavy a load is applied. These adverse conditions can usually be avoided.

The cutting speed (m/s or ft/s) is the surface speed at which the work passes the cutter. The feed is the rate at which the cutter moves along, or into, the surface of the workpiece. When the work rotates the feed is measured in mm per revolution; if the tool or the work reciprocates, then the feed is expressed in mm (in.) per stroke. For a given tool, the feed needs to be reduced if there is an increase in either the cutting speed, the hardness or ductility of the workpiece or the amount of material removed.

8.4.4 Sawing

Sawing, or cutting off, is one of the first operations performed in the workshop. This may be by hand or using one of the many machine tools available, although special machines are required for mass production. Sawing is a metal-cutting operation and for hand sawing a thin, flexible blade is often used. It is between 200 and 300 mm (8 and 12 in.) long, held in a hacksaw frame with a suitable grip. The tooth pitch is between 6 and 13 teeth per cm (15 and 32 teeth per inch); the average for handsaws is 7 teeth per cm with less required for cutting tubing and thin material. A coarse-tooth saw (larger pitch) has more space for the chip, but the selection depends upon both the thickness and the type of material to be cut. The tooth construction of metal hacksaw blades is shown in Fig. 8.36. Similar hacksaw blades are used for both power driven and hand operations. In Fig. 8.36 the straight tooth having zero rake is the most common design; the region A-B allows space for the metal chips, without which the teeth would become blocked. The undercut tooth (Fig. 8.36(b)) is used mainly for larger blades, giving faster cutting on heavy sections of material.

When cutting steel or cast iron a coarse-pitch blade is required to provide sufficient space for the metal removed; however two or more teeth should always be in contact with the workpiece. A medium-pitch blade would be used for cutting high-carbon and alloy steels and a fine-pitch blade for thin metal, tubing and brass. High-speed steel blades may vary between 300 and 900 mm (12 and 36 in.) in length and thicknesses between 1.2 and 3mm (0.048 and 0.12 in.), they have a coarse pitch between 1 and 6 teeth per cm (2.5 and 15 teeth per inch).

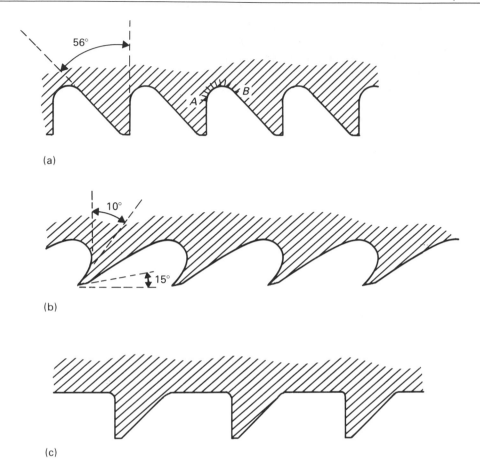

Figure 8.36 Arrangement of teeth in common metal hacksaw blades.
(a) Straight tooth. (b) Undercut tooth. (c) Skip tooth.

Hacksaw machines generally cut between 40 and 160 strokes per minute, depending upon the machinability of the metal. The surface speed is not usually specified, and it varies throughout the length of the stroke. The cut obtained is slightly wider than the blade thickness to allow clearance, and this is achieved by slightly bending certain teeth to the right or left along the blade length. This is known as the *tooth set* and may be alternate teeth to the right and left, e.g. for brass or copper, or one straight tooth alternating with two teeth set in opposite directions.

Only a small amount of heat is generated during sawing. The cutting fluid should be chosen for its lubricant properties and its ability to wash away metal chips. By consideration of the tooth construction of a hacksaw blade shown in Fig. 8.36, it can be seen that the metal chips are removed only on the forward stroke. For this reason saw cuts should only be made in a single direction—any pressure applied on the return stroke merely shortens the blade life.

The metal blades used on power-sawing machines can be either straight, circular or continuous blades depending upon the type of machine. Power sawing machines can be classified as follows:

(a) *Reciprocating saw.* This may be either a horizontal hacksaw or a vertical saw. These machines may use either positive or uniform pressure feeds. A positive feed has an exact depth of cut for each stroke and the blade pressure varies according to the number of teeth in contact with the work. When cutting a round bar, the cutting is slow at the start and finish when contact and pressure are small.

Uniform feed machines have a constant pressure at all times, the maximum pressure is determined by the load which a single tooth can withstand. This method is used with gravity or friction feeds where the pressure is exerted either by weights, or by the weight of the machine frame itself. In all cases the pressure is released on the return stroke to eliminate wear on the teeth.

(b) *Circular saw.* These machines cut with metal blades, steel friction disks and abrasive disks. It is sometimes known as cold sawing, and large-diameter saws with low rotational speeds are used. The cutting action is the same as that obtained with a milling cutter (see Section 8.4.8).

(c) *Band saw.* The blades for these machines may be either metal saws or friction or wire blades. These machines can be used for straight cuts and cutting-off operations, but they can also be used to make irregular cuts or to perform continuous filing and polishing. Most machines use a vertical blade, with the work supported on a horizontal table.

8.4.5 Broaching

Broaching is the removal of metal by an elongated tool having a number of successive teeth of increasing size. The cut is made along a fixed path and is completed in one stroke of the machine. The last teeth on the cutting tool correspond to the desired shape of the finished object. Generally the broach is moved past the work, but a stationary tool and moving workpiece can produce the same requirement. An accuracy of $\pm$ 13 μm can be obtained and a surface finish of 0.8–3 μm. This method can be used with ferrous and nonferrous materials and for relatively hard materials.

The principles of pull and push broaching operations are shown in Fig. 8.37. The use of broaching and the type of tool used to cut internal keyways are shown in Fig. 8.38. Broaching tools are usually adapted to a single operation, and the feed must be predetermined and unchanged. Certain information must be available before the tool is made, including the size and shape of both the cut and the material, tolerances, quality of finish, type of machine, method of holding the broach and the pressure which the part can withstand. Most internal broaching uses a pull method, making possible longer cuts and more metal removal. Push broaches are short to avoid buckling under load, and are mainly used for blind holes, short runs and hole sizing. For flat surfaces either straight or angular teeth can be used, the latter producing a smoother cutting action.

An internal pull-type broach is illustrated in Fig. 8.39, and the typical tooth form is shown in Fig. 8.40. The terminology that is adopted is shown in each diagram. The shapes or angles used on a broaching tool are not necessarily the same throughout its length. The clearance angle (see Fig. 8.40) is usually

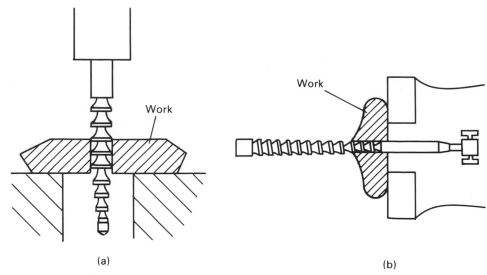

(a)   (b)

Figure 8.37   Round broaches used for push and pull broaching ((a) and (b) respectively).

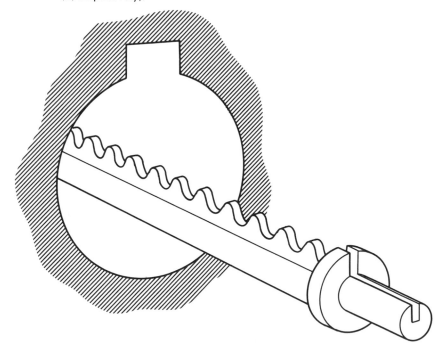

Figure 8.38   Keyway broaching.

between 1.5° and 4° on the cutting teeth and between 0° and 1.5° on the finishing teeth. The rake angle may be between 0° and 20°, depending upon the material being cut. The angle increases as the ductility increases, and between 12° and 15° is used for most steels. The force required to make the cut and the finish obtained depend upon the rake angle. The first teeth have a small rake

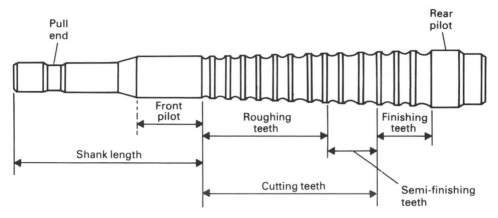

Figure 8.39   Internal pull type broach.

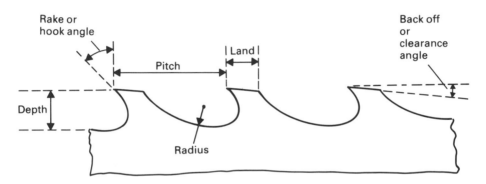

Figure 8.40   Terminology and toothform on a broach.

angle, whereas the finishing teeth have a large rake angle in order to improve the finish. The tool life will be increased if a smaller rake angle is used. Regrinding of a broaching tool involves sharpening the face or front edge of the teeth, and not the land as this would considerably change the size of the broach.

Broaching machines are usually classified according to their operation: these includes push, pull and surface broaching. The machines are of either horizontal or vertical design depending upon such factors as the size of the part, the size of the broach, the quantity and type of broaching to be performed. Continuous broaching machines are available, but only for surface machining where the work is moved continuously against stationary broaches; the path of this movement may be either straight or circular.

Broaching machines are usually classified according to their operation: these include push, pull and surface broaching. The machines are of either horizontal or vertical design depending upon such factors as the size of the part, the size of the broach, the quantity and type of broaching to be performed. Continuous broaching machines are available, but only for surface machining where the work is moved continuously against stationary broaches; the path of this movement may be either straight or circular.

8.4.6 Turning

Turning produces a cylindrical component. Many different machine tools can be used; a lathe is often employed and metal is removed by rotating the work against a single-point cutter. This method can also be used for machining plain surfaces, cutting threads or turning tapers. The principles of metal cutting or metal chip removal have been discussed in Section 8.4.3. All of the tool body (except the point) should clear the work, otherwise excessive wear results and the life of the tool is reduced. A typical single-point cutting tool is shown in Fig. 8.41(a). The ideal interaction of cutter and workpiece for different machining

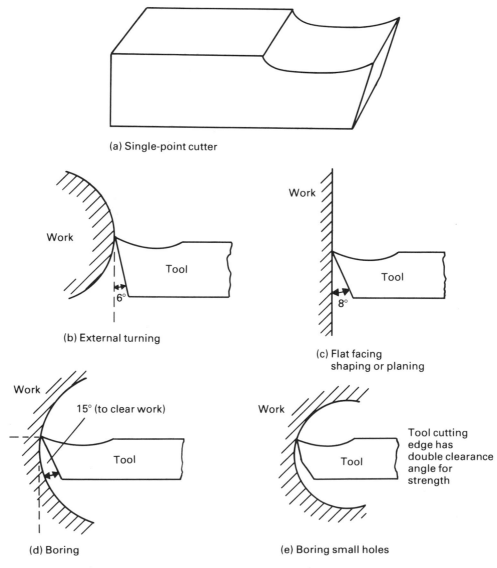

(a) Single-point cutter

(b) External turning

(c) Flat facing
shaping or planing

(d) Boring

(e) Boring small holes

Figure 8.41 Single-point cutter and positioning with work in ideal operating conditions.

operations is illustrated in Fig. 8.41(b)–(e).

Some of the machines that can be used for turning will now be mentioned briefly. The lathe is one of the most common machine tools, and there are many different types. Neither the detailed construction nor the definitions of the main parts will be given here. However, it should be mentioned that the size of a lathe is expressed in terms of the maximum diameter of the work that it can handle, and sometimes the maximum length.

The speed lathe is the simplest machine: it operates at high speed using hand tools and consequently makes small cuts. It is used for wood turning, metal spinning and centering of metal cylinders prior to further machining.

The advantages of using an engine lathe are better control of the spindle speed, additional control and support of the feed of the fixed cutting tool.

A bench lathe may be either a speed lathe or an engine lathe; it is bench mounted and only used for smaller work.

A toolroom lathe possesses additional features that make it particularly suitable for accurate tool work.

Turret lathes are particularly suitable for production work. It is possible for inexperienced operators to reproduce accurate identical parts, whereas the engine lathe requires more time and an experienced operator. The skill required when using a turret lathe is the ability to pre-set and adjust the tools prior to machining.

Automatic lathes operate so that the tools are automatically fed to the work and withdrawn after operation. These machines may only be semi-automatic in operation if an operator is required to position the work and remove the finished product. However, a continuous flow of components may be produced.

An automatic screw machine is basically a turret lathe designed to machine bar stock. The tools can be automatically fed to the work at the desired speeds, withdrawn and set to the next required position. The work feed and its positioning, clamping and releasing operations are also performed automatically. This type of machine was originally used for the manufacture of bolts and screws – hence its name.

A vertical boring mill uses stationary cutting tools, except for feed movement, with the work rotating on a horizontal table. These machines can be used for horizontal facing work, vertical turning and boring. The machine operation is that of a lathe and rotary planer. The work table diameter may be as large as 12 m (40 ft); large objects need minimum securing and the machine occupies relatively little floor space. Accurate machining can be achieved due to the stability of the work and the machine.

Many different machining operations can be performed on a lathe, including turning, boring, facing threading and taper turning. All of these operations use a single-point cutter which is fed along the revolving workpiece. Drilling and associated operations require other types of cutters. For cylindrical turning the work is often mounted between centers. The machining of a flat surface is known as facing; the cut is at right angles to the axis of rotation and it is essential that there is no axial movement during machining. Many components and tools have tapered surfaces, which may be short steep tapers, e.g. gears, or long gradual tapers on mandrels. Taper turning can be carried out on a lathe, one method being to use a special taper-turning attachment. Common commercial tapers include the Morse taper (approximately 5.2%) for drill shanks and lathe centers, and taper pins (2.08%) which are used for thread cutting, but this is usually reserved for special forms or when only a few threads are required. The cutting tool must first be ground to the correct shape. Using a series of gears, the tool is given a positive feed along the work at a rate which will produce the required number of threads per mm (in.)

* * *

Self-assessment exercises

1 List the main types of machine tools and their distinguishing features.

2 Describe how metal is removed from an object using a single-point cutting tool.

3 Describe the types of metal chip that may be formed. What is the function of a 'chip breaker'?

4 Explain the term machinability.

5 Explain the terms; tool life, the cut, speed, feed, regrind, lip angle.

6 Describe the movements of a lathe tool.

7 Explain the meaning of:
   (a) tooth pitch;
   (b) rake angle;
   (c) tooth set;
   (d) undercut tooth.

8 When using a saw blade, why is metal removed only on the forward stroke?

9 Describe how broaching is performed, and the difference between push and pull operations.

10 What is a burnishing broach? (Refer to Yankee (1979), Chapter 15, page 181.)

11 Compare the relative cutting speeds and accuracy obtained by broaching and milling (see Section 8.4.8).

12 Explain the importance of work-holding devices when performing turning operations.

13 Explain the causes of cutting-tool damage.

* * *

8.4.7 Drilling and boring

Terminology
A *drilling machine*, or *drill press*, produces a hole by forcing a rotating drill against a stationary object. The same effect can sometimes be achieved by holding and rotating the workpiece in a lathe and using a stationary drill. The drill press is one of the simplest machine tools.

Boring is a process for enlarging an existing hole. It is often used for truing a hole that has been drilled using a lathe.

Coring is used to produce large holes, mainly in castings. Machining costs are reduced and less metal is wasted. The circular saw cutters and fly cutters shown in Fig. 8.42 are both used for cutting large-diameter holes in thin metal.

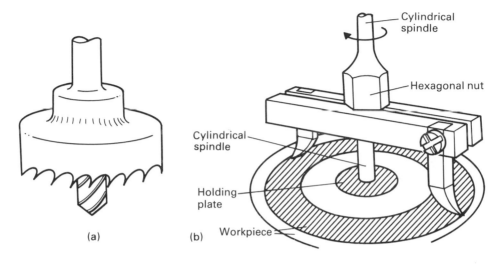

Figure 8.42   Cutters for holes in thin metal. (a) Circular saw cutter. (b) Fly cutter.

Reaming is a process of enlarging a machined hole to the required size, and at the same time producing a smooth finish. This is an accurate process and should not be employed to remove large quantities of metal.

Trepanning is a process used to produce a solid cylinder of metal using a saw cutter as shown in Fig. 8.43, or for drilling deep holes. In the latter case a straight-fluted gun drill is used (see Fig. 8.48), it has no dead center and leaves behind a solid core of metal. This solid core acts as a continuous center guide for the cutting process.

Counterboring is used to enlarge one end of a drilled hole. The enlarged hole is concentric with the original hole and is flat at the bottom. A pilot pin that

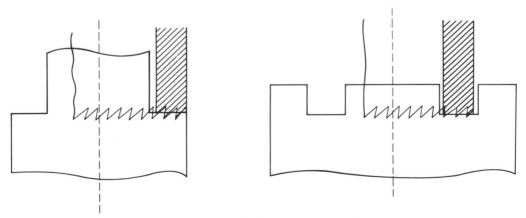

Figure 8.43   Examples of trepanning.

fits the drilled hole is used to center the cutting edges. The method is illustrated in Fig. 8.44 and is mainly used to set bolt heads and nuts below the surface.

Spotfacing is used to finish-off a small surface around a drilled hole as shown in Fig. 8.45. This is mainly used to provide a smooth seat in a rough surface.

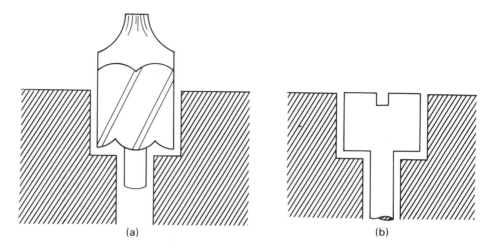

Figure. 8.44 Example of counterboring. (a) Preparation. (b) Finished product.

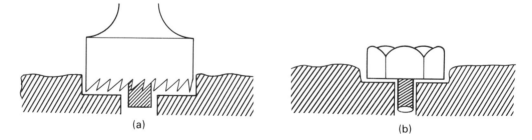

Figure 8.45 Example of spotfacing. (a) Preparation. (b) Finished product.

Countersinking produces a bevelled top to a drilled hole in order to accommodate the conical seat of a flat-head screw. This is shown in Fig. 8.46.

Holes can also be produced by *punching, flame cutting,* and using *ultrasonic* and *electric-discharge machines*. Punching is a rapid process that produces accurate holes especially in thin materials. However, the punches and dies are expensive. Flame cutting (see Section 7.10) can produce holes in a material of any commercial thickness, although the position and size of the holes produced are not particularly accurate.

Drills

A drill is a rotary-end cutting tool. It has one or more cutting edges, and corresponding flutes that continue along the length of the drill body. The flutes

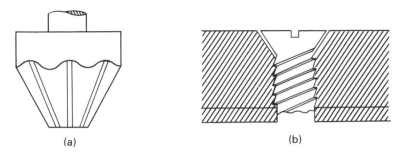

(a)   (b)

Figure 8.46 Example of counter-sinking. (a) Counter-sinking cutter.
(b) Finished product.

may be straight or helical, and they allow movement of metal chips and cutting
fluid. A two-fluted drill is traditionally used for originating and drilling holes,
whereas a single-fluted drill is used for originating and drilling deep holes. A
drill having more than two flutes is known as a core drill; it is not used for
starting a hole but for enlarging or finishing drilled and cored holes.

A twist drill having two flutes and two cutting edges, is the drill most
commonly used. The drill may have either a straight or a tapered shank. Details
of a twist drill and the associated terms are given in Fig. 8.47. A wide range of

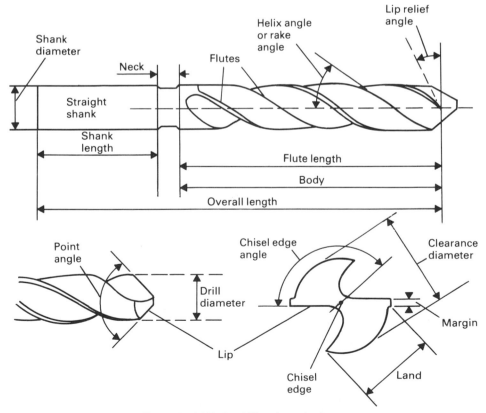

Figure 8.47 Twist drill and terminology.

drills is available for greater productivity, improved finish and for use with special alloy materials; some of these drills are shown in Fig. 8.48. Some drills, used for production drilling, have interior or exterior channels for oil flow.

Drill selection. When selecting a drill for use in particular conditions or evaluating the performance of a particular drill, various factors need to be considered. These include the type of material to be drilled. For a hard and abrasive material such as cast iron, a tungsten carbide-tipped drill may be used. On the other hand, a high-carbon, cobalt-alloy steel drill may be required for drilling stainless steels. Some drills are surface treated to provide a thin case-hardened layer, or chrome-plated to provide a hard-wearing surface.

The choice of a drill in terms of the point angle and the helix angle (see Fig. 8.47) depends mainly upon the material to be drilled. For most metals a point angle of 118° is satisfactory, although this needs to be increased with harder metals to obtain satisfactory performance. Soft materials, such as plastics, are often drilled with point angles less than 118°. In general the larger the point angle, the thicker the metal chips which are removed. Removal of thicker chips usually requires less energy than for thin chips of the same volume. Also the larger the point angle, the shorter the metal chip obtained; however, with abrasive materials less wear on the cutting edge is experienced with longer chips. The choice of a drill with the correct helix angle for the material being drilled is important; this can vary from 20° to 25° for copper alloys to approximately 30° for steel. The smaller the helix angle, the greater the torque necessary to operate at a given feed. The life of the cutting edge depends upon the use of the correct helix angle.

The amount of metal removed depends upon both the cutting speed and the feed. The cutting speed depends upon the hardness and toughness of the material; the greater these values then the slower the speed required. Speeds may vary between 0.5 m/s (1.6 ft/s) for cast iron and 1.5 m/s (4.8 ft/s) for brass. Drill feeds are expressed in mm (in.) per revolution and depend upon the drill material and the cutting speed.

To increase the life of the cutting edge and to obtain the best performance, a cutting fluid should be used. The fluid acts as a coolant, makes chip removal easier and improves the cutting action. Cutting fluids may be a dry air jet for cast iron, lard or a soluble oil for tool steel or a dry, mineral oil for magnesium.

Reamers
Reaming is an accurate process where a hole is finished by removal of small quantities of material. A reamed hole is round with a smooth surface. Reamers can have either a straight or tapered shank and the flutes can be either straight or helical. A standard reamer is illustrated in Fig. 8.49 and standard terms are also included.

Boring tools
Boring is used to enlarge holes (that have been drilled or bored) to a size suitable for reaming, or to eliminate any eccentricity. Boring may be used to finish a hole to its final dimensions. A single-point cutter is used for most boring operations; this is easy to set up and maintain. The rotating cutter is fed into the hole and a

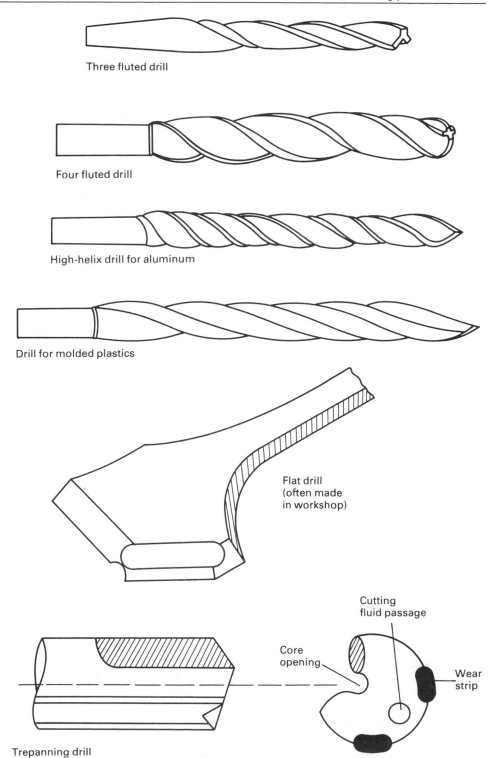

Three fluted drill

Four fluted drill

High-helix drill for aluminum

Drill for molded plastics

Flat drill
(often made
in workshop)

Cutting
fluid passage

Core
opening

Wear
strip

Trepanning drill

Figure 8.48   Types of drills.

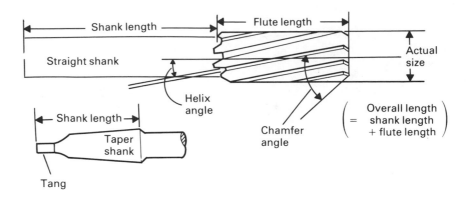

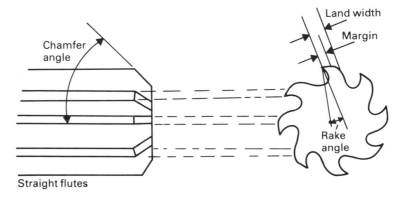

Figure 8.49 Standard reamer and terminology.

micrometer adjustment may be required. A block cutter having two cutters may be used or even multiple cutters; these tools have a longer life than single-point tools and are therefore more economical for production work.

Jigs

A jig is a device used to hold the work and guide the cutting tool. The design of the jig depends upon the shape of the workpiece and the region which is to be drilled. A jig should enable quick and easy loading and unloading, and it should be constructed so that the work can only be loaded in the correct position. With any drilling operation it is important that the work is held firmly, e.g. in a vise, which can be achieved by securing the jig. The work should be marked out using a template; this can also be used to guide the drill. Metals chips should be easily removable from the jig. Clearance is usually allowed under the drill bushings which guide the drill, so that metal chips do not have to pass through the bushing. A jig that could be used for drilling holes accurately in a metal 'tee' section is shown in Fig. 8.50. In a manufacturing process the cost of the jig has to be recovered. It is economic only if a large number of components are to be produced, and if accuracy is particularly important. Jigs are used for drilling, tapping, counterboring and reaming operations.

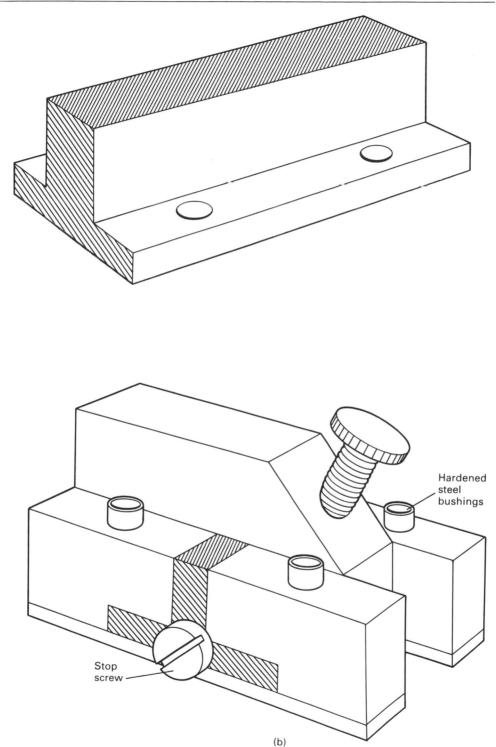

(b)

Figure 8.50 Example of a jig for drilling the component shown.
(a) Final component. (b) Assembled jig and component.

Drilling deep holes

There are various problems associated with drilling deep holes, which are normally encountered when the length of the drilled hole is more than five times its diameter. Difficulties may be experienced in feeding lubricant to the cutting edge and in the removal of metal chips. Special drills may be needed to feed the lubricant under pressure, which also carries away the metal chips. Alternatively, chips or swarf may be removed by drilling from below so that the metal falls out of the hole by gravity. It may be necessary to interrupt the drilling operation periodically to remove metal with a magnetic file (if appropriate) or compressed air, but this is not usually a satisfactory operating procedure.

The main problem to be overcome is to eliminate any movement of the work or the drill during the drilling operation. The errors that can occur are illustrated in Fig. 8.51. If the workpiece is not securely held during drilling, then the hole will be too large and tapered (Fig. 8.51(b)). If the drill moves or it is not properly aligned at the start of the operation, then the hole will be incorrectly positioned in the workpiece (Fig. 8.51(a)). In general, the effects of movement in the workpiece are less serious, as the drilled hole still follows the required axis. It is therefore preferable to rotate the workpiece (drill stationary) when drilling deep holes. In order to start the hole, the procedure to follow is: drill a hole of length equivalent to two diameters with a smaller drill, bore out this hole to the required diameter, then drill the required hole. The drilling of deep holes in materials that are not homogeneous, e.g. wrought iron, is not advisable because the presence of hard impurities or 'blow holes' (see Section 2.2) can cause the drill to deviate from its intended path.

Drilling machines

There are many drilling machines available, and the following classifications are made according to their general construction. Detailed descriptions of these machines and their operation are not given.

(a) *Portable drills*

Holes up to 6 mm (0.25 in.) diameter can often be drilled using a hard drill, and slightly larger holes using a hand-and-breast drill. Holes up

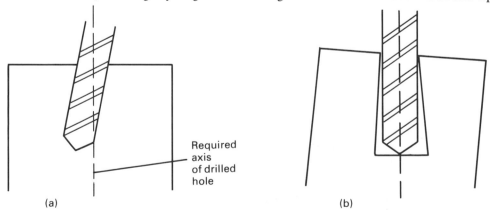

(a)   (b)

Figure 8.51 Problems associated with drilling deep holes. (a) Drill moves or incorrectly aligned. (b) Workpiece moves during drilling.

to 25 mm (1 in.) diameter can be drilled using a hand-held, power-driven drill, and for holes up to 50 mm (2 in.) diameter a drilling pillar and ratchet brace is required.

(b) *Sensitive drilling machine*

Most common is the hand-feed type (bench or floor mounted); the operator is sensitive to how fast the drill is cutting and can therefore control the cutting conditions.

(c) *Upright or vertical drilling machine*

The axis of the drilling operation is vertical; it is similar to the sensitive type but has power feed, heavier construction and can handle a wider range of work.

(d) *Radial drilling machine*

Used mainly for larger components, avoids positioning heavy work. The drilling unit is independently powered and can be operated at different positions on the workpiece surface.

(e) *Turret drilling machine*

This machine has 6, 8, or 10 tools, which can be different drill sizes or different tools, e.g. drills, reamers, etc., all contained in one head.

(f) *Automatic, production drilling machine*

Uses a multi-spindle drilling head to drill many holes at the same time. The head may be adjustable (intermediate production) or fixed (long or high production runs).

(g) *Deep-hole drilling machine*

Boring machines

There are several machines available that are designed to perform boring operations. These operations can also be performed on lathes and other machines but it is generally easier and quicker, and therefore more economic, to use a specialist boring machine. A *jig borer* is used for precision work, it is similar to a drill press and can also be used for drilling and end-milling. For large work, either a horizontal or vertical boring machine could be used. The vertical boring machine has a horizontal circular worktable, and it can also be used for facing and vertical turning operations. The horizontal boring machine differs in that the work is stationary and the tool revolves. The worktable can be adjusted longitudinally and crosswise, and the tool can be adjusted vertically. This machine is used for boring horizontal holes.

8.4.8 Milling

A milling machine removes metal when the work is fed against a rotating, circular cutter. This is the only movement of the cutter. During its rotation, each cutting edge on the circumference of the cutter acts as an individual cutter at some time. The work is held on a table which controls the feed against the cutter. This is generally either longitudinal, crosswise or vertical movement, but some machines also possess swivel or rotational movement.

The milling machine is particularly versatile and can produce an accurate product with a good surface finish. The machine can be used to machine angles, slots and gear teeth, and to recess cuts accurately due to micrometer adjustments on all table movements. Heavy cuts can also be made accurately and with an excellent finish. Shaping, drilling, broaching and gear cutting can all be carried out on a milling machine and the machining is particularly adaptable for the production of duplicate parts. The work is usually completed in a single pass, the cutters are efficient and have a long life before requiring regrinding. The average life of a cutter is approximately 25 regrinds, and as cutters are expensive items signs of wear should be continuously monitored.

In Fig. 8.52 a typical milling cutter is illustrated with the appropriate nomenclature. The choice of *rake angle* is a compromise between high strength and good cutting, and is between 10° and 15° (positive) for high-speed cutters. A larger angle can be used with softer materials, but a negative rake angle is used when steel is milled. The *clearance angle* is always positive, for cutters greater than 75 mm (3 in.) diameter angles of 4–5° are used so as not to weaken the teeth. Smaller diameter cutters have a larger clearance angle so that the teeth clear the work. This angle also depends upon the material being milled; values of 4–7° are normal for cast iron and 10–12° for softer materials. The *land* is

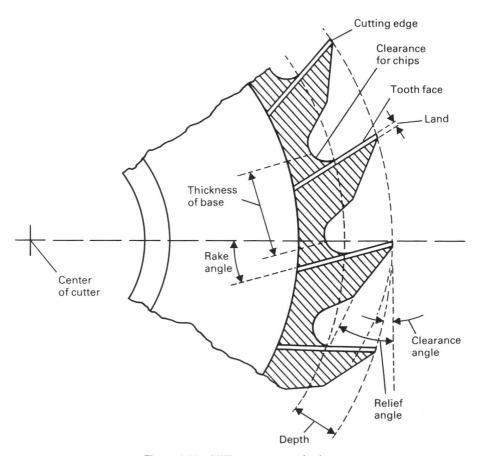

Figure 8.52 Milling cutter terminology.

usually between 0.8 mm and 1.6 mm (0.032 in. and 0.064 in.) and a secondary clearance is ground at the back of the land. Metal removal is more efficient using cutters with coarse teeth rather than fine teeth because thicker chips are removed, more clearance space is available for the chips and coarse teeth have a freer cutting action. This results in increased production and reduced power consumption. However, fine teeth are recommended for saw cutters when used for milling thin materials.

The two methods of feeding work to the cutter are shown in Fig. 8.53. When the work is fed in the same direction as the tool rotation (*down-cut milling*) larger chips are removed and the cutting action is more efficient. However, the work surface should be free from scale so that each tooth starts its cut in clean metal, otherwise *up-cut milling* is recommended. The feed on a milling machine may be expressed either as mm (in.) per revolution of the cutter or as feed of the table in mm/s (in./s). The usual range of these feed rates is 0.15–7.5 mm (0.006–0.3 in.) per revolution and 0.2–8.5 mm (0.008–0.34 in.) per second, respectively. With a plain milling cutter the rate of metal removal is mainly determined by the work feed, although the width and depth of the cut will also have some effect.

The cutting speed is determined by the surface speed of the cutter and is based upon the cutter diameter, not the work diameter. The movement of the work is not considered. The cutting speed is much higher for softer materials and the surface finish is better for high speeds and light cuts. Cutting speeds are usually quoted for high-speed steel cutters, the values being about twice those required for carbon steel cutters. The cutter life is extended if heat is removed by a coolant and if heavy cuts are carried out at slower speeds.

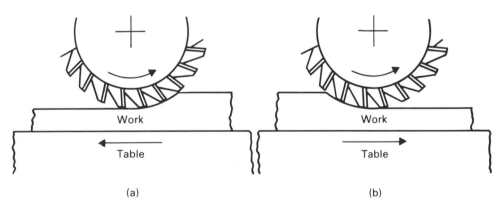

Figure 8.53 Alternative feeds for milling. (a) Conventional or up-cut milling. (b) Climb or down-cut milling.

Many different cutters are available for use on a milling machine and these can be classified in several ways. In terms of the cutter mounting, there are arbor cutters (having a central hole), shank cutters (straight or tapered) and face cutters (bolted on the end of short arbors). Cutters may also be classified according to the material from which they are made, e.g. nonferrous alloys, carbon steel, carbide-tipped or high-speed steel. High-speed steel cutters are more widely used; their cutting speeds are higher and the keen cutting edge is maintained up to 550°C (1022°F).

Milling cutters are made with two alternative types of teeth. These are fluted teeth as used on profile cutters, or relieved teeth on formed cutters; both are shown in Fig. 8.54. Profile cutters are sharpened by grinding the land at the back of the tooth cutting edge, whereas formed cutters are sharpened by grinding the face so as not to change the tooth contour. The (exaggerated) shape of the teeth of a formed cutter after many re-grinds is shown in Fig. 8.54. Cutters can also be classified according to their general shape or the type of work they perform. Some of the more common cutters are shown in Fig. 8.55.

Many different types of milling machine are available. They differ not only in their sizes but also in terms of possible table movement, the drive unit used and the method of work feed, whether by hand, by mechanical means or by a hydraulic system. The following classification of milling machines is proposed in terms of the general design, with some overlapping.

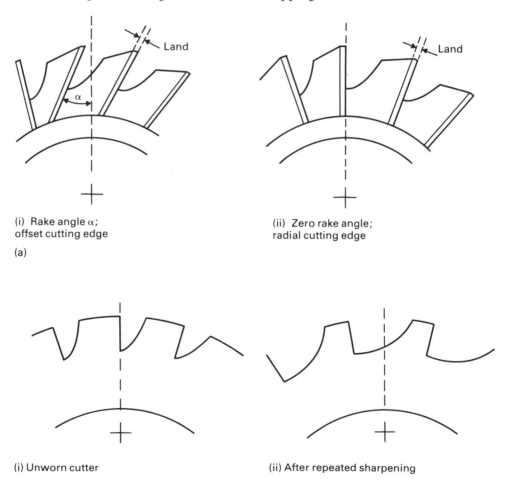

(i) Rake angle α;
offset cutting edge

(a)

(ii) Zero rake angle;
radial cutting edge

(i) Unworn cutter

(ii) After repeated sharpening

(b)

Figure 8.54 Alternative teeth forms for milling cutters. (a) Fluted teeth used on profile cutters. (b) Relieved teeth used on formed cutters.

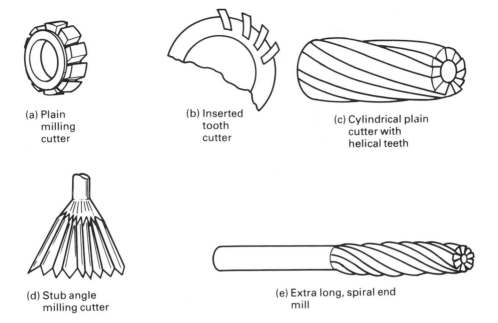

(a) Plain
    milling
    cutter

(b) Inserted
    tooth
    cutter

(c) Cylindrical plain
    cutter with
    helical teeth

(d) Stub angle
    milling cutter

(e) Extra long, spiral end
    mill

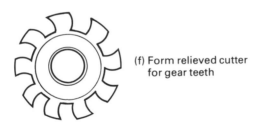

(f) Form relieved cutter
    for gear teeth

Fig 8.55   Cutters for milling operations.

(a) *Column and knee type:*
    (i)   hand miller;
    (ii) plain milling machine;
    (iii) universal milling machine;
    (iv) vertical milling machine.
(b) *Planer milling machine*
(c) *Fixed-bed type:*
    (i)   simplex milling machine;
    (ii) duplex milling machine;
    (iii) triplex milling machine.
(d) *Machining center machine*
(e) *Special types:*
    (i)   rotary table machine;
    (ii) planetary milling machine;
    (iii) profiling machine;

(iv) duplicating machine;

(v) pantograph milling machine.

In order to obtain the maximum rigidity during milling, the cutter and its support must be positioned as close to the machine body as possible. The accuracy of the machined component depends upon the stability of the work and the cutter. Milling machines are particularly suitable for mass production. When used for flat-facing, although the setting-up time is longer, the machine time per article is less on a milling machine than on a shaper. This is especially true if several cutters are used at the same time, which is known as *gang milling*. It is sometimes claimed that although a milling machine is more versatile than a lathe, it is quicker and easier to perform milling on a lathe than it is to turn on a milling machine.

8.4.9 Shaping and planing

Shaping and planing are both machining operations that are used to produce flat surfaces. The principle of these operations is shown in Fig. 8.56. A *shaper* operates by advancing the work (feed movement) and the cutting tool then traverses the work and makes a straight-line cut by a reciprocating motion. A plane surface is produced whatever the shape of the tool, and the tool does not have to be accurately prepared. By using special tools and attachments, a whole range of shapes can be produced using a shaper. A *planer* removes metal by moving the work in a straight line against a single edge tool. The tool is then advanced (feed movement) ready for the next cut. Flat surfaces are usually produced with a shaper; a planer is generally only used for large work and has now mainly been replaced by milling, broaching and abrasive machining for most production work.

Tools used for shaping and planing operations are more robust than those used on a lathe. Their construction usually consists of a toolholder with removable bits, rather than forged tools. The inserts are made of high-speed steel or cast alloys for heavy roughing cuts, and carbides for secondary roughing and finishing. Carbide-tipped tools must be automatically raised on the return stroke to avoid damage to the cutting edge. The bit should be secured near the center line of the holder or the pivot point, rather than at an angle as is customary with the lathe toolholders. Typical cutting-tool shapes for common planer operations are shown in Fig. 8.57. The cutting angles depend upon the material to be cut and the type of operation. They are similar to those used with other cutting tools, except that the end clearance should be less than 4°.

Cutting speeds depend upon the tool material and the number of tools to be used, as well as how securely the work is held and by which device. To calculate the cutting speed it is necessary to know the length of the stroke and the number of strokes per second, and also the proportion of the cycle time spent on the cutting and return strokes. Several types of quick-return mechanisms have been developed for use with shaping machines.

The following classification of shapers and planers is given in terms of their general design.

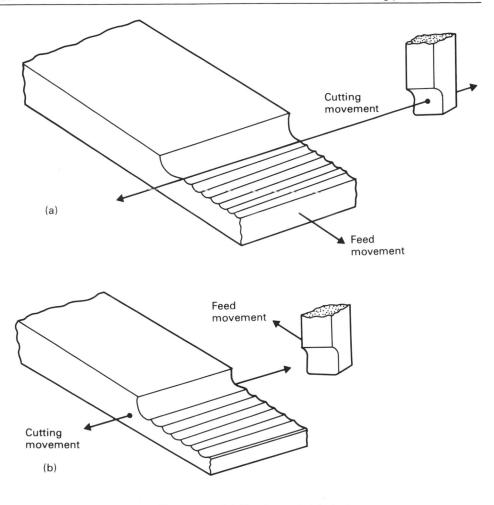

Figure 8.56 (a) Shaping and (b) planing.

Shapers
(a) Horizontal-push cut:
   (i) plain for production work;
   (ii) universal for the workshop.
(b) Horizontal - draw cut.
(c) Vertical:
   (i) slotter;
   (ii) keyseater.
(d) Special purpose, e.g. gear cutting.

Planers
(a) Double-housing.
(b) Open-side.
(c) Pit type.
(d) Edge or plate.

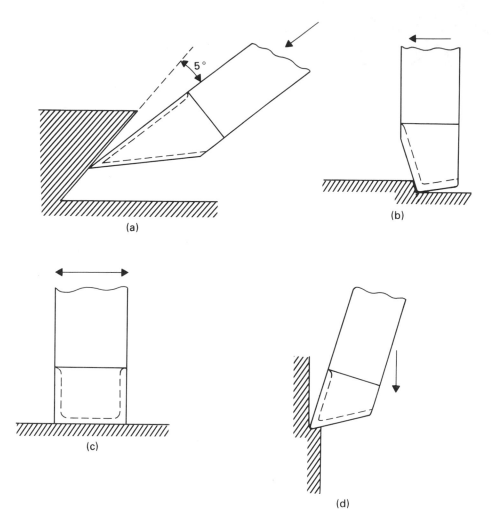

Figure 8.57 Planer cutting tools. (a) Corners. (b) Roughing flat surfaces. (c) Finishing tool. (d) Side cutting.

The size of a shaper is usually designated by the maximum length of work that can be machined. For a vertical shaper or slotter, this is modified to be the product of the maximum stroke length and the worktable diameter. The size of a planer is determined by the product of the table width, the table length and the distance between the table and rail.

* * *

Self-assessment exercises

1 Describe the following operations, and the differences between them: (a) boring; (b) coring; (c) reaming; (d) trepanning; (e) counter-boring; (f) spotfacing; (g) countersinking.

2 Describe the different types of drills that are commonly used.

3 What factors influence the selection of a drill for a particular application?

4 What factors limit the feed per revolution when drilling?

5 Explain the use of a jig for production drilling.

6 Why must metal chips be removed when drilling deep holes? How can this be achieved?

7 What is the main problem encountered when drilling deep holes? How can this be avoided?

8 Explain why a 6 mm (0.25 in.) twist drill does not drill a 6 mm hole.

9 Describe the process of metal removal by milling, emphasizing the particular aspects of this operation. Define: rake angle, clearance angle, land, up-cut and down-cut operation.

10 Explain why a milling machine is usually considered to be the most versatile machine tool.

11 How are milling cutters classified?

12 Why is down-cut milling confined to machines designed for that particular operation?

13 Describe the cutting action of a shaper.

14 How is the work held on a shaper?

15 Describe the tools used for shaping.

16 State one advantage of shaping.

17 Why is it difficult to produce curved surfaces using a shaper?

18 Answer Exercises 13 to 17 for a planing operation.

*   *   *

8.4.10 Grinding

Grinding is a process of metal removal using a rotating abrasive wheel that can act as a metal cutter or provide frictional wear; it can also be used to sharpen machine tools. The grinding process has made rapid and substantial developments in recent years. Improvements have been obtained in both accuracy and finish in terms of reduced tolerances, without increases in costs or loss in production times. This has helped to raise the quality of the products of the engineering industry. Improved techniques for grinding-wheel manufacture and better machine design are the main advances, without any major changes in the grinding process itself. Grinding wheels can be used for cutting, surfacing, shaping and other operations, and can be applied to brick, carbon, concrete, cork, glass, granite, marble, and many other materials.

The old type of whetstone performed its work by friction and abrasive wear, the residue obtained was a mixture of metal dust and sand from the wheel. Grinding with modern abrasive wheels is a metal-cutting process where the action of the wheel is similar to that of a milling cutter. The wheel contains thousands of small grains bonded together, which act as miniature cutting edges at the wheel surface. A schematic diagram of the composition of a grinding wheel is shown in Fig. 8.58; also shown are the (magnified) grindings of a ductile material obtained from a modern wheel. These resemble short turnings obtained from a cutting tool. The grain may be considered as the cutting tool and the bond as the tool holder. Grinding has advantages compared to other cutting operations, e.g. extremely smooth surfaces can be produced with surface roughnesses between 0.4 and 2000 μm, and very accurate dimensions can be produced in a short time. The low pressures required mean that fragile parts can be machined.

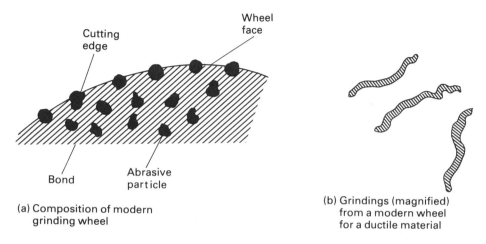

(a) Composition of modern grinding wheel

(b) Grindings (magnified) from a modern wheel for a ductile material

Figure 8.58 Modern grinding wheel and grindings.

The abrasive particles that provide the cutting action may be either natural or manufactured materials; the following classifications are used:

Natural materials
(a) Sandstone or solid quartz.
(b) Emery – 50–60% crystalline aluminum oxide (Al_2O_3) with iron oxide impurity.
(c) Corundum — 75–90% crystalline Al_2O_3 with iron oxide impurity.
(d) Diamond.
(e) Garnet.

Manufactured materials
(a) Silicon carbide (SiC).
(b) Aluminum oxide (Al_2O_3).
(c) Boron carbide (B_4C).
(d) Zirconium oxide (ZrO_2).

Natural materials such as sandstone, emery and corundum are unsuitable for production work because the bond is not uniform due to the presence of impurities, and they do not wear evenly. Diamond is a useful material that has rapid cutting ability, slow wear and free cutting action. Little heat is generated during cutting and this can be an economical material to use, despite the high initial cost. Silicon carbide is a man-made mineral of extreme hardness and sharpness with properties close to those of diamond. However, it has limited use due to its brittleness but it is suitable for grinding materials of low tensile strength such as cast iron, brass, bronze, aluminum and cemented carbides. Aluminum oxide (or fused alumina) is a manufactured material produced by the fusion of the mineral bauxite. It is tough, sharp and abrasive, although slightly softer than silicon carbide. It is suitable for grinding metals of high tensile strength such as steels, and most manufactured wheels are made of aluminum oxide. Abrasive materials are graded by sieving through screens having a certain number of holes, or meshes, per mm (in.). The *grit* signifies the number of meshes per mm (in.) used to grade any particular size. The larger the *mesh number* the finer the material passing through. The grit, or grain size, of abrasive particles is usually between 6 and 240, and this is part of a grinding-wheel specification (see Fig.8.59).

The *bond* is the substance which, when mixed with the abrasive grains, holds them together, thus enabling the mixture to be shaped in the form of a wheel. After suitable treatment this wheel can acquire the necessary mechanical

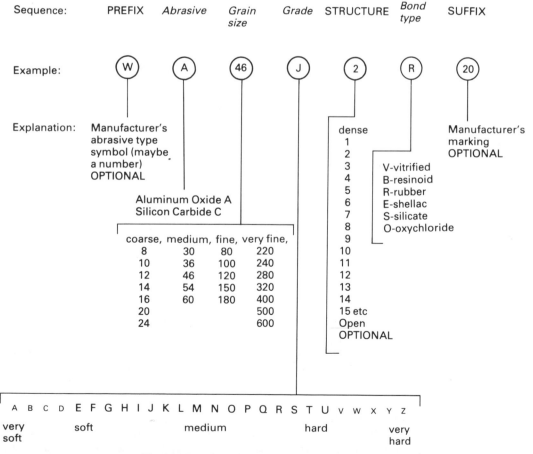

Figure 8.59 Grinding wheel classification system.

strength for the work it is required to perform. The *grade* of a wheel is a measure of the degree of hardness possessed by the bond, and indicates the ability of the bond to retain the abrasive particles. The grade is usually specified by the letters of the alphabet from A (very soft) to Z (very hard), although the normal range is E (soft) to U (hard). The appropriate letter is also given on the grinding-wheel specification (see Fig. 8.59). The grade or hardness of a wheel has no relation to the nature of the abrasive material, and a soft graded wheel may contain the hardest of particles. The choice of the correct bond is important: a soft bond allows the abrasive particles to break away more easily and should be used when grinding hard metals where the abrasive becomes rapidly dulled. The main types of bonding material that are used for making wheels are considered below.

(a) *Vitrified bonds*

About 80% of all wheels are made by a process that involves mixing the abrasive with pottery clays, pressing and shaping, and then burning in a furnace. Wheels made by this process are dense, accurately shaped, porous, strong and unaffected by water, acids, oils and atmospheric conditions. A wide range of grades can be produced by this process and recommended speeds may vary from 7 m/s (23 ft/s) for grinding hard materials up to 60 m/s (197 ft/s) with specifically reinforced wheels.

(b) *Silicate bonds*

These wheels use sodium silicate as the bonding agent, mixed with zinc oxide as a waterproofing agent. The abrasive and bonding materials are mixed, shaped and then baked at low temperatures (260°C/500°F) for two to four days. Silicate wheels have a milder action and cut with less harshness than vitrified wheels; however, their mechanical strength is less and they are more easily damaged. They are suitable for grinding fine-edged tools where heat must be kept to a mimimum.

(c) *Shellac bonds*

Flake or powdered shellac is mixed with the abrasive, and then heated in order to coat each grain. Next it is either pressed in molds or rolled in thin sheets and cut to size. The final operation is mild baking at 150°C (302°F) with the wheels packed with sand. Thin wheels less than 3 mm (0.12 in.) thick, can be produced by this process. They are strong and also possess some elasticity.

(d) *Rubber bonds*

Pure rubber, with sulfur as a vulcanizing agent, has abrasive agent forced into it by passing both materials through heated rolls. Wheels of the required thickness are then cut out and vulcanized using heat and pressure. These wheels have good elasticity and are used for high-speed grinding (45–80 m/s; 148–262 ft/s) with rapid metal removal, and for cutting off.

(e) *Synthetic resin bonds*

A very hard, strong bond is produced by mixing the abrasive with a thermosetting, synthetic resin material, e.g. bakelite, and a liquid solvent, followed by molding and baking. These wheels are operated at high speeds (45–80 m/s; 148–262 ft/s), approximately twice that of vitrified and silicate wheels. They are mainly used as thin wheels for finishing and cutting off, where small amounts of metal are removed. Despite their high mechanical strength, they are not suitable for production work where a large amount of heat is generated.

The proportion of bonding material in a wheel varies from 10% to 30% of the total volume, and the *structure* of the wheel depends upon this proportion. If the proportion of bonding material is high, then this is known as an open structure and the grinding action is similar to that obtained with soft bonds. A dense structure has a larger number of cutting edges per unit area of the wheel face. The structure or *grain spacing* of a wheel is also represented in its specification by a number usually between 1 (dense) and 15 (open) (see Fig. 8.59).

Cutting fluids are often used during a grinding operation, their main functions being to provide lubrication, which reduces the power consumption, and to cool the tool and the work, thus minimizing distortion. A cutting fluid can also wash away the removed metal, protect against corrosion, improve the surface finish and prevent welding of the chip to the tool. The three main types of cutting fluid used during grinding are detailed below.

(a) *Soluble oils or mineral oils*
These form an emulsion (slurries or suds) when added to water. They are used either neat for maximum lubrication or diluted (typical oil:water ratio is 1:30) to increase the cooling capacity. These oils leave a protective rust-resisting film on the work.

(b) *Straight oils*
These are mainly mineral oils and lard, designed to be used neat for extreme-pressure cutting. They possess good lubricating properties and are essential for slow, heavy-cutting operations.

(c) *Water-based fluids*
These are true solutions of salts and other minerals in water. They are clean and clear, and possess good cooling properties. Those containing sodium nitrite or sodium carbonate are rust-resisting.

The most efficient methods of applying cutting fluids are by a pump, an oil tray or a reservoir. The most satisfactory application provides a slow continuous stream of fluid over the cutting action. The cutting fluid most suited to a particular operation should ideally be determined from a series of trials, but this is not always possible. The following are suggested cutting fluids (listed in order of preference from left to right) for use with particular metals.

Aluminum and its alloys:	paraffin: dry; straight oils; soluble oils.
Brass, copper or bronze:	dry; soluble oils; straight oils; paraffin.
Cast iron:	dry; dry with compressed air.
Steel:	soluble oils; straight oils; water-based.

The manufacture of grinding wheels involves reducing the abrasive material to a small size in a crushing machine, then removing fines, dust and impurities by washing and iron compounds with magnetic separators. The abrasive is graded using a set of screens and the required size of particles is mixed with the bonding agent. They are molded or cut to the required shape and heat is applied. Finally, the wheel is bushed, trued, tested and inspected.

The Abrasive Industries Association has adopted a British Standard Marking System for grinding wheels, and this corresponds to that used by the American National Standards Institute. The specification includes the abrasive, the grain, the grade and the bond type; it may also include (optional) the structure, the manufacturer's abrasive identification, e.g. W for white abrasive, and the manufacturer's own identification marking. The system is shown in Fig. 8.59 (optional items are shown in capitals in the sequence line) with an example, and including the range of values used for each category.

Grinding wheels are manufactured in a wide variety of shapes; particular types of machine have special wheel adaptations. A selection of the more common shapes is shown in Fig. 8.60. The shapes (a)–(h) are used as disk wheels which grind on the periphery; they are used for cylindrical, surface and general-purpose tool grinding machines. Wheels (j) to (l) are mainly used on cup wheel surface grinders. Shapes (m), (n) and (p) are used for tool and cutter grinding, and the thin wheel (r) is for slitting and cutting off. With larger machines it is often more economical to fit a number of segments into a chuck, rather than using a solid wheel. The gaps between the segments promote freer cutting. The types of wheel faces that are used on the straight wheel-type shape (Fig. 8.60(a)) are shown in Fig. 8.61.

The selection of a suitable grinding wheel for a grinding process is difficult because of the wide range of wheels available, and the range of materials that may need to machined. If the choice is not obvious, then the manufacturer's advice should be obtained; however, the following are some of the factors which have to be considered. The correct choice of the abrasive material is important, and the properties of silicon carbide and aluminum oxide have previously been discussed. Silicon carbide is more suitable for grinding materials of low tensile strength and aluminum oxide is more efficient with hard, tough materials. This is illustrated in Fig. 8.62, which also shows the materials that can be efficiently ground with either abrasive. It can be seen that emery has only medium efficiency, mainly due to the impurities it contains. Wheels composed of large, coarse grains are used for rapid metal removal, mainly for soft materials. Smaller particles enable a finer finish to be obtained and are used with hard brittle materials.

The choice of the bond is also difficult, but in general a hard wheel is used for soft work and a soft wheel for hard work. The wheel should operate so that when the grains become dulled they either split to create a new cutting edge, or they break away from the wheel, thus exposing new grains. If the bond is too strong, then grains do not break away and the wheel becomes glazed, but if the bond is too soft then excessive wear of the wheel occurs. A vitrified bond is most commonly used but it is not particularly suitable when bending forces occur or a fine finish is required, or for thin wheels or high operating speeds. The structure (grain spacing) of the wheel must also be considered. A finer finish is obtained with a close spacing of the abrasive particles, but for soft, ductile materials a wide spacing should be used.

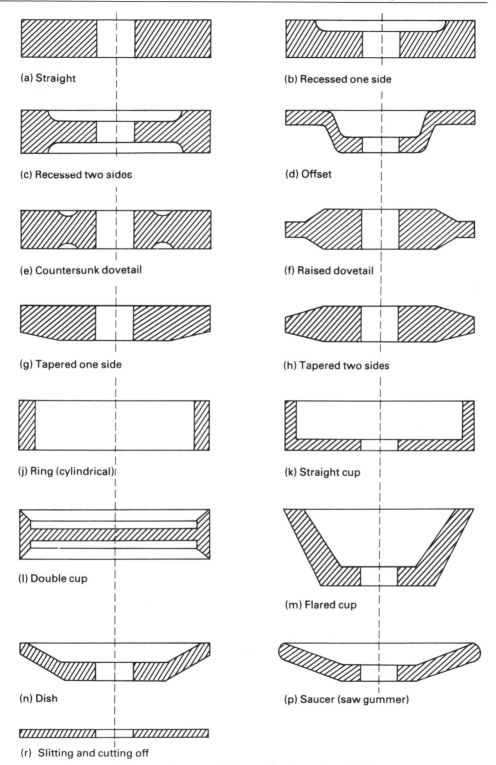

(a) Straight

(b) Recessed one side

(c) Recessed two sides

(d) Offset

(e) Countersunk dovetail

(f) Raised dovetail

(g) Tapered one side

(h) Tapered two sides

(j) Ring (cylindrical)

(k) Straight cup

(l) Double cup

(m) Flared cup

(n) Dish

(p) Saucer (saw gummer)

(r) Slitting and cutting off

Figure 8.60   Common grinding wheel shapes.

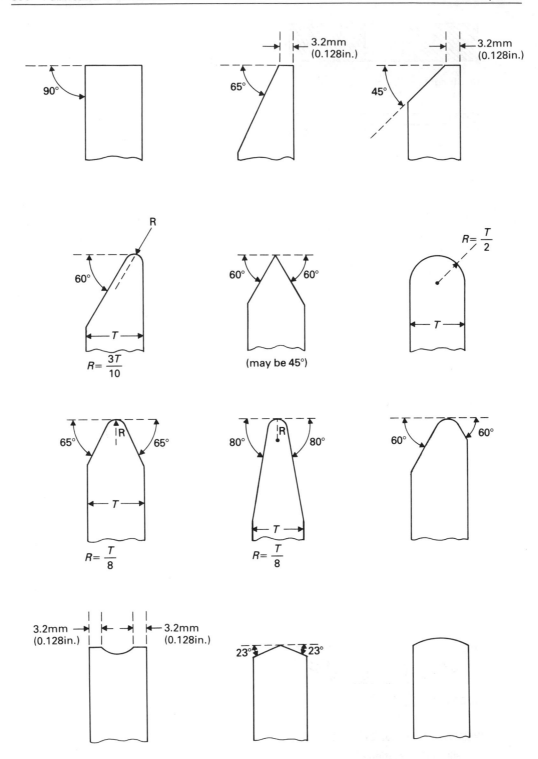

Figure 8.61 Standard faces for straight type grinding wheels.

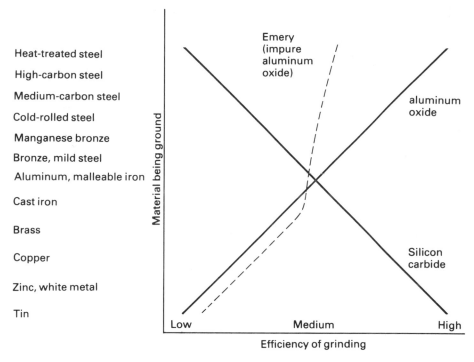

Figure 8.62 Relationship between material to be ground and abrasive material.

There are several other factors which influence the selection of a bond. The size of the region where grinding occurs is important, and softer wheels must be used for large grinding regions. This is the area of contact of a cup wheel and the arc of contact of a disk wheel; the latter is shown in Fig. 8.63. The width of the wheel face has no effect provided the pressure per unit area remains the same. The higher the work speed, the more material that is ground in a given time, and the greater the wear on the wheel. The higher the speed, the harder

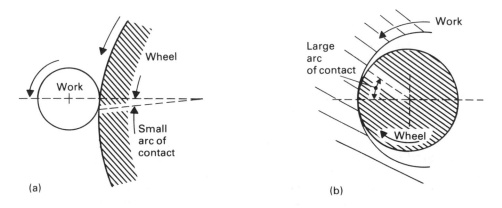

Figure 8.63 Arc of contact for cylindrical disk grinding. (a) External grinding – small diameter work. (b) Internal grinding – wheel nearly as large as hole.

the bond that is required. A harder wheel can be used for grinding if a coolant is used in the process. Heavy rigid grinding machines that are well maintained have a high cutting efficiency, and operate satisfactorily with softer wheels. If grinding is carried out by a skilled operator, it is possible to use softer wheels and operating costs are reduced.

Several problems can occur during grinding, but most of these can be easily corrected if the principles of the operation are understood. As a wheel wears down its speed of rotation must be increased in order to maintain the efficiency. Most of the other problems occur either because an incorrect bond is used or the rate of metal removal, i.e. the work rate, is incorrect. If excess wastage of the wheel occurs, this can be corrected by reducing the work speed or increasing the cut. If a wheel becomes glazed because dull grains are not being removed, then the cause is probably too hard a wheel being operated either too fast or too slowly. If a wheel becomes clogged or loaded with particles of the material being ground, then eventually the cutting edges will not project. The material may be too soft for grinding, or the wheel is used too fast or the cut too deep for the particles to escape. When used for tool grinding, the wheel rotation should always be such that pressure is against or into the top of the tool, never away from the cutting edge. The tool should be kept moving across the wheel face to avoid heavy wear in one region. Grinding is more efficient if the wheel is securely mounted and balanced, and if it is regularly trued. If a tool overheats during grinding, it is important that it is not immersed in water but is either allowed to cool naturally or the shank only is immersed.

When abrasive particles are glued to a flexible backing material the result is known as a *coated abrasive*. The term 'sandpaper' is often used to describe the entire range of products. The abrasive may be a natural material such as flint quartz, emery or corundum, or a manufactured material such as aluminum oxide or silicon carbide. The backing material is either paper, cloth, fiber, paper-cloth or cloth-fiber. For heavy duty, the particles completely cover the backing. However, for increased flexibility a more open coating is applied, provided there is no tendency for loading. The particle density and the backing material determine the characteristics of the coated abrasive and its applications. An electro-coating process is used to cover the glued-backing material with particles. The finished product is usually either belts or disks that can be used with a metal backing support.

If a metallic surface is examined microscopically, with a large vertical magnification and smaller horizontal magnification, it appears to consist of a series of 'troughs and hills' as shown in Fig. 8.64. Use of the same horizontal and vertical magnification is impractical, but if this were applied the surface would appear to be composed of smooth undulations. In order to quantify the accuracy of a surface finish, a *centre line average (CLA) index* is used. This is measured in microns and is the average height between the peaks and dips on the surface, with reference to a horizontal straight line such that the enclosed areas above and below are equal. Standards obtained depend upon the machining process used, but the following would be considered reasonable: 32–63 CLA for turning and smooth filing; 16–32 CLA for commercial grinding; 1–8 CLA for lapping and honing. Standards of surface finish become most

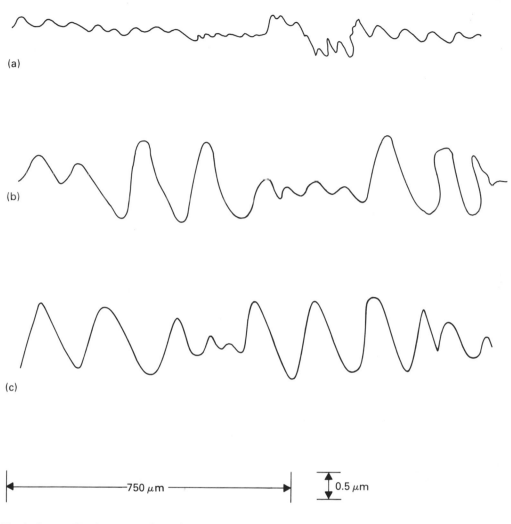

(a)

(b)

(c)

|←————————750 μm————————→| ↕ 0.5 μm

Vertical magnification approximately 8000
Horizontal magnification approximately 100

Figure 8.64 Typical surface finishes with suitable magnification.
(a) Grinding finish. (b) Turning finish. (c) Diamond boring finish.

important when two surfaces are to be in contact and movement occurs between them. Initial movement wears down the surface peaks, followed by a longer period without significant wear. It is important to allow for wear, e.g. the initial contact between nuts and bolts, and to make surfaces as smooth as possible for certain applications, e.g. motor and aircraft cylinders and pistons.

Grinding machines are used mainly for finishing operations on cylindrical, flat or internal surfaces. Machines are usually classified according to the type of surface machined or the function which is being performed. The following classification of grinding machines uses this basis.

(1) *Cylindrical grinder*
    (a) Work between centers.
    (b) A centerless grinder supports the work on a rest, and feeds the work between two wheels. The large wheel is the grinding wheel and the smaller one is the pressure or regulating wheel; both rotate in the same direction. The process is rapid and especially suited to production work.
    (c) Tool post.
    (d) Crankshaft and other special applications.

(2) *Internal grinder*
    (a) Work rotated in a chuck.
    (b) Work rotated and held by rolls.
    (c) Work stationary.

(3) *Surface grinder*
Used for plane or flat surfaces.
    (a) Reciprocating table of the planer type.
    (b) Rotating table.
Both types (a) and (b) may use a horizontal or vertical spindle.

(4) *Universal grinder*
    (a) Cylindrical work.
    (b) Thread-form work.
    (c) Gear-form work.
    (d) Oscillating.

(5) *Tool and cutter grinding*
Mainly used for sharpening single point tools and cutters, but also applicable for cylindrical, taper, internal and surface grinding. A jig grinder is similar to a jig borer but operates at far higher speeds.

(6) *Special grinding machines*
    (a) Swinging frame – snagging.
    (b) Cutting off – sawing.
    (c) Portable – offhand grinding.
    (d) Flexible shaft – general purpose.
    (e) Profiling – contouring.

(7) *Surface preparation*
    (a) *Honing*
        This is a low-speed abrasive process using minimal heat and pressure. Good accuracy is obtained and metallurgical properties are controlled. The cutting action is provided by abrasive sticks coated with alumimum oxide or silicon carbide, and mounted on a metal mandrel. The work floats (not clamped or chucked) and there is no distortion; it has a reciprocating motion as the mandrel rotates. A straight round hole is produced. Generally only 0.03 mm (0.001 in.) or less of metal is removed, although certain inaccuracies as shown in Fig. 8.65 can be corrected in amounts up to 0.5 mm (0.02 in.). Coolants are essential to remove small chips and control the temperature. Sulfurized mineral-based oil or lard oil mixed with kerosene are normally used.

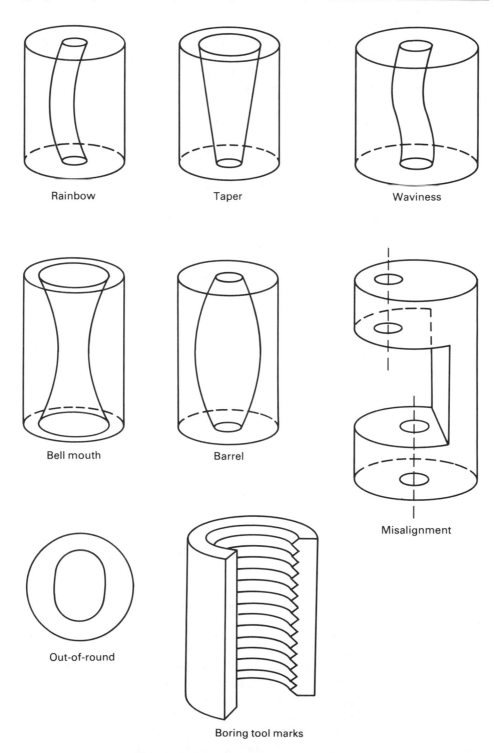

Rainbow   Taper   Waviness

Bell mouth   Barrel

Out-of-round   Boring tool marks

Misalignment

Figure 8.65 Boring errors suitable for honing.

(b) *Lapping*

This is not a true metal removal operation and usually less than 0.03 mm (0.001 in.) is removed. It is used to produce geometrically true surfaces, improve dimensional accuracy, provide a close fit between two contact surfaces or to correct minor surface defects. Lapping is used on flat, cylindrical and spherical surfaces, and specially shaped surfaces where the work surfaces are in contact with a lap. The motion between the work and the lap is such that new points of contact are constantly made. The lap must be softer than the work and gray cast iron, copper, lead or wood are suitable materials. Very fine abrasive particles such as boron carbide or silicon carbide are often embedded in the lap, or they may be suspended in oil, grease or water between the work and the lap.

(c) *Superfinishing*

This is a surface-improving process that removes noncrystalline metal fragments left behind after machining. It is not a dimensional process; it improves contact and lubrication between surfaces and reduces frictional wear. It is similar to honing in that an abrasive stone is used; this may be a rotating cup for superfinishing flat surfaces or a rectangular stone with one curved surface for cylindrical work. The features of superfinishing are shown in Fig. 8.66. An oscillating motion is given to the stone and this may traverse if necessary; also a low pressure (20–250 kN/m^2; 3–35 psi) is applied. The work is rotated at approximately 0.25 m/s (0.82 ft/s) and is continuously fed with a light oil to wash away the fine metal particles.

(d) *Wire brushing*

Rotating wire brushes are used to clean castings and to remove scratches, scale and sharp edges. Only small amounts of metal are removed.

(e) *Polishing*

Polishing operations use cloth wheels or belts coated with abrasive particles. It is mainly used to remove surface defects, and several wheels of decreasing grit size may be used to obtain the required finish.

(f) *Buffing*

This is the final operation to obtain a required surface appearance and is similar to polishing. Wheels made of cotton containing fine abrasive particles such as rouge or amorphous silica are used.

(g) *Production finishing*

Barrel finishing (or tumbling) is used to remove burrs, scale or flash from a large number of objects at the same time. A uniform surface finish is obtained and it is particularly economical for small objects. It can be used with glass, plastic, rubber and all metal objects. Parts to be finished are placed in a rotating drum or a vibrating tank containing an abrasive material, water or oil and appropriate chemical compounds. The process can be operated

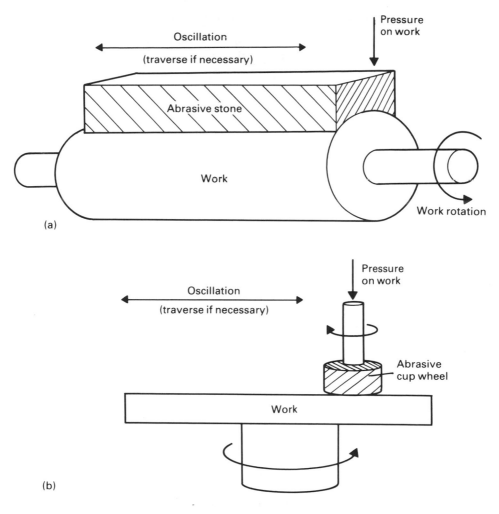

Figure 8.66   Methods of superfinishing. (a) Cylindrical superfinishing. (b) Flat superfinishing.

continuously if the drum or tank are inclined, the parts and the abrasive medium are separated at the outlet and the abrasive is recycled.

(8) *Abrasive grinding*

Abrasive machining or abrasive grinding is a metal-removal process and is sometimes termed high-energy grinding, but unlike grinding or lapping it is not a finishing operation. Using either bonded wheels or coated abrasives, material up to 12 mm (0.5 in.) thick can be removed. The process provides rapid metal removal, good surface finish and efficient size control. The machines are usually heavier and more powerful (up to 200 kW; 680 Btu/h) than conventional grinding machines. It is an economical machining process especially for forgings and castings, and can be applied to the preparation of flats, tubing and extrusions. The main types of abrasive grinding machines are:

(a) Abrasive belts, either single or multihead, with either a rotary or traversing table supporting the work.
(b) Disk.
(c) Loose grit.
(d) Flap wheel.
(e) Wire sawing.

* * *

Self-assessment exercises

1 What are the advantages of grinding compared to other metal-removal processes?

2 State the most commonly used abrasive and bond, and explain why they are selected.

3 Why is a grinding fluid used?

4 Describe the main causes of, and solutions to, the problems of excessive grinding wheel wear and inefficient metal removal.

5 How does the surface finish possible with grinding compare to that produced by milling?

6 Describe the type of cutting that is more suited to an abrasive wheel than a hacksaw.

7 Describe the function of the structure and the hardness of a grinding wheel.

8 What method is used for checking for imperfect wheels?

9 Why is surface finish important?

10 Describe the processes of honing, lapping and superfinishing. Explain the reasons for performing each operation.

11 What is barrel finishing, and why is it carried out?

12 What is the difference between polishing and buffing?

13 Which process: vibratory finishing or barrel finishing, produces the better surface finish?

14 What is the main difference between precision grinding and abrasive machining?

* * *

8.4.11 Specialized machining processes

(a) Gear-cutting machines

Gears transmit power and/or rotary motion from one shaft to another. This is accomplished by projections or teeth on the circumference of the gear. There are recesses between the teeth to prevent interference; because there is no slippage, exact speed ratios are essential. Gears can be produced by various methods including casting, stamping, powder metallurgy, extrusion, rolling, grinding or plastic molding. However, most gears are produced by machining processes and accurate work is required to obtain quiet, high-speed, long-wearing gears. The following are the main machining processes used to produce gears:

(i) shaper;
(ii) template process;
(iii) cutter generating process; uses either a cutter gear in a shaper, a rotary cutter, reciprocating cutters simulating a rack, or hobbing. *Hobbing* is a generating process consisting of rotating and advancing a fluted steel work cutter past a revolving blank.

Finishing operations are performed on gears to eliminate slight inaccuracies in the tooth spacing and profile. *Shaving* is the removal of metal using a rack or rotary cutter. Alternatively, a cold working process is used, such as *burnishing*, in which the gear is rolled under pressure with three burnishing gears. The disadvantage of burnishing is that noncrystalline metal is left on the surface of the gears. Heat-treated gears can be finished by either grinding or lapping.

(b) Thread cutting

A screw thread is a ridge of uniform section in the form of a helix, on the surface of a cylinder. External threads can be produced by the following processes:
   (i) cutting to shape on a lathe;
   (ii) die and stock (manual);
   (iii) automatic die head;
   (iv) milling machine;
   (v) threading maching (plain or automatic);
   (vi) rolling between dies (flat or circular);
   (vii) die casting;
   (viii) grinding.
Internal threads can be produced by:
   (i) cutting to shape on a lathe;
   (ii) tap and holder;
   (iii) automatic collapsible tap;
   (iv) milling machine;
   (v) screw broach.

(c) Abrasive jet machining

This is a mechanical machining process used for cutting hard brittle materials. Fine abrasives are transported by air or carbon dioxide and bombard the workpiece at high velocities (150–300 m/s; 490–980 ft/s). It can be used for cutting or finishing operations, but it is unsuitable for soft materials because the abrasive becomes embedded in the surface.

(d) Water jet machining

A high velocity (600–900 m/s; 1960–2940 ft/s) jet of water (0.25 mm/0.01 in. diameter) is used to cut wood, plastics, textiles and certain ceramics.

(e) Ultrasonic machining

Material can be removed by abrasive grains in a liquid. They are made to bombard the work at high velocities and metal is removed by the oscillating action of the tool. It is mainly used for hard brittle materials and can be used to perform drilling, tapping or coining.

(f) Electric discharge machining

This is sometimes known as spark machining or electronic erosion. Metal is removed by an electric spark passing between an electrode of the required shape and the work. It can only be used with conducting materials but can be applied to hard or soft metals. The process can perform drilling, tapping and grinding operations.

(g) Electrochemical machining

This is a process of deplating where the work is the anode and the tool is the cathode; an electrolyte is also required. Electrochemical grinding is a similar process, except that the cathode is a metal disk embedded with abrasive particles. Approximately 10% of the metal removal is due to abrasive action.

(h) Laser-beam machining

This is a thermoelectric process, in which metal is removed mainly by material evaporation caused by a high-energy beam of light. Heat affected zones are small and the process works well with hard, nonmetallic materials.

Electron-beam machining is a similar process using high-speed electrons (usually in vacuum) to bombard the workpiece. The thermal energy produced vaporizes the metal locally. The principles of the laser-beam and electron-beam processes have been discussed in connection with welding applications (see Sections 7.7 and 7.8).

(i) High-temperature machining

If the workpiece is heated the shear strength of the metal is reduced, then less power is required to produce plastic deformation ahead of the cutting tool. Continuous chips are formed, tool life is increased and cutting speeds can be doubled. The disadvantages are the cost of heating the work and any distortion or metallurgical changes that may occur.

Alternatively, if high temperatures occur at the tool-chip interface, then the tool life is reduced, and welding may occur. In this case, low-temperature machining may be advantageous and can be accomplished using a cold mist spray at approximately −80°C (−112°F).

(j) Chemical processes

Chemical milling is a selective metal-reducing process and is really a controlled etching process. The part is thoroughly cleaned and prepared for etching by masking those regions (if any) which are to be unaffected, using a chemically resistant coating. The part is then submerged in hot alkaline solution and unprotected metal is eroded. The immersion time determines the quantity of metal removed, and when this is achieved the part is rinsed and the masking material removed.

Chemical blanking is used to produce thin metallic parts by chemical action. A chemically resistant image of the required part is produced and placed on a metal sheet. This is then exposed to chemical action either by immersion or by spraying. All metal except the image and the part is dissolved. If metal is removed by immersion, then an image of the part may be required on both sides of the sheet. An electric current is not required to carry away metal in the electrolyte, and the process is sometimes known as *electroless etching*. The metal is converted chemically into metallic salts.

(k) Electroforming

This is an electrolytic metal deposition process, in which metal supplied from a solid anode is used to coat a conductive mold. This mold is the internal shape of the finished product and its quality determines the accuracy and surface finish of the product. The process differs from plating in that a solid shell is produced, but it is separated from the form upon which it was deposited. The mold may be permanent and reusable if it can be removed from the product, or it may be expendable and made of a soluble material. If the mold material is nonconducting, then it must be coated with an appropriate metal film. The prepared mold is placed in an electrolytic solution, which is usually agitated by air and has liquid level and temperature controls. The product is removed when the required thickness of metal has been deposited. The disadvantages of the process include high costs, low production rates, and only certain metals are suitable.

8.4.12 Machine lubrication

The most common application of lubricants is to machine bearings. An oil or grease film is used to prevent metallic contact between the journal and the bearings. The efficiency of an oil depends upon its viscosity; if this is too low then the film may be broken and if the viscosity is too high then the frictional resistance is increased. Gears that operate under severe conditions require extreme pressure (EP) lubricants. Oil is generally preferable to grease as a lubricant, but it requires better unit sealing. Oil can be applied either in a bath or splash system for slow and medium-speed operations, in a circulating system at moderate speeds, and by a spray or mist at high speeds. Grease is used at low temperatures and when the unit sealing is inadequate for oils. Grease has excellent self-sealing properties and is an efficient dirt excluder. Grease can be applied either as a packing for mild conditions requiring infrequent replenishment, or by a compression cup or pressure gun for the periodic addition of fresh grease. For severe conditions, mechanical lubricators or centralized pressure systems are used. These provide frequent and regular additions of fresh grease which are necessary in metal rolling mills.

* * *

Self-assessment exercises

1 Describe abrasive jet machining including the powders used, how they are projected, the types of shapes and materials suitable for use with this process.

2 What factors control or limit this machining method?

3 Explain the main difference between chemical milling and chemical blanking. Describe practical applications of each process and the main advantages of each method.

4 What is electrical discharge machining? Is the position of the electrode fixed or variable during operation? How does electrode wear affect the operation?

5 Compare and contrast electrochemical and electrical discharge machining. Explain the purpose of the electrolyte, and the reason electrode wear does not occur with the electrochemical process.

6 How is a laser used as a cutting tool? Why is a CO_2 laser used for machining nonmetals?

7 What is ultrasonic machining? Why are frictional forces small, and why does the use of a coolant improve the cutting efficiency?

8 For ultrasonic machining, describe the tool used, its direction of feed and the possible movements of the worktable.

9 Define an electron beam. Distinguish between electron-beam machining and electron-beam welding.

* * *

8.5 SURFACE TREATMENT

8.5.1 Introduction

Most metallic products require a final surface treatment, which may be to improve the appearance, to produce special surface properties such as hardness, or to protect the part against corrosion. When selecting a surface treatment process, the most important considerations are probably the cost of each alternative process and the reason for the treatment. The prevention of atmospheric corrosion is often an important reason, and the mechanisms of the various corrosion processes have been fully discussed in Chapter 5. Other important factors when selecting a process are the size, shape and composition of the part and the importance of dimensional accuracy.

Before surface treatment is applied, the surface should be thoroughly cleaned (if possible) for good adhesion. The main cleaning methods are mechanical, chemical and electrolytic. Surface treatment processes usually involve either the application of a finite coating of material, e.g. metal, polymer, paint, or changing the nature of the material at the surface, e.g. oxidizing. The more important processes are now described.

8.5.2 Metallic coating

The metallic coating on a component may prevent corrosion either directly, by forming a complete covering, or by acting as a sacrificial coating (see Section 5.7.2). Metallic coating is known generally as *electroplating* or *electrodeposition*. It is an electrochemical process using a low-voltage (4–10 V) d.c. source. The part to be plated is scoured and chemically cleaned by immersion in either dilute hydrochloric acid or dilute sulfuric acid, to remove any surface oxide film. The part acts as a cathode and is immersed in a plating bath containing a suitable electrolyte. The plating metal is provided either by the electrolyte or by a consumable metal anode. Nickel–chromium plating is particularly popular on brass, steel and zinc alloys. This consists of a nickel coating (10–30 μm thickness) under a chromium layer (0.2 μm thick). An approximately 2 μm layer of chromium provides a hard, nonporous coating. Zinc coating and tin coating are also common processes. Cadmium plating produces a ductile layer that is useful for subsequent forming operations. Copper is sometimes applied as an undercoating. Silver plating is often applied to nonferrous parts to be used in the food industry and for good electrical contact. Coatings of zinc or cadmium act as sacrificial layers (see Section 5.7.2) and complete coverage of the part is less important than when nickel or chromium are applied.

Metal coatings can also be obtained by dipping the part into a bath of molten (coating) metal. This is known as *galvanizing* for zinc coatings and is used for protecting low-carbon steels from atmospheric attack. *Hot-dip tinning* is used to produce tin coatings (up to 250 μm thick) on food-container vessels. Galvanized coatings are between 100 and 200 μm thick, which is thicker than can normally be obtained by zinc plating. Zinc coatings can also be obtained by spraying molten zinc onto steel or by tumbling the part in zinc dust at high temperatures; this is known as *sheradizing*.

Protection can be obtained by rolling a thin sheet of corrosion-resisting metal onto each side of the metal to be protected. This is know as *cladding*; typical examples are pure aluminum cladding on aluminum alloys and nickel cladding on steels.

Calorizing is a process where aluminum is diffused into the metal surface at high temperatures and forms aluminum oxide. This protects the underlying metal, and is used on steels for furnaces and oil refineries.

8.5.3 Chemical coating

Phosphating is the most common chemical coating process and is used mainly with steel and zinc-based alloys. The metal is heated in a solution of acid phosphates and the coating produced must be painted, varnished or lacquered. *Parkerizing* (using manganese dihydrogen phosphate), *bonderizing* and *granodizing* are other typical processes.

Chromating is used for magnesium alloys and zinc-based alloys, it is carried out by immersing the part in potassium bichromate and some additives.

The oxide film that forms naturally on aluminum surfaces can be thickened by immersion in an aqueous solution of sodium carbonate with either sodium chromate or potassium chromate. This layer can be hardened by subsequent immersion in a sodium solution.

Anodizing, or *anodic oxidation*, is a process used for thickening the naturally occurring oxide film on aluminum surfaces. The process differs from plating in that the coating is formed entirely by oxidation, and it is permanent and an integral part of the base metal. The part is made the anode and is immersed in an electrolyte that can liberate oxygen, such as sulfuric, oxalic or chromic acids. The cathode is either stainless steel or the lead lining for the tank. The oxide film is hard and has a porous cellular structure that can absorb dyes. Magnesium can be anodized in a similar manner.

8.5.4 Surface hardening

Hard surfaces can be produced either by applying a metallic coating or by fusion welding; the surface properties can be changed by heat treatment or by contact with other materials. *Cementation* is similar to the carburizing process used to surface harden steels (see Section 2.19.5). The metal to be heated is surrounded by a powdered protecting metal, and both metals are then heated to a temperature below their melting points. The protecting metal enters the surface of the part forming a hard casing. Typical processes are *sheradizing* (zinc case), *chromizing* (chromium case) and *calorizing* (aluminum case).

Hard surfaces can also be produced by *induction hardening* (electric heating) or *flame hardening* (torch heating), where cooling is achieved by rapid quenching.

8.5.5 Paint and polymer coating

Corrosion protection can be obtained by applying a surface coating of paint or polymer. Paints have been used for many years and they are normally classified as layers less than 250 μm thick. Thicker coatings can be obtained but these are not classified as paints. Surfaces to be treated should be thoroughly cleaned and all traces of oil, grease and rust removed. If possible the surface should be scoured with an abrasive in order to obtain better adhesion. It is important that corrosion products are removed, otherwise corrosion may continue under the surface coating. Pre-treatments and primers, e.g. zinc oxide or lead oxide, should be applied according to the manufacturer's recommendations to provide initial adhesion. Undercoats are used to provide an opaque background of uniform color before the final durable coating is applied. The way in which the coating forms depends upon its chemical composition, and not the method of application, which may be brushing, rolling, spraying, dipping, etc. Typical coating materials include lacquer, varnish, polyurethane, epoxy, silicone, water- or oil-based paints.

Many polymer coatings consist of a mixture of the polymer, some coloring material and chemicals that ensure satisfactory curing and drying of the film. The main differences between the surface coatings are generally in terms of the polymer material to be used and the way in which the surface film is formed. It is usually necessary to prepare the surface as already described for painting, scouring being especially important. Both the polymer and the part must be heated for sufficient time for the polymer coating to adhere to the metal. One method of forming the polymer coating is to heat the part to 200°C (392°F) and then immerse it in polymer powder. Excess powder is removed, the remainder is fused to the part at 160°C (320°F) for sufficient time to form an even layer. Uniform heating is important and a thermostatically controlled oven is often used.

The main factors to be considered when selecting a paint or polymer coating material, are the surface preparation that is required and the environmental conditions to which the coating will be subjected. There is plenty of literature available, especially from manufacturers, to assist in the selection process. However, it may still be necessary, if time permits, to perform field trials on alternative coatings in order to select the most suitable material.

* * *

Self-assessment exercises

1 Discuss why workpieces may require cleaning.

2 Describe the main features of impact abrasive cleaning, liquid blasting, power brushing and ultrasonic cleaning.

3 What is the difference between an oil paint, alkyd paint and an epoxy paint? Describe typical applications of silicone paints and polyesters. Try to obtain information for some of these materials from manufacturers, or refer to the handbooks in the Bibliography for Chapter 4.

4 What is the main application of ceramic coatings?

5 Describe how electroplating is performed, and the reasons why dipping may be preferred.

6 What are diffusion coating, wire metallizing and plasma spraying?

7 What is a conversion coating?

8 Explain how anodizing can be used to produce a hard, wear-resistant coating.

9 Why is a cyanide plating bath preferred to an acid bath?

10 Explain why thickness limitations exist for the production of anodized coatings.

* * *

8.6 PRESSURE VESSELS

8.6.1 Introduction

A pressure vessel is normally considered to be a closed container of limited length, whose smallest dimension is much greater than the connecting pipework.

The vessel may be subjected to internal or external pressures or both, or vacuum. Pressure vessels are distinguished from boilers which generate steam for external use.

There are two main codes which cover the design and construction of pressure vessels. These are the American Society of Mechanical Engineers (ASME) Boiler and Pressure Vessel Code and British Standard Specification BS5500. These codes do not themselves possess legal status, but they provide a specification to which a contractor may be asked to conform. In the United States and Canada, some states and provinces have adopted parts of the ASME code as a legal requirement within their jurisdiction. An ASME code stamp can be applied to a vessel only by an authorized manufacturer, and only within the United States and Canada. Outside these countries the code can still be specified but the acceptability of the final vessel must be established by an appropriate authority.

The ASME Code Committee reviews the code and periodically publishes *Code Cases*; their decisions are detailed in the journal *Mechanical Engineering*. A new edition of the code is issued every three years, but alterations and extensions are published between editions. Inspectors approved by the ASME are usually employed either by an insurance company or an appropriate city, province or state department. They can then authorize a vessel to be stamped ASME-NB (National Board). Inspectors who are employed by a vessel user can only authorize the more limited ASME stamp.

8.6.2 The ASME Boiler and Pressure Vessel Code

The code comprises eleven sections, which are as follows:

I   Power Boilers
II   Material Specifications (three parts)
III   Nuclear Power Plant Components
IV   Heating Boilers
V   Non destructive Examination
VI   Recommended Rules for Care and Operation of Heating Boilers
VII   Recommended Rules for Care of Power Boilers
VIII Pressure Vessels; Divisions 1 and 2
IX   Welding Qualifications
X   Fiberglass – Reinforced Plastic Pressure Vessels
XI   Rules for In-service Inspection of Nuclear Reactant Coolant Systems

Pressure vessels, as distinguished from boilers, are covered by Sections II, III, V and VIII–XI. Section VIII deals specifically with unfired pressure vessels and now contains the simpler Division 1 and the more technical Division 2.

A nominal safety factor of 4 is used in Division 1 and many secondary stresses acting on the vessel are ignored. Division 2 with a safety factor of 3 allows higher stresses but requires a more thorough stress analysis, and closer material quality and fabrication control. In Section VIII, Division 1 provides the specification for smaller vessels and Division 2 (representing an extension of previous ASME code issues) is concerned with large heavy vessels. The increased engineering and inspection costs for Division 2 maybe balanced by reduced thicknesses and the associated lower material costs. However, safety should not be impaired because most vessel failures are due to brittle fracture at a flaw, crack or stress concentration, and the possibility of this occurrence should have been reduced.

Section VIII, Division 1

The range of vessels covered by this part of the ASME code is shown in Table 8.3. Vessels below $0.14 \, m^3$ (5 ft^3) capacity and $1.65 \, MN/m^2$ (250 psi) pressure, or $0.042 m^3$ (1.5 ft^3) capacity and $4 \, MN/m^2$ (600 psi), should be designed according to Code rules but are exempt from Code inspection. Division 1 covers vessels operating below $20 \, MN/m^2$ (3 000 psi) pressure; above this pressure vessels must be designed according to additional principles. This is usually Section VIII, Division 2, which has no pressure limitations. Vessels that do not fall within the categories of Divisions 1 or 2 may be accepted as 'specials', if they are shown to be based on satisfactory design principles.

Division 1 contains 3 subsections (A, B and C) and 13 parts designated by letters and numbers (the first letter is U, designating unfired). There are also mandatory and nonmandatory appendices.

SUBSECTION A: DESIGN PROCEDURES AND GENERAL REQUIREMENTS
Including rules for vessel thickness, corrosion allowances (paragraph UG-25), internal (UG27 and UG32) and external (UG28, UG29, UG30, UG33) pressure design, flange design, reinforcement of openings (UG36–UG42), vessel design pressure (UG21 and UA60), pressure relief devices (UG125–UG134), inspection (UG90–UG97), hydrostatic pressure testing (UG99–UG102), stamping and reports (UG115–UG120).

SUBSECTION B: FABRICATION METHODS
Contains four parts as follows:

UW – welded vessels;
UR – riveted vessels;
UF – forged vessels;
UB – brazed vessels.

Brazing is sometimes used for small vessels in mass production, although the code does not state any size limitations. Welding is by far the most popular fabrication method. However, forging may be used for high-pressure vessels with thick shells which cannot be easily rolled and welded. Welding methods must conform to Section IX of the Code.

SUBSECTION C: MATERIALS OF CONSTRUCTION
Requires that materials subjected to stress also conform to the specifications given in Section II. Subsection C contains the following parts:

UCS – carbon and low-alloy steels;
UNF – nonferrous materials;
UHA – high-alloy steels;
UCI – cast iron;

Table 8.3 Types of unfired pressure vessels (ASME code, Section VIII, Division 1).

Contents	Size	Pressure limitations	Temperature limitations
Unrestricted	Over 15 cm (6 in.) internal diameter	Internal or external pressures above 100 kN/m² (15 psi); maximum internal pressure of 20 MN/m² (3 000 psi)	Dependent on allowable stresses
Water storage tanks	Contents exceeding 0.46m³ (16.4ft³)	as above	as above
Indirectly heated hot water storage tanks	Contents exceeding 0.46m³ (16.4ft³)	as above	as above
	over 15 cm (6 in.) internal diameter	as above	Water temperature above 95°C (203°F)
	over 15 cm (6 in.) internal diameter	as above	Heat input to the water above 5.86 kW (20 000 Btu/h)
Unrestricted	0.14 m³ (5 ft³) maximum	1.6 MN/m² (240 psi) maximum	Dependent on allowable stresses
	0.043 m³ (1.5 ft³) maximum	Internal or external pressures above 100 kN/m² (15 psi); maximum internal pressure of 20 MN/m² (3 000 psi)	as above
Steam generation	Over 15 cm (6 in.) internal diameter	as above	as above

UCL – integrally clad plate or corrosion-resistant linings;
UCD – cast ductile iron;
UHT – heat-treated ferritic steels.

Allowable stresses and other requirements that are only dependent upon the material are detailed; these apply only to the stated temperatures.

The Appendices for Division 1 are divided into mandatory (numbered I to IX) and nonmandatory (letters B to U, omitting H, I, N and O) sections. Particularly useful for the engineer using the code for the first time are Appendix III containing definitions of terms used in the code, and Appendix L containing sample problems of vessel design subject to external pressure.

Section VIII, Division 2
Division 2 comprises eight parts and 16 appendices (mandatory and nonmandatory); these are detailed as follows:

PART A GENERAL REQUIREMENTS
Specifies the user's and manufacturer's responsibilities, including the user's design specification and the manufacturer's design report. Vessels constructed according to Division 2 must remain at a fixed location during their operating life.

PART AM MATERIALS

Includes specification and properties, and also impact testing. Thorough materials inspection is required.

PART AD DESIGN

Specifies stress analysis and conditions for fatigue analysis.

PART AF FABRICATION

Contains more details than Division 1, including more extensive use of nondestructive testing.

PART AR PRESSURE RELIEF DEVICES

PART AI INSPECTION

PART AT TESTING

PART AS STAMPING, MARKING, REPORTS AND RECORDS

APPENDICES 1–10 (MANDATORY)

Cover aspects of stress analysis, fatigue analysis, flange design, weld porosity, examination techniques.

APPENDICES 11–16 (NON MANDATORY)

Cover installation, operation, temperature measurement and protection, pre-heating, thermal effects, approval of new materials.

Section III Nuclear Power Plant Components

Contains rules for pressure-vessel design and containment systems for nuclear installations.

Section X Fiberglass-reinforced Plastic Pressure Vessels

Contains rules for methods of fabrication of vessels, operating temperatures and pressures, testing procedures.

Alternative codes

Pressure vessels may be subject to different codes and regulations depending upon their contents and location. The ASME Boiler and Pressure Vessel Code is probably the most widely used code in the United States and Canada. However, pressure vessels that are a permanent part of a ship come under the rules of the American Bureau of Shipping. The standards of the Tubular Exchanger Manufacturers Association (TEMA) are often specified for the design of heat exchangers, although like the ASME Code they have no legal requirement. Vessels of unusual design or of nonstandard materials are often approved as special cases by the ASME committee.

The scope of Section VIII of the ASME Code, and the basis of its requirements, may be summarized as follows:

670 Fabrication Processes

(a) Rules cover the minimum construction requirements for design, fabrication, inspection and certification of unfired pressure vessels. Specific requirements of vessels in terms of service conditions and vessel contents are also given in the code.

(b) Includes details of combined fabrication methods: e.g. welding, brazing, riveting, etc.

(c) Does not cover all design cases, and omits special cases which should be approved by an inspector.

(d) Provides a basis for an inspector to approve a design and apply a code stamp.

(e) Codes are inadequate for complex vessels and for nonstandard situations.

Other design factors which need to be considered are as follows:

(a) Corrosion – extra material thickness is required, based upon the expected plant life.

(b) Identification and certification of materials – uncertified materials can be used if they conform to specific code requirements.

(c) Welding – materials must be demonstrated as weldable. Carbon and low-alloy steels must contain less than 0.35% carbon.

(d) Combinations of materials – must conform to Section IX of the ASME Code.

(e) Lower temperature conditions – materials must then conform to impact test requirements.

(f) Fabrication – forgings and rolled materials should be sufficiently worked to remove the coarse ingot structure. Clad plate must possess a minimum shear strength of 130 MN/m² (18 900 psi). Corrosion-resistant linings to be welded to a base metal must be of weldable quality. Pipe and tubing possessing a welding seam is limited to 75 cm (30 in.) nominal diameter. Bolting of material other than carbon steel should be threaded along its full length, or machined to the thread diameter in the unthreaded portion. Washers should be made of wrought material and nuts should be semi-finished, chamfered and trimmed.

(g) Vessel testing – all fabrication work should be examined using nondestructive methods; details of radiographic examination of butt welds are given in the code including technique, weld surface finish, thickness of reinforcement and limits of acceptable defects. Any type of weld crack, or zone of incomplete fusion or penetration, is unacceptable. Elongated slag or groups of slag inclusions can be accepted within certain limits – less stringent requirements for spot examination than full radiography. Porosity limits are also given for full radiography. Examination can also be performed by ultrasonic testing, magnetic particle or fluid penetration methods.

Pressure tests must be performed on completed vessels, and may be rejected due to permanent visual distortion. Hydrostatic testing is normally 1.5 times the normal operating pressure, although this may be less for enamelled vessels due to the possibility of distortion. Pneumatic testing requires more care due to the greater stored energy. For multichamber vessels, each chamber must be tested separately without pressure in the others.

8.6.3 British Standard specifications for pressure vessels

The following British Standard (BS) specifications are concerned with pressure-vessel design and construction:

BS5500: Unfired fusion welded pressure vessels. (Replacing BS1500:Part 1 and BS1515:Parts 1 and 2.)

BS1500: Fusion welded pressure vessels for general purposes.
Part 3 Aluminum
Covers two classes of pressure vessels made from aluminum and aluminum alloys: details include materials and design stresses; design of vessels and components; manufacture and workmanship; inspection and mechanical tests; radiographic examination of welds; pressure tests; marking and protective devices.
Design methods for flanged connections are appended.

BS2915: Domed metallic bursting disks and assemblies.

BS3274: Tubular heat exchangers for general purposes.

BS3451: Methods of testing fusion welds in aluminum and aluminum alloys.

(Note: Always check for amendments to the edition of a British Standard being used.)

Information was given in BS1500 and BS1515 (now withdrawn) which could be used to determine whether a vessel was classified as a pressure vessel. This information was presented graphically and as appropriate equations, and was similar to that given in Fig. 8.67.
The following is a summary of the information contained in BS5500.

Section 1 General
Includes definitions, references, responsibilities, etc.

Section 2 Materials
The selection of materials, design strengths (values in tables) and consideration of carbon, carbon manganese and alloy steels, and aluminum.

Section 3 Design
The major section covering corrosion, construction, external pressures, supports, flat heat exchanger tubesheets, design of welds, jacketed construction, etc.

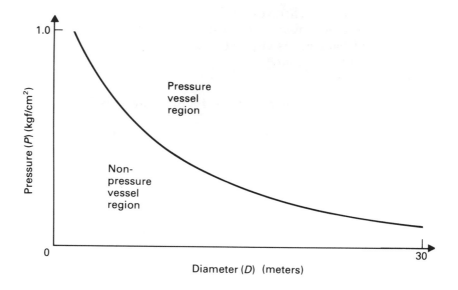

Can be calculated from the equation:

$$P = \frac{100}{3(D+6)^2} \; \text{kgf/cm}^2$$

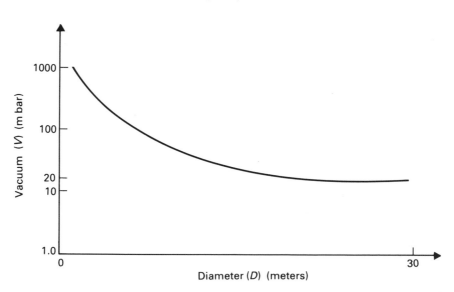

Can be calculated from the equation:

$$V = \left(\frac{310}{D} + 8.2\right) \; \text{m bar}$$

Figure 8.67 Classification of pressure vessels for conditions of pressure and vacuum.

Section 4 Manufacture and workmanship: steel

Section 4A Manufacture and workmanship: aluminum
Includes cutting, forming and tolerances, welded joints, heat treatment and surface finish.

Section 5 Inspection and testing: steel

Section 5A Inspection and testing: aluminum
Includes approval testing of fusion-welding procedures, nondestructive testing, detection of weld defects, pressure tests, etc.

The standard also contains appendices covering fatigue, welded connections, pressure-relief protective devices, etc.
Enquiry cases are also included in the standard and are notified in *BSI News*.

8.6.4 Bursting disks

This section contains details of the information presented in *BS2915: Domed metallic bursting disks and assemblies.*

Section 1 General
Including scope, definitions, applications, installation.
Bursting disks are used to protect pressure vessels, pipelines, etc., from excess pressure or vacuum; they are designed to rupture under excess pressure differences. The construction may be a simple domed metal disk with the concave side exposed to pressure, or a reverse domed metal disk with the convex surface exposed to pressure. A domed disk reverses and splits under excess pressure. Graphite disks are flat and rupture at the orifice circumference, due to the creation of excessive shear stresses.
It is often advantageous to obtain advice from a manufacturer before installing bursting disks. When installed disks should be accessible, and when burst they should not impede the fluid discharge. Consideration must be given to the effects of temperature changes and any dangerous discharges which may occur. Bursting disks can be used in conjunction with relief valves, but disks are preferable in the following situations:

(a) rapid pressure rise, valves act slower;
(b) no leakage permissible;
(c) deposits may block relief valves;
(d) severe cold may prevent valve operation.

It will be necessary to calculate the venting capacity required to discharge sufficient fluid, so that the pressure rise does not exceed 1.1 times the design pressure. For a gas or vapor, the venting capacity (W) can be calculated from an equation of the form:

$$Q = CKAP \sqrt{(M/T)}$$

where Q = rated capacity (mass flow rate);

K = coefficient of discharge (usually 0.6);

A = actual orifice area;

P = vessel pressure;

M = molecular weight;

T = inlet temperature;

C = a constant which is obtained from a table of values as a function of γ,

where $\gamma = \dfrac{\text{specific heat at constant pressure } (C_p)}{\text{specific heat at constant volume } (C_v)}$

Details of materials, testing, holders, design, inspection, temperatures and pressures are given in Sections 2–4.

Section 2 Domed disks

Section 3 Reverse domed disks

Section 4 Graphite disks
Details of two recommended types of support and holder are given in Fig. 8.68. Details of a test holder are shown in Fig. 8.69.

Section 5 Corrosion protection
Corrosion protection can be obtained by coating the disk. The choice of coating depends upon the environment, e.g. polytetrafluoroethylene (PTFE), poly-trifluoromonochlorethylene (PTFCE), epoxy resins and paints. Alternatively, composite disks can be used which are alternate layers of polymer and metal, or multiple metal foils, e.g. nickel and platinum, aluminum and lead.

Section 6 Marking and identification

Section 7 Test certificates

Section 8 Information from purchaser

Also given in the standard are tables of bolt diameter against number of bolts, in order to obtain a bolt factor. This can then be used with a graph of torque (N-m) against bolt factor for various disk sizes.

Graphs are presented of the percentage bursting pressure at elevated temperature compared to the bursting pressure at room temperature, as a function of temperature for various disk materials, e.g. nickel, graphite.

* * *

Self-assessment exercises

1 Locate and study standards or codes relating to pressure-vessel design.

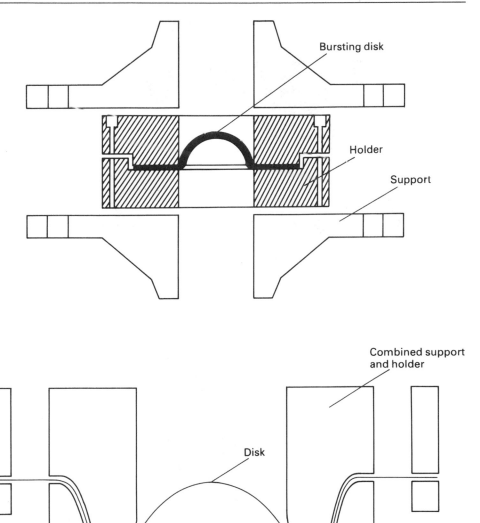

Figure 8.68 Typical bursting disk supports and holders.

2 Summarize the useful design information contained in these documents.

3 Consider the design of a propane storage tank which is to be used to supply
   the gas for welding and metal cutting operations.
          Determine the pressure required for welding and hence decide the tank

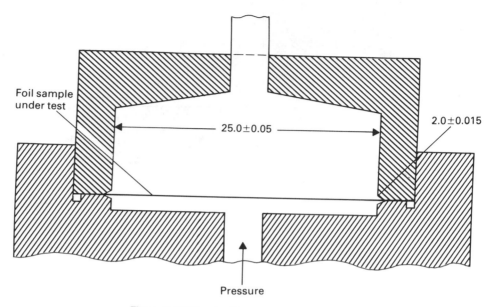

Figure 8.69 Standard bursting disk test holder.

pressure. The tank is to be supported horizontally, with domed ends. It is to be filled from approximately half full each day, and the gas is used by three operators working for six hours continuously out of an eight-hour shift. Hence, determine the size of the tank.

Using appropriate design equations and making any necessary assumptions, perform approximate calculations to determine the required wall thickness of the vessel, and any other necessary information.

Prepare a draft specification for the vessel and decide:

(a) possible materials that could be used;
(b) fabrication methods;
(c) details of any welds, etc.;
(d) the use of mechanical fasteners, e.g. bolts.

Prepare a safety plan in case the pressure rises rapidly in the tank. Could a relief valve or bursting disk be incorporated? If yes, decide appropriate specifications and precautions. If no, prepare an alternative safety strategy.

* * *

8.7 NUMERICAL CONTROL AND COMPUTER-AIDED MANUFACTURE

8.7.1 Numerical control

Numerical control (NC) is a method of automatically controlling machine tools, equipment or processes. An NC program is a logical arrangement of symbols (numbers and letters) that are organized to direct a machine tool for a specific task.

The earliest NC systems worked on a vacuum-tube principle, using servo systems to control the machine's axes. They were mainly concerned with controlling rate and motion only. Transistorized NC units were introduced in 1960 and provided more actual control functions. In 1967 integrated-circuit NC units were introduced and quickly replaced the previous NC systems. Their success was due to the miniaturization of parts and lower costs; significant improvements in data processing were obtained. Recent developments in NC have involved the use of small digital computers, e.g. minicomputers and microcomputers. The programming data can now be stored in computer memory, rather than on punched tape or cards.

Conventional NC systems contain the following elements: data input device, the tape reader, machine control unit, feedback equipment, the operating equipment such as a machine tool. A closed-loop feedback system compares the actual position of a machine tool with the programmed location, and 'feeds back' the discrepancy so that appropriate corrective action can be taken. In contrast an open-loop system obeys the programmed instruction, even if discrepancies have occurred. Possible action may be the overloading of the system so that the operation stops. The closed-loop system is more accurate, and more expensive.

The versatility of an NC machine depends upon the number of axes in which it can operate, i.e. be controlled. Simpler machines may control movement only in the x and y (and z) Cartesian coordinate axes. Other machines may include rotational control of the worktable and/or the tool. NC machine tools may operate either as point-to-point control or continuous path. Point-to-point systems are used for straight-line cuts, drilling and boring operations. Machine tools capable of continuous-path control are normally operated by computers

Instructions for NC are usually written in either the Automatic Programmed Tool (APT) or COMPACT II languages, although other languages are available. APT was developed at the outset of NC by the Aerospace Industries Association (AIA), but it has now been extended to include many other users and industries. APT is the standard programming language (software package); it can be used to control fifty different machine tools, and can be applied to five axes of machine motion. The use of APT has previously required the use of a large mainframe computer, the cost of which has restricted its use. This problem has been overcome by storing the program in a peripheral data storage bank, and using a minicomputer to access and use parts of the program as they are required. This system is known as ADAPT (Adaptation of APT).

The majority of NC applications are in metal cutting, e.g. turning, boring, drilling, milling, and the use of machining centers (multi-tools). NC is also used for the control of metal forming, but mainly for punching and shearing operations. However, many NC applications are being developed that are not related to these industries.

8.7.2 Direct numerical control

NC machines that utilize paper punch tape or similar methods of data storage are known as *hardwired controllers*. As mentioned in Section 8.7.1, the use of punch tape, etc., can be eliminated by storing the data and operating instructions in a mini- or microcomputer. This is known as *computer numerical control* (CNC). If the machine control unit is interfaced directly with the computer and is used to control the machine tool, then this is known as *direct numerical control* (DNC).

One advantage of using DNC is that many machine tools can be operated and controlled from a single computer. The computer may be a large (mainframe) machine or a minicomputer, or several minicomputers (or microcomputers) connected to a mainframe computer. Rapid technological developments and falling costs have blurred the distinctions between the various types, i.e. sizes, of computers. The main difference between the machines is the size of the memory, or logic-processing capability. Microcomputers are the least expensive computers, mainly due to the virtual elimination of core memory and the use of a *read only* memory. *Hierarchical control systems* have been developed where a mainframe computer controls the operation of a set of microcomputers, each of which controls the action of a machine tool. It is not necessary to utilize a mainframe computer entirely for NC, the use of time sharing allows other operations to be performed on the same computer. The user may not even own or lease the computer, but may simply buy computer time as it is required. This type of system is known as *remote access* and is achieved via a teletype or telephone, or both.

DNC makes quick and easy data revision and editing possible, and also interaction between the programmer and the machine tool. The installation of a *cathode ray tube display unit* (CRTDU) and a keyboard at the machine tool location enables program modifications, and changes in the machine tool operation, to be made as required. It can also be used to provide direct instructions such as coolant requirements, stopping operation, etc. Computer programs (software packages) may be written in two ways, either as a straightforward input–output operation or in an *interactive* or *conversational form*. Data are provided by the operator during the execution of a conversational or interactive computer program. The program requests data (and may specify the form) as required. The advantage of this programming method is that it can make the user 'feel' more integrated with the computer, rather than considering it merely as a plug-in machine. The user also feels in control of the operation and the machine. Requests for data can be used as a sequential check on the actual program operation, rather than supplying data in one block and then attempting to identify the source of any error messages.

The number of tools that can be controlled by a single computer depends upon the nature of the parts to be produced, and the types of machines that are used. Systems have been developed where several hundred operations or machines are controlled by a single computer. Because of the initial cost of DNC, it is usually necessary to control at least five machine tools to make production economical. However, this minimum level depends upon the complexity of the machining process and the time required.

8.7.3 Computer-aided manufacture

Computer-aided manufacture (CAM) is a general term used to describe the use of computers in manufacturing and related operations. The heart of CAM is the use of DNC (described in Section 8.7.2) for the efficient operation of production machines. CAM also includes purchasing, inventory control, testing, packaging and transportation. The influence of rapid developments in CAM has been most significant in the areas of manufacturing and assembly. DNC has enabled improvements in speed, efficiency and reliability to be achieved, especially when many functions need to be carried out in a programmed sequence using multiple tools.

Computers now have a major role in many aspects of design, such activities being termed computer-aided design (CAD). These functions are described in more detail in Chapter 9. The division between CAD and CAM has gradually narrowed as new developments have emerged, and as attempts have been made to integrate both functions. Computer-aided drafting and design (CADD) systems are also widely used. It is now possible to obtain a fully integrated CAD/CAM system where all the project functions are interfaced with a common database. Such a system is sometimes referred to as computer-aided engineering (CAE). Figure 8.70 illustrates the stages in the design of a machine part from inception to production; also shown are the CAD/CAM features that can be included in an integrated system.

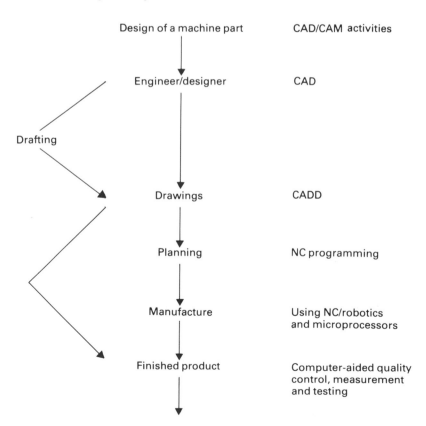

Figure 8.70   CAD/CAM activities for the design of a machine part.

Advances in CAM and NC have been achieved by using computer simulation to check the operation of a machine-tool program, rather than actual operation and subsequent adjustment. A computer can also be used to produce NC instructions directly from the geometric modelling database, although this is usually restricted to flat parts, symmetric shapes or special parts.

The importance of CAM now extends beyond DNC, and it is being utilized and developed in areas such as process planning, factory management and robotics. A robot consists of a programmed manipulator arm with grippers that move and position materials, parts, tools or other objects. Robots may be simple pick-and-place machines with limited movement, or they may be controlled by a servomechanism providing more versatility by use of a programmable controller (the 'teach' mode). There are even computer-controlled 'smart' robots. Robots are used mainly for tedious or dangerous tasks such as heavy loading, handling toxic chemicals or checking for terrorist bombs. However, advantages and applications continue to be found, e.g. welding, parts assembly, fabric cutting, material handling, etc. The automobile industry in particular has adopted robotics for spot welding and assembly functions.

Research studies are now attempting to produce robots with an increasing amount of artificial intelligence. This is not in the form of consciousness or creativity, but a problem-solving ability using sensory input. An example would be selection of components based upon their shape. The development of a robot with television camera 'eyes', so that the image received can be compared with that stored in a computer memory, is considered the most promising application.

The introduction of CAM systems into a production process should improve the productivity of that operation. There are still improvements to be made in the integration between CAD and CAM systems, and in the sophistication of CAM compared with CAD.

*   *   *

Self-assessment exercises

1 Explain the meaning of the terms:
   (a) numerical control;
   (b) machining center;
   (c) feedback;
   (d) CAM.

2 Describe the difference(s) between conventional NC, CNC and DNC. A CNC system uses a dedicated computer. What does this mean?

3 List the advantages and disadvantages of NC.

4 List the advantages and disadvantages of using a computer for NC operations.

5 Explain what is meant by adaptive NC.

6 Describe the movements required when some common machining operations are performed, e.g. milling, turning, boring, etc.

7 Describe the importance of the following operations when initiating automatic control systems:
   (a) sequencing control;

(b) position control and measurement;
(c) assembly tasks.

8 List the functions that are encompassed by CAM, other than NC.

9 Describe the functions that an industrial robot could be required to perform.

*   *   *

Exercises

1 Discuss the significance of the thermal conductivity of the mold material in casting processes.

2 Explain why the properties of a centrifugally cast part are usually better than those statically cast.

3 List the process and pattern requirements that prevent the production of sand castings to the final size.

4 Explain why brittle materials that crack in most plastic-forming operations can be safely extruded.

5 What is the main advantage of a forging, compared to a cast or welded component?

6 Explain why a metal possessing a large yield elongation is unsuitable for forming.

7 What is the main disadvantage of using a forging process?

8 Explain how the following factors influence the choice between metal spinning and metal drawing:
(a) diameter;
(b) material thickness;
(c) production quantity.

9 Explain how the following mechanical properties influence the selection of a shaping process:
(a) yield strength;
(b) ratio of yield strength:ultimate strength;
(c) elongation.

10 Explain how powder metallurgy can extend the possibilities of alloying.

11 Why is sintering performed below the melting point of the powder? How does the sintering stage influence the ultimate strength of the final part?

12 What factors restrict the size and mass of parts to be produced by powder metallurgy.

13 What type of shape is most suitable for production by powder metallurgy?

14 Describe how milling cutters can be classified. List some of the factors to be considered when selecting milling speeds and feeds.

15 Describe the cutting action of a shaper. How is the size of a shaper classified? Why is it difficult to produce curved surfaces by shaping?

16 How is the work held on a planer? How is the motion of a planer table controlled?

17 Explain the difference between a one-piece broach and a shell-type broach. What is a burnishing broach?

18 What factors affect the surface finish of a turned component?

19 What are the differences between reciprocating and band sawing machines?

20 Explain the significance of the following characteristics of a metal saw:
(a) material;
(b) blade width;
(c) pitch;
(d) set.

21 What is precision grinding? Explain why synthetic abrasives are used for production grinding. Compare the possible surface finish for grinding and milling.

22 What are the differences between abrasive machining and precision grinding? What are the advantages of abrasive machining compared to lathe or milling operations?

23 Why are very hard or very soft materials difficult to machine?

24 How does power consumption vary with speed, feed and depth of cut for machining?

25 Why is a drilled hole either bored or reamed?

*   *   *

Complementary activities

1 Perform a simple casting operation. Examine the component produced to determine any defects that are present. Perform appropriate tests on the component to determine its mechanical properties (Section 6.2).

2 Prepare small simple shapes of different materials and then shape these by;
(a) cold working;
(b) hot working.

3 Determine the mechanical properties, e.g. tensile strength, hardness, of these shapes, before and after various stages of deformation.

4 Subject the samples to appropriate heat treatment and continue the deformation processes.

5 Determine the deformation limits to which the samples can be subjected.

6 Perform or observe powder metallurgy fabrication.

7 Perform (under supervision) some of the basic machining processes on selected materials and shapes.

8 Perform finishing operations.

9 Examine some pressure vessels and pressure relief devices.

10 Observe some numerically controlled (NC) operations.

11 Obtain manufacturer's literature concerning recent developments in CAM systems.

*   *   *

KEYWORDS

semi-finished products	mold	molding box:
casting	sand casting	drag
ingot	pattern	cope

split pattern
parting powder
pouring gate
feeding gate (riser)
runner
sand
binder
contraction
taper (draft)
core
die casting:
    gravity method
    pressure method
shell-molding
centrifugal casting
investment casting
full mold process
mechanical shaping
hot-working processes
hand forging
machine forging
forge (furnace)
upsetting
drawing down
fuller
necking down
swage
reducing
flatter
drop forging
flash gutter
fin
draft
hot pressing
hot rolling
reversing mill
extrusion:
    direct process
    indirect process
tube manufacture
seamed tube
seamless tube
Mannesmann process
recrystallization
    temperature
softening
grain growth
cold-working process
cold rolling
drawing
die
solid lubricant
pickling
pressing
deep drawing
punch and die
stretch forming

coining
embossing
cold forging
spinning
flow forming
flow turning
impact extrusion
elongation
fibrous character
internal stress
stress corrosion cracking
brittle fracture
stress relief
powder metallurgy
compacting
sintering
porosity
binder
finished products
machining process
machine tools
lathe
shaper
planer
chip
single-point cutter
cutting angle
cutting edge
clearance
rake angle
coolant
machinability
tool life
cutting speed
sawing
tooth pitch
coarse-tooth saw
undercut tooth
reciprocating saw
circular saw
band saw
broaching:
    pull or push
surface broaching
turning
facing
taper turning
threading
drilling
boring
coring
reaming
trepanning
counterboring
spotfacing
countersinking
punching

flame cutting
drill
flute
shank
core drill
twist drill
jig
milling
fluted teeth
profile cutter
relieved teeth
formed cutter
shaping
planing
horizontal machine
vertical machine
grinding
abrasion
grain
bond
cutting edge
grinding wheel
glazing
clogging
coated abrasive
microscopic surface
center line average (CLA)
surface preparation
honing
lapping
superfinishing
wire brushing
polishing
buffing
production finishing
abrasive grinding
gear cutting
thread cutting
abrasive jet machining
water jet machining
ultrasonic machining
electric discharge
    machining
electrochemical
    machining
laser-beam machining
high-temperature
    machining
chemical milling
chemical blanking
electroforming
machine lubrication
oil or grease
surface treatment
metallic coating
electroplating
electrodeposition

dipping	BS2915	read only memory
cladding	pressure relief	hierarchical control
calorizing	domed metal disk	system
phosphating	flat graphite disk	time sharing
chromating	venting capacity	remote access
anodizing	numerical control (NC)	CRTDU
surface hardening	integrated circuit	keyboard
cementation	conventional NC	computer program
sheradizing	minicomputer	software package
chromizing	microcomputer	interactive program
surface coating	feedback	conversational program
paints	point-to-point	CAM
polymers	continuous path	CAD
pressure vessel	APT	CADD
boiler	COMPACT II	CAE
ASME Boiler and	mainframe computer	computer simulation
Pressure Vessel Code:	ADAPT	process planning
Section VIII;	machining center	factory management
Divisions 1 and 2	hardwired controller	robotics
BS5500	CNC	programmable
bursting disk	DNC	controller

* 　 * 　 *

BIBLIOGRAPHY

Alting, L., *Manufacturing Engineering Processes*, Marcel Dekker, Inc., New York (1982).

Amstead, B.H., Ostwald, P.F. and Begeman, M.L,. *Manufacturing Processes*, 7th Ed., John Wiley and Sons, Inc., New York (1977)

Bickell, M.B. and Ruiz, C., *Pressure Vessel Design and Analysis*, Macmillan Publishers Ltd, London (1967).

Black, B.J., *Workshop Processes, Practices and Materials*, Edward Arnold (Publishers) Ltd, London (1979).

Bram, G. and Downs, C., *Manufacturing Technology*, Macmillan Publishers Ltd, London (1975).

Dixon, R.H.T. and Clayton, A., *Powder Metallurgy for Engineers*, Machinery Publishing Co. Ltd, Brighton, Sussex (1971).

DeGarmo, E.P. and Kohser, R.A., *Materials and Processes in Manufacturing*, 6th Ed., Macmillan Publishing Co., Inc., New York (1984).

Flood, C.R., *Fabrication, Welding and Metal Joining Processes. A Textbook for Technicians and Craftsmen*, Butterworth and Co. (Publishers) Ltd, London (1981).

Gabe, D.R., *Principles of Metal Surface Treatment and Protection*, 2nd Ed., Pergamon Press Ltd, Oxford (1978).

Haslehurst, M., *Manufacturing Technology*, 3rd Ed., Hodder and Stoughton Ltd, Sevenoaks, Kent (1981).

Kazanas, H.C., Baker, G.E. and Gregor, T.G., *Basic Manufacturing Processes*, McGraw-Hill Book Co., Inc., New York (1981).

Kibbe, R.R., Neely, J.E., Meyer, R.O. and White, W.T., *Machine Tool Practices*, John Wiley and Sons, Inc., New York (1979).

Kuhn, H.A. and Lawley, A. (Eds.), *Powder Metallurgy Processing: New Techniques and Analyses*, Academic Press, Inc., San Diego, California (1978).

Lindberg, R.A., *Processes and Materials of Manufacture*, 2nd Ed., Allyn and Bacon, Inc., Boston, Massachusetts (1977).

Lissaman, A.J. and Martin, S.J., *Principles of Engineering Production*, Hodder and Stoughton Ltd,

Sevenoaks, Kent (1982).

Lowenheim, F.A., *Electroplating: Fundamentals of Surface Finishing*, McGraw-Hill Book Co., Inc., New York (1978).

Neely, J.E., *Practical Machine Shop*, with Instructor's Manual and Student Workbook, John Wiley and Sons, Inc., New York (1982).

Niebel, B.W. and Draper, A.B., *Product Design and Process Engineering*, McGraw-Hill Book Co., Inc., New York (1974).

Pollack, H.W., *Manufacturing and Machine Tool Operations*, 2nd Ed., Prentice-Hall, Inc., Englewood Cliffs, New Jersey (1979).

Rowe, G.W., *Principles of Industrial Metalworking Processes*, Edward Arnold (Publishers) Ltd, London (1977).

Schey, J.A., *Introduction to Manufacturing Processes*, McGraw-Hill Book Co., Inc., New York (1977).

Watkins, M.T., *Metal Forming I, Forging and Related Processes*, Engineering Design Guide No. 11 (1975) and *Metal Forming II, Pressing and Related Processes*, Engineering Design Guide No. 12 (1975), Oxford University Press, Oxford.

Yankee, H.W., *Manufacturing Processes*, Prentice-Hall, Inc., Englewood Cliffs, New Jersey (1979).

9
Engineering Design

─── CHAPTER OBJECTIVES ───

To obtain an understanding of:
1 the relationship between material's selection and design;
2 the work performed by a design engineer;
3 the general approach to a design problem;
4 the specific factors to be considered in order to obtain a design solution;
5 the use of computers in engineering, and the importance of computer-aided design.

─── IMPORTANT TERMS ───

engineering
design:
  adaptive
  developmental
  creative
morphology and anatomy
  of design

CAD
computer graphics
CADD
'turnkey' system
CAM

9.1 ENGINEERING MATERIALS

The life cycle of an engineering material passes through several stages, the first being the location and extraction of essential elements (often present in a combined form). This is followed by separation, concentration and purification stages. The final purity of (pure) metals usually depends upon their intended use. However, most materials for engineering applications are required as alloys, mixtures or compounds, and various elements are combined together.

If the material is available with the required composition, or purity, then materials technology can be applied. The next stages are the modification of the material properties and its behavior, followed by fabrication and joining. The aim is to produce components which can be assembled, or joined together, for industrial operation and which will operate in a satisfactory and economical manner. The last stages of the life of a material are concerned with obsolesence of the equipment, scrap and possible recycle.

This book is concerned with engineering materials from the purification stage (the condition in which materials are normally available) through to the finished component or product stage. Materials are used in engineering situations, so far we have established the basic knowledge of materials. The question now is what are we to do with this knowledge, or what use will it be to us as engineers. The intention is, and information is presented in such a way, that the reader will appreciate the integration of different topics and their interrelationship. For instance, the occurrence of dislocations in crystal structures (Section 1.3), their influence upon mechanical properties, e.g. brittle fracture (Section 6.2.3) and their importance for fabrication, e.g. cold working (Section 8.2.17), are all related. Engineers should be using and applying the knowledge they have gained, and relating different aspects of that knowledge to obtain a better understanding of the effects that occur.

The majority of the work performed by engineers is concerned with design. This is an activity that is frequently talked about and often performed, but it is not usually clear exactly what it involves. Sections 9.3 and 9.4 attempt to describe briefly the activities of engineering and design. This will be necessary in order to appreciate the applications of materials in engineering situations, and hence their function in a design. The meaning of these activities will perhaps become clearer if we consider that an engineer is required:

(a) to engineer something (efficiently) – i.e. materials and resources;
(b) to design – i.e. to plan, to devise, to specify, to build.

Materials represent two main challenges or tasks for the engineer, these are:

(a) specification of materials for a particular application;
(b) modification of the properties of particular materials in order to perform the tasks required (or to perform them better).

In order to perform tasks (a) and (b) it is necessary to have acquired the basic knowledge and understanding of materials science and technology. What the engineer needs to know (and the appropriate chapters or sections in this book) are:

(a) What situations or conditions will make a material unsuitable for its intended task?
(Chapters 2–5; Section 6.2)
(b) How and why materials fail?
(Chapter 5; Sections 6.2, 7.21 and 7.24)
(c) What tests need to be performed on materials?
(Sections 4.12, 5.6, 6.2 and 7.24)
(d) How can materials be joined (together or to other parts)?
(Chapters 2, 3 and 7)
(e) How can materials be formed into the required shape?
(Chapters 1–3 and 8; Sections 4.15 and 6.2)

It is hoped that after studying the topics presented in this book the reader will be able to answer most of these questions, both from a general viewpoint and for specific examples. If not, then the material should be reread, but this time concentrating upon the application of the information rather than merely remembering it.

9.2 MATERIALS SELECTION AND SPECIFICATION

One of the ultimate aims of an engineering materials course should be that the reader acquires sufficient knowledge and understanding to be able to select a material for a particular situation. There is no blueprint for correct materials selection, nor any complete set of procedures to be followed. Each situation is different and requires knowledge of the materials, and the conditions to which they will be subjected. However, there are some general points that can be made, providing a place to start. These are:

(1) Any material selected must possess properties consistent with the in-service conditions.
(2) It is necessary to consider the effects of changes in these conditions outside the normal limits.
(3) To select an acceptable material, first compile a list of possible materials and then eliminate some of these because of unsuitable mechanical properties, thermal properties, corrosion, brittle fracture, etc. Disregard some other materials on the basis of unpredictable availability or delivery, safety requirements, costs, etc.

If the materials remaining cannot fulfill all the requirements, then the engineer has various options. These are:

(a) change the conditions, e.g. loading, environment, etc;
(b) subject the materials to appropriate heat or chemical treatment;
(c) alloy the material with a suitable solute;
(d) add alloying additions to produce specific effects.

Material specification is concerned with providing the necessary information to a supplier in order to obtain the required materials. This is the next stage after material selection. The supplier will want to know not only which material is required but also its condition, e.g. annealed, quenched, surface hardened, etc., and the form, e.g. rod, bar, ingot, tube, etc. It is advisable to specify to the supplier any requirements for fabrication, joining, machinability, etc., and to ascertain the suitability of the material for heat treatment. The compositional limits of the individual constituents should also be checked.

* * *

Exercises

Examine some common items such as a screwdriver, car jack, chair, toothbrush, stapler, breadknife, bicycle chain, ball point pen, electric kettle, key, scissors, valve, etc.

Identify the main design feature(s) associated with selected items.
What is the main purpose or use of an item?
What task is it intended to fulfill and how is this objective achieved?
What are the engineering principles involved?
What calculations need to be performed to design an object, e.g. stresses and strains, creep, fatigue, etc.?
What conditions will normally be imposed upon the object, e.g. forces, temperature, etc.?
What will happen if unusual conditions occur?
Which materials are used to make the object?
Why were these materials used?
Which alternative materials could be used?
What fabrication methods were used?
How are the parts of the item joined?
What is the most likely cause of failure of the item?
Why will failure occur?
How can failure be prevented, if at all?
Will corrosion significantly affect performance?

Now ask and answer some questions of your own.
The aim of this exercise is to illustrate the important role of materials in all aspects of product design. Correct material specification is essential for the successful operation of a product. Many aspects of product design will be affected by the choice of materials, e.g. the cost and strength. The joining methods and fabrication processes that can be used also depend upon the materials chosen.

* * *

9.3 WHAT IS ENGINEERING?

A textbook survey would probably reveal several different definitions of the term engineering. Some would be in general terms while others would attempt a more precise definition. Most would probably contain some common theme but no single definition would satisfy everyone. The situation is not helped by the media where the terms engineering, technology and science are often used indiscriminately. A class of engineering students would probably describe engineering mainly in terms of its functions rather than defining the term absolutely. If the same class were asked what they thought was the role of engineering in society or why they chose an engineering course, the answers would be even more diverse and would illustrate more clearly the misconceptions associated with engineering.

An exact definition will not be presented here but it is hoped that the reader will formulate a personal opinion and, even more important, will consider the limitations and extensions of the definition. A starting point for discussion would be that engineering is concerned with the application of ideas and resources in order to attempt to satisfy the needs of mankind. This could provide a basis for further group tutorial discussion in order to elaborate on the main types of applications, the types of solutions which are obtained, if there are any exceptions to this general definition and whether there are applications not directed towards population needs.

Considering the scope of engineering, it can be appreciated that precise categorization, so often demanded by the scientific community, becomes less applicable. Courses are available in 'pure' sciences and various engineering disciplines. But there are also courses that take an intermediate approach, often termed engineering science. Courses in science-based subjects such as applied physics, industrial biochemistry, chemistry and economics, etc., would provide hours of discussion in search of their precise definition, but they exist to satisfy the needs of industry and the needs of students who do not wish to follow a traditional approach. The definition of engineering perhaps becomes more accessible if the broad range of scientific activity is considered as overlapping bands of 'pure' science, applications, design, engineering calculations, energy studies, manufacturing, process economics, etc., rather than exclusive categories.

9.4 WHAT IS DESIGN?

The understanding of the work performed by an engineer can assume a different significance depending upon the country of employment. It can be used to describe a mechanic concerned entirely with practical problems or a manager concerned entirely with costs and human relations. It is often thought that all engineers are engaged in design work, which is certainly not the case. However, all engineers, at least at graduate level, should have had significant grounding in the problems associated with design. It is difficult to define the term design to cover all cases; perhaps the most useful definition is in terms of the work performed by the designer or design engineer. The design process involves the application of technology for the transformation of resources, to create a product which will satisfy a need in society. The product must perform its function in the most efficient or economic manner, within the various constraints which may be imposed. The major restraint is obviously cost, although other factors such as safety, pollution, legal requirements, etc., will have to be considered. It is the ability to design an efficient product that sets apart the successful designer. It is important that the designer is not referred to as a draftsman or planner, or by other descriptions. These are specialized functions that are required in the overall design process, and they are often performed by the designer. However, the role of the designer goes beyond these particular functions. The designer should be involved in all stages of the design process; it is his intellectual ability and understanding of the problem and its solution which determines the final product.

The designer is concerned with solving a problem or producing a product. This work has resulted from a need within society and the decision to proceed to a solution. In most cases the designer has been trained either as an engineer, or with a strong engineering orientation. However, it must be remembered by the designer, as well as other members of a team, that the product is meant to satisfy a need in society. If society decides to reject a product, whether for functional reasons or merely because of habit, then the design effort is wasted. The sociological view of the product's place in society must also be considered. This will be in terms of the structure and needs of society and any changes which may occur during the lifetime of the product, e.g. occupational changes, wealth, etc. Statistics and statistical analysis are often useful in this context, although they should not be considered to the exclusion of other considerations. The designer is not expected to be a sociologist or a statistician but he should be aware of their importance to his work, and be able to appreciate the implications of their recommendations.

* * *

Exercises

Having studied Sections 9.3 and 9.4, the reader should have formulated some ideas about what engineering and design encompasses. Before engaging in design studies it will be useful to consider some general aspects of the influences of design and engineering. Try to answer the following questions; do not assume there are correct answers but try to formulate some ideas and opinions.

1 What are the main branches of engineering?

2 Is a metallurgist an engineer or a scientist?

3 What are the differences (in training and employment) between a metallurgist and a materials technologist or engineer?

4 Is a corrosion expert more likely to be a scientist with practical experience or an engineer who understands some chemistry of reaction mechanisms?

5 Who is involved in engineering?

6 Who is affected by engineering?

7 What is a 'good' design?

8 What is the difference between engineering design and industrial design?

9 Can design be taught, or is it an 'art' only appreciated from experience?

Answers, or rather ideas and suggestions, related to some of these questions can be found in Ray (1985).

* * *

9.5 TYPES OF DESIGN WORK

There are three main types of design work, each of which require different levels of intellectual ability and creativity. First there is the adaptation of existing designs. This requires only minor modifications, often in the dimensions of the product. This type of work represents a large proportion of the total design work undertaken. The designer engaged entirely in this type of work requires only the basic technical skills and does not appreciate the nature of the design process unless other work is undertaken.

Second there is development design which uses as a basis an existing product. The technical work involved can be considerable and the final product may be very different from the original. In certain cases the use of an existing design may make the solution more difficult to obtain; the level of ability and time spent by the designer may be considerable.

Finally the most demanding work involves the design of a new product that has no precedent. This requires a high level of ability and relatively few designers are engaged in this type of activity; it is sometimes referred to as creative design.

9.6 THE ROLE OF THE DESIGNER

Before considering the role of the designer in an organization or a team, it will be useful to assess some of the qualities which a designer would be expected to possess. Obviously any discussion or list of this type should not be considered exhaustive, but the following is intended to emphasize the main qualities and capabilities of a successful designer.

9.6.1 Conceptual ability

The designer should be able to understand and visualize the whole concept of the design project. This involves the imagination of simple shapes, their combination and interrelationships before the detailed design stage begins.

9.6.2 Logical thought

The ability to think logically is important and it is a skill which is acquired early in life, mainly in the school years. A logical approach to problem-solving is necessary at each individual stage of the design process, and also in the overall strategy of the design, which will be considered in Section 9.5.

9.6.3 Perseverance, concentration and memory

These are personal qualities and are also developed mainly in the school years.

However, a person who lacks these attributes is unlikely to achieve the qualifications required to begin a career in design.

9.6.4 Responsibility, integrity, willpower and temperament

These are personal or moral qualities; they determine not only the success of the designer's work but also the direction and success of the designer's career. A person who is responsible to, and for, other people must possess more than technical skills.

9.6.5 Invention or creativity

Problem solutions are obtained by hard work and logical thinking with a certain amount of inventiveness, intuition or creative ability on the part of the designer. The idea of a flash of inspiration or the existence of a creative genius is misplaced. The 'novel' solution is obtained by a clear understanding of the problem, followed by logical analysis and deduction, even though these stages may not be remembered as such when the solution is obtained.

9.6.6 Communication

The most superior design will be ineffective if the designer cannot communicate essential information to other members of an organization or to a customer. This requires presentation of information in written, visual and spoken form.

9.6.7 Scientific knowledge

Throughout his training the designer acquires basic scientific information which will be used to produce satisfactory designs. However, this knowledge is only useful when it is applied with the skills already outlined, and which the engineer should possess. The knowledge required obviously depends upon the areas in which the designer wishes to specialize. Areas of basic importance include:

    (a) mathematics – elementary and advanced levels, geometry and technical drawing;
    (b) principles of physics;
    (c) chemistry;
    (d) engineering and technology – materials, manufacture, machines, structures, production, etc.;
    (e) management studies;
    (f) economics.

    The role of a design engineer, within an engineering department or design section of an organization, can best be appreciated by reference to Fig. 9.1. In a

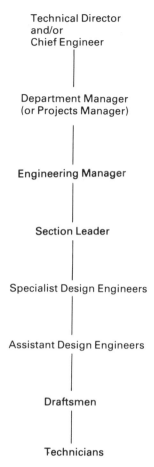

Technical Director
and/or
Chief Engineer

Department Manager
(or Projects Manager)

Engineering Manager

Section Leader

Specialist Design Engineers

Assistant Design Engineers

Draftsmen

Technicians

Figure 9.1 Hierarchy of staff within an engineering design department.

small department, all staff may be actively engaged in all aspects of design work. In larger departments, senior personnel spend a proportion of their time concerned with management tasks. Within larger companies, staff may specialize in particular design problems and sections may exist concerned primarily with these aspects. Although it is not always adhered to, it is preferable if the term designer or design engineer is reserved for persons capable of working independently and creatively toward the solution of a problem. Less qualified staff may be designated as assistants, draftsmen, technicians, etc.

For small projects the entire process of design, production and sales may be performed by the design engineer. For larger projects and within companies, the designer acts as a member of a team. In larger companies, departments or sections exist that deal primarily with different aspects of the production process, e.g. sales, commissioning, legal aspects, etc. Each department needs to cooperate with several others, and none will or should, act with complete independence. The relationship between the design department and other departments within a company is shown in Fig. 9.2. The relationship between each of the departments is not shown.

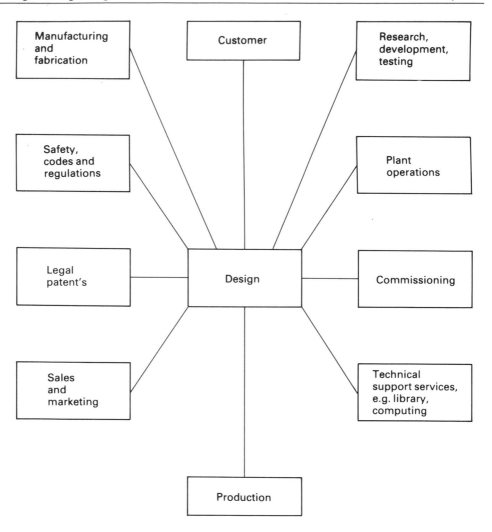

Figure 9.2 Relation between the design department and other departments.

The main areas where technical creativity occurs are design and production, and liaison should be particularly good between these departments. It is also important that there are reciprocal communications between scientific departments which provide details of recent advances, and design departments which identify areas where advancement is required. If this is not done then new developments are not exploited, and a product becomes less competitive because fundamental problems are not being investigated. Good communications are essential between all departments – e.g. customer requirements are communicated to the designer via the sales office – if the product and the company are to succeed.

9.7 THE NATURE OF THE DESIGN PROCESS

There are two methods that are normally used to describe the nature of the design process. The first approach is to examine the life cycle of the product, and to define each stage through which it passes. This is known as the morphology of the design process, and the appropriate stages are detailed in Section 9.7.1. The second approach is known as the anatomy of design, and this is described in Section 9.7.3.

9.7.1 The morphology of design

(a) Identifying the problem

The first step involves identifying a problem or a requirement of society. However good the final product there must be a need or a market for it. A full specification of this need must be obtained to avoid extra work at a later stage. Limitations and requirements of the design must be identified and stated in terms of contraints and criteria. Other relevant information should be obtained and recorded, such as the time scale, the economics, the estimated life of the problem, whether the nature of the problem will change during its life, ease of modifications, etc.

(b) Feasibility study

Having identified a need for the product, the next stage is to conduct a basic feasibility study. It must be physically possible to produce the product, it must also be economically viable and it must be acceptable. If the alternatives in the design process cannot be physically realized, either because appropriate materials or fabrication methods are not available, for example, then the project will not proceed unless these problems are overcome by prior development work. If the physical requirements appear possible, then the economics of the various alternatives can be investigated. The final stage of the feasibility study is to establish that any alternatives under consideration will be acceptable, not only to the customer for whom the need was identified, but also to society in general. If other members of the community object to the product then it may not be successful.

(c) Preliminary design

At this stage, having defined the need or the problem and ascertained that a solution may be feasible, then it becomes necessary to select the best approach to the problem. Where several alternatives exist, they should all be examined in an unbiased manner and with equal thoroughness. The personal preferences of the designer should not be allowed to influence the evaluation process. The preliminary design may be conducted entirely on paper or scale models may be used, or (if possible) experimental results may have to be obtained. At this stage it may be found that a solution which appeared possible earlier cannot now be achieved. At the conclusion of the preliminary design phase, it must be decided which approach to the final solution is to be examined further and is most likely to be adopted. This choice is not irreversible and results obtained later may require that this approach is abandoned and an alternative is developed. This may be a difficult decision to make due to the costs already incurred, but it is cheaper than producing an unsatisfactory solution.

(d) Detailed design

An approach to the solution has been chosen and a full detailed design is prepared. All components and systems are fully specified and all in-service requirements should be detailed. A complete set of drawings is also prepared. At this stage it is usual to produce a scale model of the product, if this has not been done previously, and sometimes a prototype unit. The prototype may be a small-scale version of the final product or even a full-scale replica. The prototype is tested in order to evaluate its performance and effectiveness, and to determine any modifications needed. Fewer choices are now available to the designer and a considerable economic committment is usually required. Any major revisions, or the decision to abandon this approach entirely, should be made before progressing to the next stage, otherwise considerable investment will be wasted. It may be that events outside the control or experience of the designer may cause the product to become unacceptable at a later stage. The possibility of this happening should be minimized by constant re-evaluation of the problem definition and the feasibility studies.

(e) Production

At this stage, with the detailed design finalized and approved, the product is constructed. This involves close cooperation between the designer and the production engineer. A commissioning engineer and/or plant operations engineer may also contribute specialist advice at this stage. The capital investment for the project has to be available to pay for materials, labor, fabrication, subcontractors' fees, components, etc. If the product is a machine which will be used to manufacture mass-produced parts, then the raw materials to be used by the machine will be required. The conditions concerning the repayment of the capital must also be agreed. Planning for the production phase should have been set out in the detailed design, including any special machines or devices which will be required, quality control requirements, testing procedures and standards, tolerances. The cost of the commissioning process should have been considered previously. The detailed design should have included time schedules of each phase of the operation and delivery dates; assembly times should have been carefully calculated to avoid costly waiting time.

(f) Distribution, sales, usage

The distribution stage of a design project may be concerned with the supply of raw materials and services, or the distribution of the product to suitable sales outlets. This involves storage, packaging, transportation, etc. For a design process that does not produce a marketable quantity as the product, e.g. bridge construction or a waste disposal plant, then the distribution factors will be concerned with ensuring that the inputs and outputs, e.g. vehicles or refuse in the examples quoted, have easy access and exit.

If the product is to be sold, then the sales outlets must be conveniently situated. If the market is not self-generating, then suitable advertising and marketing techniques must be employed. The project or product must be costed or priced such that optimum sales are achieved, ensuring that a profit is generated and the capital investment repaid.

The final product must be used and must satisfy the need it was intended to fulfill. If this is not achieved, then either the capital invested in the project is wasted or insufficient profits are obtained. The results of public opinion can be useful to the designer, both in terms of experience to be applied to future products and modifications which may increase consumer satisfaction. The availability of suitable spare parts and maintenance facilities are important factors which influence the final level of product use.

Although aspects of distribution, sales (if appropriate) and usage can be considered separately, it will make the overall process more efficient if close cooperation is maintained between these stages. This is due to the obvious interrelationship that exists between them.

(g) Obsolescence

The design of the product should ensure that it will not wear out or become obsolete prematurely, and at the end of its usefulness it is not still capable of prolonged operation. If either of these situations occurs, then the design is not sufficiently economic. It should be decided in the original design, although this may later be modified, whether at the end of its life the product should still be produced on a new and usually improved machine, or if the need for the product has been diminished or largely satisfied. In the latter case, continued production may be uneconomic. The scrap or resale value of the components, their disposal and any subsequent special considerations must be taken into account.

At each stage in the design process there should be a constant recycling of new information back to previous stages, so that previous decisions can be continuously reassessed. In this way unforeseen and unavoidable changes that affect the design are detected at the earliest opportunity. For example, if a product becomes obsolete prematurely due to unforeseen rapid technical advances, then this is unfortunate economically, but prolonged production only harms profitability still further.

9.7.2 Example of the morphology of design

(a) The problem
Storage, use or disposal of radioactive waste material from a nuclear power station.

(b) Feasibility study
Before embarking upon a nuclear energy program, it is necessary to decide what to do with the radioactive by-products. Assume there are no uses for these

materials and they cannot be processed into a less dangerous material (at present).

The decision is to design and build containers, either for storage until suitable uses or processing methods are discovered or for disposal by burial.

The cost of the containers represents only a very small percentage of the total budget for the nuclear program, and cost is not the main consideration. The most important requirement is that the containers conform to strict safety standards.

(c) Preliminary design

The feasibility study established that a 'solution' appeared possible and the decision was for container storage. In the preliminary design, all possible alternatives and choices are considered. For this problem the decisions to be made (immediately) relate to the selection of appropriate materials and the design of the container. The preliminary design involves finding answers to the following questions (and many others):

- Is the waste chemically reactive, either with particular materials or self-reactive, i.e. are gases produced, will the pressure increase, are liquids formed, do any reactive products have significantly different properties from the original material?
- Which materials can be used to contain the radiation within the vessel? Lead and concrete are possibilities, although it depends upon the type of radiation emitted. Neutron rays are adsorbed by hydrogen or water.
- How can external damage be prevented, or any effects minimized?
- What material thickness is required for the worst situation of external damage?
- Is it necessary to use several layers of different materials?
- If different materials are in contact, will corrosion occur?
- Will corrosion occur within the inner vessel?
- Should the vessel be designed to withstand the worst situation of external conditions, or only a 90% (say) probability of this occurring?
- What is the shape of the container, e.g. cylindrical, square, sphere?
- Is the shape based on stress considerations, ease of transportation and storage, etc.?
- What connections are required to the container, e.g. manholes, pipework, handles, etc.?
- What supports are required?
- How can the vessel be designed so that it cannot be opened by unauthorized personnel?
- Can equipment be included to monitor internal corrosion?

And so forth.

(d) Detailed design

The outcome of the preliminary design should be decisions regarding the main items: the next stage is to produce a full and detailed specification. At this stage, scale models and full-size units should be built and extensively tested. Any changes or alterations to the design or the original decisions should be made as early as possible, and before production commences. The design should also be discussed with the relevant authority and the full specification submitted for approval or modification. Items that have long lead times should be identified and appropriate planning initiated.

(e) Production
Following the approval of the detailed design, the containers are manufactured strictly in accordance with the specification. No unauthorized modifications may be made. Testing should be performed on material samples, semi-finished components and the final container. Tests should conform to approved standards, and the final container should be certified and approved for service. In this situation each vessel has to be tested, but in other situations only a certain number of finished items will require quality control checks to be performed.

(f) Delivery and use
The containers are delivered to the customer, and he may perform his own tests. The initial use of the product is closely monitored, by both the customer and the supplier.

(g) Obsolescence
This may be due to the need for an improved or alternative design, or the need for additional features. Uses may be found for the radioactive material that make storage unnecessary, or the entire nuclear program could be abandoned in favor of wind and solar energy!

9.7.3 The anatomy of the design process

The design process can also be examined by considering the steps taken by the designer, from the initial evaluation to a final solution. This is known as the anatomy of the design process, and generally contains several steps which are discussed below. These steps are carried out at each stage of the morphology approach, but in this type of analysis the work of the designer is usually concerned with the design only as far as the detailed specification. After this point specialist personnel become responsible for the project, although advice and feedback to the designer are still important.

(a) Identifying the problem and evaluating the need
At the outset of the design process, the overall objective must be defined in terms of the problem to be solved or the need which must be satisfied. The overall specification and any constraints which are imposed must be defined. However, this must be performed at each stage of the process (as outlined in the morphology approach), so that the project progresses as efficiently as possible. As the constraints and criteria are defined at each stage, the original specification may need modification. Continuous (effective) feedback of information is vital to the success of the project.

(b) Information retrieval and assessment

Information is collected and evaluated at all stages of the design process. This includes areas outside the normal work of the engineer: efficient retrieval of information as well as identifying gaps in the knowledge is necessary. The main sources of information are the customer, the user, experienced personnel, literature, patents, computer data-storage systems, observations and recording, etc. Once information is obtained it should be analyzed and assessed. The information should be categorized and referenced, and then filed for easy access. Information may be detailed and lengthy, the establishment of abstracts from each source may be valuable. If the information is not collected, assessed and stored in an efficient manner, then the design process will not proceed efficiently.

(c) Evaluating the alternatives

At each stage of the process new problems occur and there are often several alternatives available. Progression to the next stage can be made only when these alternatives have been analyzed and the best solution selected. Certain constraints affect the analysis and the selection, and the main ones are time and cost. A final design solution may be required very quickly, when only a limited amount of information has been obtained. A decision has to be made as to which areas will yield information most rapidly: technical literature, patent office, information from other companies, etc. The time to be spent on this stage will have to be decided, and also the consequences of proceeding with inadequate information. The alternatives will be assessed, and decisions made after consideration of the physical limitations and those imposed by society. At each stage the specifications and constraints imposed by the overall problem must be fulfilled, or reassessed and modified.

In order to make a decision, the alternatives available need to be tested and analysed. This may be carried out in the laboratory using scale models or prototype units, or it may be in more symbolic form using diagrams, mathematics or computer simulation. All modeling techniques require simplifying assumptions to be made; these should be clearly stated and considered when results are analyzed and decisions made. A whole field of study known as decision-theory exists, so if the alternatives are numerous or complex then specialist techniques and assistance may be valuable. A starting point for most decision theory approaches is to list the input, output and solution variables that occur during the solution of a problem. This type of analysis can be useful at each stage of the project, and helps to clarify complex situations. The approach can be illustrated by considering the variables that exist for the solution to a product distribution problem, as given below:

Input variables:	number of products,
	quantities of each product,
	arrival schedules of each product,
	shelf life of products,
	etc.
Solution variables:	storage time,

                   location of storage site,
                   method of transportation,
                   number of storage sites,
                   etc.

Output variables: storage cost,
                   transportation cost,
                   cost of product shortages,
                   fuel cost,
                   etc.

Within the decision-making process, compromises often have to be made between what is required in an ideal situation and what can be achieved in the real world. This requires optimization of the alternatives that are available. As an example, a company may require the maximum possible profit from a product. This requires production at the lowest cost and sales at the maximum price which the customer will pay. However, although smaller sales at higher prices may result in larger profits than increased sales at lower prices, this type of approach may cause other companies to enter the market attracted by these high profits. The selling price of the product may then fall but the original company may find it is unable to attract sufficient sales. In this case a compromise, or an optimal solution, to the problem of profit and price and competition should have been made.

(d) Communication and implementation

At various stages in the design process, information, results and recommendations have to be communicated to other personnel involved in the process. If this is not performed effectively, then the project will not function efficiently and wrong decisions may be made. Communications can be considered as the transfer of information between a transmitter and a receiver, generally involving feedback. In order to transfer information correctly, considerable thought is required. This involves consideration of the level of knowledge and understanding of the receiver, e.g. communicating technical information to a sales manager, in order to decide a product specification in terms of technical performance and consumer requirements. The quantity of information must be such that only essential details are presented for immediate consideration, although full project details should be available. Finally the quality of the information and its presentation should be acceptable to the receiver; otherwise it may not be fully appreciated or accepted.

It is important that decisions are made and accepted by members of the project team at each stage, and in the following stage where recommendations previously made have to be implemented. All personnel involved in the design process, from its original conception to final solution, should be kept well informed of the progress of the project and the decisions that are made at all stages, even though they may not be directly affected at that time. This will avoid decisions being made which lead to conditions which are unacceptable at a later stage. Therefore, feedforward of information is as important as feedback for the success of the project.

9.7.4 Example of the anatomy of design

(a) The problem

Before designing and manufacturing a combined can opener, bottle opener and corkscrew, the need for the product must be established. The decision to produce the item may be based upon the desire to improve the company's cash flow and profits, but if the product does not sell, the entire operation is wasted.

Can openers have been available for a long time; any new product must have distinct advantages in order to sell in significant quantities. These can be:

(a) new design or operational features;
(b) more 'appeal' compared to traditional products;
(c) cheaper.

New design features may include easier operation, less risk of accidents, fold away features into a small compact shape, etc. The appeal of a product depends mainly upon its style or appearance. A silver of gold finish provides a more sophisticated product, whereas a brightly colored polymer item appeals to a younger customer. Ultimately this product will succeed, or not, on the basis of cost and price.

(b) Information

Information is required concerning the probable sales of the product, the features which the customer will prefer, e.g size, color, mass, etc., and the expected selling price. The design can then proceed and the use of alternative materials can be considered. Information should be obtained concerning the following material requirements:

(a) strength to perform the required tasks, e.g. can opening, and without deformation;
(b) resistance to impact if dropped;
(c) mass necessary for comfortable operation;
(d) color and finish as determined by customer preference;
(e) ability to join together individual components, e.g. corkscrew, bottle opener and can opener;
(f) suitability for fabrication and machining;
(g) corrosion resistance;
(h) automating the production process with the minimum of joining or assembly operations;
(i) ability to produce a consistent and reliable product.

(c) Decision-making

The possible alternatives for the product design and the materials that could be used have been established. The next stage is to evaluate the alternatives in more detail, and make definite decisions regarding the final product. Prototypes are made and tested, for technical and consumer satisfaction. Modifications are included until a suitable product is obtained. Throughout this design stage the criteria of technical performance, consumer acceptability and competitive cost must be applied.

(d) Implementation

The final stage for the designer is the implementation of the design. This requires discussion with all personnel involved in the production and sales activities. The reasons for particular features and decisions should be fully explained, and any modifications that become necessary should always be referred back to the designer.

9.8 FACTORS INFLUENCING A DESIGN STUDY

The design process and the role of the design engineer have so far been considered in terms of the actions of the designer and the methods used to solve the problem. It now becomes necessary to consider some of the more specific aspects of design, such as planning, materials, manufacturing processes, etc., and their integration into the detailed design phase of the project.

The senior design engineer will probably spend the majority of his working time between two aspects of design work. First, the creative work required to obtain a solution to a problem (this is sometimes assumed to be the actual design process). Second, the organizational responsibilities associated with the overall design process and the functioning of the design department. The successful completion of the project requires logical sequencing and planning of the design operations, and suitable allocation of individual design tasks. These organizational aspects ensure that personnel are assigned tasks commensurate with their abilities and experience. Also parts having more intricate design requirements, and longer manufacturing times, are undertaken in sufficient time so as not to impede the progress of the project.

A junior design engineer is engaged in the creative aspects of the design process. He probably begins his career as a detail designer specializing in a narrow design field. With time and suitable training his experience broadens and he becomes competent in other design specializations. As a detail designer he is a member of a team, and aspects of the design will already have been performed. Often the problem analysis, information collection and alternative solutions evaluation will have been completed, and the detail designer is asked to prepare a detailed design for a particular specification. Some of the factors influencing a design are listed here to illustrate the diverse nature of the design problem. This is not an exhaustive list and not all these factors are applicable or of equal importance for all problems. The diversity of possible factors which the designer may have to consider at least shows why only exceptional persons are successful in the field of creative design.

The factors influencing a design problem can be listed in various categories. The factors due to the product criteria and constraints and factors caused by manufacturing methods are shown diagrammatically in Fig. 9.3(a) and (b), where it can be seen that there is a certain amount of overlapping. Although different design factors are identified, and shown separately in Fig. 9.3, during the design process no single factor can be considered in isolation. Each design factor (or component in the design) is dependent upon other considerations, and

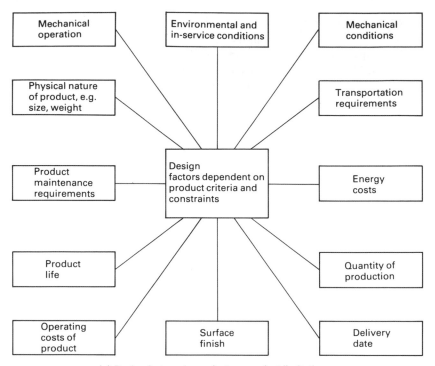

(a) *Design factors dependent on product limitations*

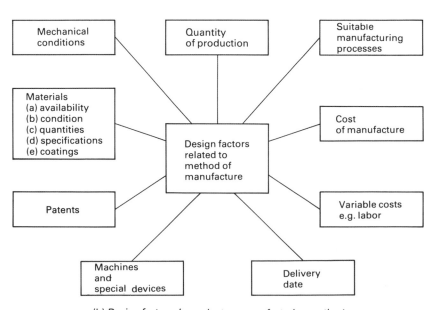

(b) *Design factors dependent on manufacturing method*

Figure 9.3 Design factors dependent on product limitations and manufacturing method.

they may need to be considered together before a decision can be made. Certain design factors are directly related, such as materials availability and design specification, it would be pointless to produce a design using specific materials which are unavailable. Other factors are related only indirectly, such as delivery date and design, where sufficient time must be available for the design process to be satisfactorily completed. However, the delivery date becomes more important during the manufacturing stage, and has a significant influence on the choice of the manufacturing process to be used.

The life cycle (morphology) of a product has been described in Section 9.5.1, and each stage within that cycle can be considered as a part of the design process. The junior designer is concerned mainly with the early stages of the process as far as producing the detailed design, but as his experience increases he may become more involved in the latter part of the cycle. Considerable interaction takes place between the designer and the production engineer, and the designer needs to be familar with the details of manufacturing processes. The problems or tasks confronting the designer early in his career, or the undergraduate embarking on a design study, are often divided for convenience into four main categories. These are the product specification (or requirements), the design specification, materials and manufacturing. These areas are interrelated, as can be clearly seen by consideration of the individual factors which are detailed as follows:

Product specification: Including operational requirements, in-service conditions, size and weight, maintenance, life, reliability, quantity, delivery, operating costs, etc.

Design specification: Including operational details, in-service conditions, size and weight, sub-components, existing designs, standards, testing, energy requirements, manufacturing requirements, etc.
(*Note:* Should comply with the product specification above.)

Materials: Including specification, standards, testing, availability, cost, condition, size and weight, quantity, delivery, manufacturing limitations, etc.

Manufacture: Including process used, time, delivery, machines and fixtures, quantity, costs, materials, reliability and maintenance, material requirements, surface finish, etc.

These are just a few of the factors to be considered, but it does indicate the relationships between various stages of the designer's work, e.g. material specification in all categories; manufacturing details mainly concerned with design, materials and manufacture.

All factors and stages in the design process must be evaluated in terms of their related costs. The overall project must either produce a profit for the company or complete the task within a specified budget; if this is not achieved, then the design has been unsuccessful.

There has been a certain amount of repetition or re-iteration of the factors within the design process. This is due to the complex nature of the design process and the alternative views which can be adopted. There is no set of rules for tackling a design problem because each problem has a new set of requirements and limitations. Suggestions for solving a problem can be advanced but each situation will have different priorities. The reader should not assume that this chapter represents a specification for a design engineer, or a blueprint for solving design problems. It will have relevance in many problems, but few problems would be satisfactorily solved merely by following these suggestions and observations. In most tasks the essence of success is experience and practice, and this is especially true of design engineering.

9.9 COMPUTER-AIDED DESIGN

9.9.1 Computers and engineering

It is expected that the reader will have some computing experience before commencing this course and should be familiar with such terms as '64K memory', 'available software and peripherals', etc. It will be worth spending some time reading through the short glossary of computing terms included at the end of this chapter, to help clarify the meaning of some of the jargon. This section will be mainly concerned with the use of computers for purposes of improved engineering design.

The computer industry was only born after the Second World War and its growth has been dramatic, if not phenomenal. Technological advances resulting in physical size reductions, improved reliability, more complex operations and shorter response times have helped to create a wider market for computers since the 1960s. The development of time-sharing computer systems and storage type, cathode ray tube (CRT) displays led to even wider applications. The integration of computers into engineering design was mainly due to the success of applying an interactive mode of operation, and significant improvements in on-line computer graphics.

The student should be aware of two important aspects of computer technology. First, the successful miniaturization of component parts and the associated improvements in computer operational ability. Second, the rapid decrease in cost and price of computers, especially the personal microcomputer.

A computer is an assembly of electronic and electromechanical devices which obeys instructions provided by an operator. It is a system comprising a *central processing unit* (CPU) and other equipment known as *peripherals*. These physical components are known as *hardware*. The computer receives its instructions in computer programs and these are known as *software*. The elements that make up a computer system are shown in Fig. 9.4.

The CPU contains three sections:

(a) the controller;
(b) the arithmetic and logic unit (ALU);
(c) the core store.

It is sometimes more economical to store data in peripheral devices using magnetic tape, cassettes or hard or floppy disks. The peripheral device is known as backing store.

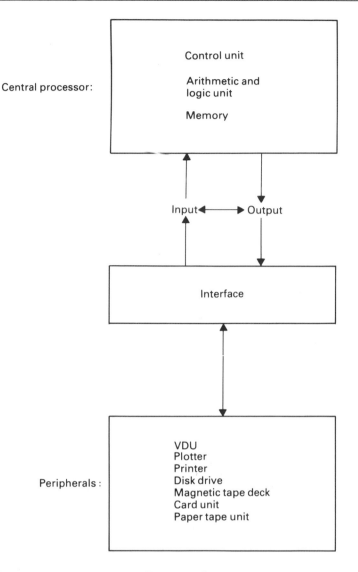

Figure 9.4   Elements of a computer system.

Before an item of data or a program instruction can be processed it must first be converted into a binary or two-digit machine code. The machine operates with the binary (base 2) arithmetic system using only the digits 0 and 1. *Bi*nary dig*its* or *bits* are the smallest units that the computer can recognize, and computer 'words' are made up of groups of bits. The *byte* is defined as:

$$1 \text{ byte} = 8 \text{ bits} = 1 \text{ alphanumeric character}$$

A computer word may represent a character, e.g. A–Z or 0–9, an instruction, e.g. 'print', and a floating point number up to a value of 10^{128}. Computers may use variable word lengths in multiples of 8 bits (1 byte) or have fixed word lengths of 16 bits or 32 bits.

Communication between the user and the computer is now frequently carried out using a teletype or alphanumeric visual display unit (VDU). Instructions are typed on a keyboard and displayed on a screen, and the answer is obtained either as printed copy from a line printer or as a VDU display.

A time-sharing or multi-access system incorporates several teletypes or VDUs so that more than one operator can use the computer at the same time. The computer may also be used by *remote access* from a telephone connection.

A computer can only provide a solution to a problem if it receives instructions in the correct form. Operations to be performed must be presented in the correct sequence, and the planning of such a sequence is known as *programming*. The actual plan is known as a *program*. The instructions are prepared and given to the computer using a specified code which is known as a *programming language. FORTRAN* and *BASIC* are two programming languages. In order to 'run' a program the computer must be able to accept (understand) the particular language used. Before running a FORTRAN program, a *program language compiler*, must be fed to the CPU. The compiler is a program (software) that the computer uses to 'translate' the instructions in a FORTRAN program into a form which the computer can accept. A system program relates all the peripherals to the CPU. BASIC is usually interpreted not compiled, i.e. the code is simply stored, and each instruction is interpreted by the computer each time it is encountered in a program.

The information given in this section is presented in general terms so that it will not be quickly dated by the rapid changes and developments within the computer industry. It is intended to provide an introduction to the subject. It is perhaps worth mentioning the difference between a mainframe computer, a minicomputer and a microcomputer.

A *mainframe computer* is a large machine housed in a purpose-built room or building, capable of performing many and varied tasks and supporting a wide range of equipment.

A *minicomputer* is a smaller machine, but the difference between it and a *microcomputer* is becoming increasingly blurred. The functions that can be performed by a microcomputer are continually being increased and improved. The power (and cost) of a computer system is mainly determined by its operating speed and the size of the internal working store. Machines are often designated by their main storage capacity, e.g. 64K (eight-bit bytes) where K is nominally 1000 (actual value 1024). In general, the larger the internal storage capacity of a machine, the greater the number of functions that can be performed and the terminals and peripheral devices which can be supported.

However, increasing the storage size of a computer is extremely costly and it may be cheaper to use backing storage, such as magnetic tape or disks.

The computers described are known as 'general-purpose, stored program, digital, electronic computers'.

9.9.2 What is CAD?

CAD is the acronym for computer-aided design.

CAD can simply mean that a computer is being used in design studies. The computer may merely be used to perform the numerical calculations quicker than they could be carried out by hand, or it may be carrying out drafting work. The term 'computer-aided design and drafting' (CADD) describes these combined functions.

It is important that the work of the engineer and the tasks performed by the computer should be integrated to provide the most efficient combination for design problem-solving. The computer and the engineer should each perform the work for which they are best suited, but the engineer is responsible for his own work and acceptance of that produced by the computer.

The stages in the design process and the possible uses of a computer are illustrated in Table 9.1. The early stages of a design are not particularly suitable for computer applications. These stages represent the more creative aspects of problem-solving and are best performed by the designer using a combination of past experience, judgement and intuition, and having the ability to locate various sources of data.

The use of computers in these preliminary stages is desirable, but the range of possibilities available is so wide that vast amounts of data would need to be stored in the computer memory. The ability of the computer to extract significant or appropriate information would require a systematic search of all data, and would be less cost-effective than the intuitive human approach.

The latter stages of a design, which require a larger amount of repetitive and iterative working as well as significant computational time, are ideally suited to computer applications. The design tasks that can be performed using a computer are:

(a) computations to be repeated many times, similar to hand calculations;
(b) computations that are complex and time-consuming, extensions of hand calculations requiring greater accuracy;
(c) the manipulation of data, including storage and retrieval of old designs and recent modifications.

Table 9.1 Potential for computer utilization in a design study.

Design stage	Computer potential
Recognition of a need and problem definition	Desirable, but few applications
Feasibility study	Limited application
Preliminary design	Certain specific applications and use of specially prepared programs or packages
Testing, evaluating and improving the design	Yes, using numerical methods and optimization
Final design and documentation	Yes, preparation of standard specification and drafting
Communicating the design	Computer graphics
Manufacture	CAM and NC
Feedback	Storage on files of subsequent results and report

Table 9.1 Potential for computer utilization in a design study.

The computer can be used as a design aid not only by the designer working on one particular aspect of a project, but by all members of the design team. Data can be made available to other engineers either by multiple access or remote access to the computer, or by storing data on devices such as disks which are portable and may be compatible with other machines.

*   *   *

Exercises

Consider the problem of designing a pipeline for the transportation of sewage. A specific solution is required, and then a general design method must be developed.

1 Identify the stages in the design process.

2 List all the variables.

3 Select (realistic) values or specifications for each unknown item except the pipe thickness.

4 Calculate the pipe thickness required.

5 Change one or more of the values specified and repeat the calculation of pipe thickness.

6 Compare the two solutions obtained.

7 What is the (relative) cost of each solution?

8 What are the effects on other aspects of the design, such as the estimated life, reliability, capacity for overload?

9 Assess the possibility of using a computer to solve this type of problem.

10 Now prepare a general design approach to this problem and assess the use of computer-aided design at each stage.

11 What factors restrict the use of a computer in solving this problem?

12 Is it possible to include such factors as customer preference or bias (e.g specifying common materials) into a computer-aided design method?

*   *   *

Computer programs or software may be written in two ways, either as a straightforward input–output operation or as an interactive or conversational mode of performance. All problems require data in order to obtain a solution, but whether these are numerical values to use in calculations or a choice between alternatives is immaterial.

The data are provided by the operator. There are three ways in which data can be provided for use in the operation of a computer program, and these are now described.

(a) Data are presented and incorporated as part of the computer program. They are available in the program when first required in the sequence of operations. The disadvantage of this method is that different applications of the program will require changes to be made at different places in the program. This may be time-consuming and liable to mistakes, especially the possibility of overlooking a value to be changed.

(b) All data are supplied as a complete 'block' or 'package' at one section of the computer program, usually the start or finish, and are provided before the program is executed (i.e. run). The data must then be in the correct order, as read (required) in operation of the program. This requirement may cause problems but they can be partially alleviated by situating all 'data read' or 'assign' statements at the beginning of the program. Also all data must be in the correct form, e.g. integer or real, and in the required units. Data provided for density may be kg/m^3 or g/cm^3 and it is important always to use a specified system. Checks can be implemented by telling the computer to print an appropriate error message if the data value is different from an expected value, e.g. if the density for liquids is less than 700 (kg/m^3) or greater than 10 (g/cm^3). An alternative is to obtain a 'hard copy' (print-out) of all data values after a program run.

(c) Data can be provided by the operator during the execution of a conversational or interactive computer program. The program requests data (and may specify the form) as required. Requests for data can be used as a sequential check on the actual program operation, rather than supplying data in one block and then attempting to identify the source of any error messages.

9.9.3 Computer graphics

The training and expectations of designers and engineers, as well as the nature of the work they perform, make it essential that design data and solutions can be represented graphically. This is also necessary to ensure the effective communication of ideas. The designer's output will be further improved if the graphics display can be easily and quickly modified by the designer.

In the 1950s all output from a computer was on paper, either printed in alphanumeric form or plotted slowly on an *xy* recorder. In the 1960s, facilities were developed that enabled results to be displayed in graphical form, although this was mainly used for data plotting. Complete systems were expensive and therefore not widely used, and also the time required to obtain a display represented a significant constraint on the design process. A further development was the use of graphics to assist the input of data, which meant that the operator required some means of indicating a position on the screen. Various devices were produced, including keyboard, light pen (a sensitive photo-electric device), 'mouse', crosshairs, joystick and digitizing tablet.

The *digitizer* is a particularly useful device for the conversion into digital form of the large amount of data which a computer may require in order to produce a drawing from a two- or three-dimensional display. Magnetic tape, floppy disks, etc., are then used to store the digital data. A pen or cursor can be used to trace details of an engineering drawing, using a flat tilting table with a programmable controller and a keyboard.

A *visual display unit* (*VDU*) presents data in a graphics or alphanumeric form. The *direct view storage tube display* uses a CRT which stores a drawing in its phosphor coating. The drawing cannot be modified unless it is first entirely erased. The *refresh graphics display* is more expensive but allows repeated modification of the image without complete erasure. A *raster scan display* uses a television monitor, although this requires scanning of the entire screen rather than a discrete line, and it produces a coarse image.

Hard-copy output from a graphics system is generally produced by a plotter, available in various forms. The *digital incremental plotter* (drum plotter) is the least expensive and is widely used. It consists of a variable speed, paper-carrying drum with a pen that moves along a line parallel to the drum's axis. It is available in different sizes and specifications, and may be operated from a peripheral unit (off-line) or directly from a computer (on-line).

An alternative machine is the *flat bed plotter* which is available in sizes from A4 to that of a drafting table (125 by 200 cm; 50 by 80 in.). It may also be operated in an off-line or on-line mode, and may be used by remote access from a telephone connection.

Finally, an *electrostatic plotter* can be used. This employs an electronic matrix-scanning technique to print high-density dots on specially prepared paper in a raster-type format. The disadvantages are that special paper and chemicals are required, the data must be preprocessed, the quality and accuracy are inferior, and multicolor or overdrawn plots cannot be produced. However, high plotting speeds are possible and the plotter can also be used as a printer. printer.

Until the mid-1970s the development of computer graphics techniques and software was relatively slow, owing to high capital costs and time-consuming data preparation. The breakthrough occurred with the availability of cheap microprocessors and cheaper computer memory. This was reinforced by a shortage of drafting skills (and hence higher salaries), more competitive markets, falling costs and prices, and the development of more flexible graphics software.

Software is available that allows the computer to draw two-dimensional objects, or three-dimensional objects in a 'wire-line' drawing form. The planes can then be smoothed out, the surface colored and hidden areas removed from view. The shape displayed can be rotated, enlarged or panned, and areas subtracted or added as they leave or come into view. This is carried out using a light pen or digitizing tablet.

The development and availability of interactive computer graphics were also major influences on the range of computer applications. The increased speed of computer response and introduction of direct input/output devices meant that the designer could see almost immediately the graphical plots of computed results, and could then amend the output as necessary. In this way the designer–computer combination became fully integrated as an interactive design team, each performing its particular expertise with support from the other.

* * *

Exercises

1 Prepare a sequence of instructions and, if possible, write a computer graphics program to perform the following tasks:

    (a) draw an octagon;
    (b) draw a regular pentagon such that all points (corners) of the shape are joined to each other;
    (c) draw a representation of a rectangular box;
    (d) draw a representation of a tetrahedron that has been cut into two parts;
    (e) show two cross-sectional views of a simple two-position flow valve.

2 Recommend appropriate computer graphics systems to be used with the following design problems:
    (a) thermal design of a heat exchanger;
    (b) mechanical design of a heat exchanger;
    (c) structural load support;
    (d) construction of a low-cost, dome-shaped house;
    (e) specification of an automobile body cross-section, including floor panel design.

3 Assess the current and future uses of computer graphics within your particular engineering discipline.

*   *   *

9.9.4 CAD systems

Computers are now widely used in the planning, design and construction stages of engineering projects. The expected benefits are greater efficiency, better standards of design and improved management planning.

CAD is not a specific term; it includes all aspects of the design process including drafting work. The present trend in CAD is away from systems performing only numerical calculations and requiring a large computer memory, toward the analysis of complex problems requiring many hours of program development. CAD is being increasingly adopted by architects, shoemakers, surgeons, industrial designers, etc.

Many requirements are for *turnkey systems* which are purchased as a complete unit: the purchaser turns the key and starts learning. The cost of such systems is declining as computer hardware costs fall and competition increases.

Before embarking on a CAD problem-solving approach it should first be ascertained that the problem is suitable for solution by a computer. Any problem which can be written as a sequence of logical steps and mathematical representations can then be expressed as a computer program. However, CAD is most economical with situations that are repetitive, and with complex problems. CAD is also useful for assessing alternative methods or possible situations, since the time required to prepare a design specification is then reduced and quotations become more accurate. Other advantages are less tangible, such as the job satisfaction associated with using modern, sophisticated techniques or the extra time available for creative work.

A long-term benefit will be the creation of a database of information covering many design situations and products. Designers and engineers then use the same set of data, thus producing consistent specifications and standards, avoiding copying errors and making improvements and modifications easier. A common database also provides the opportunity to introduce computer-aided manufacture (CAM) into a company (see Section 8.7.3).

The decision to proceed with a CAD approach to a problem means that the appropriate hardware must be available. This can be a large, time-sharing computer system or a microcomputer with access to a larger database stored by a mainframe computer. However, avoid trying to combine computer equipment from different manufacturers – the reasons should be obvious! Certain conditions must be achieved, such as convenient and easy computer access, well-defined program input, ease of data revision and error correction, results quickly available in an appropriate form and expert assistance available when required.

It may be possible to use an existing CAD system or program to solve a problem. If this option is not available, then it will be necessary to develop or obtain suitable software. There are four main types of software packages for use with CAD systems.

First, *interactive computer graphics* as described in Section 9.9.3. This should be an interactive program using CRT screen presentation and providing a complete drafting capability including labeling, dimensioning and geometrical analysis.

Second, programs that provide the engineering analysis of a design problem. This often comprises a set or 'suite' of programs that are integrated to provide a complete solution. These programs may perform a geometrical analysis, plant layout or piping diagram, stress analysis, temperature distribution, etc. *Finite element methods* may be used which divide the design problem into small components or elements, and then consider the solution in terms of the interrelationship of these elements.

Third, CAD programs that are written specifically to handle, store, retrieve and present the large amounts of data which may be involved in a design project.

Finally, software packages are available which produce final design drawings and instructions for *numerically controlled* (NC) machines (see Section 8.7.2). The communication of design information via a network of terminals is becoming increasingly popular.

The use of CAD systems in particular industries will now be considered.

The aircraft industry is probably the largest user of CAD systems. The Boeing Company is one of the leaders, having used NC equipment for production of machined parts since the 1950s, and APT technology and surface analysis programs since the late 1960s. From the mid 1970s, Boeing has utilized interactive computer graphics and now produces about 30% of the total design drawings for an aircraft using CAD systems. These drawings are for the design of parts representing about 90% of an aircraft's structural weight. All the design information is stored in a common database and the computer checks the entire assembly for clearance and fit. Individual assemblies and units may also be checked.

The automobile industry also embraced CAD at an early stage, mainly because of the wide range of interrelated variables that influence an automobile design. However, in the USA the rapid introduction of stringent requirements relating to safety, the environment and fuel economy has now made CAD a necessity. Interactive graphics can be used to compare different overall shapes for a vehicle and then make alterations and improvements. This is known as *concept surfacing*. CAD can be used for the design and analysis of component parts and to show the interaction of moving parts using animation techniques. Computer graphics have been used to simulate the shape and requirements of a human model to improve the ergonomic features of a design.

Many of the techniques developed by the aircraft and automobile industries have also been used for the design of construction equipment and agricultural machinery, such as earth movers, road rollers, combine harvesters, etc. Companies that manufacture machine tools have extended their involvement in CAM and NC to the design of the tools themselves. Computer simulation can be used to investigate tool paths, machine vibration and other operating variables, and then to modify the specification until a satisfactory tool design is achieved.

*   *   *

Exercises

Some assistance (e.g. from a tutor) will probably be required for consideration of both the design and CAD aspects of these problems. However, they should provide the basis for extensive discussion/tutorial sessions.
Make recommendations for the use of CAD systems in the following cases:

1 The design of a milk bottle (Fig. 9.5).

2 The design of two intermeshing toothed gear wheels (Fig. 9.6).

3 The arrangement or layout of several units of a chemical plant, machining tools or a production process, where the units are connected in a sequential operation (Fig. 9.7). Consider also, the use of vertical space.

4 The design of a excavator digging arm (Fig. 9.8).

5 The connection of two moving parts, e.g. two rotating shafts of different diameters.

6 Joining together individual components to form an assembly that is to be joined to, or fitted inside, another unit.

7 Selecting a body shape for a new automobile (Fig. 9.9).

8 Comparing the possible sheet car body skins for the alternatives in Exercise 7 (Fig. 9.10).

9 Deciding whether to use a pressed shell or welded panels for an automobile body.

Figures 9.5–9.10 provide illustrations relating to these exercises and may be used for clarification or as starting points.

*   *   *

The use of computers and the development of computer-aided design techniques have resulted in significant reductions in design project costs, working hours, inconsistencies and mistakes. Major technological advances in computer technology, reductions in computer costs and improvements in computer 'power' indicate that the uses of CAD will continue to increase.

The computer is a tool to be used by the designer and it should only be used for tasks where it will result in higher efficiency. It is important that CAD

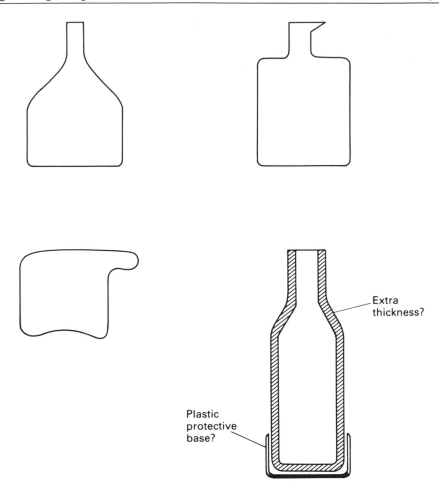

Figure 9.5 Alternative designs for a milk bottle.

becomes integrated with the activities of a design process and is accepted by all persons in the design team.

All computer applications, including CAD, CAM, NC and any other method, should be compatible with each other and with all other aspects of the design, manufacture and production process. An appropriate CAD/CAM system should provide the opportunity to reduce costs and improve profits. However, an inappropriate system will have the opposite effect. It is therefore important that the designer is familiar with CAD/CAM applications and is capable of making recommendations relating to their purchase and subsequent use.

Opposite rotations

Idler
gear

Would
they work?

Worm
gears

Figure 9.6 Design of gears.

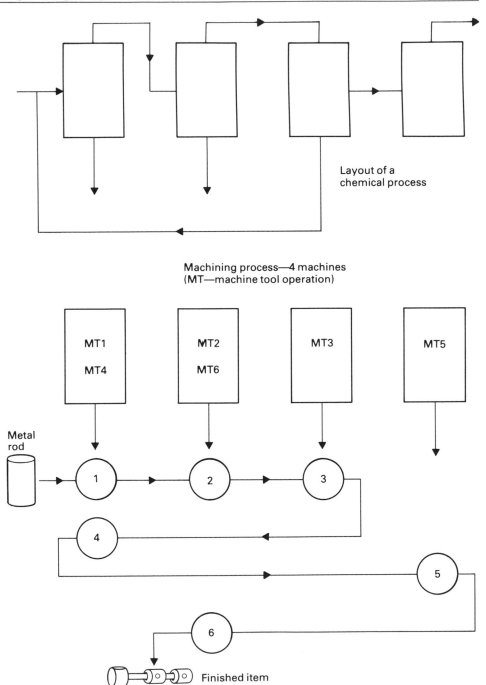

Layout of a
chemical process

Machining process—4 machines
(MT—machine tool operation)

Metal
rod

Finished item

Figure 9.7   Plant layout and operating sequence.

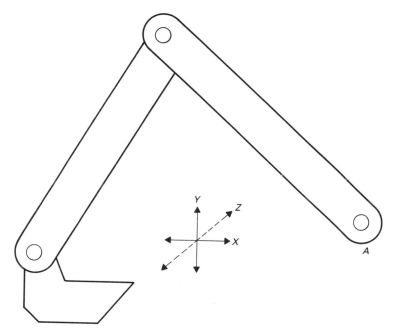

Figure 9.8 Excavator digging arm. (Note: Is the arm pneumatic or hydraulic controlled? Is Z – movement general, or is rotation about point A?)

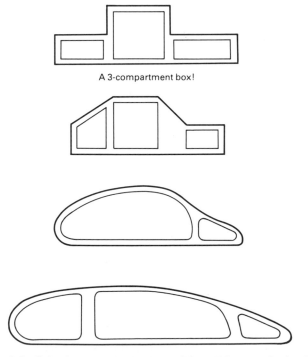

A 3-compartment box!

Figure 9.9 Selecting a car body shape. (Note: What are the implications of these designs for fuel consumption, safety, utility, consumer preference, corrosion, repair, mass production, etc., etc.?)

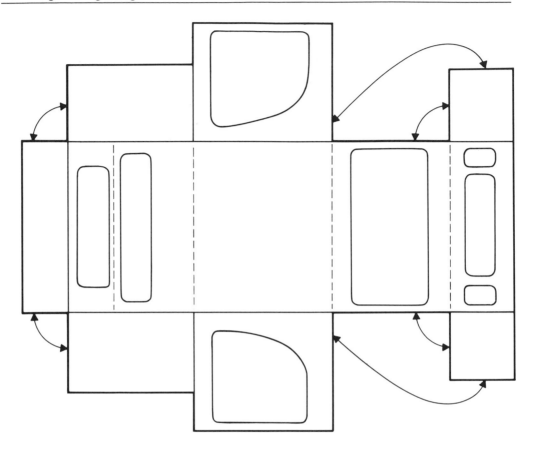

Figure 9.10 Possible automobile body shell.

9.10 THE DESIGN OF A STRUCTURAL COMPONENT

Engineers of all disciplines are frequently required to design structural components or machine parts which are required to carry a load without failure. The considerations which will be made to obtain a solution to this type of problem can be used as an example of the points discussed in this chapter. The approach will be outlined in general terms but these suggestions could be used for the design of a simple component. This would illustrate the limitations and simplifications which are made, when attempts are made to define the design process or explain it in written terms.

The steps to be taken by the designer are shown in Fig. 9.11, and the actions which the designer makes. For a structural component the main technical problems which need to be resolved are:

(a) Determination of the general shape or form of the component.
(b) Evaluation of the loads that the component must carry during its expected life.

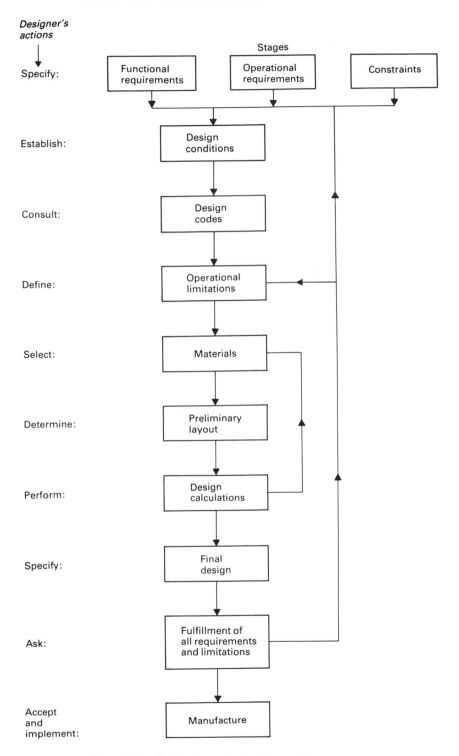

Figure 9.11 Design in terms of actions and stages.

(c) Selection of a suitable material and determination of the maximum allowable stresses.

(d) Expression of the maximum stress, or other criteria, in terms of the loads and dimensions.

(e) Modification of dimensions to comply with space, strength and fabrication requirements.

Definite procedures have been developed, and are taught, which enable solutions to these individual problems to be obtained. However, these problems are interrelated and a complete solution requires the designer to exercise judgement, experience, flexibility and innovation. The use of the component and its final appearance must be considered during the design stages, as well as the structural properties required and the cost and availability of specified materials. The component must possess sufficient hardness and rigidity to perform satisfactorily, but excessive strength may mean unnecessary weight and excessive material costs.

The relationships between the various specific design factors which need to be considered for a structural component are shown in Fig. 9.12. The shape is often determined by its required function; then the loads to which it will be subjected are estimated. The problem then is to determine the necessary size. This can be achieved only by determination of the total stresses and displacements and hence the unit stresses and strains. The maximum stress which will

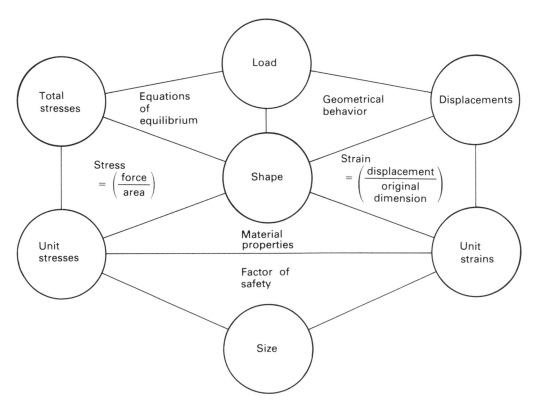

Figure 9.12 Design factors for a structural component.

occur during operation and the normal working stress must be determined. Allowances must be made for uncertainties of loading, approximations in calculations and any uncertainties in the quality of material to be used.

The effects due to failure of the component must be taken into account. Whether the component can be easily and cheaply replaced, or whether large costs will be incurred, significantly influence the design of the component. Failure usually means that the component will no longer function satisfactorily for its intended use. In certain cases failure may involve impaired performance but not necessarily require immediate action. The component must be designed so that failure will not occur as a result of either excessive distortion, cracking or rupture of the material. Therefore, calculation of the strains developed within the material becomes necessary. The most common types of failure are due to excessive elastic deformation, slip, fracture and creep.

The failure of a component depends not only upon the magnitude of the applied load, but also on the manner in which the load is applied. Loads can be steady or static loads, impact or dynamic loads and repeated loads. The action of the loading may be uniform or variable with respect to time.

9.11 CHEMICAL PLANT PROJECTS

Chemical plant design must primarily satisfy economic considerations. If a plant cannot be designed and operated so that the products are produced and sold at a profit, then the project has been a failure. Many other factors are taken into account when a chemical plant is designed, including legal aspects, patent coverage of processes, safety, reliability, flexibility, immediate investment, overall cost, etc. Certain problems may be insoluble, e.g. obtaining a license to use a patent, or the costs incurred by necessary safety features mean the project cannot then proceed beyond the design feasibility stage. However, if the major obstacles are overcome, the design of a chemical plant is concerned with incurring the minimum costs in order to obtain an optimum profit. The costs associated with building a chemical plant are the capital cost associated with each unit and the installation and commissioning costs. When the plant is operational certain running costs have to be paid, e.g. cooling water, electricity, maintenance, etc. The capital cost of the project, with interest, has to be repaid out of profits.

The design and installation of a chemical plant is often referred to as project engineering and may be divided into several stages, with a project team responsible for each stage. The overall coordination of the project is the responsibility of the chief engineer or the project manager. The design of a chemical plant involves several stages. These are:

(a) Complete product and by-product specification including purity, pressure, quantity, temperature, etc.

(b) Complete and detailed process description for each unit or section of the plant. This does not include the mechanical design of the units but the requirements in terms of the functions to be performed, e.g. heat exchanger, washing tower, etc.

(c) Prepare process flow diagrams showing flow rates of raw materials, secondary materials (e.g. wash water), products and by-products. Details of temperatures, pressures, etc., and any control features.

(d) Material or mass balances for the overall plant and for each unit. Showing the flow rates of all materials required and produced.

(e) Energy or heat balances for each unit in the plant, determination of the heat load for each unit and economic methods of utilizing the heat removed and satisfying heat requirements. The total energy requirements of the plant should also be calculated.

(f) Detailed design of each unit in the chemical plant. This should include details of the unit alternatives, and the selection on the basis of costs (both running costs and capital), efficiency, reliability, plant specification and standardization.

(g) Project description including site layout, services, safety features, back-up power, noise, roads, etc.

(h) Determine schedules for the construction, fabrication, assembly, delivery (as appropriate) of each unit. Also a schedule for the overall construction of the plant and all ancilliary items, commissioning period and length of time before the plant is fully operational. Factors which are the responsibility of the plant user, e.g. supply of raw materials or electricity costs, should be agreed.

(i) Contracts between the user and the supplier of the plant include legal liabilities, guarantees, service and repair conditions, etc. Contracts may be based on the turnkey cost, i.e. lump-sum fixed price, or percentage staged payments, or maximum price with incentives for early completion.

This description of chemical plant design is obviously an oversimplification of the actual process. No attempt has been made to consider the relative importance of each stage, or any detailed information presented for carrying out the tasks required at each stage. Books are available which deal exclusively with each aspect of the design process (a)–(i) above. The aim is to present the basic ideas of project engineering for chemical plant design, and an appreciation of the areas in which a chemical engineer may be involved.

* * *

Complementary activities

The only way to fully appreciate the activities that are involved in design, and which products can be considered examples of 'good' design, is to become involved in some design activities. The items to be designed do not have to be complex systems: ideally they should be simple common items, e.g. toothbrush, garden fork, bicycle pump.

An example; select a common item, e.g. a paper clip.

Perform the following tasks:

(a) design a 'new' item;
(b) build the designed component;
(c) test the component;
(d) modify the design;
(e) repeat 'design–build–test' process.

In practice, modifications following testing should be minimized in order to reduce costs. For the paper clip the design may be a conventional bent wire item (see Fig. 9.13), in which case determine;

(a) length of wire required;
(b) thickness of wire;
(c) number of bending operations;
(d) strength of wire;
(e) ease of fabrication;
(f) number of sheets of paper that can be secured;
(g) cost per 1000 paper clips;
(h) fabrication problems, e.g. sharp ends;

and so forth.

An alternative design is the lift-flap polymer clip shown in Fig. 9.13, in which case determine:

(a) dimensions of the clip;
(b) polymer material;
(c) color(s) preferred;
(d) length of flap, which will partly determine the strength and number of sheets that can be held;
(e) fabrication operations required;
(f) capacity for re-use of clips;
(g) ability to recycle waste material;

and so forth.

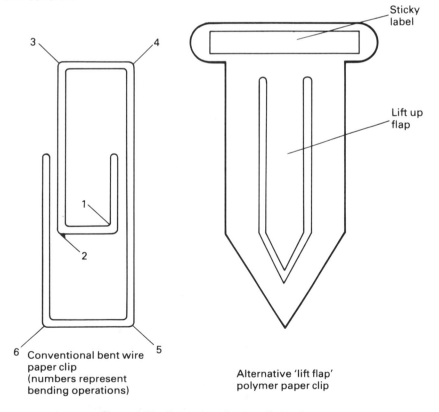

Conventional bent wire paper clip (numbers represent bending operations)

Alternative 'lift flap' polymer paper clip

Figure 9.13   Examples of paper clip design.

Having designed, built and tested the prototype item, now carry out a *critical* assessment of the product. You probably think that the design is good in the circumstances (whatever they are?). Ask for a critical assessment from someone with experience of design problems. This experienced person should be able to suggest many weaknesses in your item, and equally as many possible improvements or alternatives.

Do not get discouraged; this is only to be expected at (probably) your first design attempt. Instead of giving up, decide to make significant improvements in your next attempt, learn from previous mistakes — the sign of a 'good' engineer or designer. Be reassured by the thought that no design is ever perfect. A design is rarely the *best* solution, but is usually only *better* than the alternatives. Design work is challenging because there are always improvements to be made.

If possible arrange to visit the design or engineering departments of some large companies. Talk to the designers and engineers and ask how they perform design work, the methods they use and the problems they encounter.

*   *   *

Selected glossary of computer terminology

CAD	computer-aided design
CAD/CAM	computer-aided design and manufacture, also CAD-CAM, CADAM, CADM
CADD	computer-aided design and drafting
CAE	computer-aided engineering
CAM	computer-aided manufacture
CNC	computer numerical control
DNC	direct numerical control
NC	numerical control
APT	automatically programmed tools
BASIC	a programming language, an acronym for Beginner's All-purpose Symbolic Instruction Code
FORTRAN	a programming language, an acronym for FORmula TRANslation
ALU	arithmetic and logic unit
CPU	central processing unit or central processor
CRT	cathode ray tube
CRTDU	cathode ray tube display unit
IC	integrated circuit
VDU	visual display unit
DMA	direct memory access
RAM	random-access memory
RBT	remote batch terminal
RJE	remote job entry
ROM	read-only memory
Access time	the time taken to reference an item in storage.
Algorithm	a set of rules for performing a task or solving a problem, e.g. a formula.
Alphanumeric	characters composed of letters and numbers and sometimes arithmetic signs, punctuation marks, etc.
Binary	arithmetic system using base 2.
Bits	BInary digiTS (a 1 or 0).
Bug	computer jargon for an error in a program or system.

Bytes	a number or word comprising several (usually 8) bits.
Chip	a small piece of material (typically 6 mm square) containing electronic circuits.
Databank	a collection of stored data.
Database	an organized pool of shared data, typically a series of regularly updated applications files.
Disk	circular metal plate coated with magnetic material and used to store data on concentric tracks (a floppy disk is flexible).
File	an organized collection of records.
Graphics	visual display of data in graphical form, or as two- or three-dimensional representations.
Integrated circuit	a circuit in which all the components are chemically formed upon a single piece of semiconductor material.
Interactive	system where the operator obtains the processing results immediately so that further action can be taken, or where the program prompts the operator for data by asking questions.
Interface	a boundary between two pieces of equipment across which all the signals that pass are carefully defined.
Light pen	a photo-electric device used to detect or modify images on the surface of a CRTDU.
List	a printing operation detailing a series of records on a file or in a store
Mainframe	a large computer with a wide range of facilities.
Memory	the data storage unit (e.g. RAM, ROM).
Microcomputer	a computer utilizing a small number of chips.
Microprocessor	a CPU on a chip, part of a microcomputer.
Minicomputer	medium-sized computer.
Peripheral	device used with a computer to display, store or convert data, e.g. printer, keyboard, VDU.
Real time	system where the results of processing are produced with sufficient speed to be used for control purposes.
Software	computer programs.
Tab(ulation) character	a character used to control the format of printed output, may be required as input with data.
Terminal	device allowing communication between a computer and an operator.
Timesharing	sharing CPU time and system resources between several tasks or operators.
Wafer (or slice)	thin disk of semiconductor material.

* * *

KEYWORDS

engineering	adaptive design	willpower
design	development design	temperament
material selection	creative design	invention
material specification	conceptual ability	creativity
material properties	logical thought	communication
in-service conditions	perseverance	scientific knowledge
material condition	concentration	design engineer
material form	memory	draftsman
material requirements	responsibility	technician
compositional limits	integrity	projects manager

morphology of design
life cycle of product
problem identification
feasibility study
preliminary design
detailed design
production
distribution
sales
usage
obsolescence
anatomy of design
need evaluation
information retrieval
information assessment
evaluation of
    alternatives
scale model
prototype
computer simulation
decision theory
input variables
solution variables
output variables
optimization
implementation

feed-forward
feedback
product specification
design specification
computer-aided design
    (CAD)
time sharing
interactive mode
computer graphics
CPU
peripherals
hardware
software
bit
byte
VDU
program
mainframe computer
minicomputer
microcomputer
CADD
conversational mode
data 'block'
hard copy
digitizer

direct-view storage tube
cathode ray tube (CRT)
refresh graphics display
raster scan display
digital incremental
    plotter
flat bed plotter
electrostatic plotter
'wire line' drawing
'turnkey' systems
information database
CAM
interactive computer
    graphics
suite of programs
finite element method
NC
APT
concept surfacing
structural component
design code
chemical plant
schedule

* * *

BIBLIOGRAPHY

Beakley, G.C. and Chilton, E.G., *Introduction to Engineering Design and Graphics*, Macmillan Publishing Co., Inc., New York (1973).

Dieter, G.E., *Engineering Design; A Materials and Processing Approach*, McGraw-Hill Book Co., Inc. New York (1983).

Duderstadt, J.J., Knoll, G.F. and Springer, G.S., *Principles of Engineering*, John Wiley and Sons, Inc., New York (1982).

Farag, M.M., *Materials and Process Selection in Engineering*, Applied Science Publishers Ltd, Barking, Essex (1979).

Glorioso, R.M. and Hill, F.R. Jr., *Introduction to Engineering*, Prentice-Hall, Inc, Englewood Cliffs, New Jersey (1975).

Kapur, K.C. and Lamberson, L.R., *Reliability in Engineering Design*, John Wiley and Sons, Inc., New York (1977).

Kirby, G.N., 'How to Select Materials' 29th Biennial Report on Materials of Construction, *Chemical Engineering*, pp.86–149, 3 November 1980.

Krick, E.V., *Introduction to Engineering and Engineering Design*, 2nd Ed., John Wiley and Sons, Inc., New York (1969).

Materials Selector and Design Guide, Design Engineering, Morgan-Grampian Book Publishing Co. Ltd, London (1974).

Mayne, R. and Margolis, S., *Introduction to Engineering*, McGraw-Hill Book Co., Inc., New York (1982).

Ray, M.S., *Elements of Engineering Design: An Integrated Approach*, Prentice-Hall International (UK) Ltd, London (1985).

Spotts, M.F., *Design Engineering Projects*, Prentice-Hall, Inc., Englewood Cliffs, New Jersey (1968).

Williams, E.H., *Designing in Metals*, Iliffe Books Ltd, London (1968).

Index